Surface Chemistry in Biomedical and Environmental Science

NATO Science Series

A Series presenting the results of scientific meetings supported under the NATO Science Programme.

The Series is published by IOS Press, Amsterdam, and Springer in conjunction with the NATO Public Diplomacy Division.

Sub-Series

I. Life and Behavioural Sciences	IOS Press
II. Mathematics, Physics and Chemistry	Springer
III. Computer and Systems Science	IOS Press
IV. Earth and Environmental Sciences	Springer

The NATO Science Series continues the series of books published formerly as the NATO ASI Series.

The NATO Science Programme offers support for collaboration in civil science between scientists of countries of the Euro-Atlantic Partnership Council. The types of scientific meeting generally supported are "Advanced Study Institutes" and "Advanced Research Workshops", and the NATO Science Series collects together the results of these meetings. The meetings are co-organized by scientists from NATO countries and scientists from NATO's Partner countries — countries of the CIS and Central and Eastern Europe.

Advanced Study Institutes are high-level tutorial courses offering in-depth study of latest advances in a field.
Advanced Research Workshops are expert meetings aimed at critical assessment of a field, and identification of directions for future action.

As a consequence of the restructuring of the NATO Science Programme in 1999, the NATO Science Series was re-organized to the four sub-series noted above. Please consult the following web sites for information on previous volumes published in the Series.

http://www.nato.int/science
http://www.springer.com
http://www.iospress.nl

Series II: Mathematics, Physics and Chemistry – Vol. 228

Surface Chemistry in Biomedical and Environmental Science

edited by

Jonathan P. Blitz

Eastern Illinois University,
Charleston, IL,
U.S.A.

and

Vladimir M. Gun'ko

Institute of Surface Chemistry,
Kiev,
Ukraine

Springer

Published in cooperation with NATO Public Diplomacy Division

Proceedings of the NATO Advanced Research Workshop on
Pure and Applied Surface Chemistry and Nanomaterials for
Human Life and Environmental Protection
Kyiv, Ukraine
14-17 September 2005

A C.I.P. Catalogue record for this book is available from the Library of Congress.

ISBN-10 1-4020-4740-1 (PB)
ISBN-13 978-1-4020-4740-4 (PB)
ISBN-10 1-4020-4739-8 (HB)
ISBN-13 978-1-4020-4739-8 (HB)
ISBN-10 1-4020-4741-X (e-book)
ISBN-13 978-1-4020-4741-1 (e-book)

Published by Springer,
P.O. Box 17, 3300 AA Dordrecht, The Netherlands.

www.springer.com

Printed on acid-free paper

Professor A. A. Chuiko†, NATO ARW co-director from which these proceedings are derived, died on the 16th of January, 2006. His outstanding contributions to the fields of physical chemistry, physics, chemistry and solid surface technologies will be a lasting legacy. Professor Chuiko obtained his Doctorate in Chemical Sciences in 1972. Since 1988 he was an academician of the National Academy of Sciences of Ukraine. Some activities and awards during his career include being an active member of the Academies of Technological Sciences of Ukraine and Russian Federation, an Honored Scientist and Technologist of Ukraine, a laureate of the State Prize of UkrSSR in Science and Engineering and the Pisarzhevsky Prize, and the President of the Ukrainian Chemical Society. Professor Chuiko in 1986 was a founder and permanent Director of the Institute of Surface Chemistry of the National Academy of Sciences of Ukraine, a world renowned research organization in which he was a driving force.

A. A. Chuiko† expended considerable effort to the training of young scientists, having trained more than 100 Ph.D. and Doctors of Sciences students.

A. A. Chuiko† and his co-workers developed fundamentals of modern surface chemistry of ultra-dispersed solids, new types of functional nanomaterials, and founded a new direction in pharmacology based on nanomaterials. His compre-hensive creative activity was characterized by deep intuition and understanding of new and perspective directions in chemical science. Many of his projects led to industrial materials production.

Prof. Chuiko was a positive force in many people's lives. He is remembered as an outstanding and talented scientist, a man of deep erudition possessing a sharp and ever active mind. He was a man of inexhaustible energy, initiative, and wisdom. His sincere generosity will remain in the hearts of his colleagues.

TABLE OF CONTENTS

Preface xi
List of Contributors xiii

Fundamentals

1. Water structuring at colloidal surfaces 1
M. Chaplin

2. Microgels and core-shell particles 11
B. Vincent

3. Microencapsulation: fundamentals, methods and applications 23
D. Poncelet

4. The molecular layering method: progress in science and practical works for creation of functional nanomaterials 35
A. A. Malygin

5. Analysis of surface structures using XPS with external stimuli 45
G. Ertas, S. Suzer

6. Modification and coating of polymer surfaces 59
O. Gershevitz, P. Silickas, C. N. Sukenik

7. Is it possible to generalize the problems of porous materials formation, study and exploitation? 69
M. S. Mel'gunov, V. B. Fenelonov

8. Theoretical analysis of surface pressure of monolayers formed by nano-particles 79
V. B. Fainerman, V. I. Kovalchuk, D. O. Grigoriev, M. E. Leser, R. Miller

9. Oxide, carbon, and carbon/mineral adsorbents for toxic, explosive, and narcotic compounds – synthesis, properties, and applications 91
V. Gun'ko, R. Leboda, V. Turov, V. Zarko, A. Chuiko†

10. Porous polymeric nanocomposites filled with chemically modified fumed silicas 103
Y. Bolbukh, V. Tertykh, B. Gawdzik

11. Ion-electrostatic interaction in systems of inorganic nanoparticles and biological cells in electrolyte solution 113
L. G. Grechko, L. B. Lerman, O. Ya. Pokotylo, N. G. Shkoda, A. A. Chuiko†, K. W. Whites

12. Structural and adsorption characteristics of pyrocabon-mineral adsorbents 123
J. Skubiszewska-Zięba, R. Leboda, V. M. Gun'ko, B. Charmas

13. Synthesis and structural/adsorption characteristics of unmodified and modified hybrid carbon/mineral adsorbents 135
J. Skubiszewska-Zięba, R. Leboda, V. M. Gun'ko, O. Seledets

Biomedical Applications of Nanomaterials

14. Surface Functionalization for Biomedical Applications 145
H. J. Mathieu, X. Gao, Y. Chevolot, D. J. Balazs

15. Surface modifications to influence adhesion of biological cells and adsorption of globular proteins 159
W. Norde

16. Fundamentals of nanosilica applications for human protection 177
V. M. Gun'ko, V. V. Turov, A. A. Chuiko†

17. Medical aspects of application of highly disperse amorphous silica 191
A. Chuiko†, A. Pentyuk, E. Shtat'ko, N. Chuiko

18. The interaction of nanostructured biomaterials with human cell cultures. The choice of cell cultures for use as biocompatability probes 205
S. L. James, M. Illsley, S. E. James, E. Mendoza, S. R. P. Silva, P. Vadgama, P. Tomlins, S.V. Mikhalovsky

19. Physical characterisation of a polycaprolactone tissue scaffold 215
P. V. Grant, C. M. Vaz, P. E. Tomlins, L. Mikhalovska, S. Mikhalovsky, S. James, P. Vadgama

20. Nanostructured and nanocomposite hydrogels for biomedical applications 229
A. Spanoudaki, D. Fragiadakis, K. Vartzeli-Nikaki, P. Pissis, J. C. R. Hernandez, M. M. Pradas

21. An overview on the toxicity of inhaled nanoparticles 241
B. Fubini, I. Fenoglio, G. Martra, R. Ceschino, M. Tomatis, R. Cavalli, M. Trotta

22. Detection of V-type nerve agent degradation products using a polypyrrole/pyrroloquinoline quinone-modified electrode 253
O. V. Shulga, C. Palmer

23. Biological applications of functionalized carbon nanoparticles 265
J. Barkauskas

24. Interaction of cells with nanoparticles 277
V. A. Pokrovskiy, N. P. Galagan, A. A. Chuiko†

25. The role of nature and structure of surface sites in the biological response to silica particles 287
I. Fenoglio, M. Ghiazza, R. Ceschino, F. Gillio, G. Martra, B. Fubini

26. Magnetically sensitive nanocomposites for medical and biological applications 299
P. P. Gorbik, L. P. Storozhuk, A. A. Chuiko†, L. Yu. Vergun, V. F. Chekhun

27. Hydrophobized silica nanocomposites with immobilized antioxidants (vitamins C and E) 307
P. Kuzema, O. Stavinskaya, O. Kazakova, I. Laguta

28. Interaction of some biomolecules with modified nanosilica surfaces studied by quantum chemistry 315
O. Tsendra, A. Datsyuk, V. Lobanov, A. Grebenyuk, A. Chuiko†

Environmental Applications of Nanomaterials

29. Synthesis and characterization of ordered mesoporous materials for removal of heavy metal ions 325
M. Jaroniec, O. Olkhovyk

30. Functionalized surfaces: silica structure and metal ion adsorption behavior 337
I. P. Blitz, J. P. Blitz, V. M. Gun'ko, D. J. Sheeran

31. Heterogeneously catalyzed alkane isomerization – towards 100% selectivity 349
H. Knözinger

32. Low temperature oxidation of carbon monoxide over alkali-metal promoted palladium-tin oxide catalysts 359
B. Mirkelamoglu, G. Karakas

33. Nanoparticles of aluminium salts hydrolysis products in water treatment and disinfection 369
S. Barany, J. Gregory, A. Shcherba, I. Solomentseva

34. Co-adsorption of the low molecular carboxylic acids and cadmium ions at the metal oxide/electrolyte interface 383
W. Janusz, M. Matysek

35. Interaction of Phenolic Pollutants with PNIPA hydrogels 393
K. László, K. Kosik, E. Wilk, E. Geissler

36. Nanoscale oxides as destructive sorbents for halogenated hydrocarbons 403
A. M. Volodin, A. F. Bedilo, D. S. Heroux, V. I. Zaikovskii, I. V. Mishakov, V. V. Chesnokov, K. J. Klabunde

37. Dynamic adsorption of tert-butylbenzene, cyclohexane and water vapours on fixed activated carbon/molecular sieve beds 413
D. Palijczuk, R. Szmigielski, V. M. Gun'ko, R. Leboda

38. Influence of pre-adsorbed NaCl on breakthrough dynamics of TBB and DMMP on activated carbon at different airstream humidities 421
R. Szmigielski, D. Palijczuk, R. Leboda, V. M. Gun'ko, J. Skubiszewska-Zieba

39. Influence of the nature of active surface sites of highly disperse oxides on adsorption of heavy metal ions 429
V. I. Zarko, V. M. Gun'ko, L. S. Andriyko, E. V. Goncharuk, M. Matysek, E. Skwarek, W. Janusz

PREFACE

This volume details work presented at the NATO Advanced Research Workshop entitled "Pure and Applied Surface Chemistry and Nanomaterials for Human Life and Environmental Protection" held in Kiev, Ukraine, September 14-17, 2005. A total of 39 selected works have been compiled detailing research in three categories all related to the surface chemistry of nanomaterials: fundamentals, biomedical applications for human life, and environmental protection.

There exists great hope throughout the scientific community for the application of nanotechnologies to solve myriad technological problems. Nanomaterials and nanoparticles exhibit unique properties which are now being explored for potential uses, as well as hazards. Given the scale of nanomaterials surface chemistry inevitably plays a huge role in their properties, since most of these materials are largely comprised of surface. The synthesis of nanomaterials ranging from core-shell particles, microencapsulation techniques, molecular layering of nanoparticles, mixed oxide nanoparticles, carbon nanoparticles, hybrid carbon/mineral nanoparticulate and functionalized materials, as well as artificial polymer biomaterials filled by modified nanoparticles, catalysts, etc. are all topics that are discussed.

A common theme throughout this volume involves the adsorption and interfacial, especially biointerfacial, behaviour of all of the above mentioned nanomaterials. For environmental and human protection, the adsorption of heavy metal ions, toxins, pollutants, drugs, chemical warfare agents, narcotics, etc. is often desirable. A healthy mix of experimental and theoretical approaches to address these problems is described in various contributions. In other cases the application of materials, particularly for biomedical applications, requires a surface rendered inactive to adsorption for long term biocompatibility. Adsorption, surface chemistry, and particle size also plays an important role in the toxicological behaviour of nanoparticles, a cause for concern in the application of nanomaterials. Each one of these issues is addressed in one or more contributions in this volume.

We believe this volume holds a special niche in describing the current state of the art in the fundamentals and applications of a variety of nanomaterials. We thank all of the authors for their fine contributions, which make us proud to be editors of this book. We also thank the NATO Security Through Science Program for making the workshop and this series volume possible, for which we are most grateful. We sincerely hope you will find this volume useful.

January, 2006

Professor Jonathan Blitz – Charleston, IL, USA
Professor Vladimir Gun'ko – Kiev Ukraine
Co-editors

LIST OF CONTRIBUTORS

1. M. CHAPLIN
London South Bank University, Borough Road, London SE1 0AA, UK

2. B. VINCENT
School of Chemistry, University of Bristol, Bristol, BS8 1TS, UK

3. D. PONCELET
ENITIAA, Rue de la Géraudière BP 8225, 44322 Nantes Cedex 3, France

4. A. A. MALYGIN
Saint-Petersburg Institute of Technology (Technical University), 26 Moskovskii pr., Saint-Petersburg, Russia

5. G. ERTAS, S. SUZER
Bilkent University, Department of Chemistry and the Laboratory for Advanced Functional Materials, 06800 Ankara, Turkey

6. O. GERSHEVITZ, P. SILICKAS, C. N. SUKENIK
Department of Chemistry, Bar-Ilan University, Ramat Gan, ISRAEL 52900

7. M. S. MEL'GUNOV, V. B. FENELONOV
Boreskov Institute of Catalysis SB, RAS, Prospekt Akad. Lavrentieva 5, Novosibirsk, 630090, Russian Federation

8. V. B. FAINERMAN,[1] V. I. KOVALCHUK,[2] D. O. GRIGORIEV,[3] M. E. LESER,[4] AND R. MILLER[3]
[1]*Medical Physicochemical Centre, Donetsk Medical University, 16 Ilych Avenue, 83003 Donetsk, Ukraine*
[2]*Institute of Biocolloid Chemistry, 42 Vernadsky avenue, 03680 Kyiv (Kiev), Ukraine*
[3]*Max-Planck-Institut für Kolloid- und Grenzflächenforschung, 14424 Potsdam/Golm, Germany*
[4]*Nestec Ltd., Nestlé Research Centre, Vers-chez-les-Blanc, CH-1000 Lausanne 26, Switzerland*

9. V. GUN'KO,[1] R. LEBODA,[2] V. TUROV,[1] V. ZARKO,[1] A. CHUIKO[1†]
[1]*Institute of Surface Chemistry, 03164 Kiev, Ukraine,*
[2]*Maria Curie-Skłodowska University, 20031 Lublin, Poland*

10. Y. BOLBUKH,[1] V. TERTYKH,[1] B. GAWDZIK[2]
[1]*Institute of Surface Chemistry of National Academy of Sciences of Ukraine, Gen. Naumov Str. 17, 03164 Kyiv, Ukraine*
[2]*Faculty of Chemistry, Maria Curie-Sklodowska University, M.C. Sklodowska Sq. 3, 20031 Lublin, Poland*

11. L. G. GRECHKO,[1] L. B. LERMAN, [1] O. YA. POKOTYLO, [1] N. G. SHKODA, [1] A. A. CHUIKO†, [1] K. W. WHITES[2]
[1]Institute of Surface Chemistry, 17 General Naumov Street, Kyiv 03164, Ukraine
[2]South Dakota School of Mines and Technology, Rapid City, South Dakota 57701-3995, USA

12. J. SKUBISZEWSKA-ZIĘBA,[a] R. LEBODA,[a] V. M. GUN'KO,[b] B. CHARMAS[a]
[a]Maria Curie-Sklodowska University, Maria Curie Sklodowska Sq.3, 20-031 Lublin, Poland
[b]Institute of Surface Chemistry, 17 General Naumov Street, 03164 Kiev, Ukraine

13. J. SKUBISZEWSKA-ZIĘBA,[a] R. LEBODA,[a] V. M. GUN'KO,[b] O. SELEDETS[a]
[a]Maria Curie-Sklodowska University, Maria Curie Sklodowska Sq.3, 20-031 Lublin, Poland
[b]Institute of Surface Chemistry, 17 General Naumov Street, 03164 Kiev, Ukraine

14. H. J. MATHIEU,[1] X. GAO, [1] Y. CHEVOLOT[2] D. J. BALAZS[3]
[1]École Polytechnique Fédérale de Lausanne (EPFL)-IMX, Station 12, CH-1015 Lausanne, Switzerland
[2]Equipe Biotechnologie, UMR 5512 CNRS/ECL, Lyon/France
[3]Swiss Federal Lab. of Materials Testing and Research, EMPA-St. Gallen/Switzerland

15. W. NORDE
Laboratory of Physical Chemistry and Colloid Science, Wageningen University, Dreijenplein 6, 6703 HB Wageningen, and University Medical Centre Groningen, University of Groningen, Antonius Deusinglaan 1, Groningen, The Netherlands

16. V. M. GUN'KO, V. V. TUROV, A. A. CHUIKO†
Institute of Surface Chemistry, 17 General Naumov Street, 03164 Kiev, Ukraine

17. A. CHUIKO†,[1] A. PENTYUK,[2] E. SHTAT'KO,[2] N. CHUIKO[3]
[1]Institute of Surface Chemistry, 17 Generala Naumova Str., Kyiv 03164, Ukraine
[2]Vinnitsa State Medical University, 56 Pirogova Str., Vinnitsa 21018, Ukraine
[3]Institute for Occupational Health of the Academy of Medical Sciences of Ukraine, 75 Saksaganskogo Str., Kyiv 01033, Ukraine

18. S. L. JAMES,[1] M. ILLSLEY,[2] S. E. JAMES,[2] E. MENDOZA,[3] S. R. P. SILVA,[3] P. VADGAMA,[5] P. TOMLINS,[4] S.V. MIKHALOVSKY[1]
[1]School of Pharmacy and Biomolecular Science, University of Brighton, Moulsecoomb, Brighton BN2 4GJ, UK
[2]Blond McIndoe Centre, Queen Victoria Hospital, East Grinstead RH19 3DZ, UK
[3]Nano-electronics centre, Advanced Technology Institute, University of Surrey, Guildford GU2 7XH, UK
[4]Materials Centre, National Physical Laboratory, Queens Road, Teddington, Middlesex TW11 0LW, UK
[5]IRC in Biomedical Materials, Queen Mary, University of London, Mile End Road, London E1 4SN, UK

19. P. V. GRANT,[1] C. M. VAZ, [1] P. E. TOMLINS, [1] L. MIKHALOVSKA, [2] S. MIKHALOVSKY,[2] S. JAMES,[2] P. VADGAMA [3]
[1]Division of Engineering and Process Control, National Physical Laboratory, Hampton Road, TW11 0LW, United Kingdom
[2]Department of Pharmacy and Biomolecular Sciences, University of Brighton, Cockcroft Road, Brighton, BN2 4GJ, United Kingdom
IRC in Biomedical Materials, Queen Mary, University of London, Mile End Road, London, E1 4NS, United Kingdom

20. A. SPANOUDAKI, D. FRAGIADAKIS, K. VARTZELI-NIKAKI, P. PISSIS, J. C. RODRIGUEZ HERNANDEZ[†], M. M. PRADAS[†]
Department of Physics, National Technical University of Athens, 15780 Athens, Greece
[†]Center for Biomaterials, Universidad Politecnica de Valencia, E-46071 Valencia, Spain

21. B. FUBINI,[1] I. FENOGLIO, G. MARTRA,[1] R. CESCHINO,[1] M. TOMATIS,[1] R. CAVALLI,[2] M. TROTTA[2]
[1]Dip. di Chimica Inorganica, Chimica Fisica e Chimica dei Materiali and Inter-departmental Centre "G. Scansetti" for Studies on Asbestos and Other Toxic Particulates, University of Torino, via Pietro Giuria 7, 10125 Torino, Italy
[2]Dip. di Scienza e Tecnologia del Farmaco, University of Torino, via Pietro Giuria 9, 10125 Torino, Italy

22. O. V. SHULGA, C. PALMER
Department of Chemistry, University of Montana, Missoula, MT 59812, USA

23. J. BARKAUSKAS
Vilnius University, Naugarduko 24, 03225 Vilnius, Lithuania

24. V. A. POKROVSKIY, N. P. GALAGAN, A. A. CHUIKO[†]
Institute of Surface Chemistry, 17 General Naumov Street, 03164 Kiev, Ukraine

25. I. FENOGLIO, M. GHIAZZA, R. CESCHINO, F. GILLIO, G. MARTRA, B. FUBINI
Dip. di Chimica Inorganica, Chimica Fisica e Chimica dei Materiali and Inter-departmental Centre "G. Scansetti" for Studies on Asbestos and Other Toxic Particulates and Centre of Excellence of Nanostructured Interfaces and Surfaces (NIS) University of Torino, via Pietro Giuria 7, 10125 Torino, Italy

26. P. P. GORBIK,[1] L. P. STOROZHUK,[1]A. A. CHUIKO,[1†] L. Yu. VERGUN,[2] V. F. CHEKHUN[3]
[1]Institute of Surface Chemistry 17 General Naumov Street, Kyiv 03164, [2]Institute of Hematology and Transfusiology, 12 Berlinskiy Street, Kyiv 04060, [3]Kavetsky Institute of Experimental Pathology, Oncology and Radiobiology, 45 Vasyl'kivska Street, Kyiv 03022, Ukraine

27. P. KUZEMA, O. STAVINSKAYA, O. KAZAKOVA, I. LAGUTA
Institute of Surface Chemistry of National Academy of Sciences of Ukraine, 17 General Naumov Street, Kyiv, 03164, Ukraine

28. O. TSENDRA, A. DATSYUK, V. LOBANOV, A. GREBENYUK, A. CHUIKO†
Institute of Surface Chemistry, 17 General Naumov Street, Kiev 03164, Ukraine

29. M. JARONIEC, O. OLKHOVYK
Department of Chemistry, Kent State University, Kent, Ohio 44242, USA

30. I. P. BLITZ, J. P. BLITZ, V. M. GUN'KO[1], D. J. SHEERAN
Eastern Illinois University, Charleston, IL 61920 USA
[1]*Institute of Surface Chemistry 03164 Kiev, Ukraine*

31. H. KNÖZINGER
Department Chemie und Biochemie, Universität München, Butenandtstrasse 5-13, Haus E, 81377 München, Germany,

32. BURCU MIRKELAMOGLU, GURKAN KARAKAS
Department of Chemical Engineering, Middle East Technical University, 06531 Ankara, Turkey

33. S. BARANY,[1] J. GREGORY,[2] A. SHCHERBA,[3] I. SOLOMENTSEVA[3]
[1]*University of Miskolc, Institute of Chemistry, H-3515 Miskolc-Egyetemváros, Hungary*
[2]*University College London, Gower Street, London, WC1E 6BT, UK*
[3]*Institute for Electrodynamics, National Academy of Sciences of Ukraine, Kiev, Ukraine*

34. W. JANUSZ, M. MATYSEK
Department of Radiochemistry and Colloid Chemistry, Maria Curie Skłodowska University, pl. M. C. Skłodowskiej 3, 20-031 Lublin, Poland

35. K. LÁSZLÓ,[1] K. KOSIK, [1] E. WILK, [1] E. GEISSLER[2]
[1]*Department of Physical Chemistry, Budapest University of Technology and Economics, Budapest 1521, Hungary*
[2]*Laboratoire de Spectrométrie Physique CNRS UMR5588, Université J. Fourier de Grenoble, BP 87, 38402 St Martin d'Hères, France*

36. A. M. VOLODIN,[1] A. F. BEDILO,[1] D. S. HEROUX,[2] V. I. ZAIKOVSKII,[1] I. V. MISHAKOV,[1] V. V. CHESNOKOV,[1] K. J. KLABUNDE[3]
[1]*Boreskov Institute of Catalysis, Novosibirsk, 630090, Russia;*
[2]*Department of Natural Sciences, University of Maine at Farmington, Farmington, ME, USA;*
[3]*Department of Chemistry, Kansas State University, Manhattan, KS, USA, 66506*

37. D. PALIJCZUK,[a] R. SZMIGIELSKI,[a] V. M. GUN'KO,[b] R. LEBODA[c]
[a] *Military Institute of Chemistry and Radiometry, 00-910 Warsaw, Poland,*
[b] *Institute of Surface Chemistry, 17 General Naumov Street, 03164 Kiev, Ukraine*
[c] *Faculty of Chemistry, Maria Curie-Skłodowska University, 3 Maria Curie-Skłodowska Square, 20-031 Lublin, Poland*

38. R. SZMIGIELSKI,[a] D. PALIJCZUK,[a] R. LEBODA,[b] V. M. GUN'KO,[c] J. SKUBISZEWSKA-ZIEBA[b]
[a] *Military Institute of Chemistry and Radiometry, 00-910 Warsaw, Poland,*
[b] *Faculty of Chemistry, Maria Curie-Skłodowska University, 3 Maria Curie-Skłodowska Square, 20-031 Lublin, Poland*
[c] *Institute of Surface Chemistry, 17 General Naumov Street, 03164 Kiev, Ukraine*

39. V. I. ZARKO,[1] V. M. GUN'KO,[1] L. S. ANDRIYKO,[1] E. V. GONCHARUK,[1] M. MATYSEK,[2] E. SKWAREK,[2] W. JANUSZ[2]
[1] *Institute of Surface Chemistry, 17 General Naumov Street, 03164 Kiev, Ukraine*
[2] *Faculty of Chemistry, Maria Curie-Sklodowska University, 20031 Lublin, Poland*

WATER STRUCTURING AT COLLOIDAL SURFACES

MARTIN CHAPLIN*
London South Bank University, Borough Road, London SE1 0AA, UK

Abstract. Colloids in contact with water exert control over the arrangement of the first shell of the surrounding water by means of polar, dispersion and directed hydrogen-bonding effects. The preferred orientations of this first shell water are affected by the favored orientations of the second and more distant aqueous shells. If the colloid surface is flexible, it enables greater freedom of movement within the surface water molecules whereas if the surface is fixed, the adjacent water is more static and more extensively structured. The surface structuring of water is affected by, and will affect, the thermodynamics and kinetics for the binding of other molecules to the surface. In this paper, examples of the organization of water at colloidal surfaces are described and general conclusions drawn. These examples include (C_{60}-I_h)[5,6] fullerene, β-helix antifreeze proteins, and oxomolybdate clusters.

Keywords: antifreeze protein; β-helix; C_{60}; fullerene; hydration; hydrophilic surface; explicit water model; ice crystal; icosahedral water cluster; polyoxomolybdate; colloid

1. Introduction

Water consists of small molecules (H_2O) with significant dipole moment. The hydrogen atoms are positively charged with depleted electron density away from the electron dense oxygen atom. The charge effect alone would cause water molecules to be attracted to each other, but the attraction has added directionality due to the location of the electron depletion, away from the oxygen atom and in line with the O-H covalent bond. Increased directionality results from the partial covalency of the resultant hydrogen bonding which is most evident when the three atoms O-H···O are linearly arranged with the O-atoms close to 0.28 nm apart. Liquid water possesses a unique set of

* To whom correspondence should be addressed. Martin Chaplin, London South Bank University, Borough Road, London SE1 0AA, UK; email: martin.chaplin@lsbu.ac.uk

J.P. Blitz and V.M. Gun'ko (eds.), Surface Chemistry in Biomedical and Environmental Science, 1–10.

properties due to the high density of this hydrogen bonding, almost equivalent to the density of covalent bonding, and their preferred tetrahedral organization. Although hydrogen bonds are weak, if compared to covalent bonds, they form cooperatively with the result that most water molecules act as within clusters rather than individually. The size and density of these water clusters is determined by the temperature, pressure, solutes and the presence of surfaces; low density water with more extensive tetrahedral clustering being most evident at low temperatures and pressures. It has previously been shown that there is a sufficient and broad evidential base for the tendency of such water to ideally form icosahedral water clusters (Figure 1a), which accommodates the unique properties of water.[1,2]

2. Liquid Water Clustering at Surfaces

The colloidal solid-liquid water surface shows complex structuring that may involve several water layers. It depends on the direct surface-water interactions, polarization effects, the extent, charge and shape of the surface, rotational diffusion and interactions with secondary hydration layers and solutes. The structuring has both binding enthalpic and breaking entropic contributions, with the main difference between these surface-water interactions and bulk water-water interactions being due to the unchanging nature of the surface compared with the shifting nature of bulk water. This lowers the entropic penalty for the surface binding of water.

There are several (equivalent) ways of describing what happens to the structure of water at surfaces:

1. Water molecules at hydrophobic surfaces lose hydrogen bonds and therefore have increased enthalpy. They compensate for this by doing pressure-volume work, i.e. the network expands to form low-density water with lower entropy (e.g. see Ref. 3).
2. Water covers a hydrophobic surface with clathrate-like pentagons in partial dodecahedra. This minimizes the loss of most of the hydrogen bonds, maximizes the van der Waals contacts between the hydrophobic solute and the water, while reducing the hydrogen bonding between the surface water molecules. The rotational restriction of the hydrogen bonds at such hydrophobic surfaces may be partially responsible for their increased strength.
3. The chemical potential of water is raised due to the bonds left dangling if water cannot directly interact with the surface. The potential of water is decreased by expanding the water network. This expansion has been verified experimentally.[4]

4. Water molecules at hydrophilic surfaces form hydrogen bonds to the surface. These develop because either (a) they are enthalpically stronger than water-water hydrogen bonds and so energetically able to break water-water bonding, or (b) because otherwise water's hydrogen-bonding sites may be left 'dangling' at the surface or (c) because the entropic cost is low so allowing enthalpically weaker bonding for the same free energy change.

Clearly if more than one of these factors is operational then the bonding will appear much stronger. The primary surface-water will tend to orient to facilitate hydrogen-bonded connections amongst them. Secondary water-water hydrogen bonding then tends to mutually form the most stable preferred clustering; this extensive clustering being determined by polarization, hydrogen bond cooperativity and the favored tetrahedral stereochemistry.

If the colloid surface possesses conformational flexibility, the water structuring will have reciprocated effects on the surface structure. This enables greater freedom of movement within such surface water molecules and increases the entropic cost of hydration and so reduces the extent, duration and degree of order within the surface hydrogen bonding. Although often apparently little regarded, sometimes ignored and frequently poorly or misleadingly modeled, the surface structuring of water will affect, and is affected by, the thermodynamics and kinetics for the binding of other molecules to the surface. When there is significant rotational diffusion, such as with small proteins, then hydrogen bonds will break and water molecules will be cast off within the volume corresponding to secondary hydration.

The icosahedral cluster model[1,2] offers a structure on which large molecules can be mapped in order to investigate their interaction with water within a three-dimensional hydrogen-bonded network, and here offers new insights into the ways fullerene and polyoxomolybdate molecules interact with water in aqueous solution.

3. Fullerene Hydration

An interesting, if lesser-known fact concerning (C_{60}-I_h)[5,6] fullerene, is that it can be dissolved in water on transference from an organic solvent using sonication without the need of stabilizers or chemical modification.[5] The result is a molecular colloidal solution containing a variety of negatively charged clusters. This C_{60} fullerene is an electronegative molecule showing some aromatic behavior in its twenty six-membered (but not its twelve five-membered) rings, with the π-orbitals biased outwards.[6] The solubility of individual C_{60} molecules may be explained if the fullerene sits (ideally) within an icosahedral water cluster (Figure 1) missing its inner water dodecahedron. All twenty remaining next-to-inner water molecules are ideally situated to form

HO-H··· π hydrogen bonds to each of the twenty six-membered rings in the fullerene, by the positioning of O-H orthogonally over these aromatic rings; the optimum positioning for the hydrogen bond to a benzene molecule. Such HO-H··· π hydrogen bonds are known to possess about half the binding energy of HO-H···O hydrogen bonds with O-H atoms centrally and vertically placed and the distances from the oxygen atom to the aromatic centroid of about 3.1-3.7 Å [7] (*c.f.* 3.2 Å for all 20 such bonds to fullerene in this model, Figure 1b). These twenty water molecules can then be further linked through their remaining three hydrogen bonds to the twelve water pentagons formed from the sixty fully triply hydrogen-bonded water molecules in the outer shell. The fullerene thus presents a negatively charged surface to the environment and thus the positively charged H-atoms of water, as found experimentally. In such a structure, the carbon atoms become centers of electron-deficiency and capable of interacting with lone pair electrons donated by extra water molecules. These extra water molecules have room to sit under the outer shell water molecules without unrealistic close contacts. An increased tendency to ionize by these carbon-linked water molecules increases the negative charge on the C_{60} molecules and makes the C_{60} solution acidic, as found experimentally.[8] The resultant positioning of up to six hydroxide ions (0.184 nm and 0.28 nm from the nearest C and O atoms respectively) increases the aromatic electron density so strengthening the HO-H···π hydrogen bonding in the surrounding inner aqueous shell. The corresponding hydrogen ions may be associated with the water in the immediately surrounding shell or the bulk. It is clear that (C_{60}-I_h)[5,6] fullerene interacts with water via weak π-hydrogen bonds, which enable an unstrained surface coating with water molecules, so promoting the fullerene's unexpected aqueous solubility.

The structure is also compatible with findings by Andrievsky[8] using piezogravimetry (20-24 H_2O per C_{60}) and agrees with low temperature differential calorimetry[9] where two types of water were evident, fully hydrogen bonded water melting at 0°C (~60 H_2O per C_{60}) showing a necessary hydrogen-bonding to hydrogen-bond deficient water, melting at –2.3°C (19±1H_2O per C_{60}) with 30% less enthalpy change. In icosahedrally hydrated C_{60} (Figure 1b), there are 20 inner sphere H_2O and 60 outer (second) sphere H_2O molecules.

C_{60} molecules in water also form colloidal clusters based on 3.4 nm diameter (carbon atoms) icosahedral arrangements of thirteen C_{60} molecules.[5] Here the C_{60} molecules are necessarily separated by water molecules to form clusters with this diameter.[10] Such an arrangement is shown in Figure 2 within an expanded, but now strain-free, cluster of water icosahedral clusters. The water network is formed by tetrahedral tricyclo decamer $(H_2O)_{10}$ structures connecting groups of four C_{60} molecules. The modeled diameter of the cluster

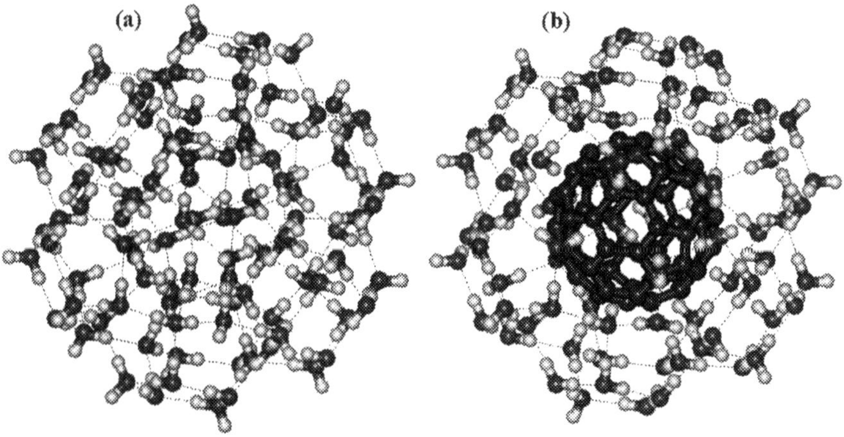

Figure 1. (a) The inner 100 molecules of water from the expanded icosahedral $(H_2O)_{280}$ network.[1] (b) A C_{60} molecule sitting inside this icosahedral water network by replacing the inner 20 molecules.

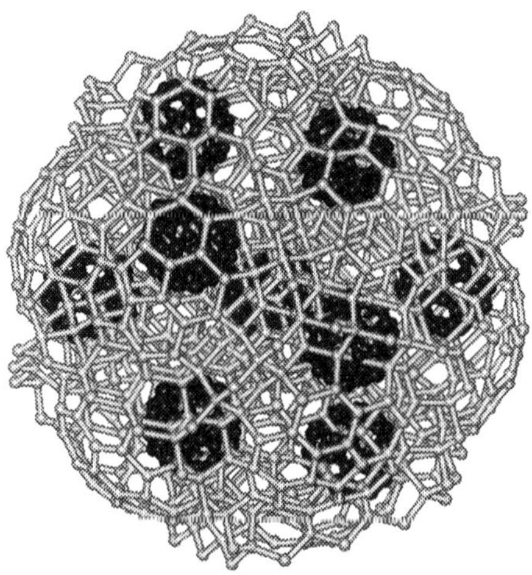

Figure 2. A tessellated network of icosahedral water clusters, $(H_2O)_{1020}$, surrounding and separating an icosahedral cluster of thirteen C_{60} molecules, shown black. The tetrahedral network of water's oxygen atoms is shown white.

(carbon atoms) is slightly larger at 3.5 nm but well within the experimental error. The required H-bond orientation toward the C_{60} molecules is allowed for this cluster size but become problematic as the condensed fully-tessellated cluster grows larger. It would, however, not offer any problem for more porous open tessellated structures. The ratio of inner sphere to outer (second) sphere water molecules varies between 1:3 for single molecules and 2:3 for infinite sized aqueous C_{60} clusters. It may be noted that ions that destroy the expanded water network also coagulate such C_{60} hydrosols.[11]

4. Water Structuring Around Ice

The growth of hexagonal ice crystals (ordinary ice) shows preference for particular faces. As such, ice formation may be considered as an extreme case of water clustering. Ice crystals may form by (slowly) growing on the basal faces in the direction of the c-axis, as inside vertical freezing pipes or where ice crystals grow down vertically from crystal platelets nucleated on still water with their c-axes vertical, and where sideways growth is prevented but axial growth allowed. Alternatively they may grow more rapidly from the prism faces, as at the disturbed randomly-directed surface of rapidly freezing or agitated lakes. Growth on the {11–20} face is even faster but forms new prism faces.[12] The hexagonal crystal structure plus the different growth characteristics of the crystal faces are behind the flat six-pointed shape of snowflakes. The relative speeds of this crystal growth, on the different faces depend on the ability of these faces to form greater extents of cooperative hydration (Figure 3).

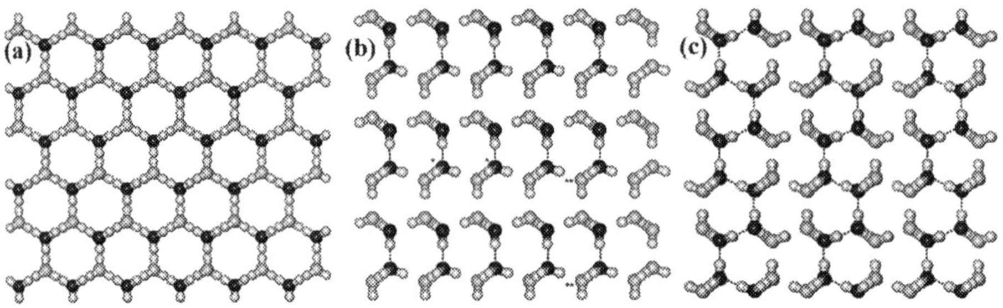

Figure 3. Views perpendicular to the faces of the ice (1h) crystal showing the next layer attached (with O-atoms black). (a) Basal face; where only isolated water molecules attach. (b) Prism face, where pairs of newly-attached water molecules may hydrogen bond to each other; one hydrogen bond/two water molecules. The distance between equivalent water molecules are 0.452 nm (marked *) and 0.738 nm (marked **). (c) {11-20} face, where chains of newly-attached water molecules may cooperatively hydrogen bond to each other; one hydrogen bond/water molecule. These form ridges which divide and encourage conversion into two prism faces.

Ice inhibitor (antifreeze) proteins achieve their objective in a number of ways. Some antifreeze proteins possess regular arrays of hydrogen-bonding and hydrophobic sites that fit well with the ice lattice. Surprisingly it is not just the hydrophilic hydrogen bonding groups but also the hydrophobic interactions that appear to be of particular relevance in preventing the growth of ice crystals. Amongst these are proteins with β-helix domains that bind to faces of the hexagonal ice crystal.[13,14] In the *Tenebrio* antifreeze protein there are tandem 12-residue repeats (TCTxSxxCxxAx) that form a β-helix with regularly spaced threonine residues (0.744 nm and 0.464 nm) that make a match to water molecules in the ice prism plane (0.738 nm and 0.452 nm, Figure 3b).[13] Importantly, not only is there a good fit with these hydrophilic hydrogen-bonding sites but also the associated methyl groups fit into the interstitial spaces in the ice's hexagonal shells, so creating multiple van der Waals contacts and reducing the free space.

5. Polyoxomolybdate Hydration

The icosahedral water cluster (Figure 1a) is a highly symmetrical idealized structure found in supercooled liquid water. The inner four shells of the cluster, consisting of 160 oxygen atoms (mostly water molecules) have been found experimentally in almost identical positions and orientations within a cavity-encapsulated icosahedral nanodrop of water in a polyoxomolybdate $\{(NH_2)_3C\}_{20}$. $\{(Mo)Mo_5O_{21}(H_2O)_6\}_{12}$. $\{Mo_2O_4(SO_4)\}_{10}$. $\{Mo_2O_4(H_2PO_2)\}_{20}$. $100H_2O$ (Figure 4a).[15] The water nanodrop is held to the molybdate by its pentagonal (Mo)Mo_5 clusters (Figure 4b).

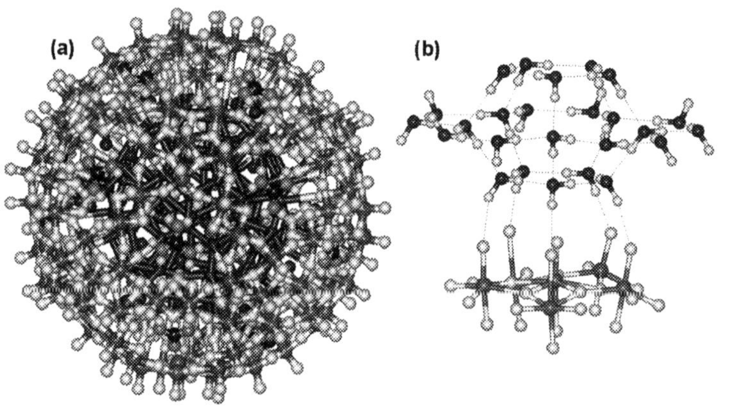

Figure 4. (a) Polyoxomolybdate nanodrop containing twelve internal pentagonal boxes, see (b). The 160 internal oxygen atoms (mainly water molecules corresponding to the icosahedral water clustering) are shown in black. (b) Linkage of water (above) to the pentagonal (Mo)Mo_5 groups (below). Five hydrogen-bonded links form between pentagonal boxes of water molecules $(H_2O)_{25}$ in the nanodrop and each of twelve such groups.

Recently some properties of a polyoxomolybdate super-cluster have been described.[16] Its most remarkable feature is that it is formed from about 1165 polyoxomolybdate $[\{(Mo)Mo_5Mo_5O_{33}(H_2O)_5\}(H_2O)_n]_{14}$ (henceforth abbreviated as $\{Mo_{154}\}$) clusters that form the surface of the 90 nm diameter spherical super-cluster. The mechanism for such clustering is initially difficult to discern as (a) each $\{Mo_{154}\}$ shows seven-fold symmetry with seven (Mo)Mo_5 groups alternating on each outside edge of a hollow tire-shaped cluster, and (b) the surface density of the $\{Mo_{154}\}$ clusters on the spherical super-cluster is such that they must be separated from each other by several layers of water.[16] The super-cluster model described here presupposes that the wheel-shaped mixed valence polyoxomolybdate $\{Mo_{154}\}$ clusters are hydrated through their pentagonal $(Mo)Mo_5$ groups as described above for the polyoxomolybdate nanodrop (Figure 4b). This then allows preferential linking between two nanowheels (Figures 5) utilizing similar water clustering arrangement to that found in supercooled water.

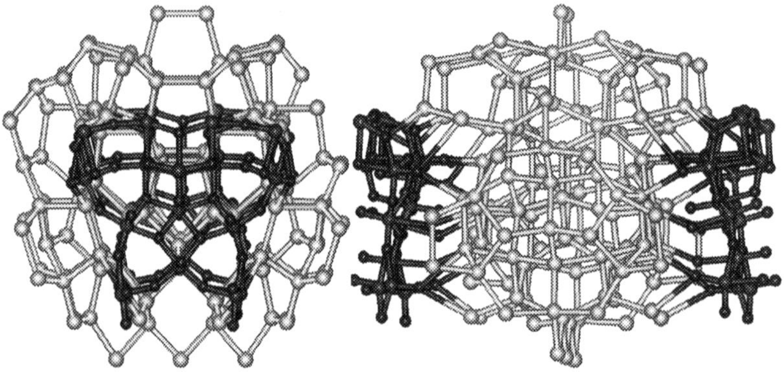

Figure 5. Two $\{Mo_{154}\}$ clusters can be held to each other through a water network involving two pentagonal boxes (one associated with each $(Mo)Mo_5$), each with five hydrogen bonded links (as Figure 4b) plus four further hydrogen bonded links from each of the local $(Mo)Mo_5Mo_5O_{33}$ clusters and associated links from neighboring clusters and including 45-plus shared molecules from the partial icosahedral water clusters. Water O-atoms are shown white with the Mo cluster atoms darker. The Mo—Mo distance between the two central surface Mo atoms of the Mo clusters is about 1.59 nm.

These $\{Mo_{154}\}$ dimers can themselves dimerize to form tetramers and then link to form octamers eventually forming the pentagons (five tetramers formed from 20 $\{Mo_{154}\}$ clusters) and hexagons (six tetramers from 24 $\{Mo_{154}\}$ clusters) on the surface of the super-cluster and further stabilized by the ring closure, as shown in Figure 6.

The route to the supercluster can thus be seen as a number of associations driven by increasing hydrogen bonded links (as Figure 5).

			Hydrogen-bonded links/$\{Mo_{154}\}$
$Mo_{154} + Mo_{154}$	$\longrightarrow$	$\{Mo_{154}\}_2$	0.5
$2\ \{Mo_{154}\}_2$	$\longrightarrow$	$\{Mo_{154}\}_4$	1.0
$2\ \{Mo_{154}\}_4$	$\longrightarrow$	$(\{Mo_{154}\}_4)_2$	1.25
$3\ (\{Mo_{154}\}_4)_2$	$\longrightarrow$	$[(\{Mo_{154}\}_4)_2]_3$	1.375
$40\ [(\{Mo_{154}\}_4)_2]_3$	$\longrightarrow$	$([(\{Mo_{154}\}_4)_2]_3)_{40}$	1.5

The overall structure of the supercluster has the following topology (Figure 6) where each vertex represents one of the two central Mo atoms (one inside the other relative to the sphere) in the linking $(Mo)Mo_5$ groups holding the water H-bonded connecting clusters.

Each vertex, therefore, corresponds to a tetrameric $\{Mo_{154}\}_4$ clusters which also stretch out along the links. The model radius is 45.1 nm which is the same as the hydrodynamic radius of the supercluster found experimentally by static light scattering (45 nm) and determined from the radius of gyration (45.2 ± 1.4 nm) and by less precise, as somewhat dehydrating, transmission electron microscopy (35-45 nm).[16] This accounts for 960 of the $\{Mo_{154}\}$ clusters, considered as 12 pentagons each made from five tetramers plus 30 hexagons each made from six tetramers. Additionally, each tetramer has two further equivalent, but exclusive, stabilized sites towards the inside of the tetramer, available for the optional binding of a further single $\{Mo_{154}\}$ cluster. If these sites are fully decorated, an average cluster size up to 1200 nanowheels may develop, $([\{(Mo_{154})_5\}_2]_3)_{40}$, in line with the Zimm plot value of ~2.54 x 10^{-7} g mol^{-1} (~1,165 nanowheels at 85% decoration).[16]

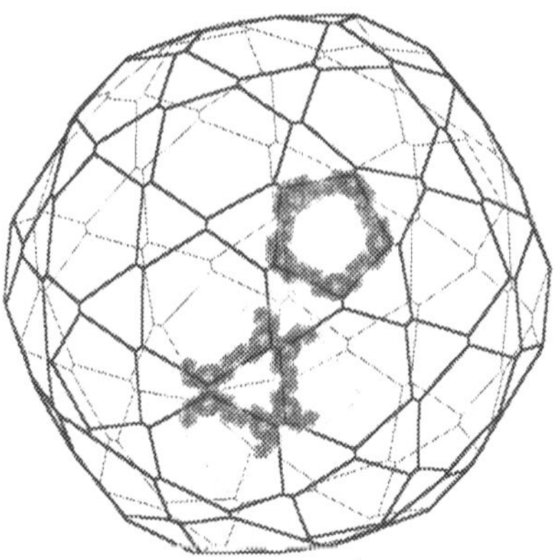

Figure 6. The polyoxomolybdate $\{Mo_{154}\}$ supercluster showing how 20 nano-wheels may form one of the 12 pentagons and how 24 nanowheels may form one of 80 distorted hexagons. The short links are the Mo—Mo linkages (~1.59 nm) as in Figure 5 and the long links (~12.36 nm) along the octamers are close to the major peak found in the SAXS experimental data (12.9 nm).[17]

Summing up, various oxomolybdates possess $(Mo)Mo_5$ ring structures that strongly hydrogen bond to water, forming almost unstrained water clusters that can extend for several aqueous shells and enable extensive colloidal clustering. The organized interfacial water surrounding $\{Mo_{154}\}$ clusters cause cluster self-association into very large spherical shells with diameter 90 nm and containing a large volume of water. The $\{Mo_{154}\}$ clusters only contact each other through the water network as evidenced by comparison of the spherical shell size, its mass and the size of its components. This is a clear, if unusual, example of how a relatively small amount of interfacial water can organize very large volumes of liquid water.

6. Conclusions

The surface properties of colloids depend on the organization and structuring of their aqueous environment, which is best modeled, at the present time, using explicit water clusters. The orientation of bound water molecules depends on a number of factors including the surface charge distribution, polarization and optimization of the secondary hydration. Access to this surface by other solutes will depend on the strength and extent of this hydration layer, which will control both the thermodynamics and kinetics of binding.

Further and updated information is available elsewhere.[2]

References

1. M. F. Chaplin, A proposal for the structuring of water, *Biophys. Chem.* 83, 211-221 (1999).
2. M. F. Chaplin, Water structure and behavior, (August 1, 2005); http//www.lsbu.ac.uk/water/.
3. T. V. Chalikian, Structural thermodynamics of hydration, *J. Phys. Chem. B* 105, 12566-12578 (2001).
4. T. R. Jensen, M. Ø. Jensen, N. Reitzel, K. Balashev, G. H. Peters, K. Kjaer, and T. Bjørnholm, Water in contact with extended hydrophobic surfaces, Direct evidence of weak dewetting, *Phys. Rev. Lett.* 90, 086101 (2003).
5. G. V. Andrievsky, V. K. Klochkov, E. L. Karyakina, and N. O. Mchedlov-Petrossyan, Studies of aqueous colloidal solutions of fullerene C_{60} by electron microscopy, *Chem. Phys. Lett.* 300, 392-396 (1999).
6. M. Bühl and A. Hirsch, Spherical aromaticity of fullerenes, *Chem. Rev.* 101, 1153-1183 (2001).
7. T. Steiner, Hydrogen bonds from water molecules to aromatic acceptors in very high-resolution protein crystal structures, *Biophys. Chem.* 95, 195-201 (2002).
8. G. V. Andrievsky, 2001, Personal communication.
9. M. V. Korobov, E. B. Stukalin, N. I. Ivanova, N. V. Avramenko, and G. V. Andrievsky, DSC study of C_{60} - water system, unexpected peaks, in: The exciting world of nanocages and nanotubes, edited by P. V. Kamat, D. M. Guldi, and K. M. Kadish, The Electrochemical Society Inc., Pennington, NJ, USA, *Fullerenes,* 12, 799-814 (2002).
10. G. V. Andrievsky, V. K. Klochkov, A. B. Bordyuh, and G. I. Dovbeshko, Comparative analysis of two aqueous-colloidal solutions of C_{60} fullerene with help of FTIR reflectance and UV–Vis spectroscopy, *Chem. Phys. Lett.* 364, 8-17 (2002).
11. N. O. Mchedlov-Petrossyan, V. K. Klochkov, G. V. Andrievsky, and A. A. Ishchenko, Interaction between colloidal particles of C_{60} hydrosol and cationic dyes, *Chem. Phys. Lett.* 341, 237-244 (2001).
12. H. Nada and Y. Furukawa, Anisotropy in growth kinetics at interfaces between proton-disordered hexagonal ice and water, A molecular dynamics study using the six-site model of H_2O, *J. Crystal Growth*, Article in press (2005).
13. Y-C. Liou, A. Tocilj, P. L. Davies, and Z. Jia, Mimicry of ice structure by surface hydroxyls and water of a β-helix antifreeze protein, *Nature* 406, 322-328 (2000).
14. E. K. Leinala, P. L. Davies, D. Doucet, M. G. Tyshenko, V. K. Walker, and Z. Jia, A β-helical antifreeze protein isoform with increased activity. Structural and functional insights, *J. Biol. Chem.* 277, 33349-33352 (2002).
15. A. Müller, H. Bögge and E. Diemann, Structure of a cavity-encapsulated nanodrop of water, *Inorg. Chem. Commun.* 6, 52-53 (2003); Corrigendum: A. Müller, H. Bögge, and E. Diemann, *Inorg. Chem. Commun.* 6, 329 (2003).
16. T. Liu, E. Diemann, H. Li, A. W. M. Dress, and A. Müller, Self-assembly in aqueous solution of wheel-shaped Mo_{154} oxide clusters into vesicles, *Nature* 426, 59-62 (2003).
17. A. Müller, E. Diemann, C. Kuhlmann, W. Eimer, C. Serain, T. Tak, A. Knöchel, and P. K. Pranzas, Hierarchic patterning, architectures beyond 'giant molecular wheels', *Chem. Commun.* 1928-1929 (2001).

MICROGELS AND CORE-SHELL PARTICLES

BRIAN VINCENT
School of Chemistry, University of Bristol, Bristol, BS8 1TS, UK
e-mail: brian.vincent@bristol.ac.uk

Abstract. An overview of the synthesis and applications of microgels and core-shell particles is provided, with emphasis on work originating from the author's laboratory. Microgels, which are cross-linked polymer latex particles, can be used for selective uptake of ions or polymers, or the controlled release of various compounds. Various methods for the synthesis of core-shell particles are described such as interfacial polymerization, layer-by-layer deposition, "colloidosomes", internal phase separation, and silica shells. The release kinetics for controlled (sustained or triggered) release purposes is discussed.

Keywords: microgel particles, core-shell particles, controlled release, triggered release

1. Introduction

Along with the rapid development over the last few decades in our theoretical understanding of colloidal dispersions and experimental techniques to study them, there have been significant advances made in the design and synthesis of colloidal particles of various kinds. The objective has usually been to design particles with a given structure and properties, having some specific function or application in mind. Some parameters which it is now possible to control well include: (a) size and shape (e.g. spheres, ellipsoids, rods, plates), including their distributions; (b) rheological properties (solid, gel, liquid particles); (c) magnetic and electrical properties; (d) porosity; (e) surface properties (charge density, hydrophobicity/hydrophilicity, grafted polymers); (f) non-uniformity, e.g. *Janus* and patchy particles; and (g) core/shell particles of various kinds.

One widespread application, which has been much researched in recent years, is the use of colloidal systems for controlled uptake and release purposes.

J.P. Blitz and V.M. Gun'ko (eds.), Surface Chemistry in Biomedical and Environmental Science, 11–22.

This process is widely-used in a range of technologies, including: pharmaceuticals (drugs); agrochemicals (pesticides, herbicides and fungicides), food processing (flavors); household and laundry products (perfumes); printing (inks). The nature of the colloidal particles used depends on the nature of the uptake/release process and the type of "active" molecules involved. Microgel particles and core/shell particles may both be used for sustained or triggered release. In addition, as described below, microgel particles also may be used for selective uptake purposes.

The purpose of this article is to review the use of microgel particles and of core shell particles, primarily in the context of selective uptake and controlled release applications. Most of the examples will be taken from the work of the author and his coworkers in Bristol, carried out over many years of researching this topic.

2. Microgel Particles

2.1. INTRODUCTION

Microgel particles are cross-linked polymer latex particles which swell in a good solvent environment for the polymer concerned.[1,2] Swelling and de-swelling of the particles can be induced, therefore, by changes in the local thermodynamic conditions for the polymer chains (e.g. temperature, solvency, osmotic pressure, and for polyelectrolytes pH or ionic strength). The maximum extent of swelling is controlled by the degree of cross-linking co-monomer introduced.

Various types of (heterogeneous) polymerization processes have been described in the literature for making microgel particles, depending on the nature of the polymer involved. Some examples are: (a) dispersion polymerization for making, e.g. poly(N-isopropylacrylamide) [PNIPAM] particles[3] or poly (vinylpyridine) particles [PVP][4] (the cross-linking (difunctional) monomers used were bisacrylamide and divinylbenzene, respectively); (b) for more water-soluble monomers (and polymers), inverse emulsion polymerization may be used, e.g. for poly (acrylic acid - co - acrylamide) particles [PAAc-co-AM];[5] (c) condensation (step-growth) polymerization, e.g. for cross-linked poly(dimethylsiloxane) [PDMS] microgel particles.[6] Of course, as well as the cross-linking monomer, other co-monomers may also be introduced, in order to introduce specific functional groups into the interior of the microgel particles. A typical example here would be cross-linked poly(N-isopropylacrylamide-co-acrylic acid) [PNIPAM-co-AAc] microgel particles. Such particles will have carboxylic acid groups distributed throughout the interior network of the microgel particles.

One important consideration concerns the distribution of (i) the cross-linker groups and (ii) any co-monomer functional groups, through out the body of the microgel particles. These distributions are controlled by the reactivity ratios of the various monomers concerned and the nature of the polymerization process (batch, "starve-fed", etc). Although some small-angle neutron scattering studies[7] have been aimed at determining the cross-link density distributions in swollen microgel particles, more work needs to be carried out to both determine and control these distributions.

Another issue to be addressed concerns the stability of the microgel particle dispersions to aggregation. A significant body of work has now been established on this topic.[8] In the swollen state, i.e. when the microgel polymer is in a good solvent environment, then the particles are *thermodynamically stable*. This is because of the van der Waals attraction between the particles is insignificant, since the Hamaker constants of the (solvent swollen) particles and of the continuous phase are more-or-less matched. Thus, there is no driving force for aggregation. In terms of the basic thermodynamic relationship governing any physico-chemical process,

$$\Delta G_{floc} = \Delta H_{floc} - T\Delta S_{floc} \tag{1}$$

in good solvent conditions, $\Delta H_{\text{floc}} << T\Delta S_{\text{floc}}$, (both negative terms) and so ΔG_{floc} is positive, (Note that ΔS_{floc} is associated with loss in translational entropy of the particles when they flocculate). However, as the solvency is made worse, and the particles de-swell, the inter-particle van der Waals attraction starts to increase, and so a point is reached where $\Delta H_{\text{floc}} > T\Delta S_{\text{floc}}$ and weak, reversible aggregation is observed.[9]

In order to resist such aggregation, classical electrostatic stabilization of the microgel particles may be introduced. This is achieved, as with normal latex particles, by having a layer of surface charge groups on the microgel particles. Such groups are usually associated with the initiator used in the polymerization process, as in a classical dispersion polymerization. In an inverse emulsion polymerization the surface charge is normally associated with the ionic surfactant molecules used to stabilize the original inverse emulsion droplets. In a condensation polymerization (e.g. of PDMS microgel particles) the surface charge groups come from the terminal, acidic silanol (Si-OH) groups at the ends of the PDMS chains. An alternative procedure to avoid aggregation under de-swollen conditions would be to introduce steric stabilization of the microgel particles. This might be achieved, for example, by co-polymerizing in, towards the end of a dispersion polymerization reaction, a suitable macro-monomer (e.g. PEO-methacrylate).

2.2. SELECTIVE UPTAKE

For selective uptake, porous solid or microgel particles are both effective, but the latter give greater flexibility in application, because of their inherent swelling/de-swelling feature. In order to effect entry/penetration of the desired molecules, some driving force (energy gain) is, in general, required to overcome the entropy loss endured by the entering molecules. Examples from the work of this group include the use of designed microgel particles for the selective preferential uptake of harmful ions (e.g. Cd, Pb and Hg species) over more benign ions (e.g. K, Na, Ca and Mg) in potable waters,[10,11] and also the uptake of polymers (e.g. poly(ethylene oxide) [PEO]), with a critical MW cut-off.[12] In both cases the microgel particles used were based on cross-linked poly(N-isopropylacrylamide-co-acrylic acid) [PNIPAM-co-AAc)] polymer latex particles. In the case of the ions, preferential adsorption of the harmful ions occurred at higher pH vales (>pH 7), when the AAc groups were in the dissociated, carboxylate ion form. The uptake of PEO, however, was most effective at low pH values (<pH 4) when the AAc moieties were in the undissociated carboxyl acid form. In the case of the cations the primary mechanism of binding was electrostatic, whilst in the PEO case it was relatively strong H-bonding between the ether oxygen atoms of the PEO and the –OH of the carboxylic acid moieties. In order to effect *desorption*, all that was required, in both cases, was a suitable change in pH to weaken the binding energy.

2.3. CONTROLLED RELEASE

With microgel particles triggered release is more usual than sustained release. Triggered release is usually effected by changing some thermodynamic property of the system which weakens the binding of the active molecule to the functional sites within the microgel particle. Alternatively, a "displacer" molecule, which interacts with the binding site more strongly, may also be used. The release rate for triggered release will normally be first-order (although it may be more complex if a variety of binding sites, having different strengths, are involved). A two-stage triggering method is also a possibility. For example, light could potentially be used as a primary trigger, if its application would induce a secondary (thermodynamic) trigger, such as a temperature rise or a local pH change.

An important aspect to bear in mind with microgel particles is that they may well also change their dimensions (swell/de-swell) when the local thermodynamic conditions are changed. With larger active molecules such as polymers or proteins (or indeed small nanoparticles), this becomes a very important consideration. Swelling/de-swelling will lead to a change in effective pore size, dramatically affecting the uptake and/or release rate. This principle may actually be utilized to "trap" large molecules (or small nanoparticles) inside the cross-linked network forming the interior of the microgel particles. This might be useful in trying to "protect" vulnerable species (e.g. enzymes or drug molecules) during a multi-stage process, where one stage involves a potentially "harmful" environment. Examples here might be an orally administered drug passing through the very low pH conditions of the stomach into the more benign intestine region, or an enzyme in a laundry formulation surviving the hot wash cycle, before entering the cooler rinse cycle in a typical washing machine process. In cases like these, it is useful to be able to invoke *two* triggering mechanisms: one to effect swelling/de-selling of the microgel network, and one to effect adsorption/desorption of the active species. For example, with the PNIPAM-co-AAc microgel particles referred to earlier, the NIPAM groups are temperature sensitive (the lower consolute solution temperature for high MW PNIPAM in water is ~32°C), and the AAc groups are pH sensitive. Hence, temperature changes could be used to trigger the swelling/de-swelling process and pH changes to trigger the adsorption/desorption process.

3. Core-Shell Particles

A typical core-shell particle is illustrated in Figure 1.

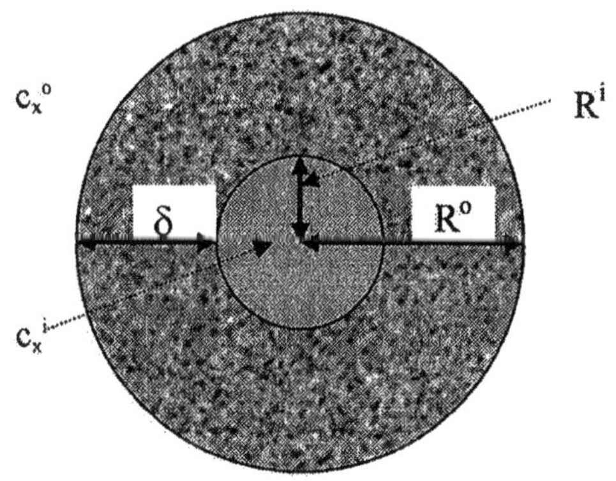

Figure 1.1 Typical core/shell particle.

The core may be a solid or a liquid, or indeed a gas but this is unusual in the context of release applications. Hollow particles have found applications in other areas such as surface coatings, where they offer a high refractive index contrast with film itself, and therefore good light scattering properties. The shell material is usually a solid, but its nature will depend on the type of release process required. Polymers or hard inorganic materials (e.g. silica or calcium carbonate) may be used.

A variety of methods has been described in the literature for preparing core/shell particles which are described below.

(a) Interfacial polymerization in emulsion systems. This is the most widely established method. A condensation reaction occurs at the oil/water interface between two reactants, one of which is soluble in the oil phase, and the other is soluble in the continuous phase. A typical example[13] would be the reaction between a polyisocyanate, dissolved in oil droplets, and an acid or base in the continuous aqueous phase. Partial hydrolysis of the isocyanate groups occurs, leading to amine groups which can react with the remaining isocyanate groups, forming a polymeric film at the interface of each emulsion droplet. The main limitation of this type of interfacial polymer film formation is that the film thickness is self-limiting, as its formation prevents further reactants in each phase meeting readily.

(b) Layer-by-layer deposition.[14] This involves the sequential deposition of alternating layers of polyelectrolytes of opposite charge onto solid particles or emulsion droplets. In this manner, many such layers can be built up. The only problem is that this method is somewhat time-consuming, as after each new layer has been deposited, the continuous phase has to be "cleaned" of any unadsorbed polyelectrolyte before the next layer is added. Nevertheless thick, robust shells of controlled porosity may be constructed.

(c) "Colloidosomes".[15] This involves the deposition of a close-packed monolayer of small, monodisperse polymer latex particles at the O/W interface of larger emulsion droplets. Heating of the particles above their T_g value results in some lateral fusion between neighboring particles in the monolayer, such that small "holes" are left in the film. The hole size will be controlled by the temperature used and the length of time heating is carried out, but this will be the main feature controlling diffusion of any active species out of the droplets. A similar method has been used in this group[16] to prepare core/shell particles with a *solid* core. This involved the deposition of positively-charged PVP microgel particles onto larger, negatively charged silica particles, to form as close-packed a layer as possible. Further vinylpyridine monomer and (oil-soluble) initiator were then added, which were taken up by the PVP layer. Polymerization was then re-initiated which caused the PVP particle layer to form into a more-or-less continuous shell.

(d) Internal phase separation. This method was developed in this group. The first publication[17] described the formation of oil core/polymer shell particles, where the oil was a hydrocarbon (typically hexadecane) and the shell was a polymer (typically poly(methylmethacrylate) [PMMA]). The method involves firstly the formation of an oil/water emulsion. The oil phase consists of two organic liquids (plus the active "molecules", if required): one liquid is a low vapor pressure/poor-solvent for the (shell-forming) polymer; the second liquid is a high vapor pressure/good solvent for the polymer (e.g. for PMMA, dichloromethane). In fact sufficient good-solvent is added to just dissolve the polymer in the liquid mixture. The aqueous phase contains a surfactant to stabilize the emulsion, which is formed by some suitable procedure (e.g. using a homogenizer or cross-flow membrane device) to give the required droplet size. The second stage involves the removal of the good solvent form the oil droplets, leading to phase separation of the polymer inside the droplets. This removal can be achieved by evaporation of the high vapor pressure solvent through the aqueous phase. However, the rate of evaporation must be carefully controlled; if it is too fast then "holes" will appear in the polymer shell which forms around the oil droplets. In fact, whether such a shell forms or not, depends on the polymer phase separation process itself, and the relative magnitudes of the various interfacial tensions involved. There are three such interfacial tensions: γ_{po}, γ_{pw}, and γ_{ow}, where p = polymer, o = oil and w = water. If one wishes to form a shell then, in effect, it is imperative that γ_{ow} is not smaller than γ_{po} or γ_{pw}. This is because, for a core/shell particle, one only wants to form p/o and p/w interfaces, and no o/w interface. If this condition is not met, then structures resembling "acorns" form, that is, two "blobs" of polymer and oil adhering to each other, both surrounded by water. Thus the choice of the emulsifying surfactant is very important. Classical surfactants, such as sodium dodecylsulfate, lead to acorns, since they lower the o/w interfacial tension too much. It was found that polyelectrolytes, such as poly(methacrylic acid), worked well for shell formation, since they adsorb at the o/w interface but do not lower the o/w interfacial tension too much.

More recently, this method has been successfully extended by us[18] to form the inverse systems, i.e. water core/polymer shell particles dispersed, initially in oil, but then transferred to an aqueous continuous phase. Clearly, whether one needs an oil or a water core depends on the nature of the active material to be released. Now one starts with a water/oil emulsion, rather than an oil/water emulsion, but the basic principles are very similar. A variety of shell polymer systems were prepared, including PMMA and poly(tetrahydrofuran) [PTHF]. The high vapor pressure liquid used in this case was in general, acetone. It turned out, however, that these water core systems are intrinsically more difficult to make than the equivalent oil core systems, because large amounts of acetone were required to dissolve the polymers initially in the water-acetone mixtures. An oil was then required which did not mix too well with acetone. In general, mineral oil worked reasonably well. In order to transfer the water core capsules into an aqueous continuous phase, the particles were centrifuged in tubes containing an upper oil phase with the dispersed capsules and a lower aqueous phase, containing a suitable surfactant to effect transfer of the capsules across the oil/water interface and into the aqueous phase.

(e) Silica shells. As well as polymer shell systems, this group has also successfully made liquid core/*silica* shell particles, with both oil cores[19] and water cores.[20] The method of making the oil core systems is again based on starting with an o/w emulsion. The emulsions systems chosen were the monodisperse, charge-stabilized silicone oil/ water emulsions, first described by Obey and Vincent.[21] These are prepared by a nucleation and growth (rather than the more traditional comminution) route. They are surfactant free systems, the droplets being stabilized by the dissociated silanol groups at the droplet/water interface.[22] It is also possible to cross-link the silicone oil droplets *in situ*,[6] to form dispersions of silicone microgel particles in water. The formation of the silicone oil/silica capsules, in water, was achieved in a two-stage process. The first stage involved the formation of a thin "skin" of silicate around the silicone droplets (or microgel particles), by deposition from a saturated sodium silicate solution, and lowering the pH to precipitate the silicate onto the droplets. This initial skin formation is necessary, so that the second stage can proceed successfully. This involves addition of tetraethoxysilane [TEOS] to the aqueous phase, now containing ethanol and ammonia; this hydrolyses the TEOS to tetrahydroxysilane, which in turn condenses to form silica; the precipitating silica deposits onto the particle, leading to the formation of a thick silica shell. The thickness is, of course, controlled by the amount of TEOS added initially. Without the initial protective "skin" of silicate around the oil droplets, the TEOS would simply dissolve into the silicone oil droplets!

The capsules based on a *water* core with a silica shell have been prepared more recently.[20] In this case a water/oil emulsion is used, where the aqueous phase contains a base (or acid) and TEOS added to the external oil phase. The TEOS is hydrolyzed at the oil/water interface and a silica layer is formed *in situ* at this interface.

3.1. CONTROLLED RELEASE

Core/shell particles exhibit a number of features which make them very useful for controlled (sustained or triggered) release purposes. The release kinetics may be zero-order (constant rate) or first order (exponentially decreasing with time. In sustained release, the release kinetics is primarily determined by the structure of the core. Solid cores (i.e. of the active molecule itself) will give rise to zero-order kinetics during sustained release through a shell. Clearly, there must, in this case, be a reverse flow of liquid from the continuous phase back into the core, across the shell, in order to maintain the integrity of the shell. If the active molecules are dispersed as nanoparticles within a *liquid* core (within which they have a finite, but limited, solubility), then zero-order release kinetics will again be observed, at least until the nanoparticles disappear from the core. However, if the active molecules are merely present in solution throughout the release process, then first order kinetics will be observed during release. The equation given below is for such a first-order, sustained release process.

$$\frac{dc_x^o}{dt} = \frac{4\pi R^o R^i P\left(c_x^o - c_x^i\right)}{\delta} \tag{2}$$

The various symbols are defined in Figure 1, except P, which is the permeability of the shell; this depends on its porosity, the solubility of the active molecules (X) in the shell material, and the diffusion constant for X in the shell material. Clearly the shell material will be selected with these properties in mind for a particular active molecule. Once the shell material has been chosen, the shell thickness (δ) becomes a primary control parameter. In two recent papers from this group[23,24] the roles of several chemical and physical properties of the polymer shells have been systematically investigated in terms of their affect on the release kinetics of a model active molecule (4-nitroanisole) from a hydrocarbon core into an aqueous continuous phase. Among the properties investigated were the chemistry and the T_g value of the polymer, post cross-linking of the polymer shell (to reduce porosity), and the shell thickness. It was shown that all of these factors have a significant effect on controlling the release rate, and also, to some extent, the release "yield" (defined as the percentage of the initial active molecule released from the capsules into the aqueous continuous phase). The capsules are stored as a very concentrated dispersion in water, since some escape of the active into the external phase is inevitable on storage, so the main factor controlling the yield is the dilution of the aqueous phase with water, but some of the active is retained in the shell, as well as the core. The final equilibrium distribution of the active material depends on the various partition coefficients involved. It is very important that there are no obvious holes in the shell through which active can escape rapidly. This is where the evaporation rate of the high vapor pressure solvent through the shell during capsule formation is critical, as referred to earlier.

As well as sustained release, core/shell particles may be used also for triggered release applications. In this case, as with the microgel systems discussed earlier, an external trigger is used to cause some change in the properties of the shell polymer, which allows the active molecules to diffuse across the shell into the exterior. With many active materials there is an optimum "dosage" that is required in the exterior, e.g. with drugs. This principle is illustrated in Figure 2. Too much drug over a given time period would be toxic, whilst too little would be ineffective.

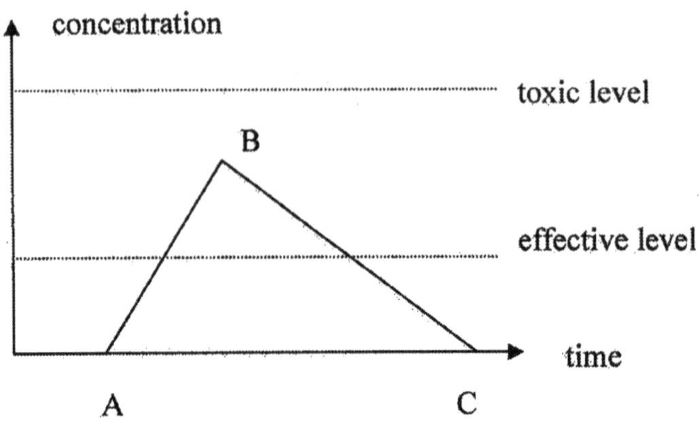

Figure 2. Triggered Release A→B is the release period and B→C is the consumption period for the active substance.

Various methods for triggering release may be used. One method is to add, say an acid or base, which causes (slow) dissolution of the polymer shell. Romero-Cano and Vincent[25] have described the release of 4-nitroanisole from oil core/polymer shell capsules into water, where the shell polymer was a polylactide, which dissolves slowly on adding acid. A second concept would be to use a cross-linked polymer shell which swells on changing pH, for example. This route has been described in a recent paper by us,[24] where PVP was used as the shell. As described earlier in the section on microgel particles, PVP protonates at low pH forming a cationic polyelectrolyte, and this leads to swelling of both PVP microgel particles and also PVP polymer shells.

The main mechanism for effecting triggered release is to use pressure to break the shell, and hence release the core contents into the external phase. This is particularly useful when the active material is a high MW molecule (e.g. a protein). The applied pressure can be in the form of simple mechanical pressure (e.g. in carbon paper copying) or high shear conditions. A more subtle method, however, is to use osmotic pressure to break the capsule. This can occur if solvent molecules from the external phase are able to diffuse across the shell into the core. If the core contains molecules (e.g. polymer) which are not able to diffuse across the shell, then clearly there will be an osmotic pressure difference across the shell, which will drive solvent to try to enter the core, leading to possible shell rupture.

One final point concerns the prevention of the inevitable release of active molecules into the external phase during storage in *sustained* release systems, referred to earlier. One potential way around this is to use a double-shell system. Here the inner shell imparts the sustained release properties, but the outer shell is impenetrable to the active molecules. Triggered breaking of this outer shell, at the required point in a given process, would then initiate the sustained release process. This topic is the focus of our current research in this subject.

4. Acknowledgements

I should like to thank the many postdocs, visitors, graduate and undergraduate students, who, over the years, have contributed to the research work in my group on both microgel and core/shell systems. The list is too long to enumerate here, but their names may be found in the reference section below. However, maybe I should just pick out the persons who started this work in each area with me: Dr (now Prof) Martin Snowden on microgels and Mr (now Dr) Andrew Loxley on core/shell systems. I should also like to thank the EPSRC, The European Commission and a range of industrial sponsors, again too many to list here, for financial support.

References

1. M. Murray and M. J. Snowden, The preparation, characterization and applications of colloidal microgels, *Adv. Colloid Interface Sci.* 54, 73-93 (1995).
2. B. R. Saunders and B. Vincent, Microgels as model colloids, *Adv. Colloid Interface Sci.* 80, 1-25 (1999).
3. M. J. Snowden and B. Vincent, The temperature-controlled flocculation of cross-linked latex particles, *J. Chem. Soc. Chem. Comm.* 42, 279-302 (1992).
4. A. Loxley and B. Vincent, Equilibrium and kinetic aspects of the pH swelling of poly(vinylpyridine) latex particles, *Colloid Polymer Sci.* 275, 1108-1114 (1997).
5. S. Nyret and B. Vincent, The properties of polyampholyte microgel particles prepared by microemulsion polymerization, *Polymer* 38, 6129-6134 (1997).
6. M. I. Goller, T. M. Obey, and B. Vincent, Inorganic silicone oil microgels, *Colloids Surf. A* 123/124, 183-193 (1997).
7. H. M. Crowther, B. R. Saunders, S. J. Mears, T. Cosgrove, B. Vincent, S. M. King, and G. E. Yu, Poly(NIPAM) microgel particle de-swelling: a light scattering and small-angle neutron scattering study, *Colloids Surf. A* 152, 327-333 (1999).
8. B. R. Saunders and B. Vincent, The aggregation of microgel dispersions, *Adv. Colloid Interface Sci.*, to be published (2006).

9. M. Rasmusson, A. Routh, and B. Vincent, Flocculation of microgel particles with sodium chloride and sodium poly(styrene sulfonate) as a function of temperature, *Langmuir* 20, 3536-3542 (2004).
10. G. M. Morris, B. Vincent, and M. J. Snowden, Adsorption of lead ions into N-isopropylacrylamide and acrylic acid copolymer microgels, *J. Colloid Interface Sci.* 190, 198-205 (1997).
11. H. M. Crowther, G. E. Morris, B. Vincent, and N. G. Wright (2003) in: *The Role of Interfaces in Environmental Protection*, NATO Science Series, IV Earth and Environmental Sciences, Vol. 24, edited by S. Barany (Kluwer, Dordrecht, 2003), pp. 169-180.
12. M. Bradley, J. Ramos, and B. Vincent, Equilibrium and kinetic aspects of the uptake of poly(ethylene oxide) by copolymer microgel particles of N-isopropylacrylamide and acrylic acid, *Langmuir* 21, 1209-1215 (2005).
13. S. Scherr, B. Rodson, and C. Lee, Encapsulation by Interfacial Polymerisation, *Pesticide Sci.* 54, 394-400, (1998).
14. E. Donath, A. Sukhornkov, F. Caruso, S. Davies, and H. Möhwald, Microcapsules from layer-by-layer polyelectrolyte deposition, *Angew. Chem. Int. Ed.* 37, 2201-2205 (1998).
15. D. Weitz, Colloidosomes, *Science* 298, 1006-1009 (2003).
16. R. Atkin, M. Bradley, and B. Vincent, Core-shell particles having silica cores and pH-responsive poly(vinylpyridine) shells, *Soft Matter* 1, 160-165 (2005).
17. A. Loxley and B. Vincent, Preparation of poly(methylmethacrylate) microcapsules with liquid cores, *J. Colloid Interface Sci.* 208, 49-62 (1998).
18. R. Atkin, P. Davies, J. Hardy, and B. Vincent, Preparation of aqueous core/polymer shell microcapsules by internal phase separation, *Macromolecules* 37, 7979-7985 (2004).
19. M. I. Goller and B. Vincent, Silica encapsulation of PDMS droplets, *Colloid Surf. A* 142, 281-285 (1998).
20. M. O'Sullivan and B. Vincent, (unpublished).
21. T. M. Obey and B. Vincent, Novel, monodisperse silicone oil/water emulsions, *J. Colloid Interface Sci.* 163, 454-463 (1994).
22. B. Neumann, B. Vincent, R. Krustev, and H.-J. Müller, Stability of various silicone oil/water emulsions as a function of surfactant and salt concentrations, *Langmuir* 20, 4336-4344 (2004).
23. P. J. Dowding, R. Atkin, B. Vincent, and P. Bouillot, Oil core/polymer shell microcapsules, I: characterization and release rates for microcapsules with polystyrene shells, *Langmuir* 20, 11374-11379 (2004).
24. P. J. Dowding, R. Atkin, B. Vincent, and P. Bouillot, Oil core/polymer shell microcapsules, II: controlling the release profile of active molecules, *Langmuir* 21, 5278-5284 (2005).
25. M. S. Romero-Cano and B. Vincent, Controlled Release of 4-nitroanisole from poly(lactic acid) nanoparticles, *J. Control. Release* 82, 127-135 (2002).

MICROENCAPSULATION: FUNDAMENTALS, METHODS AND APPLICATIONS

DENIS PONCELET
ENITIAA, Rue de la Géraudière BP 8225, 44322 Nantes Cedex 3, France, e-mail: poncelet@enitiaa-nantes.fr

Abstract. Microencapsulation is widely use in industry but remains relatively unknown from the public. The reason is that microcapsules are not an end-product, but generally a technique to overcome process limitations. Micro-encapsulation allows immobilization, protection, release and functionalisation of active ingredients. Despite the high diversity of methods, this paper proposes a classification and description of the main technologies to produce microcapsules.

Keywords: microencapsulation, immobilization, controle release

1. Introduction

In the last few years, one could see the development of commercial products based on microcapsules. However, microencapsulation has been widely used in industry for several decades. The principle of encapsulation is very old. If biochemistry is a principle of life, nothing would have been possible without its integration in membrane bound structures (cells, mitochondria...). Without immobilization and spatial organization of biochemical reactions in an internal volume and through the membrane would not be possible. The high efficiency of, for example ATP production, would not be possible.

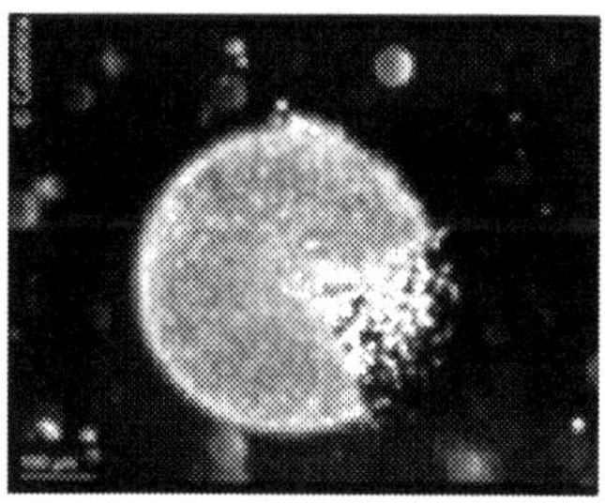

Figure 1. Multi-core microcapsules mimic biological cells and are sometimes called artificial cells. (Coletica®)

J.P. Blitz and V.M. Gun'ko (eds.), Surface Chemistry in Biomedical and Environmental Science, 23–34.

By developing encapsulation methods, scientists and engineers mimic nature to obtain innovative structures to isolate, protect, release and functionalize active ingredients.[1] However nature is not so easy to mimic, and what humans have developed are still inferior to what biological cells offer.

Encapsulation is used in many industrial and scientific domains. It is not surprising to find then diverse definitions and terminology, often directed to a specific field. However, a generic and functional definition could be

"Entrapment of a compound or a system inside a dispersed material for its immobilization, protection, controlled release, structuration and functionalization."

This definition is more oriented to objectives than on the structure of the microcapsules. It includes a very large number of systems starting from hollow molecules such as cyclodextrin, to large solid microsphers of 2 to 3 mm. It proposes a product-oriented approach, a solution that limits debate around terminologies.

If we look a little more deeply into this definition, the first question is which type of system could we encapsulate? This could range from small molecules (some try to encapsule water) to quite complex ones (peptides, drug, DNA). It could be a mix of these molecules, or complex structures like viruses, protoplasts or even complete biological cells. Inside the capsules, the active system could be in the form of a solution, a suspension or an emulsion.

Which type of structures could represent microcapsules? The "true" microcapsule is a liquid core surrounded by a membrane. However, many different structures are included under the term "microcapsules" or "nanocapsules" (Figure 2). At the smallest scale, one could use hollow molecules inside of which the active ingredient could be fixed. At a larger scale, more or less complex molecular assemblies could form nanocapsules, or nanospheres, or lipidic structures like liposomes. For sizes less than a few micrometers, one talks of nanoencapsulation. For larger sizes, one finds hydrogel beads, solid microspheres, and microcapsules. For sizes greater than 1 mm, some talk about macroencapsulation. Encapsulation could also include agglomeration of fine particles or the coating of solid particles. Finally, some include emulsions if they are stable enough to fit the above definition.

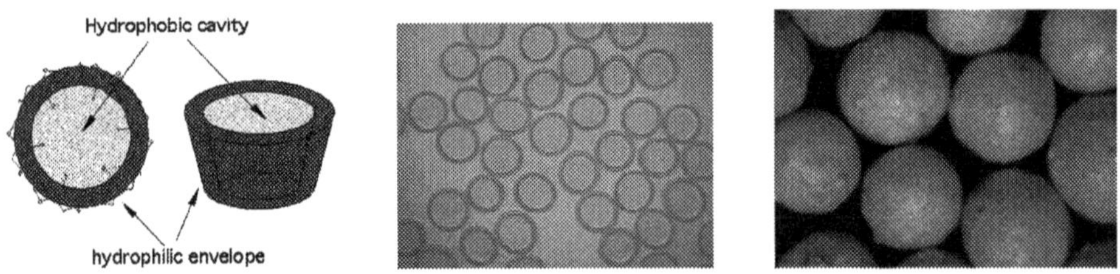

Cyclodextrin Nylon mirocapsules Alginate/amidon beads

Figure 2. Examples of microcapsule structures.

Parallel to the structural complexity, a large number of technologies exist to produce microcapsules, which is a field unto itself.

2. Why Encapsulation?

Since encapsulation is costly, the requirement must first be justified. We can classify five categories for the objectives of encapsulation.

- **Immobilization or entrapment.** To limit contact between certain parts of a system. If some ingredient must be separated, encapsulation of this ingredient and release only upon rupture of the microcapsules fills this objective. The entrapment of a flavor could create a sustained aromatic effect, or to control the release at a specific time (such as during cooking). Immobilization of batteries or enzymes allows continuous processing while avoiding washout.
- **Protection.** If some ingredients are fragile and need to be protected from their environment. For example, vitamins or polyunsaturated fatty acids are denaturized by oxygen. Many biological cells are sensitive to shear. Some drugs and probiotics are destroyed during gastric transit. When incorporated in microcapsules, all these systems will be protected to some extent against the chemical, physicochemical and mechanical environmental conditions. However, the problem may be reversed. Incorporation of iron in food promotes oxidation of fatty acids. A number of industrial additives may reduce the performance of the material itself. In this case, it is more efficient to encapsulate the minor ingredients (iron, additives). Encapsulation could then be used to protect the environment from the use of some products. Most industrial enzymes are sold in an encapsulated form to avoid allergic and professional health problems.
- **Controlled release.** For practical use the active ingredient must be released. A drug must be delivered with well defined kinetics. Sometimes it is not the encapsulated ingredient that is released but a by-product. This is the case when the encapsulated product is an enzyme or a catalyst. Encapsulation may have the objective to limit release, but in some cases to make it more rapidly available. A typical example is an instant powder consisting of aggregates made of fine particles that are insoluble, in a very soluble matrix.
- **Structuration.** Homogeneous mixing of a small liquid volume with a high volume of powder constitutes a real challenge. Microencapsulation allows converting this liquid in powder and facilitating this operation. dosage forms for pharmacy applications are readily obtained by microencapsulation. By

coating brown sugar, a quite aggregative powder, with crystalline sugar, one gets a flowing powder.

- **Functionalisation.** Finally, microencapsulation may be used to develop new functions such as regulating biocatalyst activity by controlling the membrane permeability through pH changes. Microcapsules may also offer a marketing function such as giving specific "metallic" aspects to functional food to differentiate them from food and medication.

The diversity of applications is very broad and even microencapsulation is already largely used in industry, one could expect a strong development in the next decade.

3. How to Make Capsules

Many applications from a variety of fields for diverse objectives have led to many methods of encapsulation. Moreover, terminology varies from domain to domain. The same technology may have different names in different fields. Figure 3 tries to offer an approach where most technologies fit in an unambiguous way.

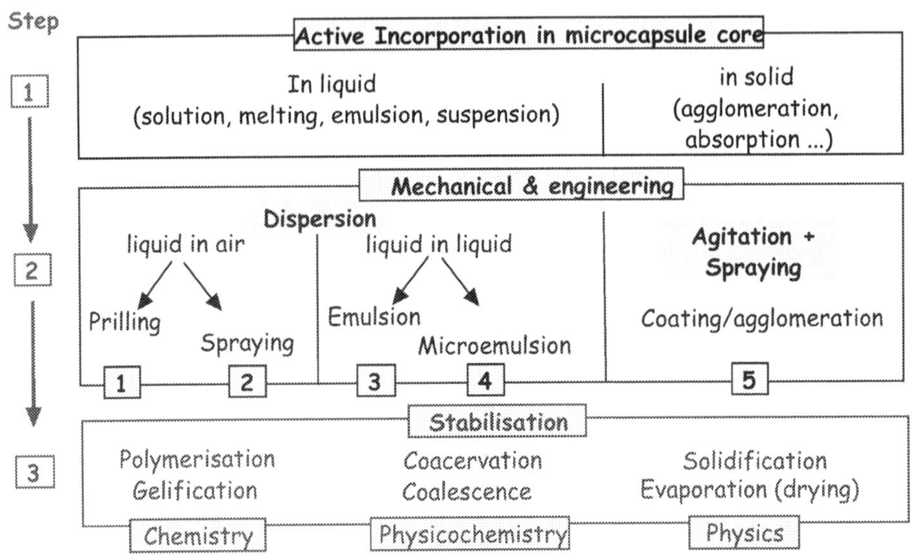

Figure 3. Technologies of encapsulation.

An encapsulation process may be generally divided into three steps.

1. The first step consists of incorporating the active ingredients in the matrix or microcapsule core. This may be in the form of a solution, emulsion or suspension, resulting in a liquid or a dispersed solid system. This could involve mixing or dispersing processes, drying, grinding and/or sieving.
2. The second stage is a mechanical operation.
   - For a liquid matrix, making a liquid–in-air (dropping or spraying) or liquid dispersion (emulsification or micro-emulsification).
   - For a solid matrix, spraying a solution on particles under agitation (fluid bed or pan coating or agglomeration).
3. The last step, is to stabilize/solidify droplets or the coating solution by a chemical process (polymerization), physicochemical process (gelation, coacervation) or physical process (drying, solidification).

These three steps may be repeated to reach the final structure. For example, microcapsules obtained by spray drying of polymer/active solution may be coated by a melted solution. The first encapsulation insures immobilization and stabilization of the active ingredient, while coating allows control of the level of protection and release.

4. Methods of Dispersion to Form Microcapsules

For incorporation of the active ingredient inside a liquid matrix (as a solution, emulsion, or suspension), the first step of encapsulation is the dispersion of this liquid as droplets, which can be classified into four categories.

4.1. PRILLING

This is simply improved methods to form droplets from a needle. The goal is to produce small droplets/microcapsules with low size dispersion (less than 10%) with a good level of production. To avoid broad size dispersion, the liquid flow must be in the laminar regime (avoiding turbulence), thus a relatively low flow rate is required compared to spraying (see below). In most cases, energy is required to reduce the droplet size (from a few millimeters with simple needle). This has led to the following systems.

- **Electrostatic generators**.[2] By application of an electrostatic potential on a pending droplet (Figure 4a), charges accumulate on its surface creating a repulsion which opposes the surface tension. The resulting droplets will then have small sizes down to 20 micrometers for high voltage (10 kV). However, the flow rate is limited to only a few milliliters per hour and the system only useful for laboratory scale or some medical applications (artificial organs).

- **Nozzle resonances technology.**[3] While forming a jet from a nozzle, this jet has a spontaneous tendency to break into droplets. By applying a vibration at a specific frequency (Figure 4b), uniform droplets are formed with a size approximately double the jet diameter. Several liters per hours can be reached for larger size droplets of 1 mm, but decreases proportionally to the droplet diameter. On the other hand, resonance is damped if the solution viscosity is too high.
- **Jet cutter.**[4] The liquid jet can also be cut by a series of wires fixed on a turntable while turning at high speed (Figure 4c). This method provides high flow rates but is mainly useful for adequate for high viscosity solutions.
- **Spinning disks.**[5] Liquid may be flowed onto a spinning disk. Jets are formed and break into droplets (Figure 4d). The flow rate can be quite high, but literature on this subject is still limited. Further research is needed to evaluate the real performance of such systems.

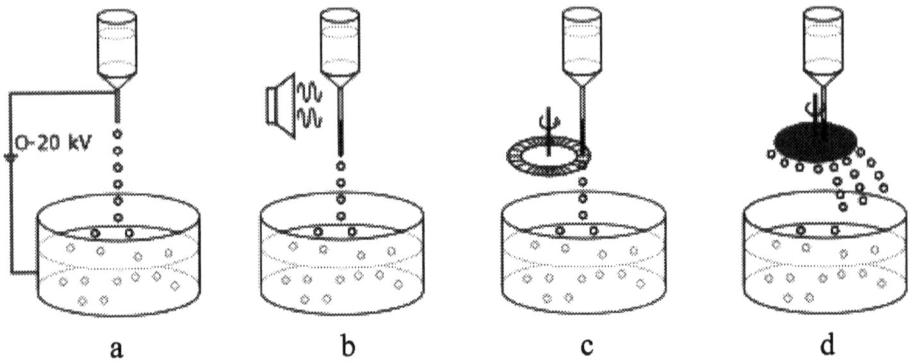

Figure 4. Scheme of different prilling methods.

4.2. SPRAY TECHNOLOGIES[6]

By flowing either through an air/liquid nozzle (Figure 5a) or on a fast rotating device (Figure 5b), a liquid can explode into fine droplets. Productivity is far superior to prilling technology, but turbulence leads to a large size dispersion (often higher than 30%).

Reactors used for spray technologies are quite similar to powder dryers, the technology of which is widely available and well understood. This technology is largely used in the food industry.

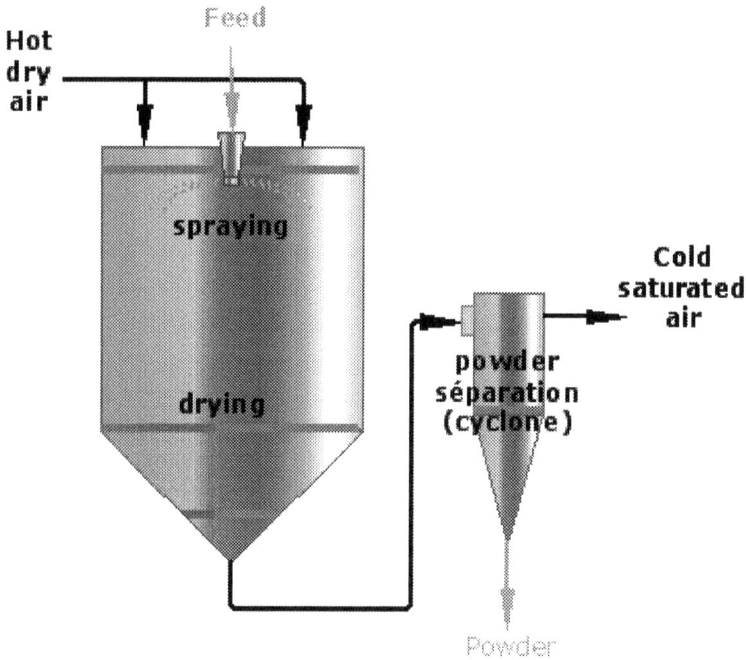

Figure 5. Spray systems.

4.3. EMULSIFICATION[7]

Liquid containing the ingredient to be encapsulated can be dispersed in an immiscible liquid to form an emulsion. For microencapsulation, one favors systems allowing dispersion at low shear to avoid denaturation of the active ingredients (for example biological cells). The simplest method is a reactor equipped with a turbine (Figure 6a). However, there is an increased interest for continuous systems, especially static mixers. This consists of a pipe where elements are inserted to promote fine division of the liquids in the static mixer (Figure 6b). Such systems allow emulsion production in a fraction of second at high flow rate (a few liters per square centimeter of section). In the majority of cases, dispersion takes place in a turbulent regime and the resulting droplet size dispersion is large (greater than 30% standard deviation).

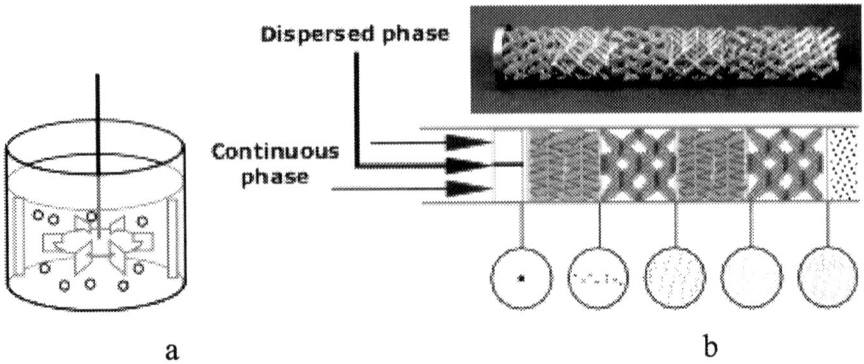

Figure 6. Liquid/liquid emulsification systems: (a) mechanical dispersion with a turbine; (b) a static mixer.

4.4. MICROEMULSIFICATION

By careful selection of the composition of a two immiscible liquid system, one can reduce the surface tension to near zero. In such conditions, with gentle agitation, a very fine dispersion (under micrometer size droplets) may be formed. This is a very stable system for an emulsion. In some conditions the size dispersion may be limited.

5. From a Liquid Dispersion to Microcapsules

All the above dispersions lead to droplets. These droplets must be transformed into solid like particles by a stabilization process. Table 1 provides the most usual encapsulation technologies by crossing dispersion methods and stabilization methods. Table 1 is not an exhaustive list.

Table 1. Usual terminology in microencapsulation technology.

	Prilling	Spraying	Emulsification	Micro-Emulsification
Solidification	Hot-melt prilling	Spray cooling		
Evaporation		Spray-drying	Solvent evaporation	Solvent evaporation
Gelation	Gelation	Spray chilling	Thermal gelation	
Polymerisation			Interfacial Polymerisation	In situ polymerisation
Coacervation	Interfacial Coacervation		Coacervation	
Micellar formation				Liposomes

- **Solidification.** The liquid forming the droplet can be melted and a reduction of temperature will result in droplet solidification. In practice, solidification is usually combined with prilling and spraying (but is easily combined with emulsification)
- **Evaporation.** As a liquid droplet is formed of a volatile solvent and a polymer, evaporation of the solvent will lead to polymer beads entrapping the active ingredients. Spray drying consists of spraying a (aqueous) polymer solution and droplet drying. Emulsification of polymer volatile organic solvent in water followed by solvent removal is called the "solvent evaporation" method.
- **Gelation.** By dropping droplet of gel forming solution in a gelation bath, hydrogel beads are formed. The gelation may be due to ionic bonding between polymer chains (such as an alginate solution dropped in calcium ions bath) or by cooling (such as an agarose solution). Gelation may also be used by spraying a thermogel (spray chilling) or through emulsification followed by cooling or pH change.
- **Polymerisation.** Emulsified droplets containing a monomer can react with a second monomer soluble in the continuous phase to form a membrane at the interface (i.e. diamine reacting with a acid dichloride). This is called interfacial polymerization. Many derivative methods can be set-up from this method, using pre-polymers in place of monomers, inversing the continuous and dispersed phases, developing a radical reaction. Covering all possible methods is not possible here.
- **Coacervation.** If an oil phase is emulsified in a polymer water solution, and the polymer is precipitated (for example) by changing the pH, the polymer precipitate (coacervate) has a tendency to accumulate at the interface. This is the coacervation process; called simple if one polymer is involved and complex if two polymers are involved. If the coacervation is obtained by dropping one polymer solution into a polymer solution of opposite charge, this is termed interfacial coacervation, or "polyelectrolyte complex formation".
- **Micellar.** By dispersing a surfactant and often a polymer, one may obtain a small assembly with diverse structures. The most well known is a liposome, generally represented by a bi-layer cell like structure. However, the structure can range from a stable emulsion to complex multi-layer vesicles.

All these capsules may need to be separated, washed and sometimes dried. They may be solid microspheres, liquid core capsules, or hydrogels beads. They may be further treated or coated to obtain the final desired properties.

6. Methods of Encapsulation by Coating, Agglomeration and Layering[8]

Coating of solid particles requires agitating the particles (to avoid agglomeration) and spraying coating solution on them. The coating must be stabilized or solidified for example by cooling or drying. The process can be repeated until a good and uniform coating is formed.

Spraying too fast, in the solidification process, results in agglomeration. However, if well controlled, the agglomeration may allow the formation of large particles from fine powders. One may also start from small particles, and spray the active ingredient to form layers of active materials. This is called layering.

Coating, agglomeration and layering can be combined to get diverse structures for the properties required by various applications.

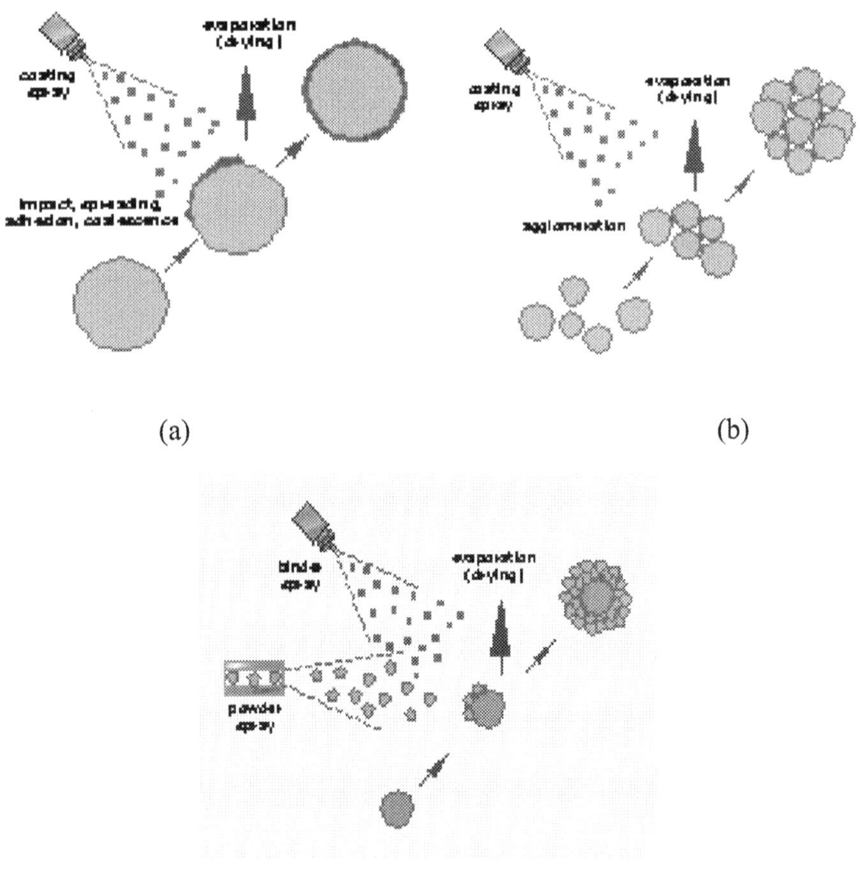

(a) (b) (c)

Figure 7. Technologies of coating (a), agglomeration (b), and layering (c). (Glatt international®)

Three actions are necessary to establish a coating:

- **Agitation.** Agitation has three main objectives: 1) to avoid unwanted agglomeration; 2) to create a spinning movement to particles for uniform coating; and 3) to circulate the particles to insure a homogenous coating over the batch. Agitation can be insured by an upper air stream leading to a suspended particle bed, having a similar behavior as a liquid and called a fluid bed (Figure 7a and b), or by a mechanical process such as rotation of a pan containing the particles (Figure 7c).
- **Spraying.** The coating or aggregative solution can be sprayed from the top (top spray, Figure 8a) or bottom (bottom spray or Wurster when a tube is inserted in the reactor, Figure 8b) or in the particle bed itself.
- **Solidification.** The most common for stabilizing the coating is to evaporate a solvent (drying in case of water) by injecting hot air (fluid bed) or heating the reactor (pan coating). However, there is an increased interest for hot melt coating, where a melted solution is sprayed on the particles. This reduces the time of processing and energy costs. The new powder dry technology consists of spraying simultaneously a very fine powder and a plastifiant. The powder will coalesce at the surface of the particles mainly at room temperature.

7. Microencapsulation and Chemistry

Microencapsulation covers a broad range of methods. A great number of variants exist around the above-described methods. The methods may also be combined. Engineers and scientists must refer to much scientific and technological knowledge, especially in chemistry. This includes:

- polymer chemistry for the synthesis and characterisation of polymers used during encapsulation;
- purification technology for extracting natural polymers, often used for hydrogel beads or coating;
- synthesis for controlling polymerization or cross-linking in diverse processes;
- physical and colloid chemistry to understand gelation and coacervation;
- and obviously formulation to get adequate support for the active ingredient and reach the optimum membrane structure.

Much progess has been made, but many challenges and opportunities remain in this field.

References

1. T. M. S. Chang and M. J. Poznansky, Microcapsules as artificial cells, *Science* 3:62-67 (1968).
2. B. Bugarski, B. Obradovic, V. Nedovic, and D. Poncelet, Immobilization of cells and enzymes using electrostatic droplet generation, in: *Fundamentals of Cell Immobilisation Biotechnology. Focus on Biotechnology Series*, Vol. 8A, edited by V. Nedovic and R. Willaert (Kluwer Academic Publishers, Dordrecht, 2004), pp. 277-294.
3. C. Heinzen, A. Berger, and I. Marison, Use of Vibration technology for jet break-up for encapsulation of cells and liquids in monodisperse microcapsules, in: *Fundamentals of Cell Immobilisation Biotechnology. Focus on Biotechnology Series*, Vol. 8A, edited by V. Nedovic and R. Willaert (Kluwer Academic Publishers, Dordrecht, 2004), pp. 257-274.
4. U. Pruesse and K.-D. Vorlop, The JetCutter technology, in: *Fundamentals of Cell Immobilisation Biotechnology. Focus on Biotechnology Series*, Vol. 8A, edited by V. Nedovic and R. Willaert (Kluwer Academic Publishers, Dordrecht, 2004), pp. 295-310.
5. E. Teunou and D. Poncelet, Rotary disk atomisation for microencapsulation applications – Prediction of the particle travel from the wheel, *J. Food Engin.* 71, 345-353 (2005).
6. G. A. Reineccius, The Spray Drying of Food Ingredients, in: *Microencapsulation of Food Ingredients*, edited by B. Per Vilstrup (Leatherhead Food RA, Leatherhead, UK, 2000), pp. 151-185.
7. R. Neufeld and D. Poncelet, Industrial scale encapsulation of cells using emulsification/ dispersion technologies, in: *Fundamentals of Cell Immobilisation Biotechnology. Focus on Biotechnology Series*, Vol. 8A, edited by V. Nedovic and R. Willaert (Kluwer Academic Publishers, Dordrecht, 2004), pp. 311-324.
8. E. Teunou and D. Poncelet, Fluid bed coating, *in Encapsulated and Food powders,* edited by C. Onwulata and R. Konstance (Marcel Dekker, New York, 2005), pp. 197-214.

THE MOLECULAR LAYERING METHOD: PROGRESS IN SCIENCE AND PRACTICAL WORKS FOR CREATION OF FUNCTIONAL NANOMATERIALS

A. A. MALYGIN
Saint-Petersburg Institute of Technology (Technical University), 26 Moskovskii pr., Saint-Petersburg, Russia, e-mail malygin@lti-gti.ru

Abstract. The Molecular Layering (ML) method is based on the chemisorption of reagents on a solid substrate surface. The main idea of the ML method consists of consecutive deposition of monolayers with structural units of set chemical composition on a solid surface.[1,2] The ML technique gives the possibility to form a wide variety of nanolayers (mono- and multilayer, multicomponent layers) on the surface of different solids (organic and inorganic powders, fibres, films etc.) with any geometrical form. Four basic effects characterise this technology.[3] They are (1) monolayer effect (ME), i.e. sharp changes in matrix properties after obtaining 1–4 mono-layers of new structural units; (2) effect of matrix overlapping (EME), i.e. the physical coverage of the solid surface (formed layer screens the surface); (3) effect of mutual co-ordination of matrix surface structure and formed nanolayer (EMC); (4) effect of the multicomponent system (MS), for example it can be a synergistic composition. Applications of the ML method can be defined from these effects in products such as adsorbents, catalysts, polymeric, metal, composition and other materials. The ML method can be important in the optimization of layer composition and thickness (for example when kernel pigments and fillers are produced), for intensification of chemical solid reactions, in sintering of ceramic powders, etc.

Keywords: nanotechnology, nanomaterials, molecular layering method, adsorbents, catalysts, nanoceramic, nanocomposition materials, pigments, nanofillers, polymers, retardants of combustibility

J.P. Blitz and V.M. Gun'ko (eds.), Surface Chemistry in Biomedical and Environmental Science, 35–44.

1. Introduction

The 21st century has been named the century of nanotechnology and nanomaterials.[4] The basic directions in the technology of solid materials are connected with new synthetic methods for obtaining these materials. Chemical approaches in the development of nanotechnologies will no doubt play a central role. One chemical approach in the field of obtaining new nanomaterials is the Molecular Layering (ML) method, the subject of this report.

2. The Molecular Layering Method as a Base of Chemical Nanotechnology

The Molecular Layering (ML) method and its main principles were formulated on the basis of solid state chemistry using the "framework hypothesis" proposed by the academician V.B. Aleskovskii in 1952.[1,2,5]

The framework hypothesis state that is any solid substance (irrespective of whether it is crystal or amorphous, organic or inorganic) consists of a skeleton (a base), on the surface of which functional groups can exist. Two basic kinds of transformations of the solid substance are possible: 1) reaction of the more inert skeleton; 2) reaction of surface functional groups which are much more reactive. Both types of reactions may occur simultaneously.

The main idea of the ML method consists of consecutive reactions of monolayers of set chemical composition on a solid surface. In performing the ML method a few basic principles must be kept in mind.

1. Synthesis is based on chemical reactions between the functional groups on a solid surface and reagent molecules brought from the outside in conditions far from equilibrium. Thus reagents and reaction products must not have chemical interactions.
2. To obtain second and subsequent monolayers, it is necessary to repeat and alternate (in the given sequence) reactions of the vapor phase reagents with functional groups of the surface. Thus each formed monolayer of new functional groups must contain reactive atoms or groups of atoms, capable of reacting with additional reagent.
3. For reactions in the ML process some structural conformity must exist between the solid surface and the reactant.

Figure 1 shows the scheme of chemical assembly of a solid surface by the ML method. On the initial surface are functional groups B. They can react with AC_4 forming a monolayer of new groups $=AC_2$. If the $=AC_2$ groups react with the reagent AB_4, a second monolayer containing A atoms in the structure of functional groups $=AB_2$ is formed. Another monolayer can be formed with reagent AC_4, then AB_4, etc. A nanolayer is formed, its thickness depends not on reaction time, but the number of chemical reaction cycles. If at different stages of synthesis different reagents are used, it is possible to synthesize nanolayers with alternating monolayers of varying chemical functionality. For example in reaction of $=AC_2$ on the solid surface with NB_4, we obtain a second monolayer containing different atoms N in NB_2 groups. Thus multi-component nanolayers can be formed.

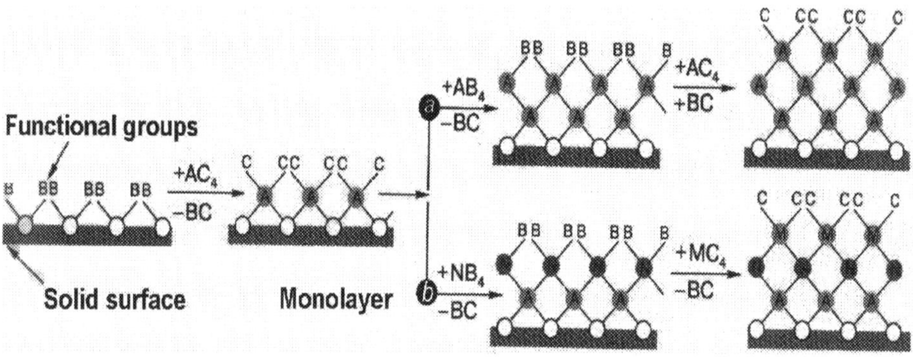

Figure 1. The chemical design of nanostructures on the surface of solids by the molecular layering method.

It is possible by the ML method to carry out chemical assembly step by step (monolayer by monolayer), by repeated surface reactions. The first nanolayers by the ML method were synthesized about 40 years ago in Saint-Petersburg Technological Institute at the Department of Chemistry of solids. There have been many publications in western countries detailing the analogs of the ML method, namely atomic layer epitaxy (ALE), and atomic layer deposition (ALD), since this techniques inception by Valentine Aleskovski and members of his science school.

Some advantages of the ML method include:

- strong (chemical) binding of sorbate with the surface of matrix;
- homogenous distribution of hetercomponents over the surface, i.e. high conformity of covering;
- a possibility to control the surface layer composition down to one monomolecular layer and to put different atomic layers one-by-one and to form multicomponent regular structures;
- low power consumption due to the use of chemical reaction heat for the synthesis;
- simplicity and flexibility of design that allows to synthesize materials for different purposes without additional investments.

A few examples of the creation of oxide nanolayers on the surface of silica gel are shown below.

$$2(\equiv Si\text{-}OH) + SiCl_4 \rightarrow (\equiv Si\text{-}O\text{-})_2SiCl_2 + 2HCl$$

$$2(\equiv Si\text{-}OH) + TiCl_4 \rightarrow (\equiv Si\text{-}O\text{-})_2TiCl_2 + 2\ HCl$$

$$3(\equiv Si\text{-}OH) + PCl_3 \rightarrow (\equiv Si\text{-}O\text{-})_3P + 3HCl$$

$$2(\equiv Si\text{-}OH) + CrO_2Cl_2 \rightarrow (\equiv Si\text{-}O\text{-})_2CrO_2 + 2HCl$$

The number of chemical bonds with the surface, and the concentration of new functional groups (FG) in a monolayer depends on the quantity and arrangement of OH groups on the initial surface, the chemical nature of reagents, and the synthesis conditions.

An important moment: in one cycle of reaction only one monolayer of a new FG is formed. Unlike traditional technologies, time does not affect mass or layer thickness. It can said that the ML method is a self-organized process.

To obtain the second monolayer it is necessary to take into account surface properties of the new monolayer. For example in the second stage of ML:

$$(\equiv Si\text{-}O\text{-})_2SiCl_2 + 2H_2O \rightarrow (\equiv Si\text{-}O\text{-})_2Si(OH)_2 + 2HCl$$

$$(\equiv Si\text{-}O\text{-})_2TiCl_2 + 2H_2O \rightarrow (\equiv Si\text{-}O\text{-})_2Ti(OH)_2 + 2HCl$$

It is thus necessary to tailor the synthesis depending on the composition and structures of an end-product. Variables include technological modes (temperature, concentration of reagents, time), set of reagents, and reaction sequence.

In the case of groups $=P$, CrO_2 can be used in an oxidation-reduction reaction:

$$(\equiv Si\text{-}O\text{-})_3P + Cl_2 \rightarrow (\equiv Si\text{-}O\text{-})_3PCl_2$$

Further reactions take place with water vapor:

$$(\equiv Si\text{-}O\text{-})_3PCl_2 + 2\ H_2O \rightarrow (\equiv Si\text{-}O\text{-})_3P(OH)_2 + 2HCl$$

To synthesize a chromium oxide nanolayer it is possible to use the following chemical reactions:

$$(\equiv Si\text{-}O\text{-})_2CrO_2 + 3/2\ H_2 \rightarrow (\equiv Si\text{-}O\text{-})_2CrOH + H_2O$$

$$(\equiv Si\text{-}O\text{-})_2CrOH + CrO_2Cl_2 \rightarrow (\equiv Si\text{-}O\text{-})_2CrOCrO_2Cl + HCl$$

$$(\equiv Si\text{-}O\text{-})_2CrOCrO_2Cl + H2O \rightarrow (\equiv Si\text{-}O\text{-})_2CrOCrO_2(OH) + HCI$$

$$(\equiv Si\text{-}O\text{-})_2CrOCrO_2(OH) + H_2 \rightarrow (\equiv Si\text{-}O\text{-})_2CrOCr\,(OH)_2 + 1/2\ H_2O$$

Further processes can be carried out by repeated alternation of the specified reactions. In Table 1, some characteristics of Cr-containing silica gel after different cycles of ML are shown. It should be noted that side reactions are also possible and must be taken into consideration.

Table 1. Chemical compound and structural characteristics of chrome-containing silica gel.

Number of ML cycles	0	I	II	III	IV	V	VI
Cr content [Cr], mmol/g	0	0.95	1.73	2.53	3.49	4.21	4.99
d[Cr] = $[Cr]_i$ - $[Cr]_{i-1}$, mmol/g	0	0.95	0.78	0.80	0.96	0.72	0.78
Specific surface S, m^2/g	246	229	209	193	169	155	136
dS, m^2/g	0	17	20	16	24	14	19
dS/d[Cr]		17.8	25.6	20.0	25.0	19.4	24.4
Volume of pores, v, cm^3/g	0,96	0.80	0.76	0.64		0.54	

The ML method has been successfully applied to the synthesis of monolayers ranging from Group II to Group VII elements onto oxides of aluminum, zirconium, magnesium, titanium, and glass. Metal, carbon fiber, and semiconductor substrates have also been used.

3. Perspectives on Using the ML Method in Chemical Nanotechnology

Nanostructures of various chemical composition can be synthesized on the surface of solid matrices by ML. Atomic chemical assembly of surface nano- and macrostructures can be also realized.

This technology is characterized by four basic effects.

- Monolayer effect (ME), i.e. sharp changes in matrix properties after obtaining 1–4 monolayers of new structural units;
- effect of matrix overlapping (EME), i.e. the physical coverage of the surface of solid (the formed layer screens the surface);
- effect of mutual coordination of matrix surface structure and formed nanolayer (EMC);

- effect of multicomponent system (MS), for example it can be a synergistic composition.

The general directions of the application of ML method can be defined from these effects.

Firstly it can be used for obtaining layers with a thickness of several monolayers to introduce and to distribute uniformly very low amounts of admixtures. This may be important for the surface of sorption and catalytic, polymeric, metal, composition and other materials. Secondly, the production of relatively thick layers, on the order of tens of nm. In this case a thickness of nanolayers is controlled with an accuracy of one monolayer. This can be important in the optimization of layer composition and thickness (for example when kernel pigments and fillers are produced). Thirdly the ML method can be used to influence the matrix surface and nanolayer phase transformation in core-shell systems. It can be used for example for intensification of chemical solid reactions, and in sintering of ceramic powders. Fourthly, the ML method can be used for the formation of multicomponent mono- and nanolayers to create surface nanostructures with uniformly varied thicknesses (for example optical applications), or with synergistic properties (for example flame retardants), or with a combination of various functions (polyfunctional coatings). Nanoelectronics can also utilize multicomponent mono- and nanolayers.

Now some results will be described utilizing the ML method in chemical nanotechnology.

3.1. SORBENTS AND CATALYSTS

Two applications of ML are heterogeneous catalyst and sorbent synthesis. In these cases it is necessary to distribute in regular intervals very small amounts of substance on a surface. It is usually enough to have from 1 to 4 monolayers. In this case each atom in a monolayer can take part in reaction. This is not obtained using traditional methods of impregnation.

Some substances change color when in contact with components of a gas phase: water vapor for example. This property has been used to create a humidity indicator in gas environments using a V-containing silica gel. Vanadium oxide changes color depending on the extent of hydration, but the hydration reaction is slow. For this reason V_2O_5 has not been used as a humidity indicator of gases. But the sample of V-containing silica has good indicator properties in contact with water vapor. It has a bright, contrast color, which changes depending on the concentration of water in gas. It is formed by reaction of silica with $VOCl_3$ and has a name in industry of IVS-1. It is important to note, that the sample with two or more V-O monolayers on silica loses its indicator properties.

It is also possible to monitor in the gas phase such components as ammonia (concentration range of 30-400 mg/m^3, color of sample is yellow), oxide sulfur (concentration range of 360-1400 mg/m^3, color is blue), hydrogen sulfide (concentration range of 20-140 mg/m^3, color is violet).[3,9]

It is known that phosphorus (+5) oxide actively reacts with water vapor. It is an effective chemisorbent of water, and it can be used for drying of gases. Phosphoric acid is formed however, and regeneration is very difficult. It was created as a sorbent on silica with P-oxide nanolayer. It is an active sorbent of water vapor and it is 5-6 times better than initial silica (at a humidity of 70%). It also adsorbs ammonia, and some organic substances. Its name in industry is P-1-3. The sorbents IVS-1 and P-1-3 are used in industry for clearing and stabilization of the gas atmosphere in devices.

An important direction of application of ML chemical nanotechnology is creation of heterogeneous catalysts. There are many publications, one of the most interesting is in membrane catalysis.

3.2. DECREASE OF COMBUSTIBILITY OF POLYMERS

The ML method can be used to synthesize composite materials for increased thermal stability and flame retardant applications.[3,10,11] To decrease polymer combustibility, a large amount of retardant is frequently not necessary. Similarly, a thick coating on the surface of materials is often unnecessary, simply replacting surface groups responsible for combustibility is all that is required. It is well known that certain form of phosphorous are excellent flame retardants. The chemistry is illustrated in Figure 2. In the case of modifying by the ML method on a polymer surface, a dense carbonic layer is formed reducing the diffusion of oxygen into the polymer, thus ending combustion.

Multicomponent nanostructures can be used to enhance flame retardant characteristics.[12] The introduction of synergistic additives is widely used to reduce combustibility of polymeric materials and composites. To enhance the effect of P-containing flame retardants, combinations of elements-synergists (such as titanium, vanadium, chromium, antimony, nitrogen, etc.) can be used. It is combustibility of initial epoxy- phenol-formaldehyde porous polymer (cellular plastic with title PEN-I) and with nanoretardants of combustion. In materials take place together with the basic P-containing mix also additives of antimony, nitrogen, chromium and vanadium. The combustibility of polymer is affected not only by the chemical nature of element - synergist, but also by a relative positioning of components in structure of retardant of combustion.

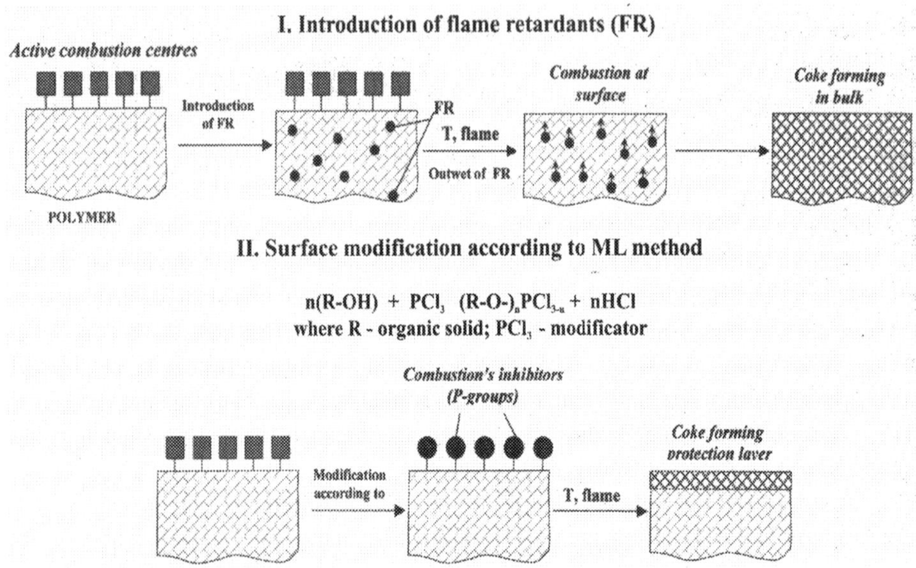

Figure 2. Reduction of combustibility of polymers.

3.3. SINTERING OF CERAMICS AND CHEMICAL SOLID REACTIONS

Nanolayers on a surface of ceramic materials lower the sintering temperature, and increase the speed of sintering.[3,10,13,14] As shown in Figure 3, with less diffusion required it is possible to speed up the sintering process.

Temperature influences the structure of TiO_2 synthesized as a nanolayer on SiO_2 and Al_2O_3. After 12 cycles at temperatures below 350°C on SiO_2, only anatase is formed. Above 600°C predominantly rutile is formed. At intermediate temperatures a mix of the two phases is formed. On aluminium oxide however, at low temperature the biphase system is formed. At 600°C 100%, anatase is formed on Al_2O_3. This illustrates the effect of the mutual influence of the matrix surface and the nanolayer on phase transformations in core-shell system.

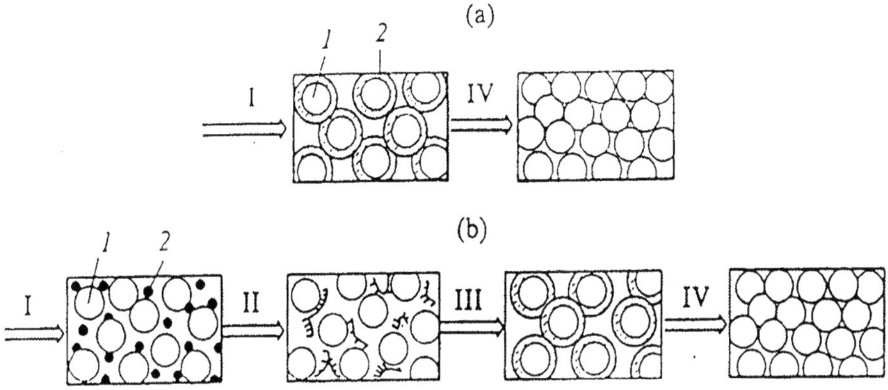

Figure 3. Sintering of doped ceramic powders on doping of ceramics powder by using (a) the ML method; and (b) mixing of components; (1) filler and (2) nanolayer.

4. Conclusion

Fundamental concepts of the molecular layering method have been developed and applied by the team headed by Professor Valentine Aleskovsky in Russia. This method, similar to atomic layer epitaxy and atomic layer deposition, has been used to create monolayers on oxides and polymers as humidity sensors, flame retardants, and agents to enhance sintering in ceramic materials.

References

1. S. I Koltsov and V. B. Aleskovskii, *Zh. Fizich. Khim.* 42, 1210 (1968).
2. V. B. Aleskovskii, *Zh. Prikl. Khim.* 47(10), 2145 (1974).
3. A. A. Malygin, A. A. Malkov, and S. D. Dubrovenskii, *Adsorption on New and Modified Inorganic Sorbents*, vol. 99, edited by Dabrowski, A. and Tertykh, V.A. (Amsterdam, Elsevier, 1996), pp. 213-237.
4. *Nanotechnology Research Directions: IWGN Workshop Report. Vision for Nanotecnology R&D in the Next Decade*, edited by M. C. Roco, R. S. Williams, and P. Alivisatos (Kluwer Academic Publishers, Dordrecht, in Russian, Moscow, Mir, 2002).
5. V. B. Aleskovskii, Stoichiometry and Synthesis of Solid Compounds (Nauka, Leningrad, 1976, in Russian).
6. T. Suntola, in: *Atomic Layer Epitaxy Materials*, Science Reports, Holland, Amsterdam 4(7) 261-312 (1989).

7. R. Puurunen, Surface chemistry of atomic layer deposition: a case study for the trimethylaluminium/water process (ALD review), *J. Appl. Phys.* 97, 121301 (2005).
8. A. A. Malygin, and V. M. Smirnov, Early work on atomic layer deposition cited, Solid State Technology 3, 14, (2002).
9. A. A. Malygin, The molecular layering method as a basis of chemical nanotechnology, in: *Natural Microporous Materials in Environmental Technology* (Kluwer Academic Publishers, 1999), pp. 487-495.
10. A. V. Beljakov, E.V. Zsharikov, and A.A. Malygin, Chemical methods in nanotechnology, in: *Chemical technologies*, edited by P. D. Sarkisov (STP «Scientific investigations of High School in priority-driven fields of Science and Techniques, RChTU Moscow, 2003) pp. 551-620.
11. S. A. Trifonov, E. A. Sosnov, and A. A. Malygin, Structure of the surface and thermal-oxidative breakdown of products that has obtained in reaction of polyethylene with vapor of PCl_3 and $VOCl_3$, *J. Appl. Chem.* (in Russian) 77(11) 1872-1876 (2004).
12. A. A Malygin, Synthesis of Multicomponent Oxide Low-Dimension Systems on Surface of Porous Silicon Dioxide Using Molecular Layering Method, *J. General Chem.* (in Russian) 72(4), 617-632 (2002).
13. A. A. Malkov, E. A. Sosnov, and A. A. Malygin, Thermal transformations of Ti-oxichloride nanostructures on the surface of silica gel, *J. Appl. Chem.* (in Russian) 78(6) 881-886 (2005).
14. Z. N. Ishutina, V. V. Gusarov, A. A. Malkov, T. V. Firsanova, and A. A. Malygin, Phase transformations in nanodimension compositions on basis of gamma-Al_2O_3-SiO_2-TiO_2, *J. Inorg. Chem.* (in Russian) 44(1) 16-19 (1999).

ANALYSIS OF SURFACE STRUCTURES USING XPS WITH EXTERNAL STIMULI

GULAY ERTAS, SEFIK SUZER*
Bilkent University, Department of Chemistry and the Laboratory for Advanced Functional Materials, 06800 Ankara, Turkey

Abstract. X-ray Photoelectron Spectroscopy, XPS, due to the perfect match of its probe length (1-10 nm) to nanoparticle size, chemical specificity, and susceptibility to electrical charges, is ideally suited for harvesting chemical, physical and electrical information from nanosized surface structures. In addition, by recording XPS spectra while applying external d.c. and/or pulsed voltage stimuli, it is also possible to control the extent of charging and extract various analytical information. In the simplest form, application of a static (d.c.) voltage stimuli enhances separation of otherwise overlapping peaks of gold nanoparticles from that of metallic gold. When the voltage stimuli is applied in the form of rectangular pulses, dynamic information is obtained from the frequency dependence of the charging shifts. This enables us to better probe the composition of nanoparticles produced (i.e. silicide formation, or whether or the extent of reduction, etc.) when platinum salt is deposited on silicon substrates. Finally, by recording the data in different time windows, XPS spectra can be recorded in time-resolved fashion. Time-resolved spectra can be used to detect, locate and quantify the charges developed in various surface structures like gold(core)/ silica(shell) nanoparticles on a copper substrate.

Keywords: XPS; differential charging; peak separation; time-resolved XPS

1. Introduction

XPS is one of the most widely used surface analysis techniques for extracting information related to the chemical nature of surface structures on the

*To whom correspondence should be addressed. S. Suzer, Bilkent University, Department of Chemistry and the Laboratory for Advanced Functional Materials, 06800 Ankara, Turkey; e-mail: suzer@fen.bilkent.edu.tr

J.P. Blitz and V.M. Gun'ko (eds.), Surface Chemistry in Biomedical and Environmental Science, 45–58.

nanometer scale. This 40+ year old technique consists of irradiation of the sample with x-rays causing photoelectrons to be emitted from the sample surface, and subsequent determination of their kinetic energies by an energy analyzer. The binding energies are obtained using the energy of the x-ray, and Einstein's equation (B.E. = hυ – K.E.), which leads to determination of elemental identities and chemical states. Surface sensitivity stems from the attenuation length of the photoelectrons (1-10 nm) within the solid samples. Except for hydrogen, all elements can be unambiguously identified.[1]

The photoelectrons ejected leave positive charges behind, which are normally replenished via the ground loop of a conducting sample. For non-conducting or poorly conducting samples this process is slow. Positive charging develops which decreases the kinetic energy of the photoelectrons, hence increasing the apparent binding energies.[2-6] For uniform samples this is not a problem, since charge correction can be accomplished using a suitable reference. For heterogeneous or layered samples with different conductivities, this problem is severe and manifests itself as differential charging. Flooding the sample with low energy electrons or ions has been successfully used for neutralizing the positive charges developed.[7-11] Complete elimination of this problem is only an ideal, and overneutralization leading to negative charging has also been encountered.

A number of publications have appeared reporting constructive use of this surface charging (mostly negative) phenomenon for elucidating chemical and/or structural parameters of various materials. Lau and coworkers have studied the structural and/or electrical properties of ultrathin dielectric films on semi-conductors utilizing the surface-charging.[12-18] Thomas et al. separated the surface spectrum (mainly silicon dioxide) from the silicon substrate spectrum (consisting of contamination and silicon dioxide on silicon), by use of surface charging.[15,16,19] Similar applications were also reported by Ermolieff et al.[20] and Bell and Joubert,[21] while Miller et al. applied the technique to separate the XPS signals of the fiber from the exposed matrix at fractured surfaces.[22] Elegant use of surface charging for lateral differentiation of mesoscopic layers, and for depth profiling in 1-10 nm thin layers, have recently been reported.[12,23,24] Surface-charging was invariably controlled/varied via a low energy electron flood gun, resulting in mostly negatively charged surfaces. We have recently demonstrated that useful information can also be extracted by analysis of positively charged surfaces. We do not attempt to neutralize the charging developed, and achieve enhanced positive charging by application of a negative voltage bias to the sample while recording the XPS spectrum.[25,26]

Application of external bias to the sample goes back 3 decades.[5] However, using an external bias for extracting chemical and/or structural information is scarce. One successful application demonstrated that biasing the sample holder with a large negative d.c. voltage (25-100V) could be used to identify chemical differences in oxide films on an aluminum alloy.[27] Similarly, we have recently reported that application of a small (1-10V) negative bias leads to enhanced differential charging by repelling stray electrons or electrons emitted from a nearby filament. A positive bias reduces charging between the oxide layers and the underlying metal substrates.[28,29]

2. Experimental

Oxide layers were grown thermally on HF-cleaned Si (100) substrates at 500°C in air. Citrate-capped gold nanoclusters were prepared according to a well-established procedure and were directly deposited from their aqueous solutions onto the SiO_2/Si substrates.[30,31]

A Kratos ES300 electron spectrometer with MgKα x-rays (nonmonochromatic) was used for XPS analysis. A typical sample was a ca. 1mm-thick silicon wafer with dimensions of 4 x 12 mm. The sample normally accepts x-rays at 45°, and emits photoelectrons at 90° with respect to its surface plane. The sample can also be rotated to decrease the emission angle to enhance surface sensitivity, while keeping the x-ray-sample-analyzer angle at 45°. Calibration of the energy scale was carried out using standard silver and gold samples, referencing to the C1s peak at 285.0 eV. Resolution of the spectrometer was better than 0.9 eV measured with the Ag3d peaks, and standard curve fitting routines were used with 0.6 eV spin-orbit parameter for the Si2p peak.

3. X-Ray Photoelectron Spectroscopy under External Stimuli

The power of the XPS technique as an analytical tool is related to both its chemical and state (i.e. ionic charge) specifity, and its applicability to all elements except H. The XPS spectrum of a silicon substrate containing ca. 4 nm oxide layer, grown thermally on top of which an aqueous solution containing 10^{-4} M $HAuCl_4$ is deposited and dried in air, is shown in Figure 1. Both the Si2p and Au4f peaks are composed of two chemically different species. For silicon, the peak at lower binding energy (99.5 eV) corresponds to the silicon substrate underneath, and the peak at 104 eV corresponds to the thermal oxide layer (SiO_2). The Au4f has a 3.6 eV spin-orbit splitting. The composite peak can be fit with 2 doublets ($Au4f_{7/2} - 4f_{5/2}$) separated by 3.0 eV, corresponding to ionic (Au^{3+}) and reduced (Au^0) forms of gold. Quantification, as well as overlayer thickness, can be obtained from peak intensities.[1] Measured binding energies contain components from both the chemical shift, related with chemical state of the elements, and charging effects. This can be very misleading if the two are not separated properly. For example, the true binding energy difference between the Si^{4+} and the Si^0 peaks is 3.2 eV, almost 1 eV smaller than measured, and can vary from –1 to about 5 eV depending on the extent of neutralization.[32,33] As will be shown below, this shift can also be influenced by application of an external bias to the sample rod. The binding energy also depends on the size/shape of the metallic clusters, offering a new analytical dimension.[34,35]

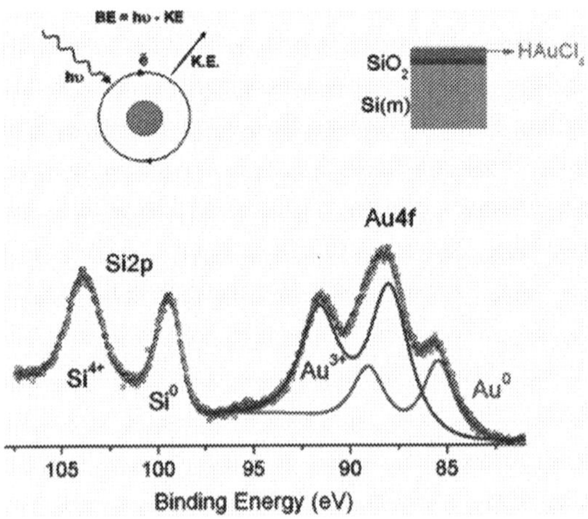

Figure 1. Part of the XPS spectrum showing the Si2p-Au4f peaks of a silicon substrate containing ca. 4 nm thermal oxide layer with one drop of 0.0001 M aqueous solution of $HAuCl_4$ deposited, and dried in air. The silicon peaks corresponding to the silicon substrate (Si^0) and the silicon oxide layer (Si^{4+}), as well as two different chemical state of gold (Au^{3+} and Au^0) are clearly resolved.

3.1. STATIC STIMULUS (D.C. BIAS)

In conventional XPS the sample is grounded. In our technique, a voltage bias to the electrically isolated sample rod is applied. Under negative bias the spectrum shifts to higher kinetic energy (blue shift), the shift is to lower energy (red shift) under positive bias. This is shown in Figure 2 for the same

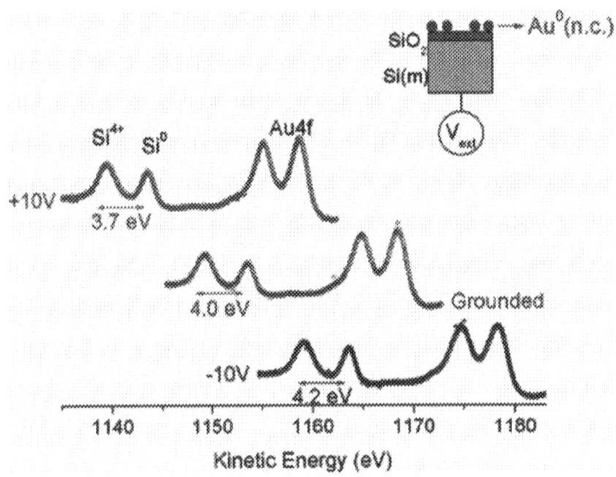

Figure 2. Figure 1 spectra under +10V, grounded, and -10V after complete reduction with x-rays. Application of an external stress results in asymmetrical peak shift.

silicon/silicon oxide sample after complete reduction by x-rays.[36] Whereas the Si^0 peak shifts exactly 10 eV, the shift in both the Si^{4+} and the Au4f peaks differ from differential charging developed in the oxide layer. Figure 3 shows spectra of a silicon sample with a 30 nm thick oxide layer. A gold metal strip is tied to the sample for referencing. The spectra are shifted with respect to Au4f peaks, where the value of the $Au4f_{7/2}$ peak is set to 84.0 eV.[1] In this case, the silicon substrate peak (Si^0) is too deep to be observed. The inset in Figure 3 shows the measured binding energy difference between the Si2p of the oxide, and the Au4f peaks as a function of the applied bias. Under negative bias all electrons which cause partial neutralization are repelled to yield a positively charged oxide layer. This leads to a larger binding energy difference between the Si2p and Au4f peaks.

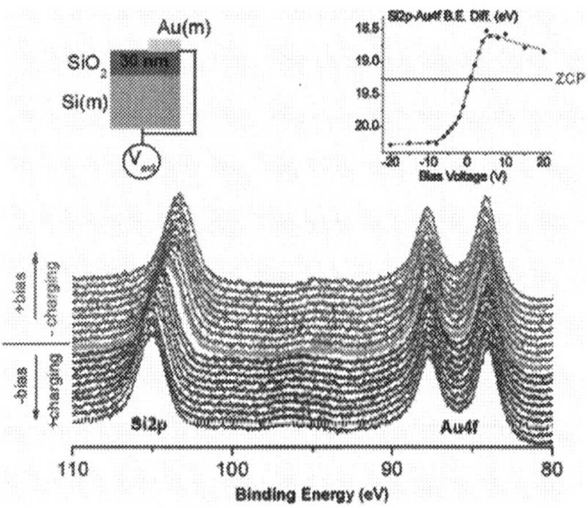

Figure 3. XPS spectra of a silicon sample containing ca. 30 nm thick oxide. A strip of gold metal is tied for referencing under different voltage stress. The inset displays the measured binding energy difference between the Si2p of the oxide layer (Si^{4+}), and the Au4f of the gold metal. Zero Charge Point (ZCP) is obtained at ca. +1V stress.

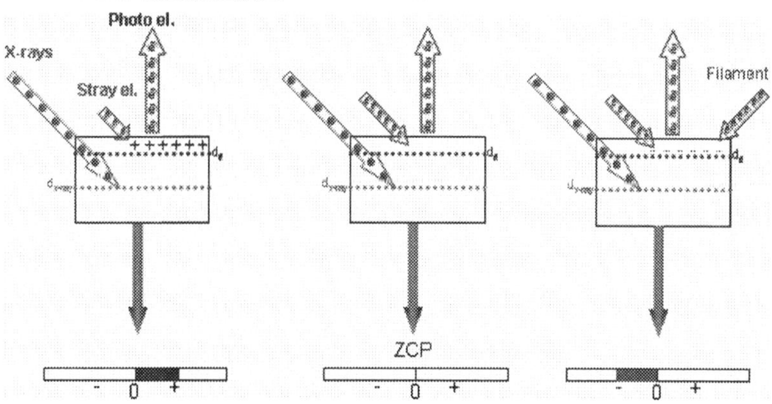

Figure 4. Schematic representation of the charging process. 3 different currents flow from the sample to the spectrometer due to; (i) photoelectrons generated by x-rays, (ii) stray electrons from the x-ray tube and/or other elements within the vacuum chamber, and (iii) low energy electrons from a nearby filament.

Zero charging point (ZCP) corresponds to a 19.4 eV difference. So the correct binding energy of Si2p (103.4) is obtained only under ca. +1 V bias. Larger positive biasing leads to negative charging of the oxide layer as shown schematically in Figure 4. External bias (d.c.) is a simple technique for controlling the extent and polarity of ultra-thin oxide layer charging. This can be used as an analytical tool for peak separation, or extracting information related to the proximity of atoms in nanometer-scale surface structures.[25,37] Representative analytical applications will be given later.

3.2. DYNAMIC (PULSED) STIMULUS

The application of external stimuli in rectangular pulses provides additional advantages. Figure 5 depicts the Si2p region of an HF-cleaned Si (100) sample containing 1 nm thick native oxide layer. During one pulse, the sample spends 50% of its time near ground potential, and 50% of its time at +10V. Therefore all peaks are twinned at all frequencies (0.1 to 1000 Hz), and with exactly the same energy difference (10 ± 0.1 eV).

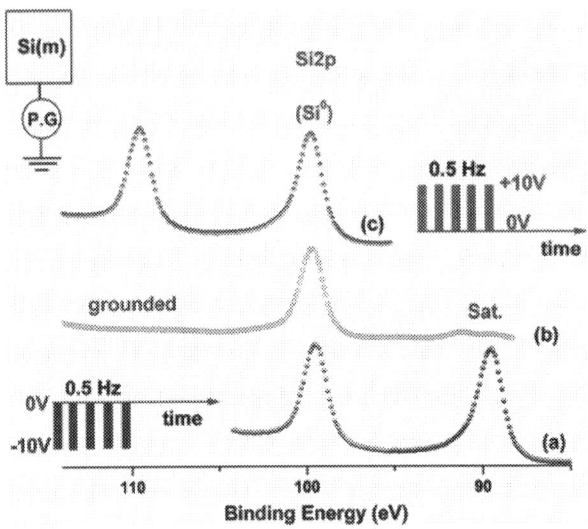

Figure 5. XPS spectra of the Si2p region of a clean silicon surface (Si^0) recorded by: (b) grounding the sample, (a) applying 0 to –10V pulses at 0.1 Hz, (c) applying 0 to +10 V pulses at 0.1 Hz. The Si^0 peak is twinned, because when pulsed the sample experiences the ground and the applied potential 50% at a time. Sat. refers to peaks of the $K\alpha_{3,4}$ x-ray satellites.

The situation is very different when the oxide layer is larger than 2 nm and the measured energy differences vary with the pulse frequency due to differential charging. To calibrate the measured energy differences a gold metal strip in electrical contact with the sample is used. In Figure 6 we show the Si2p-Au4f region of a sample containing a 30 nm thermal oxide layer. The measured binding energy difference between the Si2p of the oxide layer (Si^{4+}), and the Au4f, are different in all the three cases [i.e. (b) grounded, under (a) negative and (c) positive pulses respectively]. In the negative cycle the slow stray electrons, or electrons from the filament are repelled. The oxide layer is left positively charged to cause the measured binding energy difference to increase. In the positive cycle, the low energy electrons are withdrawn onto the sample to neutralize the oxide layer and decrease the measured binding energy difference. In this particular experiment the oxide layer is overneutralized and negatively charged since the measured ΔB.E. is 18.84 eV, which is smaller than the actual value of 19.4 eV (103.4 eV for Si2p of the oxide minus 84.0 for Au4f).

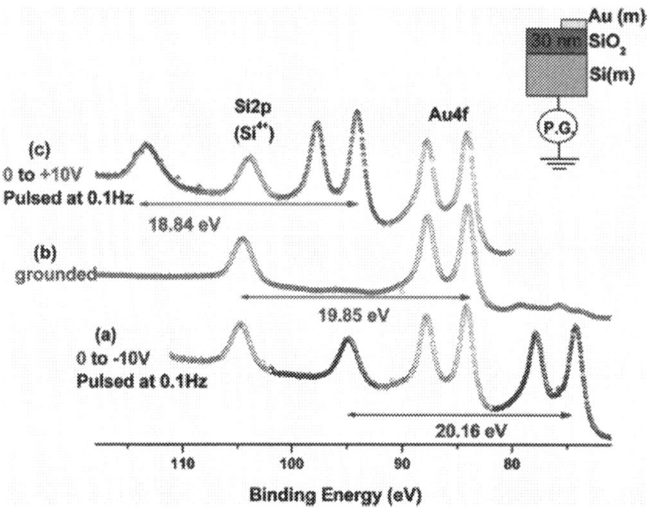

Figure 6. XPS spectra of the Si2p-Au4f region of silicon sample containing ca. 30 nm thermal oxide layer, without and with 0.1 Hz pulses as in Figure 5. Both the Au4f doublet and the Si2p (Si^{4+}) peaks are twinned when pulsed. However, the measured binding energy difference varies with the polarity of the pulses.

The measured binding energy difference changes with the pulse frequency. The difference is small at high frequencies, due to the rapid changes of the applied voltage between ground and +10V. The oxide layer cannot respond quickly enough and an average potential throughout the cycle is measured. At low frequencies the measured binding energy difference, between the Si2p and the Au4f levels under positive pulses, becomes smaller when compared to ground, since more time is allocated for the silicon oxide layer to neutralize its positive charge.

3.3. TIME RESOLVED MEASUREMENTS

It is possible to record data in different time windows after the application of a pulse to obtain time-resolved spectra. The pulses are applied to the sample rod externally using a digital-to-analog converter (DAC), and the photoelectrons are amplified and counted for a given period. The kinetic energy axis of the spectrometer is also controlled by the DAC. Each measurement starts by setting the kinetic energy to the first point with application of +10V to the sample rod.

We then apply 4 repetitive steps: (1) wait for 2.5 s for equilibration; (2) apply –10V; (3) recording data in 5 ms intervals for 1 s in separate columns for 200 data points. Return to +10V; (4) increment the energy to restart the cycle. One spectrum scan is completed after repeating these cycles for the entire XPS region. Since only 10-50 counts are collected signal averaging is necessary. Typically 50-250 scans are recorded, requiring 3-15 hours. This procedure results in a collection of 200 spectra recorded with 5 ms time resolution. In principle the time resolution is limited only by the rise-time of the applied pulses; however the 5 ms resolution limit was dictated by a combination of the counting statistics and software used.

In addition to the conventional chemical information derived from XPS analysis, time dependent measurements can now be used to extract important parameters related to dielectric properties of surface structures as will be demonstrated below.

4. Applications

Three variations of our technique for solving one different analytical problem will be shown.

4.1. STATIC STIMULUS (D.C. BIAS) - RESOLUTION ENHANCEMENT

Differentiation on the basis of chemical shifts of gold in the Au^{3+}, Au^{1+}, and Au^{0} is routine in XPS, since the binding energies are significantly different from each other.[1,38,39] Cluster size also influences the binding energy of the atom, the origin of this shift is still actively debated.[34,40-42] The binding energy of small nanoclusters of gold, prepared by physical vapor deposition, can be as much as 1.8 eV higher (85.8 eV) than metallic gold.[34] Furthermore, reports indicate that some measurable binding energy differences exist with respect to the matrix where the gold clusters reside.[40-42] Experimentally it is straight-forward to measure and quantify the contribution of each parameter separately. Severe difficulties arise when two or more parameters coexist, and their contributions complicate the spectrum due mostly to insufficient instrumental resolution. In some cases deconvolution techniques help, but other experimental tools/tricks are welcomed.

The d.c. bias controls the extent of differential charging, leading to total spectral separation of gold XPS peaks existing as metallic gold and gold nanoclusters on the same surface.

Gold nanoparticles capped with tetraoctylammonium bromide (TOAB) in toluene were prepared using well-established routes.[43] The resulting solution displays a red-wine color, and has a strong absorption peak around 530 nm.[44,45]

When the gold nanoparticles aggregate the solution becomes black and the absorption band shifts to longer wavelengths. A drop of solution containing the gold nanoclusters was deposited onto silicon substrates with various thermal oxide thicknesses (2-40 nm). The sample is tied onto the XPS sample holder together with a gold metal strip. Figure 7 depicts the Si2p-Au4f region recorded with and without external bias. An approximately 30 nm SiO_2/Si substrate is chosen so that only the Si2p of the oxide is accessible by XPS.

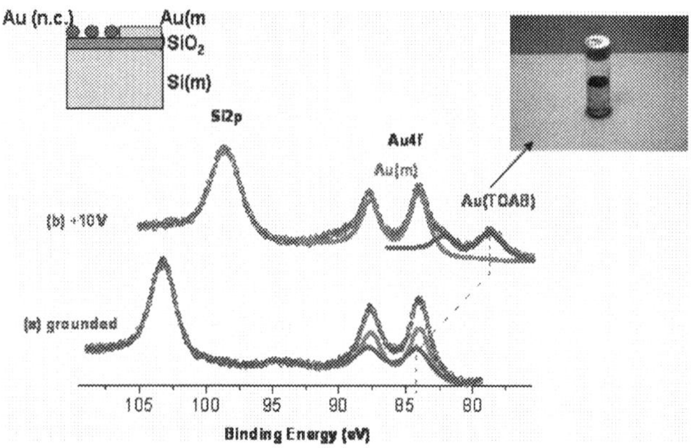

Figure 7. XPS spectra of the Si2p-Au4f region of silicon sample containing ca. 30 nm thermal oxide layer and two different forms of gold; (i) gold nanoclusters, and (ii) gold metal strip. In the spectrum recorded normally (i.e. sample rod grounded) the gold peaks overlap. However, under +10V the silicon 2p of the oxide and the Au4f of the nanoclusters shift due to charging, and the Au4f peaks become separated.

Our method is based on the ability to measure the local potential developed as a result of the different response of the gold metal and the 30 nm SiO_2 dielectric layer to the external bias applied. As shown in Figure 7a, two peaks of the Au4f spin-orbit doublet from nanoclusters and the bulk gold overlap when the normal practice is employed. When the sample is biased at +10V (Figure 7b) the Si2p of the oxide shifts to lower binding energy, and the gold peaks separate into two doublets. One of the doublets does not shift but the other shifts almost as much as the Si2p peak. In this case the oxide is overneutralized by electrons from the filament, and has an overall negative potential which is also reflected in the Au4f of the nanoclusters.

By imposing a simple external voltage stress to the sample rod we have been able to control and measure the potential(s) developed as a result of charging in and on different layers/domains. This simple procedure is a powerful tool for enhancing peak separation.

4.2. DYNAMIC (PULSED) STIMULUS

XPS analysis of a heterogeneous silicon sample with ca. 6 nm oxide layer, containing Pt particles deposited from an aqueous $PtCl_4$ solution and reduced by x-rays, is used to illustrate the pulsed technique. This involves x-ray induced production of Au nanoparticles on a SiO_2/Si system and in polymer matrices.[36] Pt behaves differently than gold on the SiO_2/Si system and questions arise such as whether or not complete reduction is accomplished, or silicide formation has taken place, etc.

Figure 8 displays a series of XPS spectra of the Si2p-Pt4f region. The Si2p region consists of two peaks corresponding to the top oxide layer (Si^{4+}) and the silicon substrate (Si^0). The Pt4f region consists of one slightly broadened spin-orbit doublet, assigned to the formal oxidation state of Pt^0 at 71.2 eV.[1] All the peaks shift after application of a d.c. voltage stress, but the shifts are different for the Si^{4+}, Si^0, and the Pt^0 peaks due to variation of the differential charging.

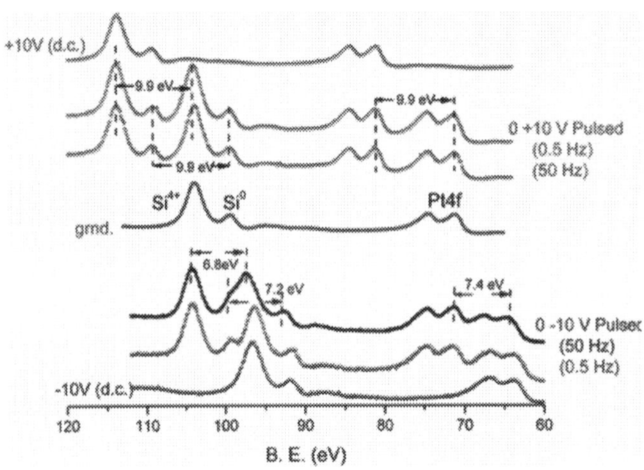

Figure 8. XPS spectra of the Si2p-Pt4f region of silicon sample containing ca. 6 nm thermal oxide layer, and also containing Pt particles deposited from an aqueous solution, and reduced by x-rays; without and under +10 and –10V external voltage stress. The figure also contains spectra recorded with: 0 to +10V pulses at 50, and 0.5 Hz, 0 to –10V pulses again at 50 and 0.5 Hz.

When the voltage stress is pulsed we observe both polarity and frequency dependent changes. For positive pulses all peaks are twinned with almost the same 9.9 eV energy difference, and almost no measurable frequency dependence. For negative pulses the measured energy difference between the twinned peaks varies and, more importantly, on the frequency of the voltage applied. As the voltage frequency decreases, the measured energy difference between the twinned peaks increase towards the asymptotic value of 9.9 eV. At low frequencies the measured binding energy difference between the twinned peaks increases since more time is allocated for equilibration.

The more important observation is that the measured difference between the twinned peaks of the Si^{4+}, the Si^0, and the Pt^0 display different functional behavior with respect to frequency. The silicon oxide layer shifts the most, the silicon substrate in the middle, and the platinum species least. Measured binding energies can change several eV, depending on parameters like particle size, matrix effects, and differential charging, etc.[1,25,28,29,36] Tabulated binding energies for the Pt4f peak of the two silicides PtSi and Pt_2Si are 73.0 and 72.5 eV, and are not very different from our observed value of 71.2 eV. However, on the basis of the frequency behavior of the peaks, silicide formation we can be ruled out, since otherwise both the Si^0 and the Pt^0 would shift together. Hence, we assign the observed Pt4f peak to separate metallic (probably nanometer size) particles on the oxide layer. Platinum's behavior is different from gold, which displays a very tight shift with the oxide layer rather than with silicon. In the case of platinum the x-ray induced reduction is incomplete (unlike gold), and the particles contain some unreduced platinum species. The concentration of unreduced platinum is not high enough to be detected as a separate peak in the spectrum, but they are effective for charging the particles.

4.3. TIME-RESOLVED XPS

In Figure 9, 200 time resolved XPS spectra are shown. The nanoparticles have ca. 15 nm gold core and ca. 6 nm silica shell as determined by TEM.[46] Both the Au4f and Si2p peaks shift to higher binding energies for the core-shell nanoparticles when recorded under –10V bias. Furthermore, their shifts are ***highly correlated in time***. Note also that the Cu2p peak does not shift at all. This shift is solely related with charge accumulation in the silica shell, the extent of which can be controlled by the magnitude and the polarity of the voltage stress, and the gold core experiences the same potential, and shifts as the shell.

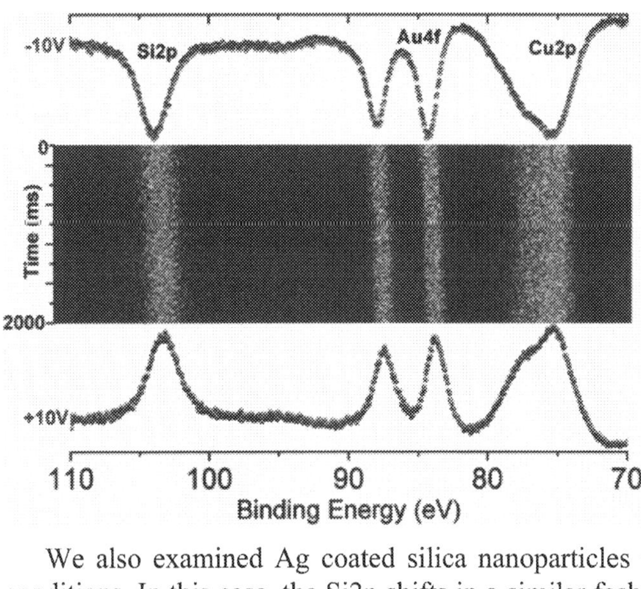

Figure 9. 200 time-resolved XPS spectra recorded with 10 ms resolution of the Si2p-Au4f-Cu3p region of gold(core) /silica (shell) nanoparticles deposited on a copper tape.

We also examined Ag coated silica nanoparticles under exactly the same conditions. In this case, the Si2p shifts in a similar fashion, but the Ag3d peaks exhibit no shift at all, since the silver atoms are now in the shell, and in contact with the copper substrate. Hence, it is possible to derive additional structural information from its charging behavior alone. Charging behavior of each peak enables us to ***detect*** and ***locate*** the charges chemically. Furthermore, from the measured shifts it is also possible to ***quantify*** the amount of charges stored on the nanoparticles.[47]

References

1. D. Briggs and M. P. Seah, *Practical Surface Analysis. Part I. Auger and X-Ray Photoelectron Spectroscopy*, Second edition (John Wiley & Sons Ltd., England, 1996).
2. T. L. Barr, Studies in Differential Charging, *J. Vac. Sci. Technol. A* 7, 1677-1683 (1989).
3. A. Cros, Charging Effects in X-Ray Photoelectron Spectroscopy, *J. Electron Spectros. Relat. Phenom.* 59, 1-14 (1992).
4. T. Dickinson, A. F. Povey, and P. M. A. Sherwood, Differential Sample Charging in ESCA, *J. Electron Spectros. Relat. Phenom.* 2, 441-447 (1973).
5. G. Johansson, J. Hedman, A. Berndtsson, M. Klasson, and R. Nilsson, Calibration of Electron Spectra, *J. Electron Spectros. Relat. Phenom.* 2, 295-317 (1973).
6. J. Cazaux and P. Lehuede, Some Physical Descriptions of the Charging Effects of Insulators under Incident Particle Bombardment, *J. Electron Spectros. Relat. Phenom.* 59, 49-71 (1992).
7. J. Cazaux, Mechanisms of Charging in Electron Spectroscopy, *J. Electron Spectros. Relat. Phenom.* 105, 155-185 (1999).
8. J. Cazaux, About the Charge Compensation of Insulating Samples in XPS, *J. Electron Spectros. Relat. Phenom.* 113, 15-33 (2000).
9. P. E. Larson, and M. A. Kelly, Surface Charge Neutralization of Insulating Samples in X-Ray Photoemission Spectroscopy, *J. Vac. Sci. Technol. A* 16, 3483-3489 (1998).
10. M. Marsi, R. Belkhou, C. Grupp, G. Panaccione, A. Taleb-Ibrahimi, L. Nahon, D. Garzella, D. Nutarelli, E. Renault, R. Roux, M. E. Couprie, and M. Billardon, Transient Charge Carrier Distribution at UV-Photoexcited SiO_2/Si Interfaces, *Phys. Rev. B* 61, R5070-R5073 (2000).
11. J. Cazaux, Electron and X-Ray Induced Electron Emissions from Insulators, *Polym. Int.* 50, 748-755 (2001).
12. W. M. Lau, Use of Surface Charging in X-Ray Photoelectron Spectroscopic Studies of Ultrathin Dielectric Films on Semiconductors, *Appl. Phys. Lett.* 54, 338-340 (1989).
13. W. M. Lau, Effects of a Depth-Dependent Specimen Potential on X-Ray Photoelectron Spectroscopic Data, *J. Appl. Phys.* 65, 2047-2052 (1989).
14. W. M. Lau and X. W. Wu, Measurements of Interface State Density by X-Ray Photoelectron Spectroscopy, *Surf. Sci.* 245, 345-352 (1991).
15. R. W. M. Chan, R. W. M. Kwok, and W. M. Lau, Modified Surface Charge Spectroscopy for the Characterization of Insulator/Semiconductor Structures, *J. Appl. Phys.* 79, 3635-3639 (1996).
16. R. W. M. Chan, R. W. M. Kwok, W. M. Lau, H. Yan, and S. P. Wong, Studies of Electrical and Chemical Properties of SiO_2/Si after Rapid Thermal Nitridation Using Surface Charge Spectroscopy and X-Ray Photoelectron Spectroscopy, *J. Vac. Sci. Technol. A* 15, 2787-2792 (1997).
17. W. M. Lau, A Surface Charging Technique in Photoemission Spectroscopic Studies of Dielectric-Semiconductor Structures, *J. Appl. Phys.* 67, 1504-1509 (1990).

18. W. M. Lau, S. Jin, X.-W. Wu, and S. Ingrey, In Situ X-Ray Photoelectron Spectroscopic Study of Remote Plasma Enhanced Chemical Vapor Deposition of Silicon Nitride on Sulfide Passivated Inp, *J. Vac. Sci. Technol. B* 8, 848-855 (1990).
19. J. H. Thomas III, C. E. Bryson III, and T. R. Pampalone, X-Ray Photoelectron Spectroscopy Surface Charge Buildup Used to Study Residue in Deep Features on Integrated Circuits, *J. Vac. Sci. Technol. B* 6, 1081-1086 (1988).
20. A. Ermolieff, S. Deleonibus, S. Marthon, B. Blanchard, and J. Piaguet, Study of SiO_2/Si Interface States in Mos Devices by Surface Charge Spectroscopy. Application to Rapid Thermal Nitridation of Silicon, *J. Electron Spectros. Relat. Phenom.* 67, 409-416 (1994).
21. F. H. Bell and O. Joubert, Polysilicon Gate Etching in High Density Plasmas. 5. Comparison between Quantitative Chemical Analysis of Photoresist and Oxide Masked Polysilicon Gates Etched in Hbr/Cl-2/O-2 Plasmas, *J. Vac. Sci. Technol. B* 15, 88-97 (1997).
22. J. D. Miller, W. C. Harris, and W. G. Zajac, Composite Interface Analysis Using Voltage Contrast Xps, *Surf. Interface Anal.* 20, 977-983 (1993).
23. H. Doron-Mor, A. Hatzor, A. Vaskevich, T. van der Boom-Moav, A. Shanzer, I. Rubinstein, and H. Cohen, Controlled Surface Charging as a Depth-Profiling Probe for Mesoscopic Layers, *Nature* 406, 382-385 (2000).
24. K. Shabtai, S. R. Cohen, H. Cohen, and I. Rubinstein, A Composite Gold-Silicon Oxide Surface for Mesoscopic Patterning, *J. Phys. Chem. B* 107, 5540-5546 (2003).
25. S. Suzer, Differential Charging in X-Ray Photoelectron Spectroscopy: A Nuisance or a Useful Tool? *Anal. Chem.* 75, 7026-7029 (2003).
26. G. Ertas and S. Suzer, XPS Analysis with External Bias: A Simple Method for Probing Differential Charging, *Surf. Interface Anal.* 36, 619-623 (2004).
27. N. J. Havercroft and P. M. A. Sherwood, Use of Differential Surface Charging to Separate Chemical Differences in X-Ray Photoelectron Spectroscopy, *Surf. Interface Anal.* 29, 232-240 (2000).
28. F. Karadas, G. Ertas, and S. Suzer, Differential Charging in SiO_2/Si System as Determined by XPS, *J. Phys. Chem. B* 108, 1515-1518 (2004).
29. B. Ulgut and S. Suzer, Xps Studies of SiO_2/Si System under External Bias, *J. Phys. Chem. B* 107, 2939-2943 (2003).
30. B. V. Enüstün and T. Turkevich, Coagulation of Colloidal Gold, *J. Am. Chem. Soc.* 85, 3317-3328 (1963).
31. S. L. Cumberland and G. F. Strouse, Analysis of the Nature of Oxyanion Adsorption on Gold Nanomaterial Surfaces, *Langmuir* 18, 269-276 (2002).
32. S. Iwata and A. Ishizaka, Electron Spectroscopic Analysis of the SiO_2/Si System and Correlation with Metal-Oxide-Semiconductor Device Characteristics, *J. Appl. Phys.* 79, 6653-6713 (1996).
33. A. Ishizaka, S. Iwata, and Y. Kamigaki, Si-SiO_2 Interface Characterization by ESCA, *Surf. Sci.* 84, 235-537 (1979).
34. K. Luo, D. Y. Kim, and D. W. Goodman, The Nucleation and Growth of Gold on Silica, *J. Mol. Catal. A* 167, 191-198 (2001).
35. M. G. Mason, Electronic Structure of Supported Small Metal Clusters, *Phys. Rev. B* 27, 749-762 (1983).
36. F. Karadas, G. Ertas, E. Ozkaraoglu, and S. Suzer, X-Ray-Induced Production of Gold Nanoparticles on a SiO_2/Si System and in a Poly(Methyl Methacrylate) Matrix, *Langmuir* 21, 437-442 (2005).
37. G. Ertas, U. K. Demirok, and S. Suzer, Enhanced Peak Separation in XPS with External Biasing, *Appl. Surf. Sci.* 249, 12-15 (2005).

38. C. Battistoni, G. Mattogno, and D. M. P. Mingos, Characterization of Some Gold Cluster Compounds by X-Ray Photoelectron Spectroscopy, *J. Electron Spectros. Relat. Phenom.* 33, 107-113 (1984).
39. C. Battistoni, G. Mattogno, and D. M. P. Mingos, Characterisation of Some Gold Clusters by X-Ray Photoelectron Spectroscopy, *J. Electron Spectros. Relat. Phenom.* 28, 23-31 (1982).
40. A. Howard, D. N. S. Clark, C. E. J. Mitchell, R. G. Egdell, and V. R. Dhanak, Initial and Final State Effects in Photoemission from Au Nanoclusters on TiO_2(110), *Surf. Sci.* 518, 210-224 (2002).
41. T. Ohgi, H. Y. Sheng, Z. C. Dong, H. Nejoh, and D. Fujita, Charging Effects in Gold Nanoclusters Grown on Octanedithiol Layers, *Appl. Phys. Lett.* 79, 2453-2455 (2001).
42. T. Ohgi and D. Fujita, Consistent Size Dependency of Core-Level Binding Energy Shifts and Single-Electron Tunneling Effects in Supported Gold Nanoclusters, *Phys. Rev. B* 66, 115410-115415 (2002).
43. M. Brust, M. Walker, D. Bethell, D. J. Schiffrin, and R. Whyman, Synthesis of Thiol-Derivatized Gold Nanoparticles in a 2-Phase Liquid-Liquid System, *J. Chem. Soc. Chem. Commun.* 7, 801-802 (1994).
44. A. N. Shipway, M. Lahav, R. Gabai, and I. Willner, Investigations into the Electrostatically Induced Aggregation of Au Nanoparticles, *Langmuir* 16, 8789-8795 (2000).
45. M. M. Alvarez, J. T. Khoury, T. G. Schaaff, M. N. Shafigullin, I. Vezmar, and R. L. Whetten, Optical Absorption Spectra of Nanocrystal Gold Molecules, *J. Phys. Chem. B* 101, 3706-3712 (1997).
46. I. Tunc, S. Suzer, M. A. Correa-Duarte, L. M. Liz-Marzan, XPS Characterization of Au (Core)/SiO_2 (Shell) Nanoparticles J. Phys. Chem. B 109, 7597-7600 (2005).
47. I. Tunc, S. Suzer, M. A. Correa-Duarte, and L. M. Liz-Marzan, Charging/Discharging of Au (Core)/Silica (Shell) Nanoparticles as revealed by XPS, *J. Phys. Chem. B* (to be published).

MODIFICATION AND COATING OF POLYMER SURFACES

OLGA GERSHEVITZ, PAULIUS SILICKAS AND CHAIM N. SUKENIK*
Department of Chemistry, Bar-Ilan University, Ramat Gan, ISRAEL 52900

Abstract. Liquid phase deposition methods are a useful way to create mineral oxide films from aqueous solution under near-ambient conditions. These approaches have been applied to the creation of ceramic coatings on polymers and on polymer-matrix composites. Control has been achieved over the adherence and crystallinity of the solution-deposited thin films based on controlling the composition of both the deposition solution and the substrate surface. The challenge of depositing such films from water, while minimizing film cracking has also been addressed. Crack-free ceramic films of up to 200 nm thickness have been achieved on a variety of polymer substrates.

Keywords: ceramic thin films, titania, polyimides, polymer matrix composites

1. Introduction

High-performance polymers and polymer-matrix composites (PMCs) are finding increasing use in demanding applications.[1] For example, PMR (Polymerization of Monomer Reactants) polyimides combine thermo-oxidative stability, processibility, good mechanical properties, and a high strength-to-weight ratio.[2] Unfortunately, they are limited to applications that avoid prolonged exposure to oxidizing atmospheres at >300°C because surface oxidation leads to weight loss and degradation of mechanical properties.

A promising approach to improving PMC performance is the use of inorganic barrier coatings to block the interaction of oxygen and oxy-radicals with the polymer surface. The effectiveness and durability of the coatings will depend on the thermal stability, compactness, and uniformity of the coating

*To whom correspondence should be addressed. C. N. Sukenik, Department of Chemistry, Bar-Ilan University, Ramat Gan, Israel 52900; e-mail: sukenc@mail.biu.ac.il

J.P. Blitz and V.M. Gun'ko (eds.), Surface Chemistry in Biomedical and Environmental Science, 59–68.

material and on the interaction between the coating and the PMC. Hard-coated PMCs should also show improved abrasion resistance.[3]

Unlike vapor-phase processes, liquid deposition methods are not line-of-sight limited and can be applied to complex substrates. Wet processing methods also have lower equipment costs but often require annealing temperatures that do not allow their application to temperature sensitive substrates.

Biomineralization,[4] where inorganic materials are formed under near ambient conditions, has inspired new approaches to coating polymers. Biominerals often show superior properties due to an elaborate microstructure of crystalline domains in an organic matrix.[5] In biomineralization the interaction of super-saturated solutions and surface chemistry is often crucial.

Liquid phase deposition (LPD) has been widely used to make silica films[6] and is being increasingly used for oxides such as titania.[7,8] It uses aqueous metal-fluoride solutions where hydrolysis is modulated by added H_3BO_3 or aluminum metal. Fluoride ligands provide for a slower and more controllable hydrolysis, while H_3BO_3 or aluminum are F^- scavengers. Small differences in pH, temperature, or solution composition can have large effects on solution supersaturation and on the morphology and growth rate of the resulting film.[9] LPD from water under mild conditions of temperature ($\leq$ 55°C) and pH (2.88-3.88) can produce a variety of stable, adherent, titania films.

The role of the substrate in aqueous depositions of oxide films can also be critical.[10] In the case of an inorganic surface, the chemical nature of the substrate surface can be controlled by the use of organic self-assembled monolayers (SAMs). A SAM is a close-packed, highly ordered array of long-chain hydrocarbon molecules anchored to a solid substrate by strong covalent or ionic bonds. The functional groups on the SAM terminus remote from the anchoring functionality are responsible for the surface properties. Using SAM-coated substrates, compact, uniform, adherent thin films of TiO_2,[8,10] SnO_2,[11] ZrO_2, and $CaCO_3$ thin films have been deposited.[12,13]

We have studied LPD titania films deposited on silicon wafers with and without SAMs and on variously sulfonated surfaces.[12] Room temperature deposition[7] from 0.3 M H_3BO_3 and 0.1 M $(NH_4)_2TiF_6$ with an initial pH of 3.88 (Method 1) gave amorphous films that adhered well to silicon wafers but not to sulfonated surfaces. Alternatively, deposition[8] at 50°C from 0.15 M H_3BO_3 and 0.05 M $(NH_4)_2TiF_6$, at an initial pH of 2.88 (Method 2) produced oriented anatase films at a faster growth rate but with poor adherence to silicon substrates and excellent adherence to sulfonated surfaces.

LPD deposition of ceramic thin films is a potentially general approach to improving the abrasion resistance and thermo-oxidative stability of polymers. Dutschke *et al.*[14,15] have deposited titania on variously treated polystyrene (PS). Continuous, adherent anatase films form on PS either after hydroxylation in aqueous sodium peroxodisulfate or after surface grafting with 2-acrylamido-2-methylpropane-1-sulfonic acid (AMPS). Non-adherent but continuous films deposit on untreated PS under identical conditions. Continuous films are not observed on polystyrene etched by sulfuric acid or SO_3 gas. On polyamide microcapsules, LPD yielded continuous anatase coatings which after heat treatment at 600°C were 200 nm thick.[16]

We report herein LPD titania films deposited on three kinds of polyimides: two high-temperature polyimide resins (PMR-15 and Kapton®) and a bismaleimide glass fiber composite (BMI). We have examined different solution conditions and different surface priming strategies. We have also learned to minimize film cracking by carefully controlling the drying process.

2. Experimental

Deposition of titania on sulfonic and benzoic acid SAMs on silicon wafers and on PMR-15 and on BMI have been described.[17,18] PMR-15 resin coupons were prepared by published procedures.[19] Glass BMI prepreg coupons F-650/120 (from Israel Aircraft Industries) were rinsed with acetone and ethanol and dried with filtered N_2. Kapton® (500HN or VN; 125 mm thick, DuPont™) was rinsed with water and ethanol and dried with filtered N_2.

Surface Activation of PMR-15 by sulfonation has been described.[17] BMI surfaces were made hydrophilic by surface oxidation with UVOCS and/or depositing a sol gel layer of SiO_2.[18] Kapton® surfaces were activated with a UVOCS cleaner or by air-plasma. Their water contact angle fell from 64° to 0°.

LPD titania deposition using TiF_4 (Method 1)[7] used 0.3 M H_3BO_3 and 0.1 M $(NH_4)_2TiF_6$ at RT for 4 to 48 h, after which they were washed with water and dried. LPD of TiF_4 with added HCl[8] (Method 2) used 0.15 M H_3BO_3 and 0.05 M $(NH_4)_2TiF_6$. HCl was used to adjust the pH to 2.88 and the substrates were kept in the solution at 50°C for 2-22 h; washed with water; and dried. Ambient drying produced cracked films. Film quality was improved by drying at 70 C while reducing the humidity in ***steps*** from 70/80% to 60% to 40% to 20% or in a steady ***ramp*** from 70% to 20%[23] over 90 h. It was also improved by rinsing samples in methanol before the variable humidity drying.

3. Results

3.1. SURFACE ACTIVATION OF THE POLYMERS

Previous work using SAM templates[12,20] suggested that sulfonated surfaces provide effective templates for the growth of a surface oxide film. For polyimide substrates such as PMR-15, the goal was to achieve maximum surface sulfonation with minimum surface damage. We investigated three approaches: 1) dipping the polymer into oleum; 2) exposing the polymer to SO_3 gas; and 3) dipping the polymer into a mixture of H_2SO_4 and $ClSO_3H$ followed by hydrolysis of the sulfonyl chloride groups. EDAX showed that relative sulfonation provided by these methods was approximately 600:3:1. While immersion in oleum provides the most extensive sulfonation, it is also the most destructive. A two-minute exposure of the PMR-15 to SO_3 vapors also gave effective sulfonation, but with substantial surface degradation - as reported for polystyrene.[14] Substantial sulfonation without significant surface damage was obtained by dipping the PMR-15 into a mixture of $ClSO_3H$ and H_2SO_4 (30:70) for 0.5 min and then soaking the sample in water for 48 hours.

Surface activation of BMI coupons was done by rinsing in organic solvents followed by UVOCS activation for 20 minutes. This was essential for good TiO_2 film adhesion.

Cleaning and activation of Kapton® was done by rinsing in ethanol and then exposing it to either UVOCS or to air-plasma. XPS analysis showed that these treatments led to the introduction of oxygen into the Kapton® surface, consistent with the decrease in the advancing water contact angle from 64° to 0°. FTIR showed only small changes and SEM indicated no significant surface damage. These pre-treatments were essential for good TiO_2 film adhesion.

3.2. TITANIA DEPOSITION AND CHARACTERIZATION

Given their modest acidity and low temperature, LPD methods are ideally suited to polymer substrates. We applied Method 1 (pH =3.8; room temperature) to unactivated PMR-15. RBS analysis showed that the deposited titania was 90 nm thick after 24 h and a 450 nm film was deposited in 48 h. Variability in the onset time for film formation may account for the seemingly slower initial growth. Method 1 titania growth on clean silicon wafers for these same time intervals gave 250 and 450 nm respectively. All samples were amorphous, as had been reported for this method on variously treated silicon wafers.[12] An adherent, amorphous, titania film (420 nm thick in 48 h) also formed on a PMR-15 surface that had been sulfonated by exposure to SO_3 gas.

Method 2 (pH 2.8, 50°C) was used to deposit (22 h) TiO_2 (750 nm thick) on $ClSO_3H$ activated PMR-15 (Figure 1). The rings in the electron diffraction pattern (inset in Figure 1a) confirmed that the film was polycrystalline anatase with some degree of texture. The high-resolution image (Figure 1b) showed crystals that were generally smaller than 10 nm in diameter (consistent with other studies using Method 2)[14] but some larger crystals are visible.

XRD analysis[17] of the sample from Figure 1 showed enhanced intensity of the (004) peak compared to the (101) and (200) peaks, suggesting significant c-axis texturing perpendicular to the substrate. A similar degree of texturing has been observed for such films on various substrates.[12,14,21] Calculations, based on the Scherrer formula[17] yielded grain sizes of 10-14 nm (similar to the electron diffraction result) regardless of the growth period (4-22 h).

Using the thickness of 750 nm obtained from TEM, we calibrated measurements of thickness done by Rutherford Back-Scattering (RBS).[17] These calibrated values appear in Table 1. Such calibration has been reported in SnO_2 films, which yielded TEM thickness values that were 55-75% of those from uncalibrated RBS measurements.[11]

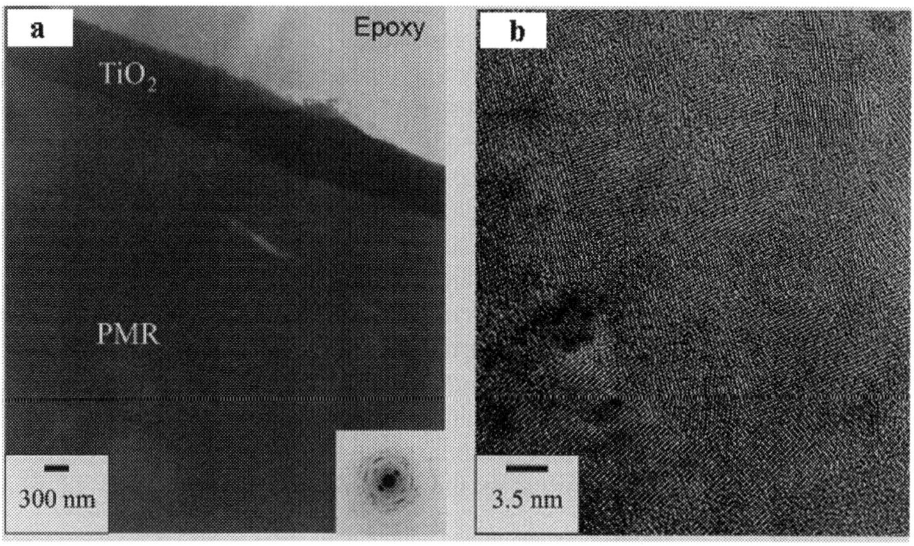

Figure 1. TEM of Method 2 LPD TiO_2 on chlorosulfonic acid-treated PMR-15; (a) cross-sectional view of the substrate and the titania layer; and (b) HRTEM of the titania shown in (a).

Though Method 2 had been reported[12] to give adherent titania films only on sulfonated surfaces, we compared PMR-15 with and without $ClSO_3H$ activation in a 22-h deposition. In both cases, adherent anatase films with significant (004) orientation formed, 660-690 nm on the sulfonated surface and 520-630 nm thick on the untreated substrate (Table 1). In a similar study on PS,[15] films formed with or without grafted sulfonic acid groups on the surface, but were only adherent on the sulfonated surface.

Table 1. Film thickness (calibrated RBS) for LPD Method 2 on PMR-15. Values for "thickness ± roughness" indicate a minimum film thickness and a point-to-point thickness variation.

Surface Type		4-h Deposition	22-h Deposition
		Thickness ± roughness (nm)/ Composition	Thickness ± roughness (nm)/Composition
Native PMR-15	sample 1	230 ± 300 / Ti, O (2.0), F (0.4)	520 ± 200 / Ti, O (2.5)
	sample 2		630 ± 300 / Ti, O (3.0)
$ClSO_3H$ treated	sample 1	230 ± 400 / Ti, O (2.0), F (0.4)	660 ± 300 / Ti, O (2.9)
	sample 2		690 ± 300 / Ti, O (3.0)

The films deposited in 4 h on untreated and on $ClSO_3H$-activated PMR-15 were both 230 nm thick (Table 1), consisting of densely packed particles ~50-100 nm in diameter that are likely agglomerates of 10 nm crystals (based on the XRD data). A few agglomerates of ~500 nm could also be seen. Open cracks ~50 nm wide were evident in both films, as compared to 270 nm thick Method 2 films on silicon that showed only closed cracks.[12]

LPD films of TiO_2 on BMI and on Kapton® were stable to sonication in water and could not be removed by a standard tape test. Figure 2 shows cross-sectional SEM of samples of BMI and Kapton® with surface oxide films ranging in thicknesses from 200-700 nm. The thickness of the titania layer is independent of the activation of the surface or the kind of polymer. The thicker films (Figure 2b and 2d) are comparable (± 20%) to those reported on sulfonate-monolayers (400 nm).[12] Thinner films (Figure 2a and 2c) were somewhat thicker than those reported on a clean silicon wafer (200-300 nm *vs.* 80 nm), likely due to variability in the time needed for the onset of film deposition.

3.3. CONTROLLED HUMIDITY DRYING AND/OR SOLVENT EXCHANGE

A good ceramic barrier layer on the surface of a polymer must be dense and crack-free. Films that are prepared by wet chemical methods (like LPD) suffer from cracking problems during drying, particularly as film thicknesses increase. An effective drying strategy is to slowly reduce the humidity around the coated samples. Using an oven with controlled temperature and programmed changes in relative humidity dramatically reduces film cracking.

Samples of BMI dried using variable humidity were analyzed by SEM and the degree of film cracking was calculated using image analysis software (Table 2). Controlled humidity drying at 60 or 70°C provided nearly crack free amorphous TiO_2 films (Method 1). Crystalline films (Method 2) needed to be dried at 70°C. Controlled humidity drying of the amorphous films at elevated temperature reduced the number of cracks. For the crystalline films, controlled humidity drying at progressively higher temperatures reduced the width of the cracks, but not necessarily their numbers. The results are summarized in Table 2 and SEM micrographs of representative samples are shown in Figure 3.

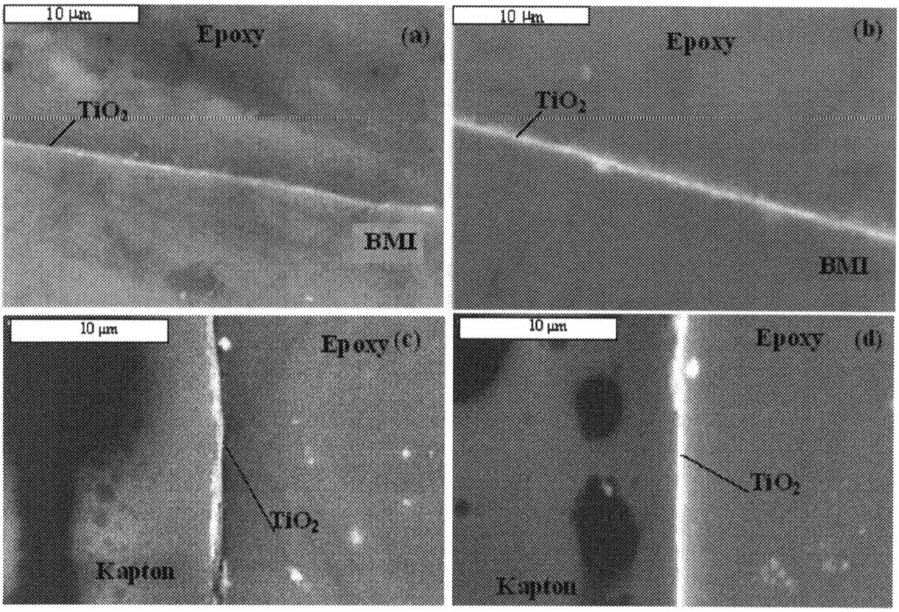

Figure 2. Cross-sectional SEM images of BMI and Kapton® coated with TiO_2 a) BMI, Method 1, 7 h coating time (~200 nm); b) BMI, Method 2, 7 h coating time (~400 nm); c) Kapton®, Method 1, 7 h coating time (310 nm); d) Kapton®, Method 2, 7 h coating time (416 nm).

We have also found that residual cracks in TiO_2 on Kapton® could be further reduced by rinsing the samples in methanol before the controlled humidity drying. Figure 4 shows examples of this improvement both for Method 1 and Method 2 TiO_2 on air plasma activated Kapton. With films up to 200 nm thick cracks can be prevented; thicker films still show nm-scale cracks.

Table 2. Crack Percentage of TiO_2 film on BMI.

Drying method: Temp, RH variation (time)		Method 2 TiO_2		Method 1 TiO_2	
		3 hr	20 hr	22 hr	44 hr
Ramp	40°C; from 80% to 20% (90 h)	4%	4.5%	1.4%	-
program	70°C; from 70% to 20% (50 h)	1%	1.4%	0%	0.6%
Step	40°C; from 80% to 20% (90 h)	1.4%	7.1%	4.2%	-
program	70°C; from 70% to 20% (90 h)	0.6%	1.8%	0.2%	-
	70°C; from 70% to 20% (50 h)	1.3%	2.8%	0.2%	-

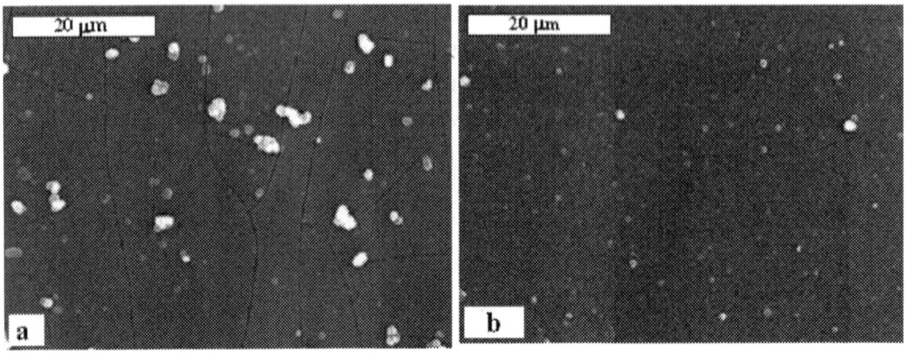

Figure 3. SEM of Method 1 titania (22 h deposition) on BMI, dried by 90 h humidity step program: a) 40ºC (4.2% cracks), b) 70ºC (0.2% cracks).

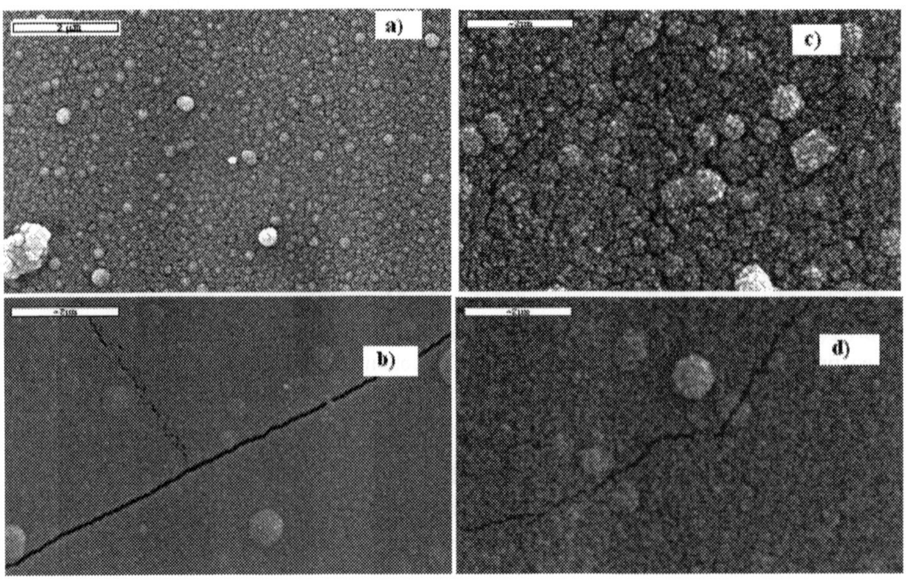

Figure 4. SEM of titania on air plasma activated Kapton®, CH_3OH washed, controlled humidity drying, (a) and (b) Method 1, 7 h and 24 h coatings; (c) and (d) Method 2, 3 h and 7 h coatings.

3.4. DISCUSSION AND CONCLUSIONS

Attaching thin adherent titania films to polyimide polymers demonstrates the promise of such coatings as barrier layers. Potential applications must examine the effect of the deposition process on the mechanical and thermal properties of the polymer and the requisite thickness of the ceramic layer. Nevertheless, the low-cost and convenience of LPD coatings, their mild conditions and their lack of line-of-site limitations, recommend their use.

Adherent, conformal LPD ceramic film either crystalline (Method 2) or amorphous (Method 1) can be obtained on polyimides. This is in contrast to silanol-bearing surfaces (good for Method 1 only) or sulfonated surfaces (good for Method 2 only). The fact that the polyimides accommodate both titania preparations may be due to the partial hydrolysis of the polymer surface under oxide deposition conditions. This provides a mix of carboxylic acid and amide sites that anchor the titania by a combination of coulombic and chelation-based effects.[22] An important lesson of this work is that the interaction of the polymer surface with the deposition solution may create oxide film anchoring sites. This does not negate activating the polymer surface. It recognizes that the polymer surface can react further under the deposition conditions.

It is clear that water-based film formation requires controlled drying protocols. While the thicker films needed for applications like scratch resistance present important challenges, scaling of such a process for large-area applications should be possible and its relative simplicity recommends its consideration. The benefits of drying at higher temperatures while carefully varying the relative humidity is likely to help other water based systems, though assessing the influence of prolonged exposure to hot humid conditions may also be an important variable in subsequent polymer/composite performance.

Finally, we note the value of rinsing the water-deposited film with methanol. As methanol is more volatile and has a lower surface tension, film cracking due to the capillary stresses exerted by the meniscus of the drying solvent on the walls of nanosized pores is reduced. Similar results have been reported by Goh.[22] Films thicker than a few hundred nm exhibit cracks due to stresses that result during drying of the film, even after methanol treatment. Eliminating these cracks will require other modifications of the drying process.

4. Acknowledgment

The authors thank the US Air Force European Office of Aerospace Research and Development, the Marie Curie Training Site for the Controlled Fabrication of Nanoscale Materials, and the Minerva Center for Microscale and Nanoscale Particles and Films as Tailored Biomaterial Interfaces for support of this work.

References

1. M. A. B. Meador, Recent advances in the development of processable high-temperature polymers, *Ann. Rev. Mater. Sci.* 28, 599-630 (1998).
2. T. T. Serafini, P. Delvigs, and G. Lightsey, Thermally stable polyimides from solutions of monomeric reactants, *J. Appl. Polymer Sci.* 16, 905-915 (1972).

3. M. Ivosevic, R. Knight, S. R. Kalidindi, G. R. Palmese, and J. K. Sutter, Erosion/oxidation resistant coatings for high temperature polymer composites, *High Performance Polymers* 15(4), 503-517 (2003).
4. S. Mann, *Biomineralization: Principles and Concepts in Bioinorganic Material Chemistry* (Oxford University Press, New York, 2001).
5. P. Calvert, Biomimetic ceramics and composites, *MRS Bulletin* 17, 37-40 (1992).
6. H. Nagayama, H. Honda, and H. Kawahara, A new process for silica coating, *J. Electrochemical Soc.* 135, 2013-2016 (1988).
7. S. Deki, Y. Aoi, O. Hiroi, and A. Kajinami, Titanium (IV) Oxide thin films prepared from aqueous solution, *Chem. Letters* 433-434 (1996).
8. K. Koumoto, S. Seo, T. Sugiyama, W. S. Seo, and W. J. Dressick, Micropatterning of titanium dioxide on self-assembled monolayers using a liquid-phase deposition process, *Chem. Mater.* 11, 2305-2309 (1999).
9. T. P. Niesen and M. R. De Guire, Deposition of ceramic thin films at low temperatures from aqueous solutions, *J. Electroceram.* 6, 169-207 (2001).
10. T. P. Niesen, J. Bill, and F. Aldinger, Deposition of titania thin films by a peroxide route on different functionalized organic self-assembled monolayers, *Chem. Mater.* 13, 1552-1559 (2001).
11. S. Supothina and M. R. De Guire, Characterization of SnO_2 thin films grown from aqueous solution, *Thin Solid Films* 371, 1-9 (2000).
12. H. Pizem, C. N. Sukenik, U. Sampathkumaran, A. K. McIlwain, and M. R. De Guire, Effects of substrate surface functionality on solution-deposited titania films, *Chem. Mater.* 14, 2476-2485 (2002).
13. M. R. De Guire, H. Shin, R. Collins, M. Agarwahl, C. N. Sukenik, and A. H. Heuer, Deposition of oxide thin films on silicon using organic self-assembled monolayers, *Proc. SPIE-Int. Soc. Opt. Eng.* 2686, 88 (1996).
14. A. Dutschke, C. Diegelmann, and P. Lobmann, Nucleation and growth of TiO_2 thin films on modified polystyrene surfaces, *Chem. Mater.* 15(18), 3501-3506 (2003).
15. A. Dutschke, C. Diegelmann, and P. Lobmann, Preparation of TiO_2 thin films on polystyrene by liquid phase deposition, *J. Mater. Chem.* 13(5), 1058-1063 (2003).
16. H. Strohm, M. Sgraja, J. Bertling, and P. Lobmann, Preparation of TiO_2-polymer hybrid microcapsules, *J. Mater. Sci.* 38(8), 1605-1609 (2003).
17. H. Pizem, O. Gershevitz, Y. Goffer, A. A. Frimer, C. N. Sukenik, U. Sampathkumaran, X. Milhet, A. Mcllwain, M. R. De Guire, M. A. Meador, and J. K. Sutter, Titania deposition on PMR-15, *Chem. Mater.* 17, 3205-3213 (2005).
18. A. Razgon and C. N. Sukenik, C. N., Ceramic Coatings for Fiber Matrix Composites: Titania thin films on bismaleimide-glass fiber composites, *J. Mater. Res.* 20, 25440-2552 (2005).
19. M. A. B. Meador, J. C. Johnston, P. J. Cavano, and A. A. Frimer, Oxidative degradation of nadic-end-capped polyimides. 2. evidence for reactions occurring at high temperatures, *Macromolecules* 30, 3215-3223 (1997).
20. M. Agarwal, M. R. De Guire, and A. H. Heuer, Synthesis of ZrO_2 and Y_2O_3-doped ZrO_2 thin films using self-assembled monolayers, *J. Amer. Ceramic Soc.* 80, 2967-2981 (1997).
21. Y. Masuda, S. Ieda, and K. Koumoto, Site-selective deposition of anatase of TiO_2 in an aqueous solution using a seed layer, *Langmuir* 19(10), 4415-4419 (2003).
22. G. K. L. Goh, K. Donthu, and P. K. Pallathadka, Cracking and orientation of solution-deposited rutile TiO_2 films, *Chem. Mater.* 16, 2857-2861 (2004).

IS IT POSSIBLE TO GENERALIZE THE PROBLEMS OF POROUS MATERIALS FORMATION, STUDY AND EXPLOITATION?

MAXIM S. MEL'GUNOV,* VLADIMIR B. FENELONOV
Boreskov Institute of Catalysis SB, RAS, Prospekt Akad. Lavrentieva 5, Novosibirsk, 630090, Russian Federation, e-mail: 2max@bk.ru

Abstract. The transition from a variety of "scientific bases of preparation" of porous materials (adsorbents, catalysts, *etc.*) to a uniform fundamental knowledge is discussed. This transition is based on allocation of two different but general levels of porous materials science: molecular (atomic) and supramolecular (textural). Fundamental relationships and laws are discussed in the application of porous materials for catalysis and adsorbents with respect to texture and structure.

Keywords: texturology, porous materials, catalysts, adsorbents

1. Introduction

Porous materials (**PM**) are applied in different situations of human life and environmental protection as adsorbents, heterogeneous catalysts, etc. However, one still can hardly formulate a definite answer to a question that appears to be very simple at first glance: is the state-of-the-art preparation of **PM** a science or art? From one side, there are well-known fundamentals for many catalytic reactions, some mechanisms of **PM** formation at some stages of their synthesis. But the complexity results in gaps between the "scientific basis of preparation" of different kinds of **PM**s, and each "basis" is considered as a separate field.
For example separate conferences on zeolites, self-assembling mesophases, carbons, sol-gel technology, supported biocatalysts, etc. are held. One can write a list of ~20 different "scientific basis of preparation".[1] The tendency is to enlarge this list, resulting in a distancing between different fields. As a result, a

*To whom correspondence should be addressed. M.S. Mel'gunov, Boreskov Institute of Catalysis SB, RAS, 5 Prospekt Akad. Lavrentieva, Novosibirsk 630090, Russian Federation, e-mail: 2max@bk.ru

J.P. Blitz and V.M. Gun'ko (eds.), Surface Chemistry in Biomedical and Environmental Science, 69– 78.

zeolite specialist does not talk to the silica gel specialist, neither one of which talks to the carbon adsorbent specialist.

However, if there are so many "scientific bases", are they fundamental? Do the ideas of the majority of technological patents and know-how appear as a result of an "Edisonian" approach of trial and error and past "art", or as a result of scientific prediction? We believe the former predominates. Thus one can consider the modern level of **PM**s preparation resulting from art, where inspiration and luck are more important than fundamental knowledge.

The widespread point of view with respect to **PM**s preparation is illustrated in the following quote (cited from Ref. 3): "All catalysts that one can imagine must be grouped by their general characteristics *mainly chemical*, which determine the properties of particular stages of synthesis, genesis and formation of catalysts in framework of a chosen method of their preparation."

We contend that this philosophy overlooks issues of fundamental importance. Not only is molecular structure important, but supramolecular structure or textural properties are also important.

In catalysis, molecular structure determines the catalytic activity in the kinetic regime, and supramolecular structure controls the degree of usage of this catalytic activity in applied catalysis, as well as heat and mass transfer, mechanical and other properties. In other words, the absence of proper molecular structure causes the absence of *catalysis*, but one is restricted in preparation of a *catalyst* by the necessity to improve the supramolecular structure.[4]

Molecular and supramolecular structures are influenced by different laws. The latter is based on effects usually not considered in traditional molecular chemistry. Rather, supramolecular chemistry is dictated by the laws of physics of dispersed matter, physical chemistry of surface phenomena, differential geometry, topology, and other "geometric" parameters. This multi-disciplinary field can be designated as the *physical chemistry of supramolecular structures*[3] or, for brevity, *texturology*.[4] Heterogeneous catalyst synthesis is usually a multistep process; some steps guided by molecular chemistry, others by texturology. When the laws of both fields apply, it is necessary to analyze both factors independently. In this way the proper set of preparation parameters for a proper catalyst can be determined. Let us consider some examples to demonstrate the typical mechanisms of **PM**s formation.

2. Capillary-Condensation Phenomena

2.1. BASIC PRINCIPLES

Classical laws of capillarity are important down to the 2-4 nm range.[5] The Laplace-Young law is fundamental. This law states that the interface of any condensed equilibrium mobile phase (e.g., liquid) should have one value of the mean curvature $\boldsymbol{g}$. This results from a gradient of intrinsic pressure (or in general, chemical potential): $\Delta P = g\sigma = 2\sigma/r$, which causes mass transfer. The direction of mass transfer is determined by the necessity to equalize the overall surface curvature. The rate of mass transfer depends on viscosity, solubility, or other properties that characterize the phase mobility.

The simplest situation (Figure 1a) shows ΔP in a simple capillary vertically

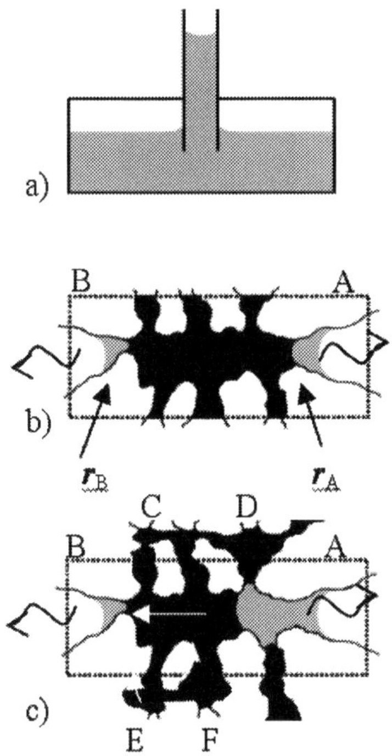

Figure 1. Scheme of liquid evaporation: a) a simple cylinder capillary connected to an infinite reservoir; b) 2D capillary of variable cross-section; c) a group of interconnected capillaries.

put into a liquid. In this situation ΔP is compensated by gravity. If the lower reservoir is infinite, or closed and kept under constant pressure evaporation occurs from the meniscus in the capillary. As a result we have a constant mass transfer of liquid to the evaporation zone. If the liquid contains dissolved non-volatile salt, under certain conditions salt will deposit in a region of evaporation, resulting in the redistribution of salt in the system.

The situation shown in Figure 1b is a coordinated evaporation from wide entrances A and B. Under continuous evaporation both menisci move inside the capillary having the same curvature. When one of the meniscus passes the narrowest section ($r_A > r_B$), the capillary equilibrium infringes, and mass transfer from A to B occurs spontaneously until the meniscus radii equalize. Such spontaneous mass transfer, known as *jumps of Haines*, occur when menisci pass through the narrowest parts of capillaries. Redistribution of deposited salt occurs.

The situation shown in Figure 1c shows mass transfer from wide to narrow ones cages, which can be found at the evaporation interface. This effect explains the redistribution of the deposited component. The less adhesion the salt has to the surface of a non-uniform capillary, the higher the heterogeneity of its distribution in a capillary. To intensify or decrease mass transfer, and thus to control the dispersion and distribution of deposited compound, one can reasonably manipulate experimental conditions such as viscosity, surface tension, rate of evaporation, and addition of compounds that compete with the main adsorber during adsorption on the surface, etc.[4]

2.2. SOL-GEL FORMATION OF NON-CRYSTALLIZING *PM*S

Control of *surface area* and *porosity* of **PMs** prepared by condensation methods is based on conditions such as condensation, aging, drying, and calcination. Let us consider the formation of the texture of amorphous **PMs** using silica gels. The surface area for silica gels depends on the size of primary particles, which form in a synthetic mixture during nucleation (gelation). The size of silica gel particles increase at successive stages, thus the surface area continually decreases.

Over-saturation is necessary to breakthrough the thermodynamic barrier of nucleation. According to the *Ostwald-Freundlich* equation, the concentration in a confined space $C/C_0 = \exp(2\,\sigma V_m/RTr)$. Figure 2 shows the dependence of

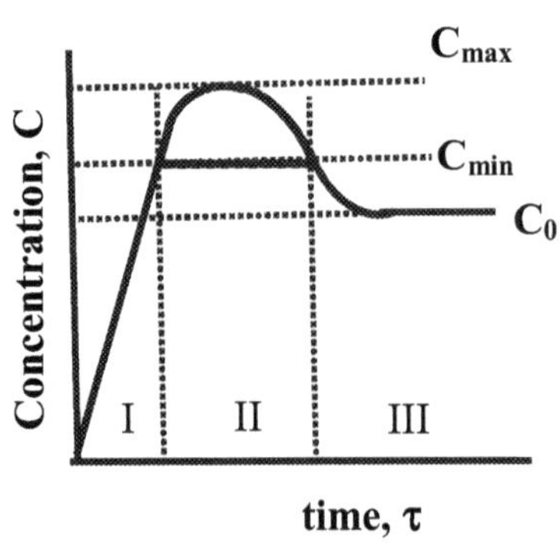

Figure 2. Typical dependence of concentration of a compound to be precipitated in a solution.[6]

concentration of a deposited compound in a solution. One can observe three typical stages: stage **I** is accompanied with increased C until C_{min} when nucleation begins; stage **II** corresponds to extreme behavior of C passing through C_{max} where the nucleation rate is maximum and returning to C_{min} again when nucleation stops; stage **III**, when concentration returns to C_0. Thus, active nucleation occurs only during stage **I**, stage **II** corresponds to an induction period. There is no nucleation during stage **III**, but the decrease in concentration is explained by two competing mechanisms: growth of particles, and *Ostwald ripening*. Poly-dispersion of the nucleated particles increases with the duration of the nucleation stage, $\Delta\tau$ due to increase in the variation of nuclei "life-time". Nuclei that appear at the beginning of stage **II** grow during a longer period than that appeared at the end of the stage. To obtain mono-disperse particles, one should increase the nucleation rate that by a decrease of $\Delta\tau$ and an increase of over-saturation.[7]

The decrease in surface area proceeds during stage **III**, which is also called the "aging period". The processes that occur during this stage are guided by the *Ostwald ripening* mechanism (Figure 3a).

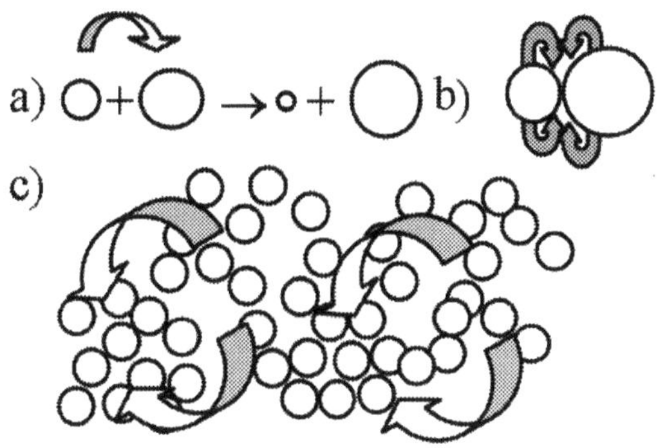

Figure 3. Scheme for explanation of mechanisms of a) *Ostwald ripening*, b) coalescence, c) overgrowing in a system of interconnected particles.

The *Ostwald-Freundlich* equation describes the dependence of particle solubility on their size. Redistribution of the solid phase follows the general trend of dissolving smaller particles and growing larger particles. However, Ostwald ripening is not the only possibility. A coalescence mechanism is important when neighboring particles come in contact. At the beginning of contact zones with negative curvature appear (Figure 3b). According to the Ostwald-Freundlich equation, over-saturation appears in negative curvature zones that result in overgrowing of the zones. This is the origin of irreversible coagulation of sol, accompanied with formation of hard interconnected aggregates in the gel that do not crash during successive drying. To intensify this process one can apply hydrothermal conditions (Figure 3c), under which overgrowing more densely packed aggregates occurs at the expense of dissolved aggregates with loose packing. In general, the Ostwald ripening mechanism is predominant for sols, and the coalescence mechanism is characteristic for gels. Both mechanisms result in a decrease of the final surface area, but have different effects on xerogel porosity during the drying step.

2.3. DRYING OF NON-CRYSTALLIZING **PM**s

Porosity formation starts with gel formation. If one decreases the influence of capillary forces during the drying step, the skeleton of the initial gel determines the maximum porosity of the xerogel, ε_{max}. But the value of ε_{max} is only achieved under supercritical drying. Drying under regular conditions results in densification of the gel and the final value of ε is always lower than ε_{max}. The minimum value, ε_{min}, can be achieved under the most favorable conditions for action of capillary forces and is usually determined by the densest random packing of particles or their hard interconnected aggregates formed during the aging step.

According to the Ostwald ripening mechanism, sol agin without particle aggregation does not change their morphology. If only particle size is changed, there is no change in the resulting packing of particles. The changes in packing due to some dispersion in size are usually not significant. However, formation of rigid aggregates by a coalescence mechanism changes the type of packing (especially when dendrite-like aggregates are formed at the beginning of coalescence). Both mechanisms result in a decrease in the surface area, but Ostwald ripening slightly influences the resulting porosity, while the coalescence mechanism results in its considerable growth.[4]

Figure 4 shows the general scheme of texture formation during drying. The evaporation interface of concave menisci is formed over the external surface (Figure 4a,b) during stage **I**. The resulting capillary forces press the globules into the bulk of the hydrogel. The action of these forces forms causes contraction of the hydrogel particle and dense packing of the external globules. A variety of situations can result depending on the rate of evaporation and the rate of tensions relaxation. In the simplest case the breakthrough of some menisci through the widest pores in the loosest zones is followed by stage **II** (Figure 4c,d). The Heines jumps mechanism appears at this stage (see, Figure 1b,c). The evaporation interface becomes more and more fractal, and the initial hydrogel splits to separate domains of higher packing density filled with liquid. The capillary forces are applied to the external surface of these domains, rather than to the external surface of the whole initial volume. Densification of the domains results in the final pore distribution. At this stage a decrease in drying rate is caused by diffusion through the dried layer. Stage **II** finishes when liquid remains only at places of contact between globules, the Heines jumps mechanism stops, the solvate shells disappear, direct contacts between the particles form, and all dissolved compounds precipitate. These complex factors guide the most intensive processes responsible for the decrease in surface area during stage **III**.

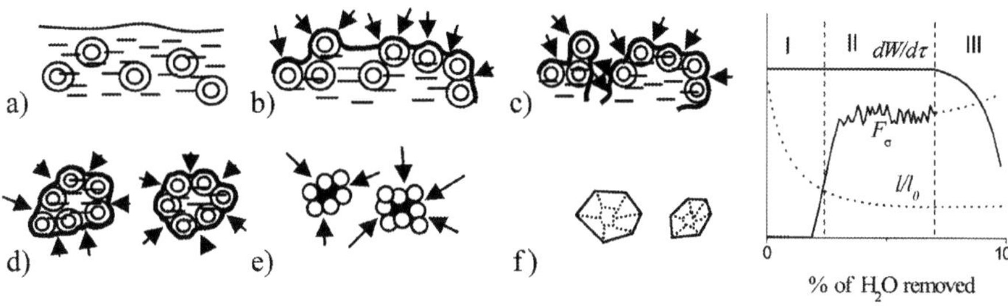

Figure 4. Typical processes that occur during texture formation during drying of hydrogel: a – d show the successive stages of formation and movement of liquid phase menisci interface (the arrows show the direction of capillary force action, the double circles are the globules of condensed phase with ionic-solvate shells), f shows high-temperature sintering. **I**, **II**, and **III** are the characteristic stages of drying, $dW/d\tau$ is the rate of drying, F_σ is the strength of capillary forces, $\Delta l/l_0$ is the linear contraction.

As a result, stage **I** directs the overall porosity of xerogel ($\varepsilon_{min} < \varepsilon < \varepsilon_{max}$), stage **II** determines the pore volume – size distribution, and stage **III** includes the decrease of surface area (the main factors here are pH, temperature and remaining liquid content).

3. Distinctions for Crystallizing PMs

From the texturological point of view, formation of crystallizing **PMs** is different because of the existence of nucleation stages at each phase transformation, usually more than one. As a result each nucleation gives a new maximum value of specific surface area, which can only decrease until the next transformation. A set of successive phase transformations including both "wet" and "dry" stages is characteristic for numerous **PMs**. Thus, each new phase transformation starts with the maximum possible surface area with its successive decrease directed by the necessity to decrease excess free energy.

There are three general principles that accompany evolution of crystallizing **PMs**: a) the *Ostwald rule of step transitions,*[7] according to which systems that can exist in different phase conditions transform from less stable to more stable phases by passing all intermediate phases; b) the *Pilling-Bedworth relation* between the volumes of starting and resulting phases;[8] c) anisotropic self-assembling.

The Pilling-Bedworth relation is not commonly known. Consider the solid-phase transformation $aA \pm C_{g1} \rightarrow bB \pm C_{g2}$. The relation of the stoichiometric volume of solid phase B to that of solid phase A can be expressed as $\Delta_{PB} = bV_B/aV_A = bM_B\rho_A/aM_A\rho_B$. When $\Delta_{PB} > 1$ the solid-phase reaction proceeds with an increase of the solid phase volume, and *vice versa*. The latter situation is illustrated in Figure 5 for "dry" and "wet" stages of synthesis.

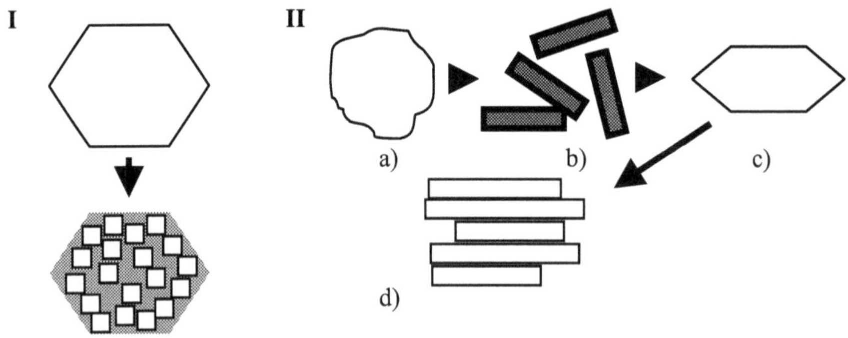

Figure 5. Textural changes during phase transformations with $\Delta_{PB} < 1.0$ in dry (**I**) and liquid (**II**) media. Scheme **II** corresponds to formation of aluminum hydroxide precipitated under pH = 6 at 473K.[9]

Phase transformations in dry media do not allow much bulk migration of the solid phase. These transformations are usually accompanied with formation of microporosity in the case of large (characteristic size of micrometers) starting particles of A, although when small nanoparticles (~3-5 nm) are transformed particle-to-particle transformation can be observed.[10] If the degree of transformation is α, and no sintering occurs (*pseudomorphous* transformation), the porosity that appears in the former case is equal to $\varepsilon_{PB,\,\alpha} = (1 - \alpha\Delta_{PB})$. The latter case results in an overall decrease of nanoparticle size, and in the absence of sintering no changes in ε are observed. Migration and redistribution of the solid phase are favorable during phase transformations in liquid media. An amorphous initial precipitate (Figure 5IIa) proceeds to self-disperse with formation of particles of a new phase (Figure 5IIb), which self-assemble in aggregates with an ordered structure (Figure 5IIc,d).

Self-assembling in liquid media results from decreased system free energy. A general mechanism of self-organization was introduced by Onzager.[11] This mechanism is based on a dependence of the change in a system's entropy ΔS of the number of building blocks in a unit volume. When N is small, entropy insignificantly depends on the neighborhood of building blocks, and is determined by arbitrary orientation. As a result at high N, the degree of freedom of each building block is lowered by neighbors, and the entropy is determined by the probability of small local movements that change the number of packing variants for each block. The probability is maximum when the blocks are parallel oriented. In this case $\sim Nn$, where $n > 2$ and depends on local self-orientation of neighboring blocks.

4. Other Possibilities

That part of texturology which corresponds to the formation of **PM**s with simple composition has been considered. Variations in chemical and phase composition of **PM**s form another dimension. Important here is the additivity principle and possible distinctions from this principle are discussed elsewhere.[4,10,12] This principle is based on the assumption that morpho-independent (independent of shape) textural characteristics of complex **PM**s (surface areas, volumes, etc) can be considered as the sum of the characteristics of individual components. This principle allows prediction and/or control of the behavior of texture during formation, aging, and application of **PM**s.

5. Conclusion

These concepts can be considered a basis for the creation and development of a uniform fundamental theory of formation of various porous (nano-)materials, including heterogeneous catalysts and adsorbents. However, texturology can also be applied to similar problems in physics, geology, biology and biochemistry, and other natural sciences.

NOTATIONS

P – pressure;
PM – porous material;
g – mean curvature, local g is equal to the half sum of two main curvatures, integral g is equal to the integral of curvature by volume of considered PM;
r – radius of surface curvature;
C – concentration in solution;
V_m – molar volume;
R – universal gas constant;
T – temperature;
τ – time;
ε – porosity;
Δ_{PB} – Pilling-Bedworth relation;
Subscript indexes: ***0*** – correspondence to saturation in infinite bulk

References

1. *Handbook of Heterogeneous Catalysis*, edited by G. Ertl, H. Knözinger, and J. Weitkamp (VCH, Weinheim, 1997), Vol. 1 and 2.
2. Yu. D. Tretyakov, Self-organization processes in the materials chemistry, *Russ. Chem. Rev.* 72, 731-763 (2003).
3. R. A. Buyanov, Scientific basis of catalysts preparation and technology, *Siberian Chem. J.* 1, 5-14 (1991).
4. V. B. Fenelonov, *Introduction to the Physical Chemistry of Supramolecular Structure of the Adsorbent and Catalysts Forming*, 2nd edition (The Publisher House of Siberian Branch of Russian Academy of Sciences, Novosibirsk, 2004, in Russian).
5. J. N. Israelashvili, *Intermolecular and Surface Forces*, 2nd edition (Academic Press, London, 1991).
6. F. Schuth and K. Unger, in: *Handbook of Heterogeneous Catalysts*, etited by G. Ertl, H. Knözinger, and J. Weitkamp (VCH, Weinheim, 1997), Vol. 1, p. 72.
7. W. Ostwald, Studien uber die bildung und umwandlung fester korper, *Z. Phys. Chem.* 22, 289-302 (1897).
8. W. M. Pilling and R. E. Bedworth, The oxidation of metals in high temperature, *J. Inst. Met.* 1, 529-533 (1923).
9. O. P. Krivoruchko, B. P. Zolotovskii, L. M. Plyasova, R. A. Buyanov, and V. I. Zaikovskii, Boehmite formation mechanism in precipitate aging, *React. Kinet. Catal. Lett.* 21, 103-109 (1982).
10. M. S. Mel'gunov, V. B. Fenelonov, E. A. Mel'gunova, A. F. Bedilo, and K. J. Klabunde, Textural changes during topochemical decomposition of nanocrystalline $Mg(OH)_2$ to MgO *J. Phys. Chem. B.* 107, 2427-2434 (2003).
11. L. Onzager, Orientational order of hard rods, *Ann. N. Y. Acad. Sci.* 51, 627 (1949). Cited by W. M. Gelbart and A. Ben-Saul, The "new" science of "complex fluids", *J. Phys. Chem.* 100, 13169-13189 (1996).
12. M. S. Mel'gunov, E. A. Mel'gunova, V. I. Zaikovskii, V. B. Fenelonov, A. F. Bedilo, and K. J. Klabunde, Carbon Dispersion and Morphology in Carbon-Coated Nanocrystalline MgO, *Langmuir* 19(24), 10426-10433 (2003).

THEORETICAL ANALYSIS OF SURFACE PRESSURE OF MONOLAYERS FORMED BY NANO-PARTICLES

V.B. FAINERMAN,[1] V.I. KOVALCHUK,[2] D.O. GRIGORIEV,[3] M.E. LESER,[4] AND R. MILLER[3,*]

[1] *Medical Physicochemical Centre, Donetsk Medical University, 16 Ilych Avenue, 83003 Donetsk, Ukraine*

[2] *Institute of Biocolloid Chemistry, 42 Vernadsky avenue, 03680 Kyiv (Kiev), Ukraine*

[3] *Max-Planck-Institut für Kolloid- und Grenzflächenforschung, 14424 Potsdam/Golm, Germany*

[4] *Nestec Ltd., Nestlé Research Centre, Vers-chez-les-Blanc, CH-1000 Lausanne 26, Switzerland*

Abstract. Surface pressure/area isotherms of monolayers of micro- and nanoparticles at fluid/liquid interfaces can be used to obtain information about particle properties (dimensions, interfacial contact angles), the structure of interfacial particle layers, interparticle interactions as well as relaxation processes within layers. Such information is important for understanding the stabilisation/destabilisation effects of particles for emulsions and foams. For a correct description of Π-A isotherms of nanoparticle monolayers, the significant differences in particle size and solvent molecule size should be taken into account. The corresponding equations are derived by using the thermodynamic model of a two-dimensional solution. The equations not only provide satisfactory agreement with experimental data for the surface pressure of monolayers in a wide range of particle sizes from 75 μm to 7.5 nm, but also predict the areas per particle and per solvent molecule close to the experimental values. Similar equations can also be applied to protein molecule monolayers at liquid interfaces.

Keywords: surface pressure isotherm, surface monolayers, nanoparticles, thermodynamic model

*To whom correspondence should be addressed. R. Miller, Max-Planck-Institut für Kolloid- und Grenzflächenforschung, 14424 Potsdam/Golm, Germany, e-mail: miller@mpikg.mpg.de

J.P. Blitz and V.M. Gun'ko (eds.), Surface Chemistry in Biomedical and Environmental Science, 79–90.

1. Introduction

Fine solid particles adsorb at interfaces and can provide long-term kinetic stability of emulsions and foams.[1-5] For more effective stabilization the particles must be much smaller than the dispersed droplets.[1,2] For the production of microemulsions, nano-sized particles are therefore of particular interest.

The stabilization mechanism of liquid films, emulsions, and foams by microparticles and nanoparticles are yet not completely understood. It is usually assumed that particles produce a steric barrier against coalescence. The efficiency of particles in stabilizing emulsions is often related to the particle adsorption energy. When the interface is not completely covered by particles, their lateral displacements along the interface are important.[6] On the other hand, the stability of solid-stabilized emulsions and foams is determined by the capillary pressure in the emulsion/foam films.[5,7] Therefore emulsions of highest stability can be obtained with particles with contact angles that provide both a sufficiently high adsorption energy, and threshold capillary pressure in the film. These contact angles are not too close to 0° or 90°.[8-11] Adsorbed particles also stabilise emulsions by modification of the interfacial rheology.[12-16] At sufficiently high interfacial concentrations the interface covered by particles can exhibit visco-elastic behaviour.[16,17]

Because of the high adsorption energy of particles, and very slow adsorption/desorption kinetics, spread particle monolayers at fluid interfaces often behave as though they are insoluble. They can thus be treated in a similar way as insoluble surfactant or polymer monolayers, e.g. through surface pressure/area isotherms. Early studies of this kind were performed with spherical monodisperse polystyrene particles and information on particle sizes and particle-particle repulsive interactions was obtained.[18-22] It was shown that the collapse pressure is greatest if the particles are monodisperse.[18] A method for determining the contact angle of particles at interfaces by measuring the collapse pressure, and equating it with the work required to remove particles into one of the two bulk phases was proposed by Clint and co-authors.[23,24] Though the potential applicability of this method was demonstrated, the problems of determining contact angles of particles from the surface pressure/area isotherms have also been described.[25-27]

The applicability of Volmer and van der Waals equations of state for a description of particle monolayers found that the shape described by the van der Waals equation of state is similar to the behaviour observed experimentally for repulsive particles within monolayers.[3] Nevertheless, both the Volmer and van der Waals equations give a dependence of surface pressure on particle size unsuitable for a quantitative analysis of experimental data. It has been recently shown[28] that for monolayers of nanoparticles, the equations of state should take into account the significant size differences of particles and solvent molecules.

The aim of the present study is to demonstrate that the corresponding equations can be derived similarly to expressions for the surface tension of mixed solutions of molecules with different sizes.[29] The same approach was also applied to protein solutions.[30] The derived equations are compared with some experimental Π-A isotherms of monolayers of micro- and nanoparticles and proteins reported in the literature.

2. Equation of State

The chemical potentials of components in a surface layer μ_i^s depend on the composition of the layer and its surface tension γ. The dependence of μ_i^s is given by a known relation[31]:

$$\mu_i^S = \mu_i^{0S}(T,P) - \int_0^{\gamma} \Omega_i d\gamma + RT \ln f_i^S x_i^S \quad (1)$$

Here $\mu_i^{0S}(T,P)=\mu_i^{0S}$ are the standard chemical potentials dependent on temperature T and pressure P, $x_i^S = m_i / \Sigma m_i$ are the molar fraction of the i-th component, m_i the number of moles, f_i the activity coefficients, and Ω_i the molar area. The superscript 's' refers to the surface (interface). Assuming that the value Ω_i is independent of γ, and integrating Eq. (1) one obtains the expression:

$$\mu_i^S = \mu_i^{0S} + RT \ln f_i^S x_i^S - \gamma\Omega_i \quad (2)$$

Equation (2) is the well-known Butler equation[32], often used to derive equations of state for surface layer and adsorption isotherm equations. In the solution bulk the chemical potentials μ_i^{α} obey the equation:

$$\mu_i^{\alpha} = \mu_i^{0\alpha} + RT \ln f_i^{\alpha} x_i^{\alpha} \quad (3)$$

with $\mu_i^{0\alpha} = \mu_i^{0\alpha}(T,P)$ being the standard chemical potentials. The superscript 'α' refers to the solution bulk. If the particles or molecules are insoluble, that is a Langmuir monolayer forms, then equation (3) is suitable only for the solvent.

An equation of state for the surface layer can be derived from corresponding equations for chemical potentials of components in the bulk and within the surface layer. The most common approach, suitable for soluble and insoluble molecules as well as for particles, equates the chemical potentials of the solvent (i = 0) in the solution bulk and in the surface layer. Eqs. (2) and (3) yield:

$$\mu_0^{0S} + RT\ln f_0^S x_0^S - \gamma\Omega_0 = \mu_0^{0\alpha} + RT\ln f_0^\alpha x_0^\alpha . \quad (4)$$

The standard state must now be formulated. For the solvent (i = 0) usually a pure component is assumed. This means $x_0^S = 1$, $f_0^S = 1$, $x_0^\alpha = 1$, $f_0^\alpha = 1$, and $\gamma = \gamma_0$. From Eq. (4) we obtain

$$\mu_0^{0S} - \gamma_0\Omega_0 = \mu_0^{0\alpha} . \quad (5)$$

Eqs. (4) and (5) yield the equation of state for the surface layer for any number of components with any geometry[30]:

$$\Pi = -\frac{kT}{\omega_0}\left(\ln x_0^S + \ln f_0^S\right). \quad (6)$$

Here $\Pi = \gamma_0 - \gamma$ is the surface pressure, k the Boltzmann constant, and ω_0 the solvent molecule molecular area. An important feature of Eq. (6) is that it involves solvent characteristics only. The ω_0 value depends on the choice of the position of the dividing surface. Assuming the solvent adsorption to be positive, the equation was proposed[30] which relates the surface excesses Γ_i of the solvent (subscript i = 0) and dissolved species (i ≥ 1) with any molecular area ω_i:

$$\sum_{i=0}^{n}\Gamma_i = 1/N\omega_0 + (1 - \omega/\omega_0)\sum_{i=1}^{n}\Gamma_i , \quad (7)$$

where N is Avogadro's number. The average molecular area of dissolved species ω is determined from the expression:

$$\omega_0 \neq \omega = \frac{\omega_1\Gamma_1 + \omega_2\Gamma_2 + \ldots}{\Gamma_1 + \Gamma_2 + \ldots} . \quad (8)$$

Equation (6) can be transformed into a more convenient form by expressing the mole fraction of surface layer components via the fraction of surface θ_i occupied by the jth component [as follows from Eq. (7)],

$$x_j^S = \frac{\theta_j}{n_j \sum_{i\geq 0}(\theta_i/n_i)} , \quad (9)$$

and also taking into account that the total value of f_i^S can now be obtained from the additivity of enthalpy (H) and entropy (E) in the Gibbs free energy:

$$\ln f_i^S = \ln f_i^H + \ln f_i^E . \tag{10}$$

In Eqs. (9) and (10) $n_i = \omega_i/\omega_0$, and

$$\ln f_0^H = a\theta^2 , \ \ln f_j^H = a\, n_j \theta_0^2 , \tag{11}$$

$$\ln f_j^E = 1 - n_j \sum_{i \geq 0} (\theta_i / n_i) + \ln\left[n_j \sum_{i \geq 0} (\theta_i / n_i) \right]. \tag{12}$$

By introducing Eqs. (9)-(12) into (6), an equation of state for a two-dimensional surface layer which contains the solvent and any other component (e.g., protein or solid particles) was obtained,[30] valid for $n = \omega/\omega_0 > 1$:

$$-\frac{\Pi\omega_0}{kT} = \ln(1-\theta) + \theta(1-\omega_0/\omega) + a\theta^2 , \tag{13}$$

where θ is the fraction of the surface covered by molecules or solid particles with average area per one entity ω. Comparing Eq. (13) with the expression (13.24) which was derived in Ref. 29 for the surface tension in the mixture of molecules of different sizes, and assuming the fraction of dissolved component in the bulk phase is almost zero [which enables one to simplify significantly Eq. (13.24)[29]] one can see that these equations are identical to each other. It should be noted that Eq. (13.24)[29] was derived using the approach entirely different from that employed to derive Eq. (13)[30]: the statistical mechanics formulation applied in Ref. 29 involved the calculation of the configuration energy of a mixture, the free energy, the chemical potentials of the components at the surface, and then the surface tension of mixed monolayer was calculated in the framework of the so-called parallel model.

Introducing the area of surface layer per one micro- or nanoparticle A via the relation $\theta = \omega/A$, and using, instead of Eq. (11) which expresses the enthalpy contribution to the solvent activity coefficient, the assumption that this contribution is independent of θ (i.e., taking $\ln f_0^H$ to be constant, which corresponds better to the liquid-expanded monolayer[33]), with the condition $n = \omega/\omega_0 >> 1$ which is true for micro- and nanoparticles, one obtains from Eq. (13) the expression for the Π-A isotherm valid for a monolayer of particles:

$$\Pi = -\frac{kT}{\omega_0}\left[\ln\left(1 - \frac{\omega}{A}\right) + \left(\frac{\omega}{A}\right) \right] - \Pi_{coh} \tag{14}$$

Π_{coh} is the cohesion pressure. As one can see, the surface pressure does not depend on the particle size but is determined by the monolayer coverage ω/A, and the parameters ω_0 and Π_{coh}. As Eq. (14) involves the ratio ω/A, then corresponding areas can refer not only to individual particles, but also to the areas occupied by a certain mass of particles or, for example, to the entire monolayer in the Langmuir trough.

For Π-A isotherms of insoluble monolayes of amphiphilic molecules, assuming the association or dissociation of these molecules in the surface layer, a generalised Volmer equation was derived (based on Butler's and Gibbs' equations[33-35]), which has the form

$$\Pi = \frac{kT(\omega/A)}{\omega_0(1-\omega/A)} - \Pi_{coh} = \frac{nkT}{A-\omega} - \Pi_{coh}\,. \qquad (15)$$

It is seen that Eq. (15), which follows approximately from Eq. (14) (assuming low monolayer coverage and neglecting entropy non-ideality), can also describe the behaviour of monolayers which comprise particles of any size. Similarly to Eq. (14), this equation involves not the geometric parameters of amphiphilic molecules (or particles), but only monolayer coverage by these entities. Equation (15) provides a good description of the experimental results obtained for various systems. For example, for some insoluble proteins in the liquid-expanded monolayer range, the value n = 20–100 was obtained.[35]

3. Comparison with Experiments

The experimental Π-A isotherms[27] obtained for silanized glass particles (75 ± 5 μm diameter) in a broad hydrophobicity range (40-90°) are presented in Figure 1. In this Figure the area A value refers to the entire monolayer in the Langmuir trough, and different experimental curves correspond to particles with different degrees of hydrophobicity. The theoretical isotherms calculated from Eqs. (14) and (15) are also shown in Figure 1, and the parameters of the equations are summarised in Table 1, along with the measured wetting angles ϕ.[27] The two theoretical models agree well with experimental data for all the microparticles studied, however, one can see that Eq. (14) gives more realistic values (Table 1). For example, the ω values (for the entire monolayer) calculated from Eq. (15) are 10-15% lower than experimental values, while values calculated from Eq. (14) correspond exactly to those obtained experimentally.[27] The ω_0 values obtained for the two models are of correct order; however, the values calculated from Eq. (14) are closer to the real values characteristic for water (ca. 0.1 nm^2). It should be noted that as the position of the dividing surface is chosen according to Eq. (7), the area per solvent molecule could be somewhat different from the true molecular area. A very interesting relationship between the non-ideality of enthalpy (given by the parameter Π_{coh}) and the wetting angle (cosϕ) results: Π_{coh} increases almost linearly with cosϕ (see Table 1) as expected: the more hydrophilic the particles, the higher the interparticle interaction at the water/air interface.

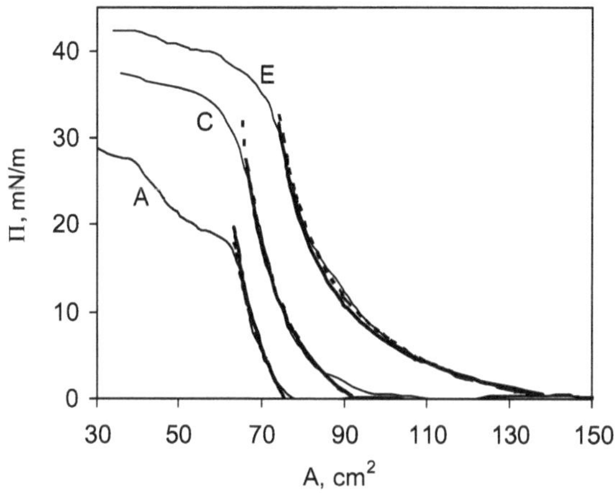

Figure 1. Dependence of surface pressure on total Langmuir trough area for silanized glass particles 75±15 μm in diameter with different hydrophobisation degree (thin lines) according to Horvolgyi et al.;[27] theoretical calculations used Eq. (14) (bold solid line) and Eq. (15) (bold dashed line) with model parameters given in Table 1.

Table 1. Parameters of theoretical models for silanized glass particles 75 μm in diameter of different hydrophobicity according to Horvolgyi et al.[27] (values shown as numerator/denominator correspond to Eq. (15)/Eq. (14), respectively).

Parameter	A	C	E
ω (cm^2)	50/57	54/61	60/68
ω_0 (nm^2)	0.4/0.16	0.43/0.17	0.44/0.18
Π_{coh} (mN/m)	19.7/17.0	13.6/9.5	6.9/3.5
Contact angle ϕ (degr.)	39	71	87

Let us consider now the Π-A isotherms (here A is the area per unit particle mass) for polymeric spherical particles 113 nm in diameter.[36] The results obtained for a monolayer of these particles with dispersant (copolymer) and a monolayer of particles without dispersant are shown in Figure 2. Table 2 contains the model parameters for the two types of monolayers. The theoretical curves presented in Figure 2 were calculated from Eq. (14). The calculations using Eq. (15) result in almost the same curves (as in Figure 1), but corresponding values of ω_0 are somewhat higher than those from Eq. (14), while the values of ω are somewhat lower than those obtained from Eq. (14). Note that for both the microparticles 75 μm in diameter (cf. Figure 1), and for the polymeric particles 113 nm in size, the ω_0 values predicted by theory are almost the same, and are almost equal [taking into account reservations which follow from Eq. (7)] to the actual area of a water molecule. It is seen from Figure 2 that for an individual monolayer of polymeric particles, the theory fits the experimental curve well throughout a very wide range of surface pressures (up to the monolayer collapse). At the same time, for a mixture of polymeric particles with dispersant, agreement with experiment is seen only at low surface pressure values. Presumably, for such systems a more complicated theory must be used, which assumes the presence of the second dissolved component in the mixture. Such models for protein/surfactant mixtures were recently developed.[37,38]

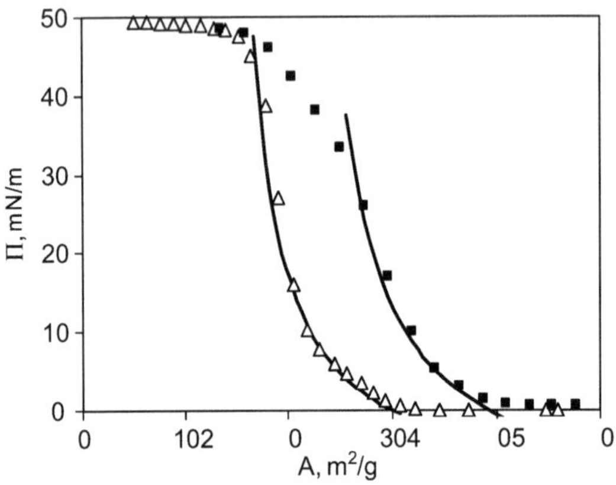

Figure 2. Dependence of surface pressure on monolayer coverage for polymeric particles 113 nm in diameter without dispersant (△) and with dispersant (■) according to Wolert et al;[36] theoretical calculations used Eq. (14) with model parameters given in Table 2.

Table 2. Parameters of theoretical model (14) for polymer particles 113 nm in diameter.[36]

Model parameters	Without dispersant	With dispersant
ω (m^2/g)	15.5	23
ω_0 (nm^2)	0.12	0.12
Π_{coh} (mN/m)	7.0	10.0

Let us consider now the Π-A isotherms (here A is the area per unit particle mass) for polymeric spherical particles 113 nm in diameter.[36] The results obtained for a monolayer of these particles with dispersant (copolymer) and a monolayer of particles without dispersant are shown in Figure 2. Table 2 contains the model parameters for the two types of monolayers. The theoretical curves presented in Figure 2 were calculated from Eq. (14). The calculations using Eq.(15) result in almost the same curves (as in Figure 1), but corresponding values of ω_0 are somewhat higher than those from Eq.(14), while the values of ω are somewhat lower than those obtained from Eq. (14). Note that for both the microparticles 75 μm in diameter (cf. Figure 1), and for the polymeric particles 113 nm in size, the ω_0 values predicted by theory are almost the same, and are almost equal [taking into account reservations which follow from Eq. (7)] to the actual area of a water molecule. It is seen from Figure 2 that for an individual monolayer of polymeric particles, the theory fits the experimental curve well throughout a very wide range of surface pressures (up to the monolayer collapse). At the same time, for a mixture of polymeric particles with dispersant, agreement with experiment is seen only at low surface pressure values. Presumably, for such systems a more complicated theory must be used, which assumes the presence of the second dissolved component in the mixture. Such models for protein/surfactant mixtures were recently developed.[37,38]

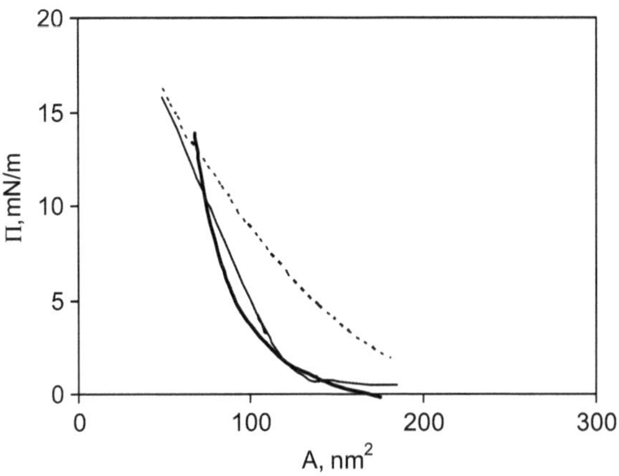

Figure 3. Dependence of surface pressure on area per particle in the monolayer for monodisperse magnetic particles 7.5 nm in diameter,[39] thin dashed line – first monolayer compression; thin solid line – third monolayer compression cycle; thick solid line – theoretical calculation.

Figure 4 illustrates the data obtained for the pulmonary surfactant protein SP-B[35] here the A value was calculated per protein molecule). The theoretical curve for this isotherm was calculated from Eq. (14) with the parameters: ω = 8.5 nm^2; ω_0 = 0.115 nm^2; and Π_{coll} = 20.0 mN/m. Again we can see the good agreement between theory and experiment.

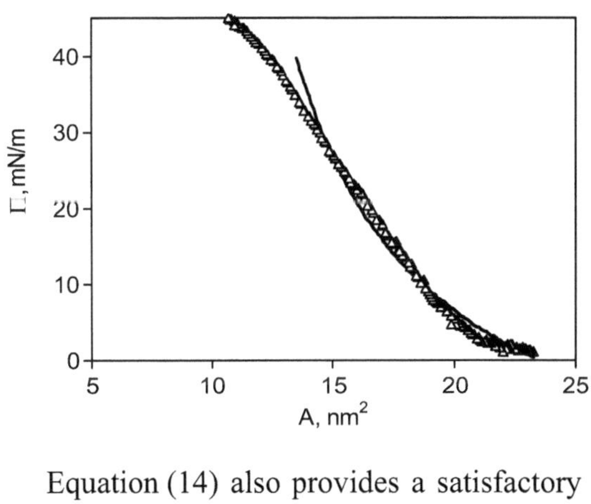

Figure 4. Dependence of surface pressure on area per molecule for the pulmonary surfactant protein SP-B according to Wüstneck et al.;[35] points – experimental data; thick solid line – theoretical calculation.

Equation (14) also provides a satisfactory description of the Π-A isotherm for monodisperse spherical polystyrene particles 2.6 μm in diameter at the water/octane interface.[40,41] For this system, fitting parameters using Eq. (14)

yields $\omega_0 = 0.12\ \text{nm}^2$, $\omega = 32\ \text{cm}^2$ (calculated for the complete area in the Langmuir trough) and the Π_{coh} value is close to zero.

4. Conclusion

Monolayers of micro- and nanoparticles at fluid/liquid interfaces can be described in a similar way as surfactants or polymers, easily studied via surface pressure/area isotherms. Such studies provide information on the properties of particles (dimensions, interfacial contact angles), the structure of interfacial layers, interactions between the particles as well as about relaxation processes within the layers. Such type of information is important for understanding how the particles stabilize (or destabilize) emulsions and foams. The performed analysis shows that for an adequate description of Π-A dependencies for nanoparticle monolayers the significant difference in size of particles and solvent molecules has be taken into account. The corresponding equations can be obtained by using a thermodynamic model developed for two-dimensional solutions. The obtained equations provide a satisfactory agreement with experimental data of surface pressure isotherms in a wide range of particle sizes between 75 μm and 7.5 nm. Moreover, the model can predict the area per particle and per solvent molecule close to real values. Similar equations were applied also to protein monolayers at liquid interfaces.

Acknowledgements

The work was financially supported by projects of the European Space Agency (FASES MAP AO-99-052), the DFG (Mi418/14) and the Ukrainian SFFR (Project No. 03.07/00227).

References

1. S. Levine, B. D. Bowen, and S. J. Partridge, Stabilization of emulsions by fine particles. I. Partitioning of particles between continuous phase and oil/water interface, *Colloids Surf.* 38, 325-343 (1989); S. Levine, B. D. Bowen, and S. J. Partridge, Stabilization of emulsions by fine particles. II. Capillary and van der Waals forces between particles, *Colloids Surf.* 38, 345-364 (1989).
2. N. Yan and J. H. Masliyah, Characterization and demulsification of solids-stabilized oil-in-water emulsions. Part 2. Demulsification by the addition of fresh oil, *Colloids Surf.* 96, 243-252 (1995).
3. B. P. Binks, Particles as surfactants - similarities and differences, *Current Opinion Colloid Interface Sci.* 7, 21-41 (2002).
4. R. Aveyard, B. P. Binks, and J. H. Clint, Emulsions stabilised solely by colloidal particles, *Adv. Colloid Interface Sci.* 100-102, 503-546 (2003).

5. P. Kruglyakov and A. Nushtayeva, Emulsions stabilized by solid particles: the role of capillary pressure in the emulsion films, in: *Emulsions: Structure Stability and Interactions*, edited by D. N. Petsev (Elsevier, Amsterdam, 2004), pp. 641-676.
6. S. Simovic and C.A. Prestidge, Nanoparticles of varying hydrophobicity at the emulsion droplet-water interface: adsorption and coalescence stability, *Langmuir* 20, 8357-8365 (2004).
7. N. D. Denkov, I. B. Ivanov, P. A. Kralchevsky, and D. T. Wasan, A possible mechanism of stabilization of emulsions by solid particles, *J. Colloid Interface Sci.* 150, 589-593 (1992).
8. N. Yan and J. H. Masliyah, Characterization and demulsification of solids-stabilized oil-in-water emulsions. Part 1. Partitioning of clay particles and preparation of emulsions, *Colloids Surf. A* 96, 229-242 (1995).
9. Th. F. Tadros and B. Vincent, in: *Encyclopedia of Emulsion Technology, Basic Theory* (Marcel Dekker, New York, 1983), Vol. 1, p. 129.
10. B. P. Binks and S. O. Lumsdon, Influence of particle wettability on the type and stability of surfactant-free emulsions, *Langmuir* 16, 8622-8631 (2000).
11. R. Aveyard and J. Clint, Liquid droplets and solid particles at surfactant solution interfaces, *J. Chem. Soc., Faraday Trans.* 91, 2681-2697 (1995).
12. J. Ding, H. E. Warriner, and J. A. Zasadzinski, Viscosity of two-dimensional suspensions. *Phys. Rev. Lett.* 88, 168102-(1-4) (2002).
13. E. J. Stancik, M. J. O. Widenbrant, A. T. Laschitsch, J. Vermant, and G. G. Fuller, Structure and dynamics of particle monolayers at a liquid-liquid interface subjected to extensional flow, *Langmuir* 18, 4372-4375 (2002).
14. E. J. Stancik, G. T. Gavranovic, M. J. O. Widenbrant, A. T. Laschitsch, J. Vermant, and G. G. Fuller, Structure and dynamics of particle monolayers at a liquid-liquid interface subjected to shear flow, *Faraday Discuss.* 123, 145-156 (2003).
15. D. Rousseau, Fat crystals and emulsion stability - a review, *Food Res. Int.* 33, 3-14 (2000).
16. D. E. Tambe and M. M. Sharma, Hydrodynamics of thin liquid-films bounded by viscoelastic interfaces, *J. Colloid Interface Sci.* 147, 137-151 (1991); Factors controlling the stability of colloid-stabilized emulsions. 1. An experimental investigation, *J. Colloid Interface Sci.* 157, 244-253 (1993); Factors controlling the stability of colloid-stabilized emulsions. 2. A model for the rheological properties of colloid-laden interfaces, *J. Colloid Interface Sci.* 162, 1-10 (1994); Factors controlling the stability of colloid-stabilized emulsions. 3. Measurement of the rheological properties of colloid-laden interfaces, *J. Colloid Interface Sci.* 171, 456-462 (1995).
17. J. Lucassen, Capillary forces between solid particles in fluid interfaces, *Colloids Surf.* 65, 131-137 (1992); Dynamic dilational properties of composite surfaces, *Colloids Surf.* 65, 139-149 (1992).
18. H. Schuller, Modellversuche zur spreitung von kolloid-partikeln. *Kolloid Z. Z. Polym.* 216-217, 380-388 (1967).
19. E. Sheppard and N. Tcheurekdjian, Monolayer studies. 4. Surface films of emulsion latex particles, *J. Colloid Interface Sci.* 28, 481-486 (1968).
20. E. Sheppard and N. Tcheurekdjian, Monolayer studies. 3. Spreading of polystyrene latexes at water/air interface, *Kolloid Z. Z. Polym.* 225, 162-170 (1968).
21. A. Doroszkowski and R. Lambourne, Measurement of strength of steric barriers in non-aqueous polymer dispersions, *J. Polym. Sci. C* 34, 253-264 (1971).
22. M. J. Garvey, D. Mitchell and A. L. Smith, Compression studies on a monolayer of polymer stabilized lattices at the air-2 molar sodium-chloride solution interface, *Colloid Polym. Sci.* 257, 70-74 (1979).
23. J. H. Clint and S. E. Taylor, Particle size and interparticle forces of overbased detergents: a Langmuir trough study, *Colloids Surf.* 65, 61-67 (1992).

24. J. H. Clint and N. Quirke, Contact Angles On Particles From Langmuir Trough Studies, *Colloids Surf.* 78, 277-278 (1993).
25. R. Aveyard, B. P. Binks, P. D. I. Fletcher, and C. E. Rutherford, Measurement of contact angles of spherical monodisperse particles with surfactant solutions, *Colloids Surf. A* 83, 89-98 (1994).
26. M. Mate, J. H. Fendler, J. J. Ramsden, J. Szalma, and Z. Horvolgyi, Eliminating surface pressure gradient effects in contact angle determination of nano- and microparticles using a film balance, *Langmuir* 14, 6501-6504 (1998).
27. Z. Horvolgyi, M. Mate, A. Daniel, and J. Szalma, Wetting behaviour of silanized glass microspheres at water-air interfaces: a Wilhelmy film balance study, *Colloids Surf. A* 156, 501-508 (1999).
28. V. B. Fainerman, V. I. Kovalchuk, D. O. Grigoriev, M. E. Leser, M. Michel, R. Miller and H. Möhwald, Surface pressure isotherms of monolayers formed by microsize and nanosize particles, *Langmuir*, submitted for publication.
29. R. Defay and I. Prigogine, *Surface Tension and Adsorption* (Longmans, Green, London, 1966).
30. V. B. Fainerman, E. H. Lucassen-Reynders and R. Miller, Description of the adsorption behaviour of proteins at water/fluid interfaces in the framework of a two-dimensional solution model, *Adv. Colloid Interface Sci.* 106, 237-259 (2003).
31. A. I. Rusanov, *Phasengleichgewichte und Grenzflächenerscheinungen* (Akademie-Verlag, Berlin, 1978).
32. J. A. V. Butler, The thermodynamics of the surfaces of solutions, *Proc. Roy. Soc. Ser. A* 138, 348-375 (1932).
33. V. B. Fainerman and D. Vollhardt, Equations of State for Langmuir Monolayers with Two-dimensional Phase Transition, *J. Phys. Chem. B* 103, 145-150 (1999).
34. V. B. Fainerman and D. Vollhardt, Equation of state for monolayers under consideration of the two-dimensional compressibility in the condensed state, *J. Phys. Chem. B* 107, 3098-3100 (2003).
35. R. Wüstneck, V. B Fainerman, N. Wüstneck, and U. Pison, Interfacial behavior of spread SP-B and SP-C layers and the influence of oligomerization and secondary structure, *J. Phys. Chem. B* 108, 1766-1770 (2004).
36. E. Wolert, S. M. Setz, R. S. Underhill, R. S. Duran, M. Schappacher, A. Deffieux, M. Hölderle, and R. Mülhaupt, Meso- and microscopic behaviour of spherical polymer particles assembling at the air-water interface, *Langmuir* 17, 5671-5677 (2001).
37. V. B. Fainerman, S. A. Zholob, M. Leser, M. Michel, and R. Miller, Competitive adsorption from mixed non-ionic surfactant/protein solutions, *J. Colloid Interface Sci.* 274, 496-501 (2004).
38. V. B. Fainerman, S. A. Zholob, M. E. Leser, M. Michel, and R. Miller, Adsorption from mixed ionic surfactant/protein solutions - analysis of ion binding, *J. Phys. Chem. B* 108, 16780-16785 (2004).
39. S. Lefebure, C. Menager, V. Cabuil, M. Assenheimer, F. Gallet, and C. Flament, Langmuir monolayers of monodispersed magnetic nanoparticles coated with a surfactant, *J. Phys. Chem. B* 102, 2733-2738 (1998).
40. R. Aveyard, J. H. Clint, D. Nees, and V. N. Paunov, Compression and structure of monolayers of charged latex particles at air/water and Octane/Water interfaces, *Langmuir* 16, 1969-1979 (2000).
41. R. Aveyard, J. H. Clint, D. Nees, and N. Quirke, Structure and collapse of particle monolayers under lateral pressure at the octane/aqueous surfactant solution interface, *Langmuir* 16, 8820-8828 (2000).

OXIDE, CARBON, AND CARBON/MINERAL ADSORBENTS FOR TOXIC, EXPLOSIVE, AND NARCOTIC COMPOUNDS – SYNTHESIS, PROPERTIES, AND APPLICATIONS

V. GUN'KO,[1,*] R. LEBODA,[2] V. TUROV,[1] V. ZARKO,[1]
A. CHUIKO[1†]
[1]*Institute of Surface Chemistry, 03164 Kiev, Ukraine,*
[2]*Maria Curie-Skłodowska University, 20031 Lublin, Poland*

Abstract. Structural and adsorption characteristics of various adsorbents such as fumed silicas, silica gels, activated carbons and carbon/silicas were analyzed. The adsorption of a variety of compounds reveals the effects of adsorbent grain size, specific surface area, pore volume, pore size distribution, surface chemistry, conditions of adsorbent synthesis and pre-treatment. Both dynamic (nonequilibrium) and static (equilibrium) adsorption conditions are addressed.

Keywords: fumed oxides; silica gels; activated carbons; carbon-mineral adsorbents; toxic and narcotic compounds; explosives; dynamic and static adsorption

1. Introduction

Fast and effective adsorption of harmful compounds of different origin and molecular size is of importance for environmental and human protection. Clearly the structural and adsorption characteristics of adsorbents play an important role for their use in various applications. Certain adsorption characteristics can be optimized for adsorbates depending on the application. For instance, specific surface area and adsorption capacity are important because if they increase then the adsorption effectiveness increases. However,

*To whom correspondence should be addressed. V. M. Gun'ko, Institute of Surface Chemistry, 17 General Naumov Street, 03164 Kiev, Ukraine; e-mail: gun@voliacable.com

J.P. Blitz and V.M. Gun'ko (eds.), Surface Chemistry in Biomedical and Environmental Science, 91–102.

other adsorbent characteristics can be very different for different adsorbates, e.g. the kind of surface sites, pore size distribution, and grain size.[1-32] These aspects of the adsorption of harmful compounds on various adsorbents will be analyzed here using several experimental methods. The aim of this work is to elucidate regularities in the adsorption of different toxic, explosive and narcotic compounds on a variety of adsorbents.

2. Experimental

Fumed oxides (silica and mixed oxides based on silica synthesized at the Pilot plant of the Institute of Surface Chemistry, Kalush, Ukraine),[1,2,25-28] unmodified commercial and modified silica gels,[3-5,29] carbons,[6-17] and carbon-mineral adsorbents[18-24] (synthesized at the Maria Curie-Sklodowska University, Lublin, Poland) were used in the adsorption investigations. Synthesis, activation, modify-cation, pretreatment, structural and adsorption characteristics were described in detail elsewhere.[1-26] Features of the adsorption of aromatics (benzene, tert-butylbenzene (TBB), phloroglucinol (PG)),[6,7,14,25] drugs and macromolecules,[11,12] and metal ions (Pb(II), Sr(II), Ni(II), Cd(II), Cs(I))[2] as models of toxic compounds; explosives ((i) nitrate esters (NE): nitroglycol (ethylene glycol dinitrate, EGDN), nitroglycerine (glycerol trinitrate, NG) and pentaerythritol tetranitrate (PETN); (ii) cyclic nitroamines (CNA): octogen (1,3,5,7-tetranitro-1,3,5,7-tetrazacyclooctane, HMX) and hexogen (1,3,5-trinitro-1,3,5-triazacyclohexane, RDX); and (iii) nitroaromatics (NA): 1,3,5-trinitrobenzene (TNB), 2,4,6-trinitrotoluene (TNT), tetryl (2,4,6,N-tetranitro-N-methylaniline, TNMA), hexyl (2,2', 4,4', 6,6'-hexanitrodiphenylamine, HNDPA) and 2,2', 4,4', 6,6'-hexanitrodibenzyl (HNDB)),[9] and narcotics (amphetamine (1-phenyl-2-propanamine) (A) and its alkyl derivatives such as methamphetamine (N-methyl-1-phenyl-2-propanamine, MA), N-ethyl- (EA), N-propyl- (PA), and N-butyl-1-phenyl-2-propanamine (BA), dimethylamphetamine (N,N-dimethyl-1-phenyl-2-propanamine, DMA), N,N-methylethyl- (MEA), N,N-methylpropyl- (MPA), and N,N-methylbutyl-1-phenyl-2-propanamine (MBA))[13] on different adsorbents were analyzed. Equilibrium (static) adsorption of nitrogen, water, benzene, PG, metal ions, explosives, and narcotics, as well as dynamic (nonequilibrium) adsorption of TBB, were analyzed using different adsorption apparatuses and techniques described in detail elsewhere.[1-32] The structural characteristics (specific surface area, S_{BET}, pore volume, V_p, incremental pore size distributions (IPSD)) were calculated from nitrogen adsorption-desorption isotherms.[1-25]

3. Results and Discussion

Comparison of IPSDs and pore volume V_p as functions of the S_{BET} value of fumed silicas (S_{BET} = 52-492 m^2/g, 30 samples),[1,25-28] carbons (3-2789 m^2/g, 130 samples),[6-17] silica gels and aerogels (26-2135 m^2/g, 30 samples),[3-5,29] and representative carbon-mineral adsorbents[18-24] at S_{BET} values close (typically slightly smaller) to that of the oxide matrices, are shown in Figures 1 and 2.

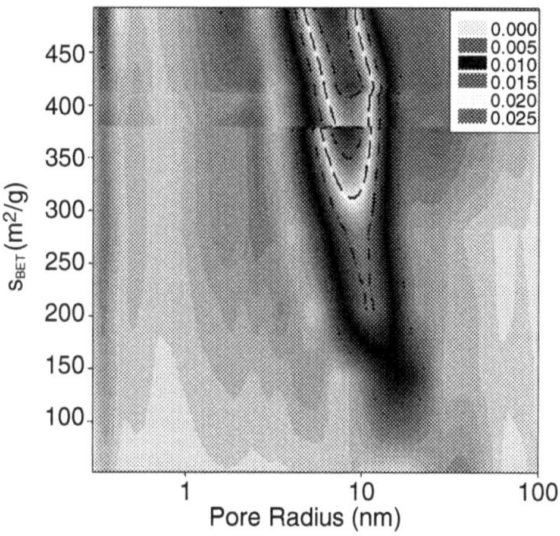

(a)

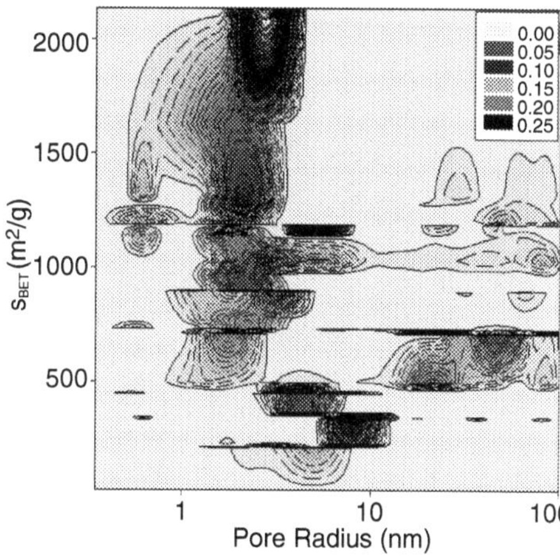

(b)

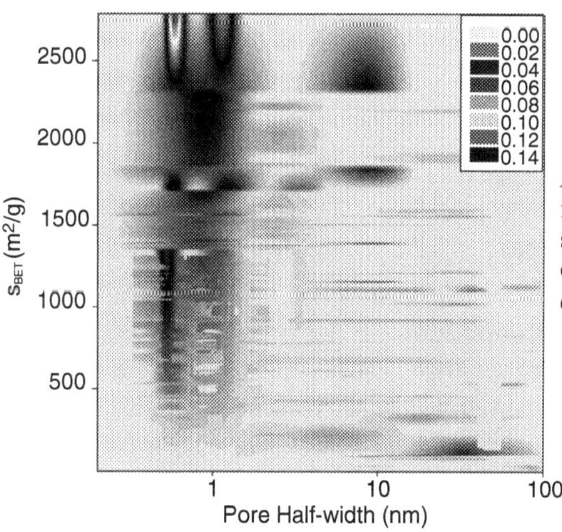

(c)

Figure 1. Contour plots of IPSD as a function of S_{BET} for (a) fumed silicas, (b) silica gels and aerogels, and (c) activated carbons, carbon blacks, and graphitized carbons.

These data reveal a strong correlation between changes in the IPSDs (position of the main peak and its intensity proportional to pore volume)[1-25] and the S_{BET}

value for all nanosilicas, and only for narrow series of the other adsorbents (e.g. porous silicas at S_{BET} between 200 and 500 m^2/g), because of variations in adsorbent synthesis conditions. Many samples are mesoporous or mesoporous/macroporous (all silicas and some carbons). Only some activated carbons are predominantly nanoporous with a minor contribution of mesopores that affect their adsorption capability with respect to the studied adsorbates.

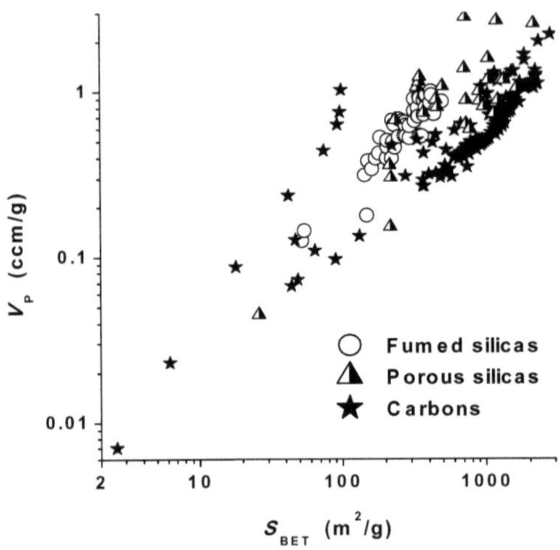

Figure 2. Relationship between the S_{BET} and V_p values for fumed and porous silicas and carbon adsorbents.

Structural features of adsorbents play an important role in both static and dynamic adsorption, as well as in the desorption and elution of organic and inorganic compounds in different media. For fast and complete dynamic adsorption of gaseous adsorbates (e.g. harmful organics in an airstream) there are several bottlenecks caused by the structural (S_{BET}, V_p, PSD), morphological (granule size, bed depth), and chemical (e.g. content of O, N, S, P, and Cl-containing surface functionalities) properties of adsorbents.[6,16] For instance, changes in the granule size even in relatively narrow ranges, e.g. d = 0.5-0.7 mm and 0.7-1.0 mm for the same activated carbon, affect the breakthrough time t_b (corresponding to outlet TBB concentration $c_x = 10^{-4}c_0$ where c_0 is the inlet TBB concentration). Typically the larger the d value, the shorter the t_b value because adsorbate molecules can penetrate in a certain surface layer of granules. For nano/mesoporous carbon PS2 t_b = 102 and 67 min at d = 0.5-0.7 mm and 0.7-1.0 mm, respectively. For nanoporous commercial carbon D55/2 t_b = 45 and 6 min, respectively.[16] However, for static equilibrium adsorption of probe compounds (e.g. benzene or nitrogen), the effect of granule size is much smaller because adsorption occurs in the entire interior volume of accessible pores. In other words the 'dynamic' accessibility of pores depends on the granule size, in contrast to the 'static' accessibility of pores. Therefore, the PSD, i.e. contributions of nano-, meso- and macropores, is an important factor in dynamic adsorption. Appropriate contributions of meso- and macropores provide a faster penetration of adsorbates into the interior volume of a thicker surface layer of adsorbent granules (Figure 3).[6,16]

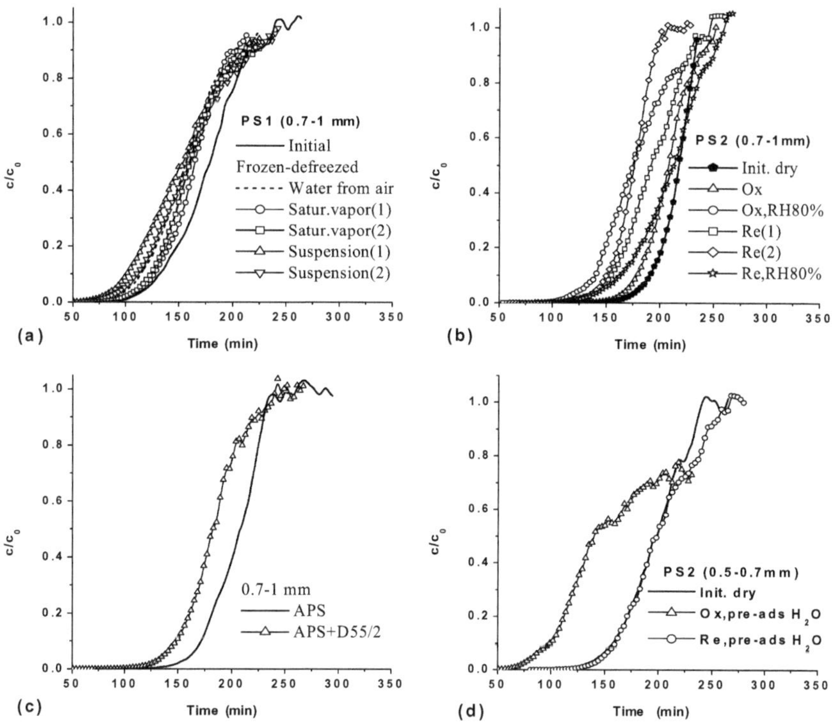

Figure 3. Breakthrough plots of TBB (dry or wet airstream) on different carbons after different treatments such as freezing-thawing with water adsorbed in different media or pre-adsorbed water, oxidizing (labelled Ox) and reduction (Re).

Another important factor is linked to the presence of water vapour in air. Pre-adsorption of water and its competitive adsorption with target adsorbates, especially when significant amounts of O-containing surface functionalities are present which interact more strongly with water than with organics (Figure 3d), can significantly hinder adsorption. Similar effects of the structural characteristics of adsorbents are observed in solid phase extraction, SPE, (i.e. adsorption then elution) of explosives and narcotics from liquid media.[9,13] However, solvent effects can cause unpredictable results (Figures 4 and 5). The Gibbs free energy of solvation and desolvation on adsorption/elution depends not only on adsorbent/adsorbate interactions, but also solvation effect in pores which can differ quite a bit from on the structural characteristics of both bulk solution.[1,7,14,26]

If $S_{BET} < 200\ m^2/g$ and $V_p < 0.2\ cm^3/g$ then the recovery rate (η)is low (10-50%) (Figure 5) because of a low adsorption capacity. The presence of mesopores is necessary for effective elution of adsorbed compounds, especially aromatics, because desorption from narrow nanopores is more difficult compared to broad mesopores or macropores.

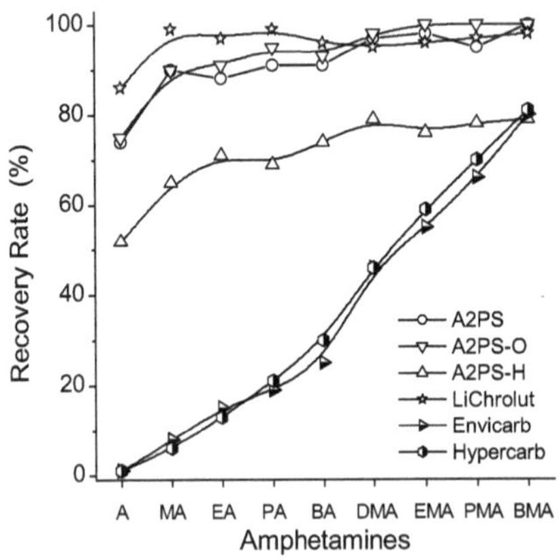

Figure 4. Recovery rate on the SPE of amphetamines on different carbons: A2PS, A2PS-O and A2PS-H oxidized and reduced A2PS; commercial graphitized carbons Hypercarb (ThermoHypersil, UK) and Envicarb (Supelco) and polymeric LiChrolut EN (Merck).

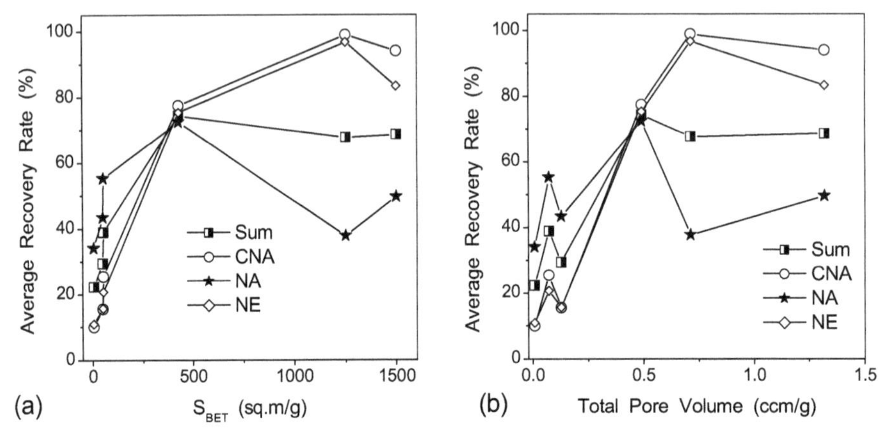

Figure 5. Average recovery rate on the SPE of three kinds of explosives: nitrate esters (NE), cyclic nitroamines (CNA), and nitroaromatics (NA) and summarized η value as a function of (a) specific surface area and (b) pore volume of carbon adsorbents.

Therefore, pure nanoporous carbons exhibit low η values when used for SPE of explosives (Figure 5, NA)[9] and narcotics (Figure 4).[13] The polarity and molecular size of solutes and solvents affect adsorption/desorption, as well as the distribution of different liquids in different pores, because these parameters govern changes in the Gibbs free energy in the adsorption layers. For instance, water adsorbed on activated nanoporous carbon Norit R 0.8 Extra in air can be displaced to varying extents by nonpolar and polar solvents (Figure 6).

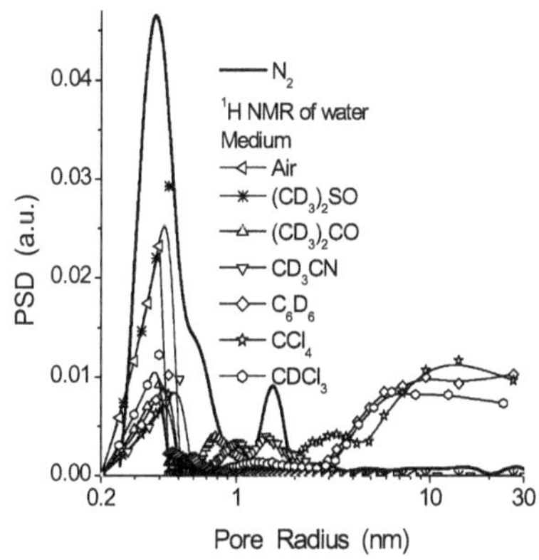

Figure 6. Carbon Norit R 0.8 Extra: PSD determined by nitrogen desorption data (N_2), and PSDs of pores filled by water unfrozen in different environments at T < 273 K.

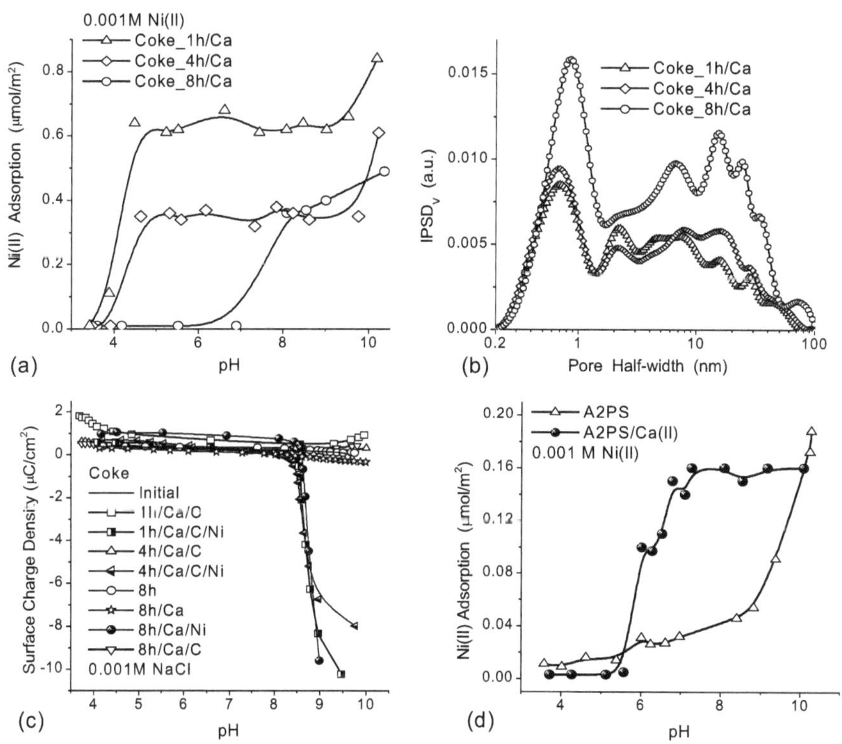

Figure 7. Adsorption of Ni(II) on activated (a) coke and (d) carbon A2PS; (b) PSD of the coke; and (c) surface charge density on the coke with and without adsorbed Ni(II).

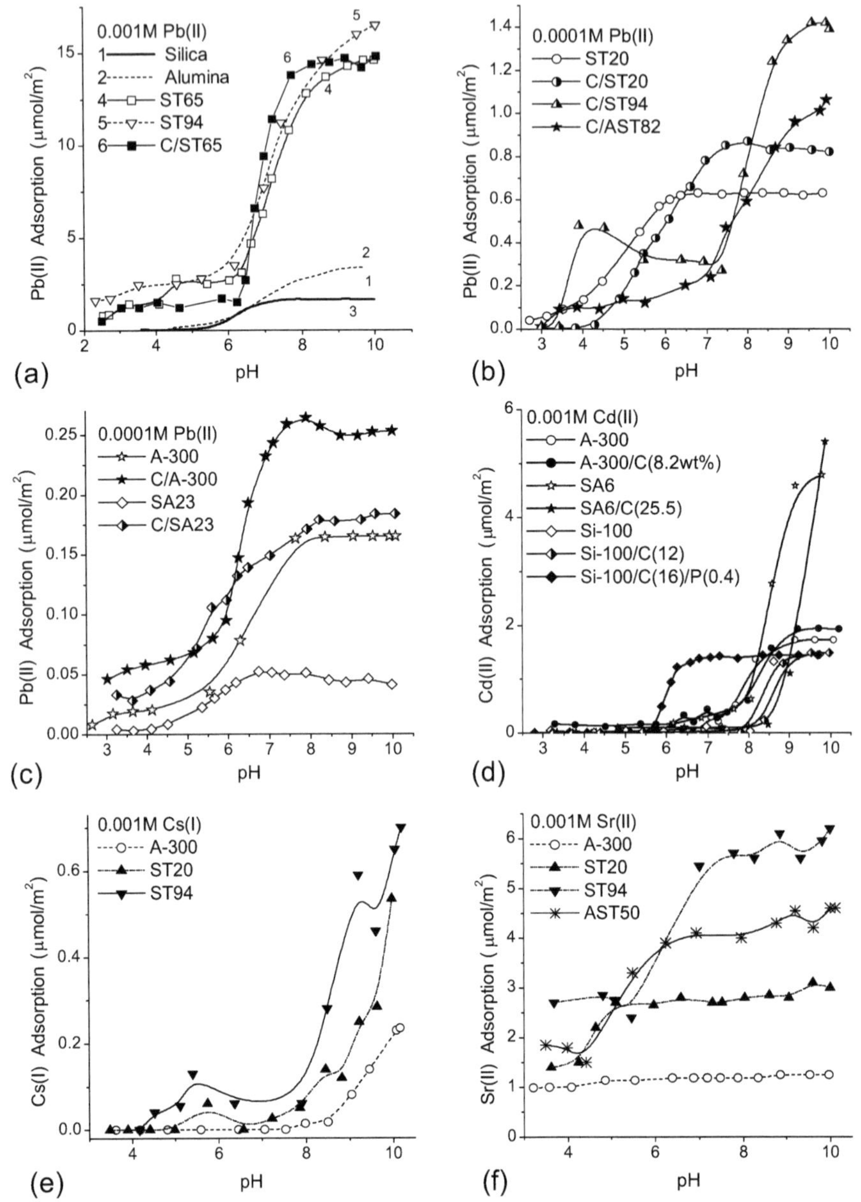

Figure 8. Adsorption of (a, b, c) Pb(II), (d) Cd(II), (e) Cs(I), and (f) Sr(II) onto fumed oxides: silica A-300, alumina, silica/alumina (c) SA23 and (d) SA6 (23 and 6 wt.% of Al_2O_3 respectively), silica/titania ST20, ST65, ST94 (C_{TiO2} = 20, 65, and 94 wt.%), and alumina/silica/titania AST82 and AST50 (C_{TiO2} = 82 and 50 wt.%), silica gel Si-100 unmodified and carbonized; C_C = 9.1-9.8 wt.% in C/ST and C/AST and 25.6 wt.% in C/SA23; carbon (and phosphorus) content on Si-100 and SA6 is shown in the legend.

Maximal displacement of water into larger pores is provided by benzene, carbon tetrachloride, and acetonitrile. A minimal effect is observed for dimethyl sulfoxide, which cannot penetrate into narrow nanopores. Similar effects are observed on the breakthrough dynamics of organics in the case of pre-adsorption of water, which should be displaced by organics. However, this displacement can be too slow to change the breakthrough dynamics, if the granule size is large or the surface contains significant quantities of oxidized groups. Adsorption of metal ions on carbon (Figure 7) and oxide (Figures 8 and 9) adsorbents depends on surface chemistry (e.g. the presence of sites capable

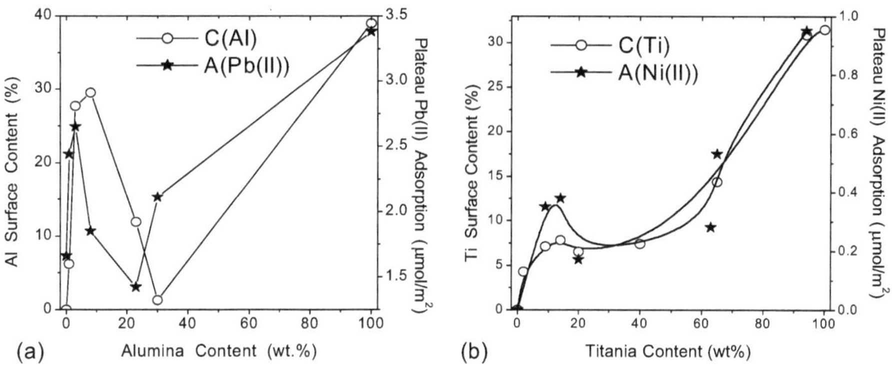

Figure 9. Surface content of (a) Al in fumed silica/alumina and (b) Ti in titania/silica and the maximum adsorption of (a) Pb(II) and (b) Ni(II) as a function of the total (a) alumina or (b) titania content in mixed oxides.

of taking part in ion-exchange reactions), pH value (affecting formation of hydroxy species with metal ions), and adsorbent structural characteristics (PSD, surface area). The adsorption of Ni(II) on a coke activated with Ca(II) for 8 h (labelled Coke_8h/Ca) (giving the best results for SPE of explosives because of good elution of adsorbates from developed nano-, meso- and macropores (Figure 7b)[9]) shows relatively poor results (Figure 7a). However, shorter activation (sample Coke-1h/Ca activated for 1 h) provides better adsorption of Ni(II) even at low pH values (> 4). For a nanoporous carbon, A2PS, activation with Ca(II) results in increased adsorption of Ni(II) at pH > 6 (Figure 7d).

Different activations of the coke adsorbent give rise to smaller changes in surface charge density than adsorbed highly charged hydroxy species of Ni(II) at pH > 8.5 (Figure 7c). Therefore, marked changes in the adsorption of Ni(II) on A2PS and Coke_8h/Ca with increasing pH may be caused by the adsorption of Ni(II) hydroxy species because the contribution of ion-exchange reactions is small compared with A2PS/Ca(II) and Coke_1h/Ca samples. Similar effects of adsorbed metal ions and their hydroxy species on surface charge density can play an important role on the adsorption of polar or charged organic compounds (especially macromolecules, e.g. proteins) from aqueous media.

The effect of surface sites capable of taking part in ion-exchange reactions is also observed on the adsorption of different metal ions onto fumed silica, as well as unmodified and carbonized mixed oxides (Figures 8 and 9). For instance, the adsorption of Pb(II) and Ni(II) correlates with the surface content of alumina or titania in mixed oxides (Figure 9), because bridges MO(H)Si strong acid sites take part in ion-exchange reactions unlike SiOH groups. Carbon deposits on fumed oxides or silica gel enhance the adsorption of metal ions at lower pH values compared with unmodified oxides (Figure 8).

4. Conclusion

There are several structural factors governing static (equilibrium) and dynamic (non-equilibrium) adsorption of organics and inorganics in various media. Pure nanoporosity of adsorbents as well as a large granule size negatively affect the breakthrough dynamics, or the SPE of explosives and narcotics. Adsorbents with significant meso- and macroporosity provide improved parameters for breakthrough dynamics and SPE; however, this may cause a reduction in the adsorption of metal ions because of enhanced solvation effects in larger pores. O-containing surface functionalities can cause both positive (metal ion adsorption through ion-exchange reactions) and negative (water role on the adsorption of nonpolar origins) effects.

References

1. V. M. Gun'ko, I. F. Mironyuk, V. I. Zarko, E. F. Voronin, V. V. Turov, E. M. Pakhlov, E. V. Goncharuk, Yu. M. Nicheporuk, T. V. Kulik, B. B. Palyanytsya, S. V. Pakhovchishin, N. N. Vlasova, P. P. Gorbik, O. A. Mishchuk, A. A. Chuiko, J. Skubiszewska-Zięba, W. Janusz, A. V. Turov, and R. Leboda, Morphology and Surface Properties of Fumed Silicas, *J. Colloid Interface Sci.* 289, 427-445 (2005).
2. V. M. Gun'ko, V. I. Zarko, I. F. Mironyuk, E. V. Goncharuk, N. V. Guzenko, M. V. Borysenko, P. P. Gorbik, O. A. Mishchuk, W. Janusz, R. Leboda, J. Skubiszewska-Zięba, W. Grzegorczyk, M. Matysek, and S. Chibowski, Surface Electric and Titration Behaviour of Fumed Oxides, *Colloids Surf. A* 240, 9-25 (2004).
3. V. M. Gun'ko, J. Skubiszewska-Zieba, R. Leboda, and V. V. Turov, Impact of Thermal and Hydrothermal Treatments on Structural Characteristics of Silica Gel Si-40 and Carbon/Silica Gel Adsorbents, *Colloids Surf. A* 235 (2004) 101-111.
4. V. M. Gun'ko, V. V. Turov, J. Skubiszewska-Zięba, B. Charmas, and R. Leboda, Structural and Adsorptive Characteristics of Pyrocabon/Silica Gel Si-60, *Adsorption* 10, 5-18 (2004).
5. J. Skubiszewska-Zięba, R. Leboda, O. Seledets, and V. M. Gun'ko, Effect of Preparation Conditions of Carbon-Silica Adsorbents Based on Mesoporous Silica Gel Si-100 and Carbonised Glucose on Their Pore Structure, *Colloids Surf. A* 231(1-3), 39-49 (2003).

6. D. Palijczuk, V. M. Gun'ko, R. Leboda, J. Skubiszewska-Zięba, and S. Ziętek, Porous Structure of Activated Carbons and Tert-butylbenzene Breakthrough Dynamics, *J. Colloid Interface Sci.* 250, 5-17 (2002).
7. V. M. Gun'ko, V. V. Turov, J. Skubiszewska-Zięba, R. Leboda, M. D. Tsapko, and D. Palijczuk, Structural Characteristics of a Carbon Adsorbent and Influence of Organic Solvents on Interfacial Water, *Appl. Surf. Sci.* 214(1-4), 178-189 (2003).
8. R. Leboda, J. Skubiszewska-Zięba, W. Tomaszewski, and V. M. Gun'ko, Structural and Adsorptive Properties of Activated Carbons Prepared by Carbonization and Activation of Resins, *J. Colloid Interface Sci.* 263(2), 533-541 (2003).
9. W. Tomaszewski, V. M. Gun'ko, R. Leboda, and J. Skubiszewska-Zięba, Structural Characteristics of Modified Activated Carbons and Adsorption of Explosives, *J. Colloid Interface Sci.* 266, 388-402 (2003).
10. V. M. Gun'ko and S. V. Mikhalovsky, Evaluation of Slitlike Porosity of Carbon Adsorbents, *Carbon* 42, 843-849 (2004).
11. M. Melillo, V. M. Gun'ko, L. I. Mikhalovska, G. J. Phillips, J. G. Davies, A. W. Lloyd, S. R. Tennison, O. P. Kozynchenko, D. J. Malik, M. Streat, and S. V. Mikhalovsky, Ibuprofen Adsorption Affected by Bovine Serum Albumin, *Langmuir* 20, 2837-2851 (2004).
12. T. A. Alexeeva, N. I. Lebovka, V. M. Gun'ko, V. V. Strashko, and S. V. Mikhalovsky, Characteristics of Interfacial Water Affected by Proteins Adsorbed on Activated Carbon, *J. Colloid Interface Sci.* 278, 333-341 (2004).
13. W. Tomaszewski, V. M. Gun'ko, R. Leboda, and J. Skubiszewska-Zięba, Interaction of Amphetamines with Micro- and Mesoporous Adsorbents in Polar Liquids, *J. Colloid Interface Sci.* 282 (2), 261-269 (2005).
14. V. M. Gun'ko, V. V. Turov, R. Leboda, J. Skubiszewska-Zięba, M. D. Tsapko, and D. Palijczuk, Influence of Organics on Structure of Water Adsorbed on Activated Carbons, *Adsorption* 11, 163-168 (2005).
15. M. M. Seredych, V. M. Gun'ko, and A. Gierak, Structural and Energetic Heterogeneities and Adsorptive Properties of Synthetic Carbon Adsorbents, *Appl. Surf. Sci.* 242, 154-161 (2005).
16. V. M. Gun'ko, D. Palijczuk, R. Leboda, J. Skubiszewska-Zięba, and S. Ziętek, Influence of Pore Structure and Pretreatments of Activated Carbons and Water Effects on Breakthrough Dynamics of tert-Butylbenzene, *J. Colloid Interface Sci.* 294(1) 53-68 (2006).
17. V. M. Gun'ko, R. Leboda, J. Skubiszewska-Zięba, B. Charmas, and P. Oleszczuk, Carbon Adsorbents from Waste Ion Exchange Resins, *Carbon* 43, 1143-1150 (2005).
18. V. M. Gun'ko and R. Leboda, Carbon-Silica Adsorbents, in *Encyclopedia of Surface and Colloid Science*, edited by A.T. Hubbard, pp. 864-878 (Marcel Dekker, 2002).
19. V. M. Gun'ko, R. Leboda, V. V. Turov, B. Charmas, and J. Skubiszewska-Zięba, Structural and Energetic Heterogeneities of Hybrid Carbon-Mineral Adsorbents, *Appl. Surf. Sci.* 191, 286-299 (2002).
20. V. M. Gun'ko, R. Leboda, V. I. Zarko, J. Skubiszewska-Zieba, W. Grzegorczyk, E. M. Pakhlov, E. F. Voronin, O. Seledets, and E. Chibowski, Fumed Oxides Modified due to Pyrolysis of Cyclohexene, *Colloid. Surf. A* 218(1-3), 103-124 (2003).
21. V. M. Gun'ko, J. Skubiszewska-Zięba, R. Leboda, E. F. Voronin, V. I. Zarko, S. I. Levitskaya, V. V. Brei, N. V. Guzenko, O. A. Kazakova, O. Seledets, W. Janusz, and S. Chibowski, Pyrocarbons Prepared by Carbonisation of Polymers Adsorbed or Synthesised on a Surface of Silica and Mixed Oxides, *Appl. Surf. Sci.* 227, 219-243 (2004).
22. O. Seledets, V. M. Gun'ko, J. Skubiszewska-Zieba, R. Leboda, M. Musiatowicz, P. Podkoscielny, and A. Dabrowski, Structural and Energetic Heterogeneities of Pyrocarbon-Silica Gel Systems and Their Adsorption Properties, *Appl. Surf. Sci.* 240, 222-235 (2005).
23. R. Leboda, B. Charmas, J. Skubiszewska-Zięba, S. Chodorowski, P. Oleszczuk, V. M. Gun'ko, and V. A. Pokroskiy, Carbon-Mineral Adsorbents Prepared by Pyrolysis of Waste Materials in the Presence of Tetrachloromethane Vapours, *J. Colloid Interface Sci.* 284(1), 39-47 (2005).

24. R. Leboda, B. Charmas, S. Chodorowski, J. Skubiszewska-Zięba, and V. M. Gun'ko, Carbon-Mineral Adsorbents Prepared from Spent Palygorskite – Crosslinking Effect, *Microporous and Mesoporous Materials*, in press.
25. V. M. Gun'ko, V. V. Turov, V. N. Barvinchenko, V. M. Bogatyrev, A. V. Turov, O. Shulga, O. V. Stebelska, V. A. Pokrovsky, R. Leboda, V. G. Sukretny, Y. M. Nicheporuk, B. A. Chuikov, and Y. G. Ptushinskii, Characteristics of Interfacial Water at Nanosilica Surface with Adsorbed 1,3,5-Trihydroxybenzene over Wide Temperature Range, *Colloids Surf. A*, in press.
26. V. M. Gun'ko, V. V. Turov, V. M. Bogatyrev, V. I. Zarko, R. Leboda, E. V. Goncharuk, A. A. Novza, A. V. Turov, and A. A. Chuiko, Unusual properties of water at hydrophilic/ hydrophobic interfaces, *Adv. Colloid Interface Sci.* 118, 125-172 (2005).
27. V. M. Gun'ko, V. I. Zarko, R. Leboda, and E. Chibowski, Aqueous Suspensions of Fumed Oxides: Particle Size Distribution and Zeta Potential, *Adv. Colloid Interface Sci.* 91(1), 1-112 (2001).
28. V. M. Gun'ko, V. I. Zarko, E. F. Voronin, V. V. Turov, I. F. Mironyuk, I. I. Gerashchenko, E. V. Goncharuk, E. M. Pakhlov, N. V. Guzenko, R. Leboda, J. Skubiszewska-Zięba, W. Janusz, S. Chibowski, Yu. N. Levchuk, and A. V. Klyueva, Impact of Some Organics on Structural and Adsorptive Characteristics of Fumed Silica in Different Media, *Langmuir* 18(3), 581-596 (2002).
29. V. M. Gun'ko, D. J. Sheeran, S. M. Augustine, and J. P. Blitz, Structural and Energetic Characteristics of Silicas Modified by Organosilicon Compounds, *J. Colloid Interface Sci.* 249(1), 123-133 (2002).
30. V. M. Gun'ko, V. I. Zarko, D. J. Sheeran, J. P. Blitz, R. Leboda, W. Janusz, and S. Chibowski, Characteristics of Modified Cab-O-Sil in Aqueous Media, *J. Colloid Interface Sci.* 252(1), 109-118 (2002).
31. V. M. Gun'ko, V. I. Zarko, V. V. Turov, E. F. Voronin, I. F. Mironyuk, and A. A. Chuiko, Structural and Adsorptive Characteristics of Fumed Silicas in Different Media, in: *Colloidal Silica: Fundamentals and Applications*, edited by H.E. Bergna (Taylor & Francis LLC, Salisbury, 2005), pp. 499-530.
32. V. M. Gun'ko and A. A. Chuiko, Chemical Reactions at Fumed Silica Surfaces, in: *Colloidal Silica: Fundamentals and Applications*, edited by H.E. Bergna (Taylor & Francis LLC, Salisbury, 2005), pp. 465-497.

POROUS POLYMERIC NANOCOMPOSITES FILLED WITH CHEMICALLY MODIFIED FUMED SILICAS

YULIA BOLBUKH, VALENTIN TERTYKH*
Institute of Surface Chemistry of National Academy of Sciences of Ukraine, Gen. Naumov Str. 17, 03164 Kyiv, Ukraine
BARBARA GAWDZIK
Faculty of Chemistry, Maria Curie-Sklodowska University, M.C. Sklodowska Sq. 3, 20031 Lublin, Poland

Abstract. Porous copolymers have been prepared by suspension-emulsion polymerization of divinylbenzene with styrene or some methacrylic monomers: di(methacryloyloxymethyl)naphthalene, methacrylic ester of p,p′-dihydroxydiphenylpropane diglycidyl ether, and dimethacrylglycolethylene in the presence and absence of chemically modified fillers (fumed silicas with grafted methyl and silicon hydride groups). The results of investigations of the unfilled and filled polymeric systems by IR and ^{13}C NMR spectroscopies combined with AFM are presented.

Keywords: porous copolymers, divinylbenzene, styrene, methacrylic monomers, chemically modified silicas, fillers

1. Introduction

Porous polymeric sorbents are of interest for applications in ion-exchange and size-exclusion chromatography,[1] but their thermal stability and swelling properties require improvement. In normal and reversed-phase high-performance liquid chromatography, silica-based packings are dominant due to their high thermal and mechanical stability, and low swelling. Improved porous polymer

*To whom correspondence should be addressed. V. Tertykh, Institute of Surface Chemistry, 17 General Naumov Street, Kyiv, 03164, Ukraine, e-mail: tertykh@public.ua.net

J. P. Blitz and V. M. Gun'ko (eds.), Surface Chemistry in Biomedical and Environmental Science, 103–111.

characteristics can be obtained with methacrylic monomers,[1,2] the preparation of copolymers, and composites filled with nanosized silica particles.

We have shown that grafted surface silicon hydride groups can influence the structure of filled polymers and polymerization of unsaturated monomers owing to formation of polymer-filler covalent Si-C bonds.[3,4] The presence of both methylsilyl and chemically active silicon hydride groups on silica may provide improved compatibility, to obtain a more uniform filler distribution of the filled composite.

This study is devoted to the investigation of porous methacrylate polymeric systems filled with chemically modified fumed silicas. IR and ^{13}C NMR spectroscopies combined with AFM was applied to characterize changes in the material structure, and also the effect of surface functional groups of inorganic particles on the polymer-filler interaction.

2. Experimental

2.1. MATERIALS

Divinylbenzene, styrene, α,α'-azoisobutyronitrile and methacrylic monomers: di(methacryloyloxymethyl)-naphthalene (DMN), methacrylic ester of p,p'-dihydroxydiphenylpropane diglycidyl ether (MEDDE), and dimethacrylglycol-ethylene (DMGE) were received from Fluka and used without further purification. Poly(vinyl alcohol), *n*-decanol, ethanol, acetone and toluene (Merck), fumed silica with specific surface area of 300 m^2/g (Kalush), triethoxysilane (Kremnepolimer) and hexamethyldisilazane (Fluka) were applied.

2.2. SYNTHESIS OF COMPOSITES

The porous composites were obtained using copolymers of divinylbenzene (DVB) with styrene (St), methacrylic ester of *p,p*'-dihydroxydiphenylpropane diglycidyl ether (MEDDE), di(methacryloyl-oxymethyl)naphthalene (DMN) and dimethacrylglycolethylene (DMGE) by combined suspension-emulsion polymerization as previously described.[5,6] The filled nanocomposites were prepared by introduction of the methyl- or methyl,hydride-containing silicas in mixture of monomers with solvent and initiator before polymerization. The degree of filling was 15 wt%.

After polymerization the products were filtered, washed with hot water and ethanol, and dried with sequential extraction using a conventional Soxhlet technique with boiling acetone. Finally, the samples were dried at 150°C.

The modification of fumed silica was performed at room temperature by treatment with hexamethyldisilazane vapor for 48 h at room temperature (for introduction of methylsilyl groups into the surface layer) and then with a solution of triethoxysilane in ethanol for 24 h at room temperature (for a surface with grafted silicon hydride groups). After modification the fillers were dried on a water bath for 5 h, and then were treated for 15 min at 150°C.

2.3. MEASUREMENTS

Infrared spectra of the unfilled and filled copolymers were measured using a Perkin-Elmer model 1700 FTIR spectrometer. The ^{13}C CP/MAS NMR measurements were conducted on a Bruker 300 instrument operating at 75.51 MHz. The samples were rotated with a spectra width of 40.0 Hz, the CP time was 5 ms. $^{13}C\{^1H\}$ distortionless enhancement by polarization transfer (DEPT) technique was applied for analysis of monomers. The process was performed at 75.51 MHz, rotated with a spectral width of 0.75 Hz and a CP time of 15 ms. Atomic force microscopy measurements were carried out using a Nanoscope IIIa controlled Dimension 3000 AFM (Digital Instrument, Santa Barbara, CA).

3. Results and Discussion

IR-spectra of pure and filled copolymers are shown in Figure 1. The most visible changes in spectra of the filled copolymers were for compositions based on DVB-St. Absorption bands at 1070 and 1030 cm^{-1} in the spectrum of the unfilled copolymer indicates a heterotactic pseudocrystalline structure.[7]

The copolymer formed in the presence of modified silica had a pseudocrystalline isotactic well-ordered structure,[7] and is characterized by absorption bands at 1030 and 1113 cm^{-1} in the spectrum (Figure 1*a*). These bands can be detected only at significant sample dilution with KBr. Though a band at 1700 cm^{-1} characteristic for aromatic groups is not manifested in the spectrum, there is a band at 1584 cm^{-1} which may be attributed to a C = C vibration of the benzene rings orientated near the silica surface.

Usually in the presence of silica filler bands of macromolecular conformation[8] at 1370, 468 and 1309-1312 cm^{-1} are not observed because of interference with bands of the silica skeleton Si-O-Si group vibrations around 1100 and 475 cm^{-1}. However, filling with methyl,hydride-containing silica the low displacement of the bands at 827 and 541 cm^{-1}, and the appearance of a band at 988 cm^{-1} takes place. These bands indicate conformational changes in the polymer with an increase of styrene block length, and degree of crystallinity degree. Also note the presence of a band from aromatic groups at 1701 cm^{-1} in the composite spectrum (with the same band position as in the case of unfilled copolymer).

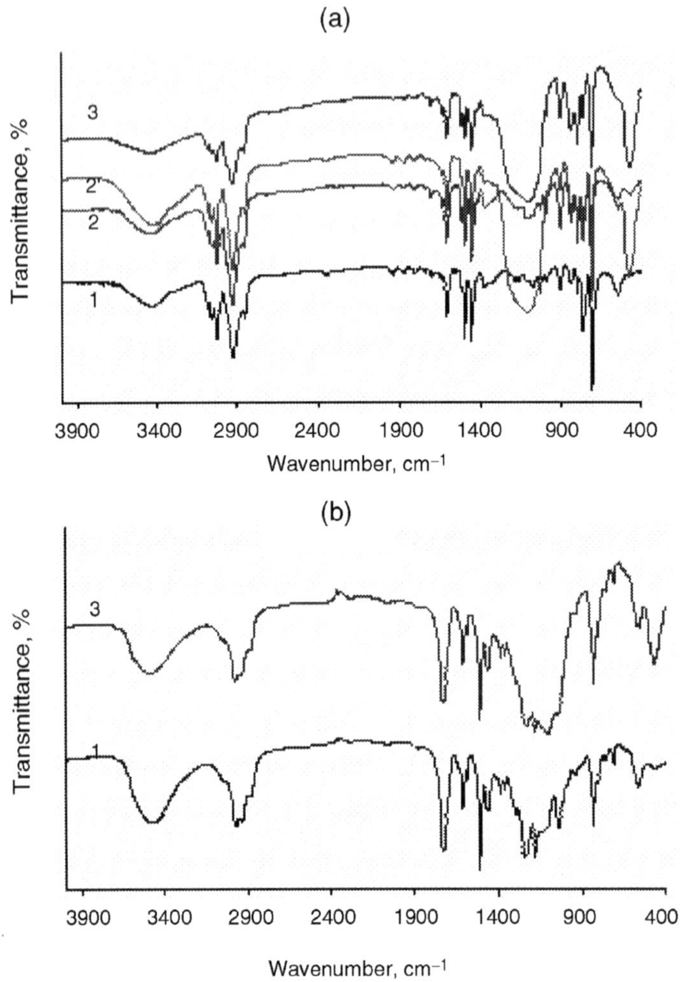

Figure 1. FTIR spectra of the DVB-St (*a*) and DVB-MEDDE (*b*) copolymers without filler (*1*) and filled with methyl- (*2,2'*) or methyl,hydride-containing fumed silicas (*3*). The spectrum marked as 2' was obtained at significant dilution of a sample with KBr.

In the spectrum of the filled DVB-MEDDE copolymer (Figure 1, *b*) the displacement of a C-O band from 1234 to 1220 cm^{-1} may indicate a change in orientation. The disappearance of 1012 and 986 cm^{-1} bands suggest a loss of order in the structure. The band at 556 cm^{-1} in spectra of unfilled and filled DVB-MEDDE copolymers indicates the absence of a change in the amount (1-3) of DVB monomer linked in blocks.[7-9]

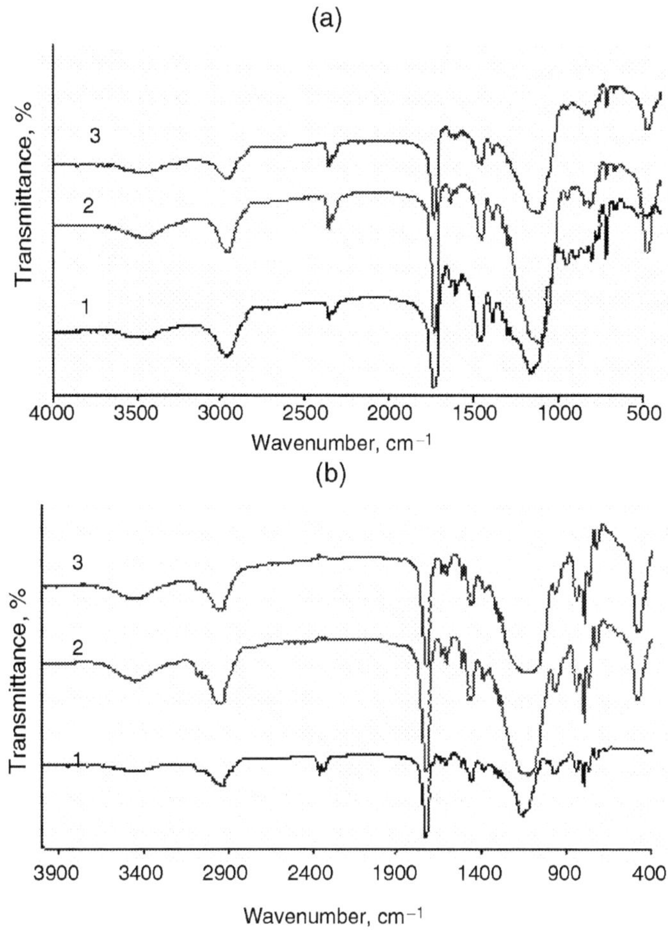

Figure 2. FTIR spectra of the DVB-DMGE (*a*) and DVB-DMN (*b*) copolymers without filler (*1*) and filled with methyl- (*2*) or methyl,hydride-containing fumed silicas (*3*).

The formation of amorphous structures has been detected for the filled DVB-DMGE copolymers. This is demonstrated (Figure 2*a*) by the disappearance of the methyl-containing silica bands at 986 and 882 cm^{-1}, which normally characterizes regular and ordered structures. Also, the band at 1044 cm^{-1} corresponding to gauche C-O-C groups (in the unfilled copolymer) was not detected in the composite spectrum, but may be masked by interference with absorption bands of Si-O-Si groups.

The presence in the DVB-DMN copolymer spectrum (Figure 2, *b*) of an absorption band at 574 cm^{-1}, and a shift of the 708 cm^{-1} band in the filled composite, indicates an increase of material cross-linking.

After filling of the DVB-DMN copolymer with methyl-containing silica, a band at 1144 cm^{-1} was detected in the spectrum indicating increased crystallinity. Crystallinity was not detected in the spectrum of copolymer filled with methyl,hydride-containing silica.

The increased crystallinity in the presence of methyl-containing silica was co

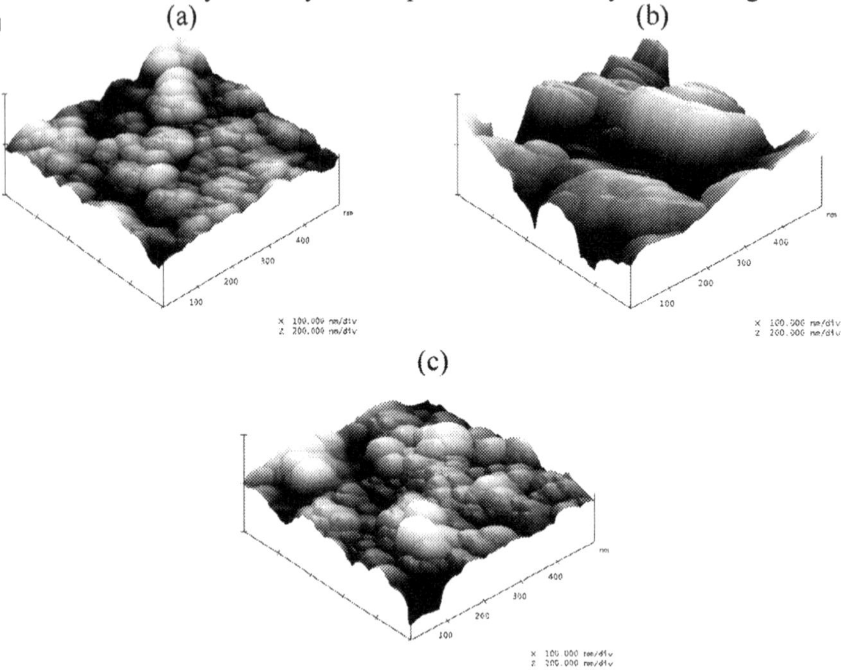

Figure 3. AFM micrographs of DMN-DVB copolymers without filler (*a*) and filled with methyl- (*b*) or methyl,hydride-containing fumed silicas (*c*).

According to AFM micrographs, the surface roughness of porous spheres of DMN-DVB copolymer increases in the presence of methyl-containing silica. At the same time, the availability of methylsilyl and silicon hydride groups on the silica surface promotes surface smoothing upon filling, similar to an unfilled system.

According to high-resolution solid-state ^{13}C CP/MAS NMR data we have observed the following changes in the presence of methyl-containing silica for DVB-St copolymer (Figure 4, *1*): decreased intensity of a 136.3 ppm (carbon atom of vinyl groups –CH =)[10] resonance, and disappearance of a 112.6 ppm (= CH_2)[10] resonance. This shows that there is a high degree of polymerization.

After filling with methyl,hydride-containing silica (Figure 4, 2), these chemical shifts were detected but with lower intensity. Resonances were

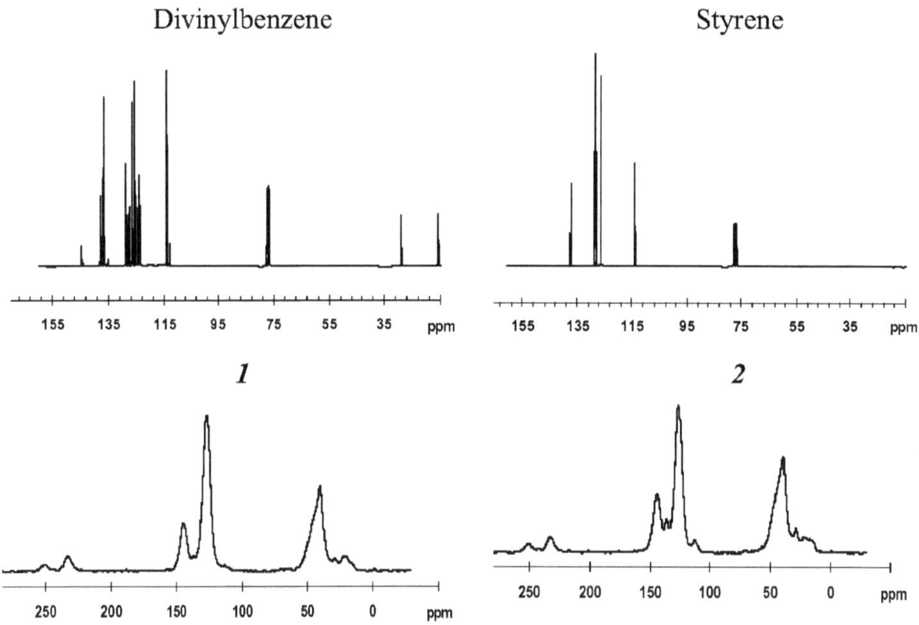

Figure 4. $^{13}C\{^1H\}$DEPT NMR spectra for DVB and St monomers and ^{13}C CP/MAS NMR spectra for DVB-St copolymers filled with methyl- (*1*) and methyl,hydride-containing silicas (*2*).

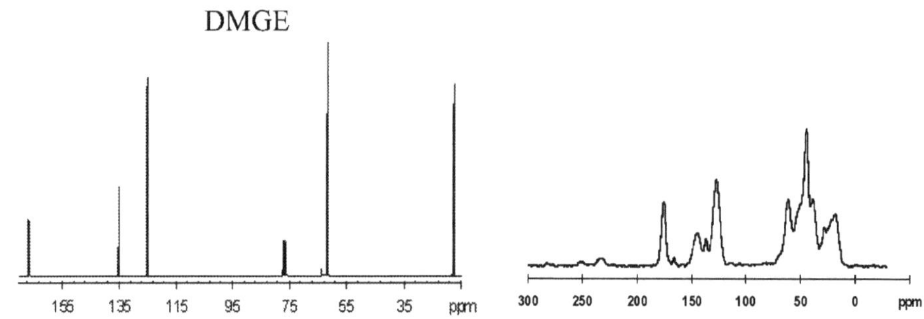

Figure 5. $^{13}C\{^1H\}$DEPT NMR spectra for DMGE monomers and ^{13}C CP/MAS NMR spectra for DMGE-DVB copolymer filled with methyl,hydride-containing silica.

assigned to carbon atoms of styrene and divinylbenzene immobilized on the silica surface. This was confirmed by the appearance of a 15 ppm resonance of carbon in Si-C bonds.

After filling of DMGE-DVB copolymer with methyl,hydride-containing silica, the appearance of a resonance at 39.9 ppm indicated the formation of Si-C bonds between the silica surface and DMGE (Figure 5). Comparison with the chemical shift of carbon in the ester carbonyl groups of the monomer at 167.0 ppm, suggests the formation of ≡Si-O-CHR'-OR" or ≡Si-C(CH_3)(CH_2R')OR" bonds.

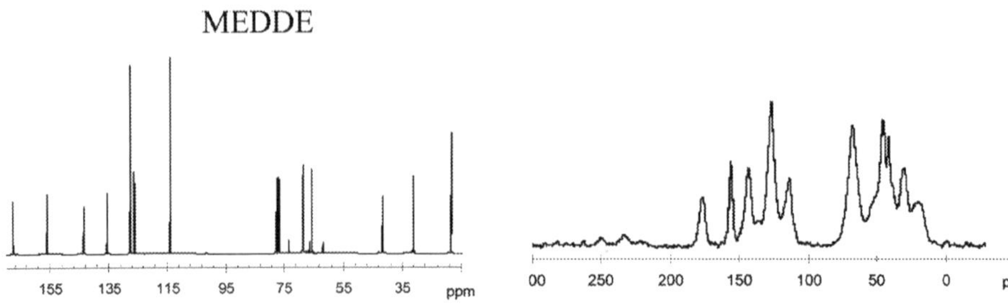

Figure 6. ^{13}C{^{1}H}DEPT NMR spectra for MEDDE monomers and ^{13}C CP/MAS NMR spectra for MEDDE-DVB copolymer filled with methyl,hydride-containing silica.

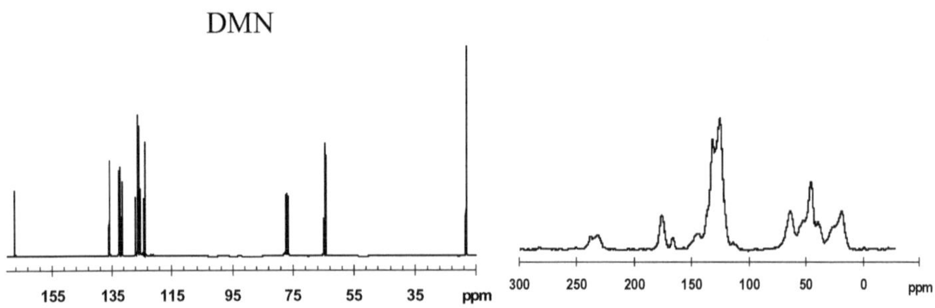

Figure 7. ^{13}C{^{1}H}DEPT NMR spectra for MEDDE monomers and ^{13}C CP/MAS NMR spectra for MEDDE-DVB copolymer filled with methyl,hydride-containing silica.

An intense resonance at 45.5 ppm in the spectrum of MEDDE-DVB copolymer filled with methyl,hydride-containing silica suggests the formation of new ≡SiCH_2R bonds (Figure 6). A similar resonance at 45 ppm was observed in the spectrum of DMN-DVB copolymer filled with methyl,hydride-containing silica (Figure 7). Thus a similarity of immobilization reactions of macromolecules in the surface layer is proposed. A high degree of polymerization was also detected.

4. Conclusion

Chemically modified silica fillers with grafted methyl groups or methyl and silicon hydride groups, influenced the micro- and macrostructures of various copolymers. Changes in cross-linking, orderliness, crystallinity, microtacticity and conformation of macromolecules have been detected in the presence of fillers. Surface functionality of the silica filler determines the disposition of macromolecular chains at the interface.

The character of the filler effect depends on the affinity of the silica surface for the copolymer, and the rigidity of macromolecules. For MEDDE-DVB and DMGE-DVB copolymers, the introduction of both methyl-, and methyl,hydride-containing silicas results in the formation of amorphous structures. A greater degree of disorder was detected in the presence of silicon hydride groups on the filler surface.

Filling of DVB-St copolymer with methyl,hydride-containing silica results in material cross-linking, and an increase in crystallinity was observed. For the DMN-DVB copolymer, an increase in crystallinity was significant with methyl-containing silica. The presence of silicon hydride groups on the filler surface promotes material cross-linking and the formation of an amorphous structure.

References

1. B. Gawdzik and J. Osypiuk, Reversed-phase high-performance liquid chromatography on porous copolymers of different chemical structure, *J. Chromatogr. A* 898(1), 13-21 (2000).
2. T. Jablonska-Pikus, W. Charmas, and B. Gawdzik, Synthesis and characterization of methacrylate polymeric packing based on bisphenol-S, *J. Appl. Polym. Sci.* 75(1), 142-148 (2000).
3. V. Tertykh, Yu. Bol'bukh, and V. Yanishpolskii, Effect of unmodified and surface treated fumed silica on the polymerization of HEMA, *Macromol. Symp.* 221(1), 145-152 (2005).
4. Yu. N. Bolbukh, Ye. P. Mamunya, and V. A. Tertykh, Thermomechanical investigation of composites based on 2-hydroxyethylmethacrylate and high disperse initial and modified silicas, *J. Therm. Anal. Cal.* 81(1), 15-20 (2005).
5. B. Gawdzik and T. Matynia, Thermal properties of porous copolymers of 1,4-di(metha-cryloyloxymehyl)naphthalene with divinylbenzene, *React. Polym.* 5, 197-202 (1987).
6. B. Gawdzik, T. Matynia, and J. Osypiuk, Chemical modification of the polymeric sorbent containing hydroxyl functional groups, *J. Chromatogr.* 44(1-2), 25-30 (1997).
7. J. Dechant, R. Danz, W. Kimmer, and R. Schmolke, *Ultrarotspektroskopische Untersuchungen an Polymeren* (Akademie, Verlag, Berlin , 1972).
8. C. Chiang, H. Ishido, and J.L. Koenig, The structure of γ-aminopropyltriethoxysilane on glass surface, *J. Colloid Interface Sci.* 74(2), 396-404 (1980).
9. H. Ishido and J. L. Koenig, Fourier transform infrared spectroscopic study of the structure of silane coupling agent on E-glass fiber, *J. Colloid Interface Sci.* 64(3), 565-576 (1978).
10. F. A. Bovey, *High Resolution NMR of Macromolecules* (Academic Press, New York and London, 1977).

ION-ELECTROSTATIC INTERACTION IN SYSTEMS OF INORGANIC NANOPARTICLES AND BIOLOGICAL CELLS IN ELECTROLYTE SOLUTION

L. G. GRECHKO,* L. B. LERMAN, O. YA. POKOTYLO,
N. G. SHKODA, A. A. CHUIKO†
Institute of Surface Chemistry, 17 General Naumov Street, Kyiv 03164, Ukraine

K. W. WHITES
South Dakota School of Mines and Technology, Rapid City, South Dakota 57701-3995, USA

Abstract. The solution of the Debye-Huckel equation for a system of spheres with arbitrary radii and surface charge in electrolyte solutions is described. The general theoretical approach to describe such systems is elaborated. The practically important case of two spheres is considered in detail. Finite closed formulae to calculate the interaction energy of two spherical particles with constant surface charges are obtained from general expressions in zero approximation. Known relationships follow from our formulae in limiting cases.

Keywords: small particles; nanoparticles; biological cells; electrostatic interactions; electrolyte; Debay-Huckel equation

1. Introduction

The interaction of small mineral particles with microorganisms and biological cells is one of the most general ways of contact between living and inorganic substances in nature. The basic problem can be defined by calculating the

*To whom correspondence should be addressed. L.G. Grechko, 17 General Naumov Street, Kyiv 03164, Ukraine, e-mail: user@surfchem.freenet.kiev.ua

J.P. Blitz and V.M. Gun'ko (eds.), Surface Chemistry in Biomedical and Environmental Science, 113–122.

energy of interparticle forces between cells and particles in an electrolyte solution. This problem is closely connected to the determination of the electric double layer energy on interaction between two charged spherical particles suspended in an aqueous electrolyte dispersion medium. Considerable attention has been paid to this problem.[1-7] Interaction of diffuse double layers of particles with constant surface charge density is usually calculated on the basis of Deryaguin's approximation. Unfortunately, this approach can lead to incorrect results in some cases, as has been noted.[2,4]

In the paper we consider the general problem of the mutual interaction of an arbitrary number of spherical particles in electrolyte solutions. It is assumed that particles are arranged in space at random, they can have different radii and different surface charge densities/potentials. For the determination of potentials, the potential energy of interaction and forces for ensembles of spherical particles in an electrolyte can be reduced to a solution of the corresponding boundary problems for the Debye-Huckel equation. We propose an analytical solution using expansions of a series of spherical functions (the scalar spherical harmonics) and modified spherical Bessel functions. The expansions give solutions of external problems for the Debye-Huckel equation in the case of distinct spheres, and to solve the boundary problem we use an additional theorem.[8] From boundary conditions, the infinite algebraic systems for determination of the unknown potentials' coefficients can be obtained. A similar approach has been used[9] for analyzing the propagation of diffuse light in a medium with spherical inhomogeneities, but different boundary conditions were assumed. The practically important case of two spheres is considered in detail. From general expressions, closed formulae to calculate the interaction energies of two particles with constant surface charges are derived without approximation.

2. Statement of Problem

A system of N spherical particles in an electrolyte solution with permittivity ε_m is considered. Particle radii are denoted as a_k, and their permittivities are denoted as ε_k ($k = 1, 2, ..., j, ..., N$). We link the local polar spherical coordinates $(r_k, \theta_k, \varphi_k)$ with the particle centers (r_k is a polar radius, θ_k is an azimuth angle, φ_k is a polar angle). The arrangement of two arbitrarily chosen particles from the ensemble is shown in Figure 1 with corresponding coordinates indicated. Global coordinates (x, y, z) of an observation point $P(x, y, z)$ are determined by vectors $\mathbf{r}_k, \mathbf{r}_j$ in the local coordinates, and a distance between centers of the spheres is R_{kj} (Figure 1).

Potentials corresponding to the internal and external domains comparing the spheres' surfaces are marked relatively by overscripts "<" and ">". In the external domain the potential $\varphi^{>}$ appears to be a sum of potentials $\phi_k^{>} = \phi_k^{>}(r_k, \theta_k, \phi_k)$, created by each sphere, providing that there is no external field.

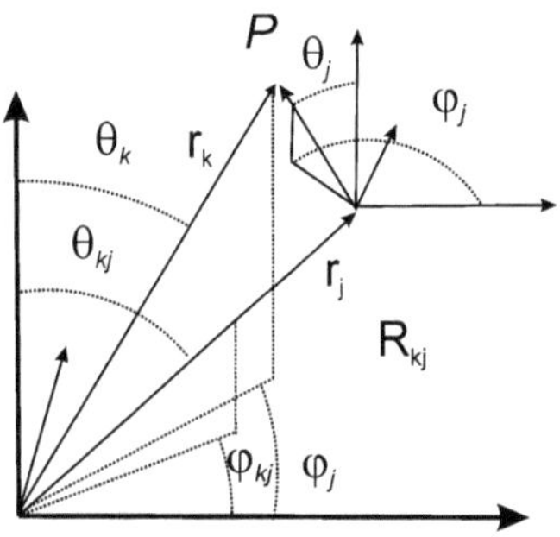

Figure 1. The local coordinates connection.

In the electrostatic approximation every potential $\phi_k^{>}$ ($k = 1, 2, ..., N$) is a solution of the Debye-Huckel equation (1), and potentials inside the spheres $\phi_k^{<} = \phi_k^{<}(r_k, \theta_k, \phi_k)$ are the solutions of the Laplace equation (2) respectively

$$\Delta\varphi_k^{>} - \kappa^2\varphi_k^{>} = 0, \qquad (1)$$

$$\Delta\varphi_k^{<} = 0. \qquad (2)$$

Boundary conditions on the surface of the k-th sphere at $r_k = a_k$ can be formulated in different ways. We consider the case when the densities of surface charges are adjusted. The boundary conditions reflect a continuity of potentials and electric inductions on the surfaces of spheres, so we have

$$\varphi_k^{<} = \varphi^{>}, \qquad \varepsilon_k \frac{\partial\varphi_k^{<}}{\partial r_k} - \varepsilon_m \frac{\partial\varphi^{>}}{\partial r_k} = 4\pi\sigma_k, \qquad (3)$$

where surface charge densities σ_k can be functions of local coordinates $\sigma_k = \sigma_k(\theta_k, \phi_k)$ generally.

As usual, it is necessary to add conditions of the potentials' limits:

$$\varphi_k^{>} \to 0 \text{ at } r_k^{>} \to \infty \text{ and } \varphi_k^{<} < \infty \text{ at } r_k^{>} \to 0. \qquad (4)$$

3. The Solution for a System of N Spheres

To solve the problem we used the expansion of solutions by series in the spherical functions $Y_{lm}(\theta_k, \phi_k)$, at $l = 0, 1, 2, ...$ and $m = -l, -l+1, ..., 0, 1, 2, ..., l$.[11] We assume that the system of the spherical functions is normalized. Inside and outside the spheres the expansions are respectively:

$$\phi_k^< = \sum_{l,m} A_{lm}^{(k)} r_k^l Y_{lm}(\theta_k, \phi_k); \tag{5}$$

$$\phi_k^> = \sum_{l=0}^{\infty} \sum_{m=-l}^{l} B_{lm}^{(k)} k_l(\kappa r_k) Y_{lm}(\theta_k, \phi_k). \tag{6}$$

In Eq. (6) the modified spherical Bessel functions of third kind $k_l(z)$ [10] are used. The total potential in the surrounding media can be written as follows

$$\phi^> = \sum_{l=0}^{\infty} \sum_{m=-l}^{l} B_{lm}^{(k)} [k_l(\kappa r_k)] Y_{lm}(\theta_k, \phi_k) +$$

$$+ \sum_{j=1}^{N}{}' \left[\sum_{l_j=0}^{\infty} \sum_{m_j=-l_j}^{l_j} B_{l_j m_j}^{(j)} [k_{l_j}(\kappa r_j)] Y_{l_j m_j}(\theta_j, \varphi_j) \right]. \tag{7}$$

The primes near the sum mean the term with subscripts $j = k$ is excluded. The sum subscripts l_j, m_j mean that they can vary independently from subscripts l, m, which correspond to the k-th sphere.

The determination of unknown coefficients $A_{lm}^{(k)}$, $B_{lm}^{(k)}$ in the expansions of potentials in Eqs. (5) and (6) from boundary conditions in Eq. (3) must be obtained. The boundary conditions expressions for potentials and its derivates in different local coordinates are needed. Using the addition theorems[8], we transform the product of the spherical Bessel functions by scalar spherical functions, to a product of the modified spherical Bessel functions by the scalar spherical functions. Moreover, the properties of 3-j Wigner symbols are taken into account and the transition to coefficients of Klebsh-Gordon $C_{l\,m\,l'\,m'}^{l''\,m''}$ is fulfilled.[8,10,11] As a result, we get the following formula for the products $k_l(\kappa r_j) Y_{lm}(\theta_j, \varphi_j)$ at the condition $r_k < |\mathbf{r}_j - \mathbf{r}_k|$.

$$k_l(\kappa r_j) Y_{lm}(\theta_j, \varphi_j) =$$

$$\sum_{l'=0}^{\infty} \sum_{m'=-l'}^{l'} (-1)^{l'-m'} Y_{l'm'}(\theta_k, \varphi_k) i_{l'}(\kappa r_k) \sum_{l''=|l-l'|}^{l+l} Y_{l'',m-m'}(\theta_{jk}, \varphi_{jk}) k_{l''}(\kappa R_{jk}) \varphi_{m\,-m'\,m-m'}^{l\,l'\,l''}, \tag{8}$$

where the $i_l(z)$ is the modified spherical Bessel function of the first kind,[10] θ_{jk}, φ_{jk}, $\mathbf{R}_{jk}$ are shown in Figure 1, and

$$\varphi_{m\,-m'\,m-m'}^{l\,l'\,l''} = \left[4\pi(2l+1)(2l'+1)/(2l''+1)\right]^{1/2} C_{l\,0\,l'\,0}^{l''\,0} C_{l\,m\,l'\,-m'}^{l''\,m-m'}. \tag{9}$$

The symbol $C_{l\,0\,l'\,0}^{l''\,0'}$ is not zero if $l + l' + l''$ is an even integer.

Now we can write the expression for total potential using the addition theorem[8] in local coordinates linked with the k-th sphere. Because the variables are separated we can find derivatives directly, then calculate their value on surface of the k - th sphere.The expansions of the potentials and their derivatives we substitute to the boundary conditions (Eq. 3). As a result, we have a system of $2N$ functional equations. Then we multiply the obtained equations by the complex conjugate functions $Y^*_{lm}(\theta_k, \varphi_k)$ and integrate over the sphere surface. This leads to a family of infinite systems of algebraic linear equations. As the spherical harmonics are orthogonal functions, the terms in summation by indices l', m' with subscripts $l'=l, m'=m$ only remain, and we get the following systems

$$a_k^l A_{lm}^{(k)} = B_{lm}^{(k)} k_l(\kappa a_k) + (-1)^{l-m} i_l(\kappa a_k) \sum_{j=1}^{N}{}' \sum_{l_j=0}^{\infty} \sum_{m_j=-l_j}^{l_j} B_{l_j m_j}^{(j)} \times$$

$$\left\{ \sum_{l''..} Y_{l'',m-m'}(\theta_{jk}, \varphi_{jk}) k_{l''}(\kappa R_{jk}) \varphi_{m-m'\,m-m'}^{l\,l'\,l''} \right\}, \tag{10}$$

$$B_{lm}^{(k)} + \alpha_l^{(k)} (-1)^{l-m} \sum_{j=1}^{N}{}' \sum_{l_j=0}^{\infty} \sum_{m_j=-l_j}^{l_j} B_{l_j m_j}^{(j)} \times$$

$$\left\{ \sum_{l''_j.} Y_{l''_j, m_j-m'_j}(\theta_{jk}, \varphi_{jk}) k_{l''_j}(\kappa R_{jk}) \varphi_{m_j-m'_j\,m_j-m'_j}^{l_j\,l'_j\,l''_j} \right\} = f_{lm}^{(k)}, \tag{11}$$

where $\sigma_{lm}^{(k)}$ are the expansions' coefficients of the surface charges in the spherical functions, and the notations are introduced

$$\alpha_{lm}^{(k)} = \frac{\varepsilon_k l i_l(\kappa a_k) - \varepsilon_m \kappa a_k i'_l(\kappa a_k)}{\varepsilon_k l k_l(\kappa a_k) - \varepsilon_m \kappa a_k k'_l(\kappa a_k)}, \qquad f_{lm}^{(k)} = \frac{4\pi a_k \sigma_{lm}^{(k)}}{\varepsilon_k l k_l(\kappa a_k) - \varepsilon_m \kappa a_k k'_l(\kappa a_k)}. \tag{12}$$

The primes mean a differentiation of functions by their arguments. It should be noted that for the constant densities of surface charges σ_k we have $\sigma_{lm}^{(k)} = \sqrt{4\pi}\sigma_k \delta_{l0}\delta_{m0}$, where δ_{ij} is the Kronnecker symbol.

When performing algebraic transformations we exclude the coefficients $A_{lm}^{(k)}$ from the second boundary condition in Eq. (3) using Eq. (10), and the number of unknown coefficients is halved. After solution of Eq. (11), coefficients for inner potentials are found by summing the series in Eq. (10).

As a result, we have the aggregate of N connected infinite systems of linear algebraic equations. The systems described by Eq. (11) contain only the coefficients $B_{lm}^{(k)}$ of external potentials, and the problem of the interaction of N spheres is completely solved.

4. Ion Electrostatic Interaction Energy of Two Particles

Now we consider in detail the interaction of two spheres starting from general relationships. The line crossing the centers of spheres is taken as the z-axis. The shortest distance between spheres we denote as H, so the distance between sphere centers is $d = H + a_1 + a_2$. The Debye-Huckel approximation to the double layer free energy $F \equiv F_{ij}$ for the interaction of the i-th and j-th spheres of known surface charge density can be found using the following formula[3,4]

$$F = \frac{1}{2}[\int_{s_i} \sigma_i(P_i)\varphi_i^{<}(P_i)dS_i + \int_{s_j} \sigma_j(P_j)\varphi_i^{<}(P_i)dS_j]. \tag{13}$$

For the ensuing discussion we take $i, j = 1,2; i \neq j$. The potential energy V of the double layer interaction is given by the equality $V = F - F_0$,[4] where F_0 is the free energy for two single spheres. If the densities σ_1 and σ_2 are constants then

$$F_0 = \frac{8\pi^2 a_1^3 \sigma_1^2}{\varepsilon_m (1 + \kappa a_1)} + \frac{8\pi^2 a_2^3 \sigma_2^2}{\varepsilon_m (1 + \kappa a_2)}. \tag{14}$$

Integrating Eq. (13) is executed over the sphere surface. Since the potentials on the surfaces of the spheres are equal inside and outside, we can use either potentials' representations. If the surface charges are constant, integration leads to calculations of the spherical functions integrals over the total surfaces of the spheres. After integrating Eq. (13), taking into account the expansions (Eq. 5), we obtain for the free energy

$$F = \sqrt{\pi}[\sigma_1 a_1^2 A_{00}^{(1)} + \sigma_2 a_2^2 A_{00}^{(2)}] \tag{15}$$

To find the energy of pair interaction, only the first coefficients of series expansions (5)-(6) $A_{00}^{(i)}, B_{00}^{(i)}$ are needed, but their values are to be determined from the infinite systems.

In this case, the problem is axis-symmetrical, and the system for determination of the potentials' coefficients looks like

$$A_{00}^{(i)} = B_{00}^{(i)} k_0(\kappa a_k) + i_0(\kappa a_i) \sum_{l'} B_{l'}^{(j)} (-1)^{l'} (2l'+1)^{1/2} k_{l'}(\kappa d), \; i, j = 1,2; i \neq j, \tag{16}$$

$$B_{00}^{(i)} + \alpha_0^{(i)} \sum_{l'} B_{l'0}^{(j)} (2l'+1)^{1/2} k_{l'}(\kappa d) = f_0^{(i)}, \tag{17}$$

$$B_{l0}^{(i)} + \alpha_l^{(i)} (-1)^l (2l+1)^{1/2} \sum_{l', l''} B_{l'0}^{(j)} (2l'+1)^{1/2} (-1)^{l''} k_{l''}(\kappa d) \left(C_{l'0l0}^{l''0}\right)^2 = 0. \tag{18}$$

We take into account that the spheres are placed on the z-axis, and in Eqs. (10) and (11) the functions Y_{lm} at $\theta = \pi$ have the value[11]

$$Y_{lm}(\pi, \varphi) = \delta_{m0} (-1)^l \sqrt{(2l+1)/(4\pi)}.$$

The system obtained allows further simplification, because there is an opportunity to separate the coefficients $B_{l0}^{(k)}$, $B_{l0}^{(j)}$ to get independent systems for every sphere. For all that, only the right parts of the systems define the connection between spheres.

5. Zero Approximation for Two Spheres

Now we assume that the surface charge density is constant. The simplest case is to take only one term in the expansions of potentials, i.e. if $l = l' = 0$ (a zero approximation). It should be noted that in the zero approximation solutions do not depend on sphere permittivity, as it follows from Eq. (12) for $\alpha_0^{(k)}$. If terms of higher order in comparison with the quantity $k_0^2(\kappa d)$ are neglected, the formula for the potential energy of interaction $V(d)$ can be derived

$$V(H) = F - F_0 =$$

$$\frac{8\pi^2 a_1^3}{1+\kappa a_1}\frac{\sigma_1\sigma_2}{\varepsilon_m}\frac{k_0(\kappa d)}{(\kappa a_1)^2 k_0(\kappa a_1)k_1(\kappa a_2)}\left[(\kappa a_2)^2 i_0(\kappa a_2)k_1(\kappa a_2) + (\kappa a_1)^2 i_1(\kappa a_1)k_0(\kappa a_1)\right]+$$

$$\frac{8\pi^2 a_2^3}{1+\kappa a_2}\frac{\sigma_1\sigma_2}{\varepsilon_m}\frac{k_0(\kappa d)}{(\kappa a_2)^2 k_0(\kappa a_2)k_1(\kappa a_1)}\left[(\kappa a_1)^2 i_0(\kappa a_1)k_1(\kappa a_1) + (\kappa a_2)^2 i_1(\kappa a_2)k_0(\kappa a_2)\right]. \quad (19)$$

In Eq. (19) the modified spherical Bessel functions of the first kind $k_1(z) = -k_0'(z)$, $i_1(z) = i_0'(z)$, ($k_0(z) = (\pi/2)\exp(-z)/z$, $i_0(z) = \sinh z/z$) [10] are used.

When the distances between spheres are large, as $d \to \infty$ and $k_0(\kappa d) \to 0$, Eq. (19) gives the same results as previously described.[3,4]

Now we consider the spheres with $a_1 = a_2 = a$ and $\sigma_1 \neq \sigma_2$. To compare we write the Ohshima formulae.[4] We also assume that $\sigma_1 = \sigma_2$. Then the resulting formula takes the form:

$$V_{HO}(H) = -F_0 \frac{(1+\kappa a)}{(\kappa a)^2}\cdot\frac{H+a}{H+2a}\ln\left(1 - \frac{a}{H+a}e^{-\kappa H}\right). \quad (20)$$

Derjaguin's method[1] gives the following simpler expression

$$V_D = -F_0 \frac{(1+\kappa a)}{(\kappa a)^2}\ln(1 - e^{-\kappa H}), \quad (21)$$

where $F_0 = 16\pi^2 a^3\sigma^2/[\varepsilon_m(1+\kappa a)]$.

From our general formula Eq. (19) we have the expression for an improved zero approximation ($d = H + 2a$, $F_0 = 16\pi^2 a^3 \sigma_1^2/[\varepsilon_m(1+\kappa a)]$)

$$V(H) = F_0 \frac{\sigma_2/\sigma_1}{1+\kappa a}\frac{a}{H+2a}e^{-\kappa H} +$$

$$F_0 \frac{(\sigma_2/\sigma_1)^2}{4(1+\kappa a)^2}\left[(\kappa a - 1) + (\kappa a + 1)e^{-2\kappa a}\right]\left(\frac{a}{H+2a}\right)^2 e^{-2\kappa H}. \quad (22)$$

Neglecting the second term in Eq. (22), it takes the form at $\sigma_1 = \sigma_2 = \sigma$

$$V(H) = F_0 \frac{1}{1+\kappa a}\frac{a}{H+2a}e^{-\kappa H}. \quad (23)$$

When $\kappa H >> 1$ and $\kappa a >> 1$, after expanding the logarithms to the corresponding series and keeping the first terms, we get the same expressions as Eqs. (20) and (21).

6. Numerical Results

We obtain closed formulae consistent with the results described elsewhere[3,4] in limiting cases. However our results are more general and can be improved with an additional approximation. With the help of different formulae we determine dependences of the non-dimensional interaction energy $V^* = V / V_0$, where $V_0 \equiv F_0 = 16\pi^2 a^3 \sigma^2 / (\varepsilon_m (1 + \kappa a))$, for two identical spheres with constant and equal charges versus the parameter κH (Figure 2). The value $\kappa a = 1$ is adopted.

It follows from this data that at $\kappa H > 3$ Ohshima formulae, and our formula for zero approximation give similar values, but Derjaguin's formula gives too high a value. All the results begin to agree at large values, when $\kappa H > 5$. The results from Eqs. (22) and (23) differ slightly and only at small κH values. When $\kappa H < 1$, our results differ greatly from the results after Ohshima and Derjaguin, because $V_D \to \infty$ and $V_{HO} \to \infty$, when $\kappa H \to 0$.

In the second case, we assume that the radii are identical, but the surface charge densities may differ, and the ratio $q = \sigma_2 / \sigma_1$ is varied (Figure 3). Calculations are worked out by Eqs. (22) and (23). In this case, the expression $V_0 \equiv F_0 = 16\pi^2 a^3 \sigma_1^2 / (\varepsilon_m (1 + \kappa a))$ is used.

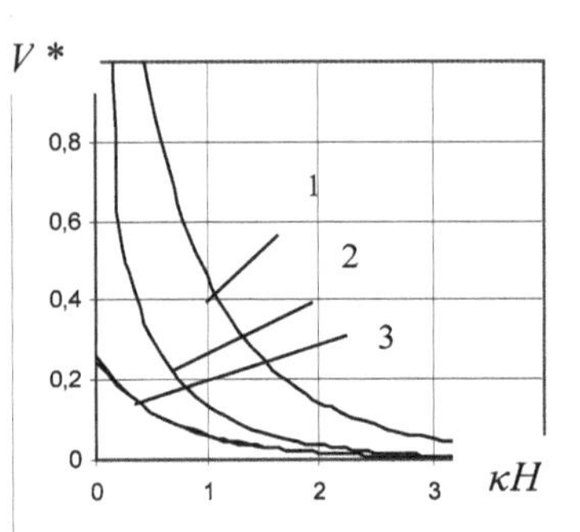

Figure 2. Interactions' energy $V^* = V / V_0$ of two identical spheres with the constant charges versus κH at $\kappa a = 1$; 1 – *Derjaguin* approximation; 2 – *Ohshima* approximation; 3 – our simplest zero approximation.

Results obtained from the improved zero approximation and the simplest zero approximation begin to differ at values $\kappa H << 1$. The difference depends on the value parameter κa. For example, at $\kappa a = 10$ the maximum difference is about 30% when charges have the same signs at $H = 0$ and $q = 5$. At large q and κa values it is necessary to use the improved zero approximation. In the case of different surface charge densities and different radii when $a_2 / a_1 >> 1$, the energy values tends to a limit. This value can be founded with the help of the limit transition in Eq. (19).

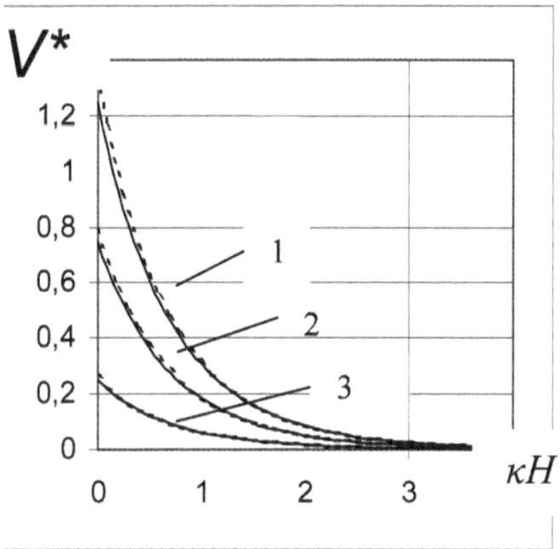

Figure 3. Interactions' energy $V^* = V / V_0$ of two identical spheres with identical radii but the different charges versus κH at $\kappa a = 1$; 1 – $q = 1$; 2 – $q = 3$; 3 – $q = 5$; _____ our simplest zero approximation; --------- our improved zero approximation.

7. Conclusion

The exact solution for interaction of a system of small spherical particles in an electrolyte is obtained. On the basis of the exact solutions, closed formulae for calculating ion-electrostatic energy of two spheres are derived. Our zero approximation corresponds to results of other authors in simple cases, and generalizes ones in the range of small values where the parameter $\kappa H < 2$. In this paper we consider the case when surface charges are given, but the problem of spherical particles interaction with given surface potentials can be solved similarly.

Results show the effectiveness of our method. It should be noted that the formulae can be improved if subsequent terms are included in the series for potentials. It opens the opportunity to calculate the potential energy of an arbitrary number of particles with great accuracy. Moreover, at insignificant modification using previously elaborated algorithms[12, 13] we can take into account a lamellar structure of particles. This circumstance is important when calculating the interaction energy of mineral particles with different biological objects such as cells.

References

1. B. V. Deryaguin, N. V. Churaev, and V. M. Muller, *Surface forces* (Nauka, Moscow, 1985). (In Russian).
2. G. M. Bell, S. Levine, and L. N. McCartney, Approximate Method of Determining the Double-Layer Free Energy of Interaction between Two Charged Colloidal Spheres, *J. Colloid Interface Sci.* 33 (3), 335-359 (1970).
3. E. J. W. Verner and J. Th. G. Overbeek, *Theory of the Stability of Lyophobic Colloids* (Elsevier, Amsterdam, 1948).
4. H. Ohshima, Diffuse double layer interaction between two spherical particles with constant surface charge density in an electrolyte solution, *Colloid Polymer Sci.* 263, 158-163 (1975).
5. H. Zonntag, Some remarks of the cell-cell adhesion from physical point of view, *Stad. Biophys* 56, 209-217 (1976).
6. J. P. Hsu and S. H. Lin, Electrical Interaction between Two Planar, Parallel Dissimilar Surfaces in a General Electrolytic Solution, *Langmuir* 19, 10610-10616 (2003).
7. L. G. Grechko, L. B. Lerman, O. Ya. Pokotylo, and N. G. Shkoda, Interaction of small spherical particles in a electrolyte, *Bulletin of Kyiv University, Series Physics & Mathematics* 3, 203-210 (2005). (In Ukraine).
8. W. C. Chew, *Waves and Fields in Inhomogeneous Media. App. D* (IEEE Press, New York, 1995).
9. V. N. Pustovit and V. A. Markel, Propagation of diffuse light in turbit medium with multiple spherical inhomogeneities, *Appl. Optics* 43 (1), 104-112 (2004).
10. *Handbook of mathematical functions with formulas, graphs and mathematical tables*, edited by M. Abramowitz and I.A. Stegun, (National Bureau of Standards Applied Mathematics Series, 1964).
11. D. A. Varshalovich, A. N. Moskalyov, and V. K. Hersons, *Quantum theory of angles moments*, (Moscow, Nauka, 1975). (In Russian).
12. L. G. Grechko, L. B. Lerman, and N. G. Shkoda, Scattering of electromagnetic waves on multi-layered sphere, *Bulletin of Kyiv University, Series: Physics & Mathematics* 3, 376-385 (2004). (In Ukraine).
13. L. B. Lerman, L. G. Grechko, and V. V. Gozhenko, Electromagnetic waves interaction with a lamellar spherical lens, in: *Proceedings of the 5th International Conference on Antenna Theory and Techniques,* (National technical university 'KPI', Kyiv, Ukraine, 2005), pp. 234-237.

STRUCTURAL AND ADSORPTION CHARACTERISTICS OF PYROCABON-MINERAL ADSORBENTS

J. SKUBISZEWSKA-ZIĘBA,[a,*] R. LEBODA,[a] V. M. GUN'KO,[b] B. CHARMAS[a]
[a] *Maria Curie-Sklodowska University, Maria Curie Sklodowska Sq.3, 20-031 Lublin, Poland*
[b] *Institute of Surface Chemistry, 17 General Naumov Street, 03164 Kiev, Ukraine*

Abstract. A variety of pyrocarbon/silica gel adsorbents were prepared using commercial mesoporous silica gels Si-40, Si-60, and Si-100 as matrices modified by carbon deposits from pyrolysis of several organic precursors. The second type of hybrid carbon-mineral adsorbents was synthesized using spent natural palygorskite utilized in paraffin purification. The adsorbents were then heated, hydrothermally treated, or modified by additional deposition of carbon. Changes in the structural and adsorption characteristics of hybrid adsorbents before and after treatments were analyzed by microscopy, *p*-nitrophenol and nitrogen adsorption isotherms, and TG, TEM, XRD, and XRF methods.

Keywords: carbon-mineral adsorbents, pyrolysis, hydrothermal treatment, CCl_4 vapors, waste material, utilization, structural characteristics, pore size distribution

1. Introduction

Carbon-mineral adsorbents are of interest because of their potential applicability in practice, and as model adsorbents in research. Such composite materials can be applied for the adsorption of both polar and nonpolar compounds.[1,2] These adsorbents have been utilized in chromatography, trace analysis, environmental protection, technology of water and sewage purification, and in other processes.[1]

*To whom correspondence should be addressed. J. Skubiszewska-Zięba, Maria Curie-Sklodowska University, Maria Curie Sklodowska Sq.3, 20-031 Lublin, Poland; e-mail: jskubisz@hermes.umcs.lublin.pl

J.P. Blitz and V.M. Gun'ko (eds.), Surface Chemistry in Biomedical and Environmental Science, 123–134.

Additionally, they are potential polymer fillers and condensation agents of various dispersed media. These materials are also used as intermediates in the preparation of carbon adsorbents[3,4] as they can provide desirable and original properties to the finished adsorbents. Adsorption features of such hybrid adsorbents depend on the nature of the support, the carbonized precursor, as well as pyrolysis conditions. These factors determine the morphology of adsorbents, the accessibility of oxide and carbon patches, and the availability of different acidic and basic surface sites. These hybrid adsorbents can work at high temperature, in the presence of water vapor at high pressures, as well as in the presence of other aggressive substances. Certain changes in the structural and adsorption properties of these materials can occur under such conditions.[1-4]

There is an important problem related to the utilization of spent adsorbents and catalysts, especially natural zeolites, used in chemical, petrochemical and food industries after adsorbing large amounts of organic substances.[5-7] As a whole they can be used as raw materials to prepare carbon-mineral adsorbents.[8] Therefore, the aim of this work was to elucidate the influence of different factors such as the kind of organic precursors and inorganic matrices, including spent adsorbents, and the preparation and modification techniques on the structural and adsorption properties of carbon-mineral adsorbents.

2. Experimental

2.1. PREPARATION OF ADSORBENTS

Different silica gels (Merck) Si-40 (specific surface area S_{BET} = 732 m^2/g, total pore volume V_p = 0.542 cm^3/g, average pore radius R_p = 1.48 nm, granule size d = 0.2-0.5 mm), Si-60 (S_{BET} = 369 m^2/g, V_p = 0.753 cm^3/g, R_p = 4.1 nm, d = 0.1-0.2 mm), Si-60 (S_{BET} = 447 m^2/g, V_p = 0.82 cm^3/g, R_p = 3.7 nm, d = 0.2-0.5 mm), and Si-100 (S_{BET} = 350 m^2/g, V_p = 1.23 cm^3/g, R_p = 7.0 nm, d = 0.1-0.2 mm) were used as inorganic matrices for preparation of carbon-silica adsorbents. Various organics such as acenaphthene (corresponding carbon-silica samples were labeled as AN), acetylacetone (AC), glucose (GL), and dichloromethane (DC) were applied as carbon deposit precursors.

Before pyrolysis silicas were dried at 200°C and cooled to room temperature. Different amounts of organic precursors were deposited on dry silica (weight 5 g) to obtain carbon-silica adsorbents (carbosils) with different amounts of carbon deposits. Samples based on acenaphthene (Tables 1 and 2), acetylacetone and glucose (Table 3), were pyrolysed under static conditions in a stainless steel autoclave (0.3 dm^3) at 773 K for 6 h. After reaction, all the prepared carbosils were washed in a Soxhlet apparatus with N,N-dimethylformamide and acetone, and then dried at 200°C.

Two other series of carbosils were prepared under dynamic conditions by using a rotary flow reactor in nitrogen atmosphere fed at a rate of 100 cm^3/min. Dichloromethane was pyrolysed on silica gel Si-60 at a liquid feeding rate (through a glass evaporator heated at 373 K) of 0.6 cm^3 min^{-1} using a Masterflex (Cole Parmer) pump.

To elucidate the influence of the preparation procedure (dynamic or static) two series of gluocose based adsorbents were pre-deposited on the surface of silica gel Si-100. The proportions of glucose to silica gel were the same for the static (S) and dynamic (D) series of carbosils. Additionally, the influence of water (formed during glucose decomposition) on the porous structure of silica was analyzed.

Initial silica samples and prepared carbosils were subjected to hydrothermal modification (HTT). A sample (2 g) of silica gel or carbosil was placed in a quartz vessel in an autoclave containing 20 cm^3 of water heated at 150°C or 200°C for 6 h. These conditions ensured a saturated water vapor pressure in the autoclave. After modification, the samples were dried at 200°C for 6 h. To investigate the pyrolysis effects of different precursors on the silica structure, pyrocarbon was removed by heating at 500°C for 24 h in air (labeled T).

Spent natural zeolite palygorskite (Cherkawskiy District, Ukraine), after use for paraffin purification, was utilized as the initial material to prepare carbon-mineral adsorbents. CCl_4 (Merck) was used as a cross-linking agent of the organic paraffin phase during pyrolysis. This series of adsorbents was prepared by several procedures. Thermal treatment (T) was carried out in a rotary quartz reactor in a nitrogen stream (flow rate 100 cm^3/min) ramping the temperature from 20-200°C in 1 h, then from 200-700°C in 1 h, and maintaining at 700°C for 1 h. Hydrothermal (H) modification was performed under static conditions in a steel autoclave at 200°C with a water vapor pressure of 13.5 atm for 8 h. Thermal treatment in CCl_4 atmosphere (procedure C) was carried out in a rotary quartz reactor under the same temperature conditions as with the thermal treatment (T). CCl_4 vapors were supplied to the reactor at a rate of 0.6 cm^3/min by heating from 200-700°C for 1 h. Modification (M) of the porous structure of complex carbon-mineral adsorbents was performed by partial gasification of carbon deposits with water vapor in the fluidal quartz reactor (nitrogen flow 300 cm^3/min). After heating up to 800°C (~2 h) the water vapor stream was supplied into the reactor by evaporation of liquid water from an evaporator at 300°C.

To determine the effect of the procedures C and M on the properties of carbon-mineral adsorbents, reference carbon adsorbents were prepared using paraffins treated under the C and C + M procedures.

2.2. TESTING METHODS

Low-temperature (77.4 K) nitrogen adsorption-desorption isotherms were recorded using a Micromeritics ASAP 2405N adsorption analyzer. The specific surface area S_{BET} was calculated using the standard BET equation[9] at p/p_0 (where p and p_0 denote the equilibrium and saturation pressures of nitrogen, respectively) between 0.06 and 0.2. The pore volume V_p was estimated at $p/p_0 \approx 0.98$ converting the volume of adsorbed nitrogen gas to the volume of fluid. Pore size distributions (PSDs) (differential PSD $f_V(R) \sim dV_p/dR$) of the studied adsorbents were calculated using a modified overall adsorption equation in the form proposed by Nguyen and Do (ND method)[10] for slitshaped pores of carbons, and modified for other materials with cylindrical pores and pores as gaps between spherical particles.[11]

The specific surface area of only the pyrocarbon (S_C) was determined by *p*-nitrophenol (PNP) adsorption (assuming preferable adsorption of PNP onto carbon deposits) from an aqueous solution of hydrochloric acid, studied using a Specord M-40 (Karl Zeiss, Jena) UV/vis spectrophotometer at 400 nm. This technique was described in detail elsewhere.[12]

The carbon content (C_C) in carbon-mineral samples was determined using a Derivatograph C (Paulik, Paulik & Erdey, MOM, Budapest) at a heating rate of 10 K/min, in air from 293 to 1273 K.

Transmission Electron Microscopy (TEM) micrographs were obtained using a BS 540 (Tesla) apparatus. Microscope samples were prepared using the platinum-carbon replication method, with evaporation of platinum and a low amount of carbon onto the adsorbents, then treated in HF to dissolve silica gel.

Powder X-ray diffraction (XRD) patterns were recorded using a HZG 4A2 X-ray diffractometer with nickel-filtered Cu K_α ($\lambda = 0.15418$ nm) radiation with a 0.04 degree step, a step time of 2 s, and a measurement range $2\Theta = 3\text{-}85^0$.

The metal concentrations in certain samples were determined using a XRF (Canberra) X-ray fluorescence spectrometer with radioactive sources of ^{109}Cd, ^{55}Fe and ^{241}Am.

3. Results and Discussion

It is known[13,14] that both thermal (T) and hydrothermal treatments (HTT) of silica gels change their pore structure. The specific surface area diminishes and the pore size increases depending on treatment temperature. This occurs due to hydrolysis of Si-O-Si bonds, transferring of $Si(OH)_4$ and larger complexes with water molecules, and re-condensation ≡SiOH + HOSi≡ →≡Si–O–Si≡ + H_2O. These processes cause significant reduction of the total porosity and the formation of spongy particles with large mesopores and macropores. However, carbon deposits can shield the silica surface against structural changes on HTT.[11,15,16]

Table 1. Structural characteristics of initial, heated (T) and hydrothermally treated at 150°C (HTT) silica gel Si-40 and carbosils prepared with acenaphthene pyrolysis under static conditions.[15]

Sample	C_C (% w/w)	S_{BET} (m^2/g)	V_p (cm^3/g)	V_{nano} (cm^3/g)	R_p (nm)
Si-40	-	732	0.542	0.011	1.48
Si-40-HTT	-	309	0.512	0.014	3.31
Si-40-HTT-T	-	305	0.504	0.006	3.30
Si-40-T	-	613	0.476	0.045	1.55
Si-40-T-HTT	-	290	0.445	0.005	3.07
Si-40-T-HTT-T	-	284	0.435	0.004	3.08
$AN_{40}1$	5.6	313	0.375	0.004	2.40
$AN_{40}1$-HTT	not determ.	211	0.343	0.003	3.26
$AN_{40}1$-HTT-T	-	239	0.418	0.003	3.50
$AN_{40}1$-T	-	364	0.425	0.004	2.33
$AN_{40}2$	8.3	231	0.317	0.002	2.74
$AN_{40}2$-HTT	not determ.	198	0.312	0.002	3.16
$AN_{40}2$-HTT-T	-	233	0.405	0.003	3.49
$AN_{40}2$-T	-	328	0.423	0.007	2.59
$AN_{40}3$	19.2	128	0.163	0.002	2.53
$AN_{40}3$-HTT	not determ.	126	0.158	0.001	2.51
$AN_{40}3$-HTT-T	-	221	0.378	0.003	3.42
$AN_{40}3$-T	-	323	0.388	0.011	2.40
$AN_{40}4$	21.8	92	0.149	0.001	3.24
$AN_{40}4$-HTT	not determ.	91	0.132	0.001	2.90
$AN_{40}4$-HTT-T	-	154	0.351	0.001	4.56
$AN_{40}4$-T	-	260	0.386	0.003	2.94

Note. C_C is the carbon deposit concentration, S_{BET} is the specific sureface area, V_p is the total pores volume, V_{nano} is the nanopores volume, and R_p is the pore radius.

Table 1 presents the structural characteristics of adsorbents prepared with Si-40 and acenaphthene.[15] Larger changes in the Si-40 structure are caused by hydrothermal treatment, despite a relatively low temperature (150°C), than by heating at 500°C. However in the case of carbon-silica adsorbents, both hydrothermal modification and high-temperature pyrolysis changes the pore structure to a large extent.

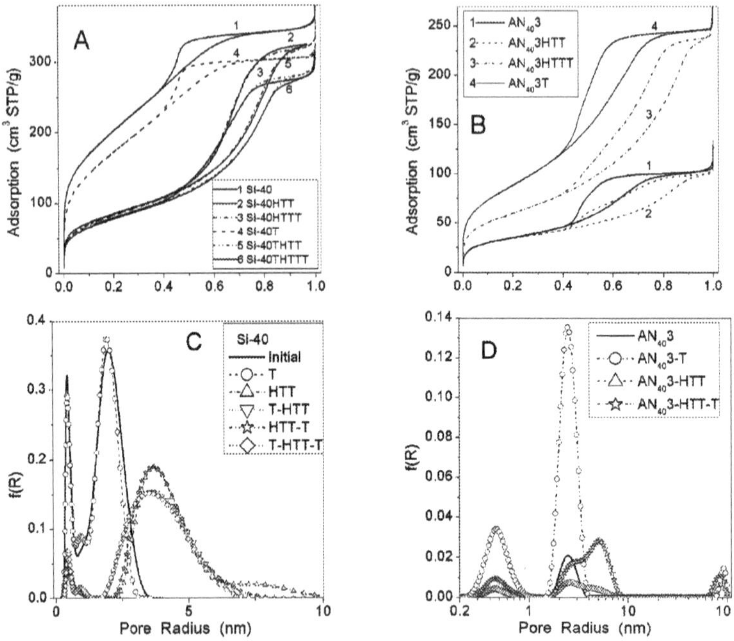

Figure 1. Nitrogen isotherms (A,B) and pore size distributions (C,D) for Si-40, carbosil $AN_{40}3$ initial and treated under various conditions.[15]

The shapes of nitrogen adsorption-desorption isotherms strongly differ for silica and carbon-silica adsorbents before and after hydrothermal treatment (Figure 1). The main changes are a reduction of porosity and an enhancement of the pore size (Figure 1A, C). Carbon deposits correlate with a diminution of the porosity nearly linearly with increasing carbon concentration (Table 1). Significant changes in the pore structure (Figure 1B, D) result as the products of acenaphthene carbonization can block narrow pores and partially fill larger mesopores of silica gel. These changes in the surface morphology of initial and modified adsorbents are clearly seen in TEM micrographs presented elsewhere.[15]

The pore volume and specific surface area of carbosils decrease with C_C values independent of pyrocarbon origin (Table 2, Figure 2).[16] However, the precursor type as well as the initial pore structure of silica gels, e.g. Si-40 (Table 1) and Si-60 (Table 2), affect the pore characteristics of carbosils (Figures 1 and 2). The narrower the pores of the pristine silica gel, the larger the specific surface area reduction under the same pyrolysis conditions.

Table 2. Structural characteristics of initial, heated (T) and hydrothermally (HTT) treated at 150 and 200°C silica gel Si-60 and carbosils prepared from pyrolysis of acenaphthene (AN), acetylacetone (AC) and glucose (GL) under static conditions.[16]

Sample	C_C (% w/w)	S_{BET} (m²/g)	V_p (cm³/g)	V_{nano} (cm³/g)	R_p (nm)
Si-60	-	369	0.753	0.19	4.1
Si-60-T	-	362	0.747	0.02	4.1
Si-60-HTT-150	-	121	0.731	0.08	12.1
Si-60-HTT-150-T	-	122	0.738	0.07	12.1
Si-60-HTT-200	-	47	0.272	0.03	11.8
Si-60-HTT-200-T	-	46	0.305	0.03	13.0
$AN_{60}1$	7.0	327	0.650	0.18	4.8
$AN_{60}1$-T	-	365	0.736	0.20	4.0
$AN_{60}1$-HTT-150	not determ.	189	0.613	0.13	6.5
$AN_{60}1$-HTT-150-T	-	212	0.711	0.14	6.7
$AN_{60}2$	15.8	244	0.480	0.13	3.9
$AN_{60}2$-T	-	358	0.725	0.19	4.0
$AN_{60}2$-HTT-150	not determ.	167	0.478	0.09	5.7
$AN_{60}2$-HTT-150-T	-	181	0.714	0.13	7.9
$AN_{60}3$	22.7	233	0.453	0.15	3.9
$AN_{60}3$-T	-	337	0.726	0.09	4.3
$AN_{60}3$-HTT-150	not determ.	150	0.385	0.11	5.1
$AN_{60}3$-HTT-150-T	-	173	0.698	0.10	8.0
$AC_{60}1$	4.0	339	0.717	0.18	4.2
$AC_{60}1$-T	-	345	0.743	0.19	4.3
$AC_{60}1$-HTT-200	not determ.	68	0.634	0.03	19.5
$AC_{60}1$-HTT-200-T	-	60	0.709	0.04	23.5
$AC_{60}2$	9.1	296	0.606	0.18	4.1
$AC_{60}2$-T	-	340	0.730	0.18	4.3
$AC_{60}2$-HTT-200	not determ.	90	0.648	0.05	14.5
$AC_{60}2$-HTT-200-T	-	85	0.704	0.04	16.6
$AC_{60}3$	14.5	275	0.566	0.19	4.1
AC3-T	-	327	0.728	0.19	4.5
$AC_{60}3$-HTT-200	not determ.	108	0.545	0.08	10.1
$AC_{60}3$-HTT-200-T	-	83	0.698	0.05	16.9
GL_{60}	16.5	174	0.433	0.09	5.0
GL_{60}-T	-	288	0.704	0.16	4.9
GL_{60}-HTT-150	not determ.	224	0.470	0.12	4.1
GL_{60}-HTT-150-T	-	200	0.667	0.11	6.6

Additionally, the pyrolysis of oxygen containing organics leads to a greater decrease in the specific surface area, because of the hydrolysis of Si-O-Si bonds and disruption of the pore walls (Table 2, AC and GL). This is particularly apparent in the case of adsorbent GL. Comparing samples $AN_{60}2$ and GL_{60} with similar carbon deposit concentrations (15.8 and 16.5% w/w), one can see that after thermal treatments the specific surface area of the samples $AN_{60}2$-T and GL_{60}-T differ significantly (358 m²/g and 224 m²/g, respectively). Subsequent thermal treatment of previously HTT modified carbosils $AN_{60}2$-HTT and

GL_{60}-HTT gives silica samples of different porous structure (S_{BET} = 181 m^2/g for AN_{60}2-HTT-T and 200 m^2/g for GL_{60}-HTT-T). TEM microphotographs (Figures 3-5) confirm the above data.

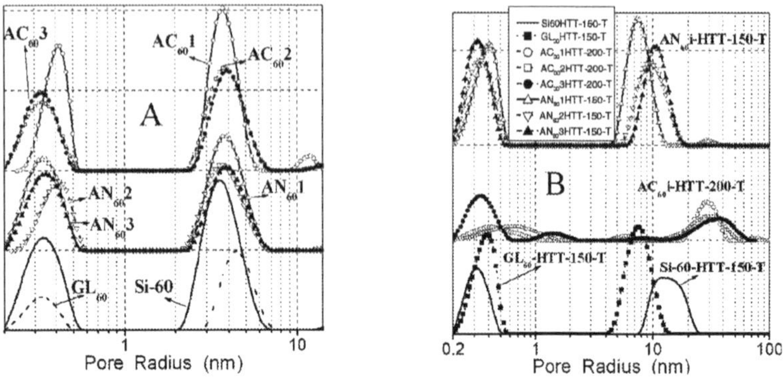

Figure 2. Pore size distribution for (A) initial silica Si-60 and carbosils and (B) after their hydrothermal at 150 (AN, GL) and 200°C (AC) and thermal at 500°C treatments.[16]

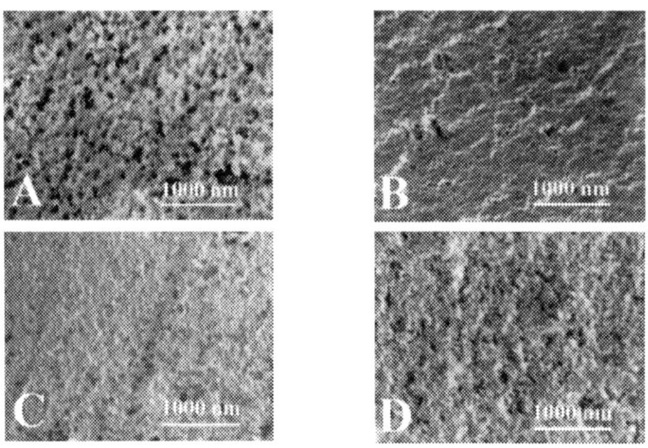

Figure 3. TEM micrographs of (A) Si-60; (B) Si-60-T; (C) Si-60-HTT-150; and (D) Si-60-HTT-200.

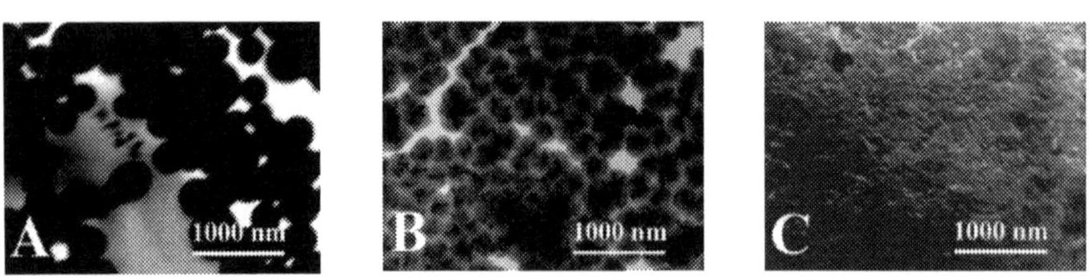

Figure 4. TEM micrographs of (A) AC_{60}3; (B) GL_{60}; and (C) AN_{60}2.

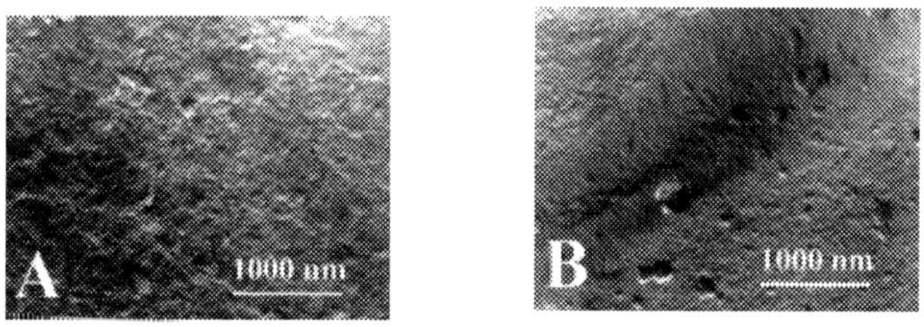

Figure 5. TEM micrographs of (A) $AN_{60}2$-T and (B) GL_{60}-T.

The Si-60 surface seems rough (Figure 3A) with large transport pores. After heating at 500°C for 24 h in air, the silica gel surface appears slightly smoother, and the globular structure is less visible (Figure 3B). After hydrothermal treatment (HTT) of Si-60 at 150°C for 6 h, the silica surface becomes spongy (Figure 3C). After HTT at 200°C, the spongy character of the surface becomes more clearly visible (Figure 3D). Pyrocarbon grafting on the silica gel surfaces results in formation of carbon particles (Figures 4A and 4B), and a carbon layer with smaller graphene particles on the silica surfaces (Figure 4C). A shielding effect of the pyrocarbon is responsible for a smaller reduction of the specific surface area on HTT and HTT-T compared to Si-60-HTT and Si-60-HTT-T (Table 2). However, glucose carbonization is accompanied, in fact, by HTT of silica gel, and the GL_{60}-T surface seems close to that of Si-60-HTT (Figure 5B).

Previous work[11] has shown the structural characteristics of adsorbents prepared on the basis of Si-60 ($S_{BET} = 447$ m^2/g, $V_p = 0.82$ cm^3/g, $R_p = 3.7$ nm, $d = 0.2$-0.5 mm) and dichloromethane (DC_{60} series) in dynamic conditions followed by hydrothermal treatment at 200°C in an autoclave. Carbon concentrations in this series of adsorbents are from 1.5% w/w for $DC_{60}1$ to 19.5% w/w for $DC_{60}6$. The pyrocarbon deposits reduce the pore structure of hybrid adsorbents nearly linearly with increasing C_C ($S_{BET} = 431$m^2/g and $V_p =$ 0.74 m^2/g for $DC_{60}1$ and $S_{BET} = 375$ m^2/g and $V_p = 0.535$m^2/g for $DC_{60}6$). Pyrocarbon strongly shields the silica gels against pore wall disruption during the hydrothermal treatment process, and results in significant differences in the shapes of the isotherms of initial and HTT adsorbents.[11] The structural parameters of carbon deposits in this series of carbosils, determined on the basis of the p-nitrophenol adsorption method[12] were as follows: 152, 158, 109, 86, 99, and 70 m^2/g accordingly for $DC_{60}1$ and $DC_{60}6$. The removal of carbon deposits from hydrothermally treated carbosils by heating at 500°C results in silica samples with a slightly reduced specific surface area compared to initial Si-60 ($S_{BET} = 413$m^2/g for $DC_{60}1$-HTT-T and 418 for $DC_{60}6$-HTT-T). However, the total pore volume remains almost unchanged compared to initial Si-60.

Interesting data concerning the impact of the applied preparation technique (in autoclave A, or in reactor R) and the amount of carbon precursor (glucose) on the structural characteristics of the adsorbents has been previously reported.[17]

Table 3. Structural characteristics of initial and heated at 500°C Si-100 and carbosils.[17]

Sample	C_C (wt.%)	S_{BET} (m^2/g)	V_p (cm^3/g)	V_{nano} (cm^3/g)	R_m (nm)
Si–100	–	350	1.225	0.008	7.0
Si–100T	–	358	1.233	0.008	6.9
GL_{100}–A1	5.6	321 (315)	1.119 (1.133)	0.013 (0.002)	7.0 (7.2)
GL_{100}–A2	14.1	309 (318)	0.916 (1.153)	0.027	5.9 (7.2)
GL_{100}–A3	23.3	308 (314)	0.729 (1.085)	0.040	4.8 (6.9)
GL_{100}–A4	24.6	256 (271)	0.694 (1.077)	0.030	5.4 (6.7)
GL_{100}–R1	5.2	343 (344)	1.094 (1.147)	0.009 (0.001)	6.4 (6.7)
GL_{100}–R2	13.8	345 (357)	0.947 (1.198)	0.028	5.5 (6.7)
GL_{100}–R3	21.4	365 (347)	0.796 (1.130)	0.047	4.4 (6.5)

Note. Parameters in abbreviations concern samples after carbon removing at 500°C.

Comparing data in Table 3 one can see that the carbonization technique affects not only the pyrocarbon amounts, but also structural parameters of the samples. Water molecules evolved during pyrolysis modify the silica matrix to a marked extent. This is clearly seen in the case of adsorbents prepared in an autoclave, particularly for samples GL_{100}-A3 and GL_{100}-A4. In these samples the proportion of glucose to silica was the same, but in sample designated –A4 there was triple the amount of glucose in the system, thus three time higher water pressure.

In the utilization of spent adsorbents, it is very important to obtain materials with developed porous structure, high adsorption capacity, and good mechanical strength. Results of research for a series of adsorbents prepared on the basis of palygorskite spent in paraffin purification have been reported.[18] The specific surface area of pure palygorskite was 197 m^2/g, and the pore volume 0.59 cm^3/g. Thermal treatment at 700°C of the initial spent sample, as well as after its previous hydrothermal modification, gave carbon-silica adsorbents with S_{BET} of 108 and 104 m^2/g accordingly and, unfortunately, small carbon deposit concentrations (1.8 and 1.47% w/w). It follows that normal sample heating causes the loss of carbon matter to a great extent. However thermal treatment in CCl_4 atmosphere results in less carbon loss on the palygorskite surface, and produces carbon-mineral adsorbents with a larger carbon deposit concentration (10.4% w/w, S_{BET} = 287 m^2/g and V_p = 0.52 cm^3/g). The best results were obtained when thermal treatment in CCl_4 vapors was performed after the hydrothermal modification process. Such a prepared sample had 29.83% w/w of carbon deposit, and its S_{BET} was 333 m^2/g and V_p 0.54 cm^3/g.[18] Partial gasification of carbon deposits with water vapor is a commonly used method to develop porous structures. Applying such modification to this adsorbent gave a material of slightly lower carbon deposit concentration (28.99% w/w) and pore structure parameters. It should be added that XRD data for the carbon-mineral samples under discussion don't contain lines characteristic of palygorskite. However, lines corresponding to quartz were observed.[18] Additionally, HCl as a by product of cross-linking pyrolysis occurring at 700°C can react with different metal ions (Al, Ti, Mn, Cu, and Zr) present in palygorskite. This leads to a reduction of the content of these metals in the adsorbent after carbonization.

4. Conclusion

The origin and amounts of pyrocarbon impact the main mesopore structure of carbon-mineral adsorbents. Pyrocarbon forms relatively large globules on the external surfaces of silica gel particles, and covers the oxide surfaces by tiny graphene particles. The presence of oxygen atoms in pyrolyzed precursors (glucose, acetylacetone) has a substantial influence on the carbon deposit character as well as structural parameters of carbosils (both pyrocarbon and silica) in comparison with those for carbosils prepared by CH_2Cl_2 and acenaphthene carbonization. For spent adsorbents with large amounts of such adsorbed organics as paraffins, drastic pyrolysis conditions had to be used. The use of tetrachloromethane as a cross-linking agent for accelerating condensation and carbonization of organics was suggested as a new method with distinct advantages.

Acknowledgement

R. L. is grateful to the Foundation for Polish Science for financial support.

References

1. V. M. Gun'ko and R. Leboda, Carbon-Silica Adsorbents in: *Encyclopedia of Surface and Colloid Science*, edited by A. T. Hubbard (Marcel Dekker, New York, 2002), pp. 864-878.
2. R. Leboda, A. Gierak, B. Charmas, and Z. Hubicki, Complex carbon-silica adsorbents: preparation, properties and some applications as model adsorbents in: *Fundamentals of Adsorption*, edited by M. D. Le Van (Kluwer Academic Publishers, Boston, 1996), pp. 497-504.
3. T. J. Bandosz, K. Putyera, J. Jagiełło, and J. A. Schwarz, Study od nanocomposites obtained by carbonization of different organic precursors within taeniolite matrices, *Clays & Clay Miner.* 44, 23-51 (1996).
4. W. Holstein, The roles of ordinary and soret diffusion in the metal-catalysed formation of filamentous carbon, *J. Catal.* 152, 42-51 (1995).

5. M. E. Fernandez, J. A. Ascencio, D. Mendoza-Anaya, V. Rodriguez Lugo, and M. Jose-Yacaman, Experimental and theoretical studies of palygorskite clays, *J. Mater. Sci.* 34, 5243-5255 (1999).
6. E. Galan, Properties and applications of palygorskite, sepilite clays, *Clays Minerals* 31, 443- 445 (1996).
7. H. H. Muray, Traditional and new applications for kaolin, smectite, and palygorskite: a new overwiew, *Appl. Clay Sci.* 17, 207-221 (2000).
8. R. Leboda, B. Charmas, J. Skubiszewska-Zięba, S. Chodorowski, P. Oleszczuk, V. M. Gun'ko, and V. A. Pokrovskiy, Carbon-mineral adsorbents prepared by pyrolysis of waste materials in the presence of tetrachloromethane vapours, *J. Colloid Interface Sci.* 284, 39-47 (2005).
9. S. J. Gregg and K. S. W. Sing, *Adsorption, Surface and Porosity*, 2nd ed. (Academic Press, London, 1982).
10. C. Nguyen and D. D. Do, A New method for the Characterization of Porous Materials, *Langmuir* 15, 3608-3615 (1999).
11. V. M. Gun'ko, V. V. Turov, J. Skubiszewska-Zięba, B. Charmas, and R. Leboda, Structural and Adsorptive Characteristics of Pyrocabon/Silica Gel Si-60, *Adsorption* 10, 5-18 (2004).
12. K. Kamegawa and H. Yoshida, A Method for Measuring Surface Area of carbon of Carbon-Coated Silica gel, *Bull. Chem. Soc. Japan* 63, 3683-3685 (1990).
13. M. P. McDaniel and T. D. Hottovy, Total porosity of high-pore volume silicas by liquid nitrogen, *J. Colloid Interface Sci.* 78, 31-36 (1980).
14. R. Leboda, E. Mendyk, A. Gierak, and V. A. Tertykh, Hydrothermal modification of silica gels (xerogels) 1. Effect of treatment temperature on their porous structure, *Colloids Surf. A* 105, 181-189 (1995).
15. V. M. Gun'ko, J. Skubiszewska-Zięba, R. Leboda, and V. V. Turov, Impact of thermal and hydrothermal treatments on structural characteristics of silica gel (Si-40) and pyrocarbon/silica gel adsorbents, *Colloids Surf. A* 235, 219-243 (2004).
16. V. M. Gun'ko, R. Leboda, J. Skubiszewska-Zięba, V. V. Turov, and P. Kowalczyk, Structure of silica gel Si-60 and pyrocarbon/silica gel adsorbents thermally and hydrothermally treated, *Langmuir* 17(11), 3148-3161 (2001).
17. J. Skubiszewska-Zięba, R. Leboda, O. Seledets, and V. M. Gun'ko, Effect of preparation conditions of carbon-silica adsorbents based on mesoporous silica gel Si-100 and carbonized glucose on their pore structure, *Colloids Surf. A* 231(1-3), 39-49 (2003).
18. R. Leboda, B. Charmas, S. Chodorowski, J. Skubiszewska-Zięba, and V. M. Gun'ko, Improved carbon–mineral adsorbents derived from cross-linking carbon-bearing residues in spent palygorskite, *Micropor. Mesopor. Mater.* 87, 207-216 (2006).

SYNTHESIS AND STRUCTURAL/ADSORPTION CHARACTERISTICS OF UNMODIFIED AND MODIFIED HYBRID CARBON/MINERAL ADSORBENTS

J. SKUBISZEWSKA-ZIĘBA,[a,*] R. LEBODA,[a] V. M. GUN'KO,[b] O. SELEDETS[a]

[a] *Maria Curie-Sklodowska University, Maria Curie Sklodowska Sq.3, 20-031 Lublin, Poland*

[b] *Institute of Surface Chemistry, 17 General Naumov Street, 03164 Kiev, Ukraine*

Abstract. Several series of pyrocarbon/silica adsorbents were prepared using fumed oxides of different specific surface areas, and mesoporous silica gel Si-100, as inorganic matrices. Different synthetic and natural polymers as well as glucose were used as carbon precursors. Solutions of phosphoric acid at various concentrations were utilized to prepare functionalized hybrid carbon-silica adsorbents. Nitrogen, p-nitrophenol and Cd(II) adsorption isotherms as well as AFM, XRD and XRF methods were used to estimate the structural and adsorption characteristics of the adsorbents.

Keywords: carbon-silica adsorbents, fumed silica, pyrolysis, glucose, starch, cellulose, phosphoric acid, polyvinylpyrrolidone, polystyrene, structural characteristics.

1. Introduction

Carbon/silica adsorbents with pure or functionalized carbon deposits, or functionalized silica surfaces, are of interest for many purposes, An improvement of the structural and adsorption characteristics of carbon deposits is desirable.[1] Pyrocarbon deposits formed by carbonization of low-molecular organic precursors (dichloromethane, cyclohexene, alcohols, acetylacetone, acenaphthene, *etc.*) at oxide surfaces typically possess a low inner specific

*To whom correspondence should be addressed. J. Skubiszewska-Zięba, Maria Curie-Sklodowska University, Maria Curie Sklodowska Sq.3, 20-031 Lublin, Poland; e-mail: jskubisz@hermes.umcs.lublin.pl

J.P. Blitz and V.M. Gun'ko (eds.), Surface Chemistry in Biomedical and Environmental Science, 135–144.

surface area. The outer surface of these deposits is the main contributor to the specific surface area of the pyrocarbon.[1-3] For many applications of carbon or carbon-mineral adsorbents, appropriate contributions of both nano- and mesopores are necessary.[1-4] This problem is difficult to solve in the case of hybrid carbon-mineral adsorbents, because the pyrocarbon synthesis results in deposition of practically nonporous carbon particles (similar to carbon black) onto the support. Additionally, grafted carbon particles can block a significant portion of the pores on the substrate, particularly nanopores. Porous matrices with silica or other oxides, and grafted porous carbon particles, can provide a very wide spectrum of adsorption properties for polar and nonpolar adsorbates.[5-7] The aim of this paper is to prepare hybrid adsorbents with functionalized carbon deposits possessing inner porosity, and to investigate the properties of these complex materials in gaseous (nitrogen adsorption) and liquid (p-nitrophenol and metal ion adsorption) media.

2. Experimental

2.1. PREPARATION OF ADSORBENTS

A series of fumed silicas A-150, A-200, A-300, and A-380 (Institute of Surface Chemistry, Kalush, Ukraine), OX-50 (Degussa), and silica gel Si-100 (Merck) were used as the initial oxide matrices. Starch (St), methylcellulose (MC), polyvinylpyrrolidone (PVP), polystyrene (PS), and cellulose (C), were deposited onto fumed silicas matrices and carbonised under dynamic conditions to prepare carbosil samples (Tables 1 and 2). Before carbonization polymers were deposited from aqueous solution with subsequent drying of the residue at 320-335 K (Table 1). Sample C/A_{380}-PVP_{20}-1 was prepared using a mixture of silica and PVP/ethanol stirred and heated at 313 K for 5 h in a reactor, and sample C/A_{380}-PVP_{20}-2 was prepared using silica and 25 wt.% solution of PVP in water/ethanol (1:1) treated in a ball mill for 5 h. Precursors of samples shown in Table 2 were prepared by deposition of starch (St) and cellulose (C) by mechanical mixing of weighted portions of A-200 and polymer, and saturated with water vapour in a dessicator for 48 h. Glucose was deposited onto porous silica Si-100 and carbonized under dynamic (rotary reactor, R) or static (autoclave, A) conditions to obtain carbosil samples (Table 3). Selected samples from the carbosil series were modified by phosphorus compounds using phosphoric acid solution before carbonization (Table 4).

Carbonization of grafted polymers and glucose (Tables 1-4) was performed under dynamic conditions in a flow rotary quartz reactor (with a nitrogen stream of 100 cm^3/min). The sample was heated from 293 K to 773 K for 2 h, then kept at 773 K for 5 h, and finally cooled to 293 K for 1 h. A series of carbosil samples (Table 3, labelled A) was carbonized in a steel autoclave (volume 0.3 dm^3) using the same temperatures and times as in the rotary reactor. To remove pitch substances, the carbosils were also heated in the rotary reactor in a nitrogen stream at 773 K for 2 h.

2.2. TESTING METHODS

Low-temperature (77.4 K) nitrogen adsorption-desorption isotherms were recorded using a Micromeritics ASAP 2405N adsorption analyzer. The specific surface area S_{BET} was calculated using the standard BET equation[8] at p/p_0 (where p and p_0 denote the equilibrium and saturation pressures of nitrogen, respectively) between 0.06 and 0.2. The pore volume V_p was estimated at $p/p_0 \approx$ 0.98 converting the volume of adsorbed nitrogen gas to the volume of fluid. The pore volume of nanopores and narrow mesopores V_{DA} were estimated using the Dubinin-Astakhov equation.[4] Pore size distributions (PSDs) (differential PSD $f_V(R) \sim dV_p/dR$) of the adsorbents were calculated using a modified overall adsorption equation in the form proposed by Nguyen and Do (ND method)[9] for slitshaped pores of carbons, and modified for other materials with cylindrical pores and pores as gaps between spherical particles.[10]

The specific surface area of the pyrocarbon (S_C) was determined from p-nitrophenol (PNP) adsorption (assuming preferable adsorption of PNP onto carbon deposits) from an aqueous hydrochloric acid solution using a Specord M-40 (Karl Zeiss, Jena) UV/vis spectrophotometer at 400 nm. The data analysis was described in detail elsewhere.[11]

The carbon content (C_C) in carbon-mineral samples was determined by heating in air from 293 to 1273 K using a Derivatograph C (Paulik, Paulik & Erdey, MOM, Budapest) at a heating rate of 10 K/min.

Atomic Force Microscopic images were obtained with a NanoScope III (Digital Instruments, USA) apparatus working in Tapping Mode. Powder samples were prepared by hand pressing using a glass plate.

Powder X-ray diffraction (XRD) patterns were recorded using a HZG 4A2 X-ray diffractometer with nickel-filtered Cu K_α ($\lambda = 0.15418$ nm) radiation with a 0.03 degree step, a step time of 8 s, and measurement range $2\Theta = 5\text{-}70^\circ$.

The phosphorus content was determined using a XRF (Canberra) X-ray fluorescence spectrometer with ^{109}Cd and ^{55}Fe radioactive sources.

Adsorption of Cd(II) on the silica gel and carbosils surfaces (sample concentration 0.2 wt.%) was studied with an aqueous solution of $Cd(ClO_4)_2$ at an initial Cd(II) concentration of 10^{-3} M. Adsorption experiments used ^{109}Cd radioisotope (DuPont) with a neutral electrolyte (10^{-3} M $NaClO_4$) in a Teflon cell (50 cm^3) temperature-controlled at $T = 25 \pm 0.2$ °C under a nitrogen stream.[7]

3. Results and Discussion

Data presented in Table 1 show that depending on the kind of oxide matrices (S_{BET}, V_p, structure of aggregates and agglomerates), polymer type (with polar or nonpolar groups), and deposition technique, we obtain carbon-silica materials of different porous structure and carbon content.

Table 1. Structural characteristics of initial fumed silicas and hybrid pyrocarbon/silica adsorbents prepared from liquid solutions of polymers and subsequent carbonization.[5]

Sample	C_{pol} (%w/w)	C_C (%w/w)	S_{BET} (m²/g)	V_p (cm³/g)	V_{DA} (cm³/g)	R_p (nm)
OX-50	-	-	54	0.108	0.014	4.05
C/OX$_{50}$-St$_{20}$	20	2.7	68	0.168	0.019	4.90
A-150	-	-	142	0.261	0.039	3.70
C/A$_{150}$-St$_{20}$	20	2.9	126	0.415	0.028	6.60
C/A$_{150}$-MC$_{20}$	20	2.0	128	0.528	0.024	8.30
A-300	-	-	294	0.442	0.089	3.00
C/A$_{300}$-PS$_{20}$	20	5.6	288	0.855	0.061	5.95
A-380	-	-	383	0.536	0.074	7.90
C/A$_{380}$-St$_{10}$	10	1.9	314	1.388	0.053	8.85
C/A$_{380}$-St$_{20}$	20	4.2	289	0.794	0.063	5.50
C/A$_{380}$-St$_{30}$	30	5.8	292	1.179	0.051	8.05
C/A$_{380}$-St$_{40}$	40	8.2	268	0.969	0.054	7.25
C/A$_{380}$-MC$_{20}$	20	3.4	307	1.253	0.051	8.20
C/A$_{380}$-PVP$_{20}$-1	20	2.1	300	0.785	0.068	5.25
C/A$_{380}$-PVP$_{20}$-2	20	1.0	329	0.862	0.057	5.25

St denotes starch, MC is methyl cellulose, PS is polystyrene, and PVP is polyvinylpyrrolidone. C_{pol} polymer concentration, C_C carbon deposit concentration, S_{BET} specific surface area, V_p total pores volume, V_{DA} nanopores volume, and R_p pore radius.

The same amounts of polymers grafted on different matrices lead to distinct structural changes. This is clearly seen for samples C/OX$_{50}$-St$_{20}$, C/A$_{150}$-St$_{20}$ and C/A$_{380}$-St$_{20}$ prepared from the same organic precursor (starch), under the same polymer/oxide proportion (20% w/w), but with silicas of different morphology. The porous structure of the initial oxide influences the carbon yield (Table 1, C_C). The larger the S_{BET} and V_p values of the initial silica, the greater the carbon deposit content (2.7 and 2.9% w/w for C/OX$_{50}$-St$_{20}$ and C/A$_{150}$-St$_{20}$ and 4.2% w/w for C/A$_{380}$-St$_{20}$). This observation is affirmed when samples C/A$_{150}$-MC$_{20}$ and C/A$_{380}$-MC$_{20}$ prepared on different oxide matrices (A-150 and A-380), but from the same polymer (methylcellulose), are compared. This effect is due to increased aggregation of primary particles for fumed silicas with increasing S_{BET} value. However, this trend does not hold for samples based on A-380 and PVP (C/A$_{380}$-PVP$_{20}$-1 and C/A$_{380}$-PVP$_{20}$-2). These carbosils have relatively low amounts of carbon deposits (2.1 and 1% w/w) compared to other samples with the same proportion (20% w/w) of polymer to oxide. However, deposition of the same polymer (starch) in different proportions to the oxide matrix (from 10 to 40% w/w on A-380), gives adsorbents with increasing carbon deposit content (Table 1). Applying polystyrene (PS) as a carbon precursor (sample C/A_{300}-PS_{20}) allows the synthesis of carbosil with comparable carbon content to C/A_{380}-St_{30}. However, during polymer immobilization, polystyrene adsorbs in the folded state due to the relatively weak adsorption interaction of nonpolar –CH–CH_2– chain and aromatic side groups with surface silanols (in this case aryl-aryl interactions are preferable). The adsorption of polar polymers (starch, cellulose, and PVP) forms relatively strong hydrogen bonds with SiOH groups. The FTIR spectra[12] of A-300 with adsorbed polystyrene (C_{pol} = 50% w/w) show that a significant portion of silanols remain undisturbed, compared to the sample with adsorbed PVP (17.5 wt.%) where practically all the SiOH groups are disturbed.

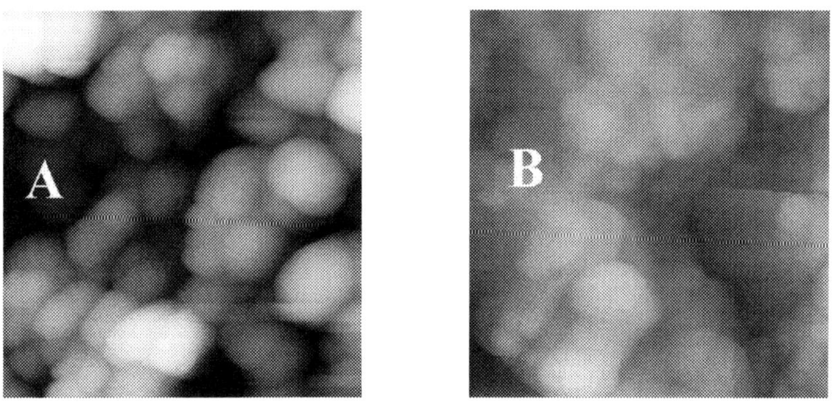

Figure 1. 2D AFM images (1 μm × 1 μm) of OX-50 (A) and carbosil C/OX_{50}-St_{20} (B).[5]

Carbon deposition on an oxide matrix typically leads to loss of porosity. But here these changes depend on the kind of organic precursors used, the method of immobilization, and carbonization method. Differences in the morphology of the initial oxide matrix and carbon-silica adsorbents are clearly seen in AFM micrographs of OX-50 and carbosil C/OX_{50}-St_{20} (Figure 1). One can see that primary particles of OX-50 form regular loose aggregates and agglomerates (Figure 1A), but deposition of the carbon phase leads to more dense packing of this material (Figure 1B).

Table 2. Structural characteristics of initial fumed silica A-200 and hybrid pyrocarbon/silica adsorbents prepared by mechanical mixing of oxide with starch (St) or cellulose (C), saturation with water vapour in dessicator for 48 hours and subsequent carbonization.[5]

Sample	C_{pol} (%w/w)	C_C (% w/w)	S_{BET} (m^2/g)	V_p (cm^3/g)	V_{DA} (cm^3/g)	R_p (nm)
A-200	-	-	232	0.483	0.057	4.15
C/A$_{200}$-St$_{35}$	35	9.6	249	0.436	0.075	3.50
C/A$_{200}$-St$_{47}$	47	16.8	253	0.385	0.082	3.05
C/A$_{200}$-St$_{64}$	64	35.8	281	0.293	0.105	1.15
C/A$_{200}$-C$_{35}$	35	10.9	239	0.452	0.067	3.80
C/A$_{200}$-C$_{47}$	47	20.5	256	0.433	0.078	3.40
C/A$_{200}$-C$_{64}$	64	30.2	284	0.385	0.097	2.70

From the data shown in Table 2, one can conclude that the method of polymer deposition (mechanical mixing) before pyrolysis allows one to prepare carbosils with a greater carbon content, compared with carbosils prepared from liquid solutions of polymers and subsequent carbonization (Table 1).

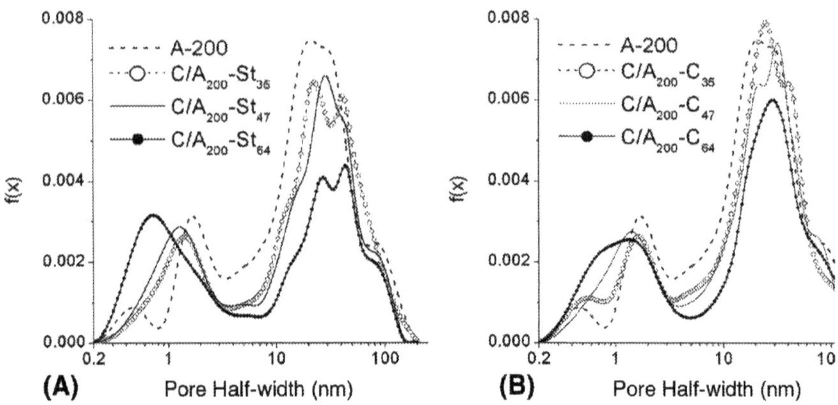

Figure 2. Pore sizes distributions for initial A-200 and carbosils prepared on the basis of starch (A) and cellulose (B) (Table 2).[5]

In these adsorbents an increase in carbon deposit content leads to reduction of the total pore volume, but an enhancement of the specific surface area (S_{BET}) and contribution of nanopores because the V_{DA} value increases. A noticeable increase in nanoporosity of these carbosils is accompanied by significant changes in the pore size distributions (PSDs) at $R_p < 2$ nm (Figure 2).

Table 3. Structural characteristics of initial silica gel Si-100 and carbosils prepared by pyrolysis of glucose (GL) under static (A) and dynamic (R) conditions.[6]

Sample	C_C (%w/w)	S_{BET} (m^2/g)	V_p (cm^3/g)	V_{DA} (cm^3/g)	R_p (nm)	a^{cs}_{mon} (mg/m^2)	S_C (m^2/g)	D (nm)
Si-100	-	349	1.225	0.023	7.0	-	-	-
C/GL_{100}-A_1	5.4	339	1.071	0.019	6.3	0.019	225	14.8
C/GL_{100}-A_2	15.6	349	0.924	0.038	5.3	0.060	260	12.8
C/GL_{100}-A_3	23.5	373	0.762	0.057	4.1	0.064	191	17.4
C/GL_{100}-R_1	4.1	339	1.089	0.011	6.4	0.016	253	13.1
C/GL_{100}-R_2	11.8	338	0.938	0.026	5.6	0.039	212	15.7
C/GL_{100}-R_3	19.0	359	0.810	0.040	4.5	0.063	224	14.0

Note: C_C is the percentage of carbon, a^{cs}_{mon} is the amount of monolayer adsorption of PNP on carbosil, S_C is the specific surface area of carbon deposit itself (from PNP adsorption), and D is the diameter of carbon particles.

Comparing data presented in Table 3, one can see that pyrocarbon grafting in silica gel pores (or blocking of the pore entrances by relatively large carbon particles at $D > 10$ nm), leads to a reduction of the V_p and R_p values. Despite this effect, the S_{BET} and V_{DA} values increase for C/GL_{100}-A_3 and C/GL_{100}-R_3, characterized by maximal C_C values for the two series. Consequently, pyrocarbon deposits affect the pore size distribution over a wide range.[6] Carbonization of glucose at the surface of silica gel Si-100 under static conditions in the autoclave (series A) provides greater amounts of pyrocarbon deposits, larger nanoporosity and S_{BET} values compared to samples prepared in the rotary reactor.

PNP poorly adsorbs onto silica surfaces.[6,11] Therefore, it can be used to estimate the structural characteristics of the carbon phase in carbosils. An increase in the C_C value leads to a significant enhancement of PNP adsorption per m^2 of carbosil surface (Figure 3A), which is greater for the A series (Table 3).

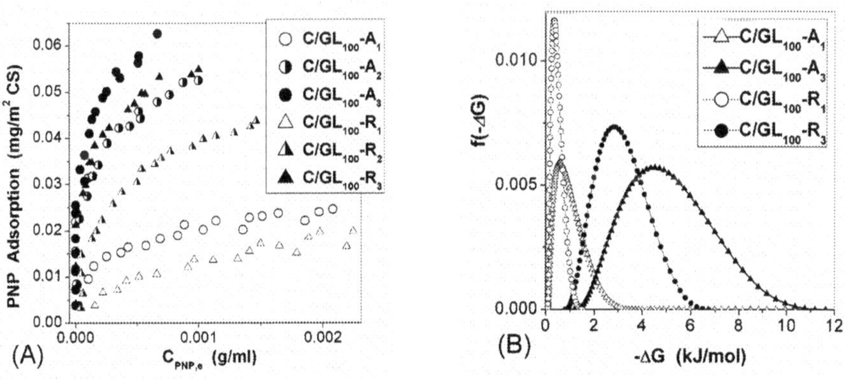

Figure 3. p-Nitrophenol (PNP) adsorption isotherms (A) and distribution functions of changes in the Gibbs free energy on the PNP adsorption (B) on carbosils prepared from glucose and Si-100.[6]

This is confirmed by the distribution functions of changes in the Gibbs free energy on PNP adsorption (Figure 3B). It is known[13] that PNP molecules can displace water molecules not only from pyrocarbon pores with hydrophobic walls, but also from oxygen-containing sites at the pore entrances. It is thus proposed that such materials as C/GL$_{100}$-A$_3$ and C/GL$_{100}$-R$_3$ could be used in the solid phase extraction (SPE) of aromatics and other organics.[14] To enhance the specific adsorption properties of carbosils for heavy metal ions a series of carbosils functionalized by phosphorus were prepared. These hybrid materials were synthesized using silica gel Si-100 and glucose (0.03 mol of glucose per 10 g of silica gel as for C/GL$_{100}$-R$_2$, Table 3) as a carbon precursor, and H_3PO_4 (as a functionalization agent) at different concentrations (5, 20 and 65% v/v).[7] Samples were carbonized in a rotary reactor in nitrogen at 500°C (5 h) at a heating rate of 2 or 4°C/min. After pyrolysis the samples were washed with distilled water to remove residual acid. Structural characteristics of these samples are shown in Table 4.

Table 4. Structural characteristics of silica gel Si-100 and carbosils with phosphorus.[7]

Sample	HR	C_{H3PO4} (%v/v)	C_C (%w/w)	C_P (%w/w)	S_{BET} (m^2/g)	V_p (cm^3/g)	V_{DA} (cm^3/g)	R_p (nm)
Si-100	-	-	-	-	349	1.225	0.023	7.0
C/GL$_{100}$-R$_2$	4	-	11.8	-	338	0.938	0.026	5.6
C/GL$_{100}$-P$_{4-65}$	4	65	24.4	0.85	544	0.491	0.057	1.8
C/GL$_{100}$-P$_{2-65}$	2	65	31.7	0.50	610	0.788	0.017	2.6
C/GL$_{100}$-P$_{4-20}$	4	20	16.5	2.06	188	0.312	0.022	3.3
C/GL$_{100}$-P$_{2-20}$	2	20	16.9	2.24	204	0.437	0.026	4.3
C/GL$_{100}$-P$_{4-5}$	4	5	16.1	0.43	314	0.836	0.006	5.3
C/GL$_{100}$-P$_{2-5}$	2	5	16.1	0.45	331	0.854	0.005	5.2

Note: HR is the heating rate from 20 to 500°C (°C/min), C_{H3PO4} concentration of H_3PO_4, C_P – concentration of phosphorus, C/GL$_{100}$-R$_2$ reference sample without P.

The use of phosphoric acid during glucose pyrolysis provides a greater yield of carbon deposits compared to C/GL$_{100}$-R$_2$ (Table 4). Additionally, higher phosphoric acid concentrations and low (2°C/min) heating rates give higher carbon contents (e.g. C/GL$_{100}$-P$_{2-65}$). It should be noted that the samples with maximal C_C values are characterized by larger specific surface areas (544 and 610 m^2/g, samples C/GL$_{100}$-P$_{4-65}$ and C/GL$_{100}$-P$_{4-65}$). However samples prepared with medium phosphoric acid concentration have a higher phosphorus content (2.06 and 2.24% w/w for C/GL$_{100}$-P$_{4-20}$ and C/GL$_{100}$-P$_{2-20}$ respectively).

To study the effect of phosphoric acid on the structure of complex adsorbents XRD measurements were performed. All of the adsorbents were amorphous with one exception (C/GL$_{100}$-P$_{4-65}$). This sample was partially crystalline because there are XRD peaks characteristic for $Si_5P_6O_{25}$.[7]

The increased phosphorus content in the carbosils samples (Table 4) enhances their acidic properties: the greater the C_P value, the lower is the curve of the surface charge density.[7] These changes with increasing C_P values influence the adsorption properties of carbosils in aqueous media. For instance, the adsorption of Cd (II) was essentially greater (between pH 6 and 8) on C/GL_{100}-$P_{4\text{-}20}$ and C/GL_{100}-$P_{2\text{-}20}$ than for other adsorbents at lower C_P values, or for C/GL_{100}-R_2 without phosphorus.[7]

4. Conclusion

Structural and adsorptive characteristics of pyrocarbon-silica samples prepared by carbonization of polymers adsorbed on fumed oxides depend on the technique of polymer grafting, substrate morphology, and the kind of oxide matrix surface. Carbonization of glucose at the surface of mesoporous silica gel Si-100 under dynamic conditions, and in the high-pressure autoclave, gives pyrocarbon/silica gel adsorbents characterized by different texture. The autoclave process provides greater amounts of pyrocarbon deposits, larger nanoporosity and total specific surface area, and a greater adsorption of p-nitrophenol onto the carbon phase. The morphology of the substrates significantly changes the carbonization of oxygen-containing carbon precursors (such as glucose, starch and cellulose), because of hydrothermal treatment of the pore walls by water eliminated as a product of pyrolysis. Addition of phosphoric acid, variation of its concentration and heating rate on glucose carbonization at the silica gel Si-100 surface, allows the preparation of a series of carbosils with different textural and adsorption characteristics appropriate for adsorption of heavy metal ions from aqueous solution.

Acknowledgement

R. L. is grateful to the Foundation for Polish Science for financial support.

References

1. V. M. Gun'ko and R. Leboda, Carbon-Silica Adsorbents in: *Encyclopedia of Surface and Colloid Science*, edited by A. T. Hubbard (Marcel Dekker, New York, 2002), pp. 864-878.
2. V. M. Gun'ko, R. Leboda, J. Skubiszewska-Zięba, V. V. Turov, and P. Kowalczyk, Structure of silica gel Si-60 and pyrocarbon/silica gel adsorbents thermally and hydrothermally treated, *Langmuir* 17(11), 3148-3161 (2001).
3. V. M. Gun'ko; R. Leboda, W. Grzegorczyk, J. Skubiszewska-Zieba, M. Marciniak, A. A. Malygin, and A. A. Malkov, *Langmuir* 16, 3227-3243 (2000).
4. V. B. Fenelonov, *Porous Carbon* (Nauka, Novosibirsk, 1995).
5. V. M. Gun'ko, J. Skubiszewska-Zięba, R. Leboda, E. F. Voronin, V. I. Zarko, S. I. Levitskaya, V. V. Brei, N. V. Guzenko, O. A. Kazakova, O. Seledets, W. Janusz, and S.

Chibowski, Pyrocarbons prepared by carbonisation of polymers adsorbed or synthesised on a surface of silica and mixed oxides, *Appl. Surf. Sci.* 227, 219-243 (2004).
6. O. Seledets, V. M. Gun'ko, J. Skubiszewska-Zięba, R. Leboda, M. Musiatowicz, P. Podkościelny, and A. Dąbrowski, Structural and energetic heterogeneities of pyrocarbon/ silica gel and its adsorptive properties. *Appl. Surf. Sci.* 240, 222-245 (2005).
7. V. M. Gun'ko, O. Seledets, J. Skubiszewska-Zięba, V. I. Zarko, R. Leboda, W. Janusz, and S. Chibowski, Phosphorus-containing pyrocarbon deposits on silica gel Si-100, *Micropor. Mesopor. Mater.* 87(2), 133-145 (2005).
8. S. J. Gregg and K. S. W. Sing, *Adsorption, Surface and Porosity*, 2nd ed. (Academic Press, London, 1982).
9. C. Nguyen and D. D. Do, A New Method for the Characterization of Porous Materials, *Langmuir* 15, 3608-3615 (1999).
10. V. M. Gun'ko, V. V. Turov, J. Skubiszewska-Zięba, B. Charmas, and R. Leboda, Structural and Adsorptive Characteristics of Pyrocabon/Silica Gel Si-60, *Adsorption* 10, 5-18 (2004).
11. K. Kamegawa and H. Yoshida, A Method for Measuring Surface Area of carbon of Carbon-Coated Silica gel, *Bull. Chem. Soc. Jpn.* 63, 3683-3685 (1990).
12. V. M. Gun'ko, V I. Zarko, E. F. Voronin V. V. Turov, I. I. Gerashchenko, E. V. Goncharuk, E. M. Pakhlov, N. V. Guzenko, R. Leboda, J. Skubiszewska-Zięba, W. Janusz, and S. Chibowski, Impact of some organics on structural and adsorptive characteristics of fumed silica in different media, *Langmuir* 18, 3 581-596 (2002).
13. V. M. Gun'ko, V. V. Turov, J. Skubiszewska-Zięba, R. Leboda, M. D. Tsapko, and D. Palijczuk, Structural characteristics of a carbon adsorbent and influence of organic solvents on interfacial water, *Appl. Surf. Sci.* 214, 178-189 (2003).
14. W. Tomaszewski, V. M. Gun'ko, J. Skubiszewska-Zięba, and R. Leboda, Structural Characteristics of Modified Activated Carbons and Adsorption of Explosives, *J. Colloid Interface Sci.* 266, 388-402 (2003).

SURFACE FUNCTIONALIZATION FOR BIOMEDICAL APPLICATIONS

H. J. MATHIEU,* X. GAO
École Polytechnique Fédérale de Lausanne (EPFL)-IMX
Station 12, CH-1015 Lausanne, Switzerland
Y. CHEVOLOT
Equipe Biotechnologie, UMR 5512 CNRS/ECL, Lyon/France
D. J. BALAZS
Swiss Federal Lab. of Materials Testing and Research, EMPA-St. Gallen/Switzerland

Abstract. This paper describes the functionalization of surfaces against non-specific protein adsorption. For surface modification photo-immobilization, γ–activation or a RF physical plasma is used which changes the chemical surface composition within the first 10 nm region. The surface chemistry is controlled by the use of Time-of-Flight Secondary Ion Mass Spectrometry and X-ray Photoelectron Spectroscopy.

Keywords: Biomaterials surface functionalization, XPS, Tof-SIMS, photo-immobilization, glycoengineering, bacterial adhesion, endotracheal tubes, PS biosensors fluorescence immunoassays, $AgNO_3$, PVC, Pseudomonas aeruginosa

1. Introduction

Biomaterials are non-viable materials used in medical devices, which are biocompatible with minimal non-specific protein adsorption. This paper describes some functionalization techniques of surfaces against non-specific protein adsorption, such as (1) photo-immobilization, (2) γ–activation or a rf plasma modification and (3) a wet-chemical treatment. The modification changes the chemical surface composition within the first 10 nm.

Surface chemical composition is controlled by Time-of-Flight Secondary Ion Mass Analysis (ToF-SIMS) and X-ray Photoelectron Spectroscopy (XPS).[1] The analysis under ultra-high vacuum (UHV) conditions allows characterization

* To whom correspondence should be addressed. H. J. Mathieu, Ecole Polytechnique Fédérale de Lausanne, Materials Science, Station 12, CH-1015 Lausanne, Switzerland, email: Hansjoerg.mathieu@EPFL.ch

J.P. Blitz and V.M. Gun'ko (eds.), Surface Chemistry in Biomedical and Environmental Science, 145–158.

with secondary ions or electrons down to femto-mol sensitivity. The information obtained is either structural information or the binding energy, respectively. The lateral resolution is several micrometers in the case of XPS compared to sub-micrometer resolution for ToF-SIMS. Description of the analytical tools is found elsewhere.[2-4] This review presents several examples of surface functionalization applying various surface modifications (pt. 2-4).

2. Photo-Immobilization for Biosensors

A. Control of bio-molecular architecture on surfaces in conjunction with the retention of biological activities of bio-molecules offers a wide range of applications in bio-sensing, cell guidance and molecular electronics.[5] Semiconductor microlithography can be extended to surface bioengineering with appropriate molecular tools enabling light addressable bio-molecule immobilization. Evidence is provided for the controlled binding of the bifunctional reagent shown in Figure 1 to material surfaces by light-dependent reactions. *N*-[*m*-(3-(trifluoromethyl) diazirine-3-yl)phenyl]-4-maleimido-butyramide (MAD) carries a diazirine function that will be lost during light activation (350 nm) leading to carbene-mediated binding to solid supports and a maleimide function with thiolated reagents.[6,7] MAD is stable and should not be degraded during the 350 nm illumination. Surface grafting was controlled by use of XPS and ToF-SIMS. Thanks to its molecular sensitivity and its surface sensitivity ToF-SIMS is the ideal complementary tool even if absolute quantitative information is not available.

The photobonding of the molecule will be characterized by comparing the influence of molecule deposition, intensive washing and illumination processes. The carbene-mediated grafting is versatile and can lead to binding to very different substrates. The grafting of the basic reagent MAD to various inert substrates is compared to identify the substrate with the highest grafting efficiency.

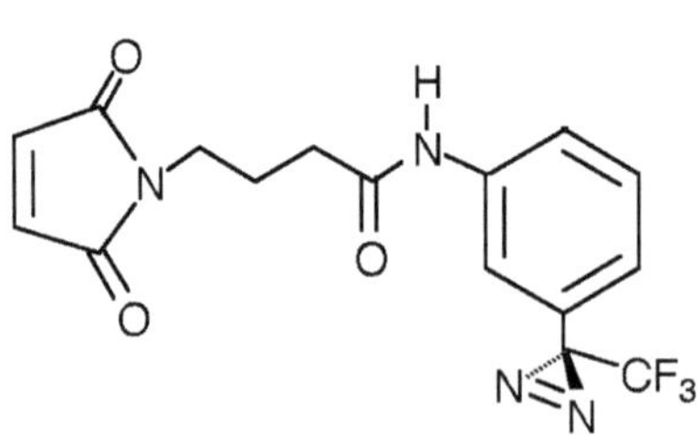

Figure 1. Chemical structure of the MAD molecule.

MAD was synthesized according to Collioud *et al.*[8] Silicon wafers with a naturally grown oxide layer (referenced as silicon), CVD deposited Si_3N_4 on silicon wafers (silicon nitride) and hot filament CVD (HFCVD) deposited diamond were tested.[9] Si_3N_4 layers were prepared by low pressure CVD on silicon wafers. Diamond layers were grown on 100 mm diameter Si/Si_3N_4 (200 nm) wafers at a deposition rate of 0.3 μm/h using a mixture of 1% CH_4 and 99% H_2.

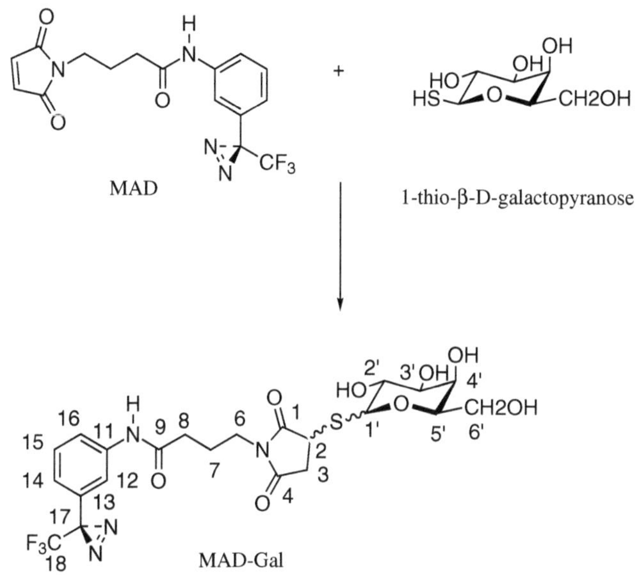

Figure 2. Reaction of 1-thio-β-D-galactopyranose with MAD yielding *N*-[*m*-[3-(trifluoromethyl) diazirin-3-yl] phenyl]-4-(-3-thio (-1-D-galactopyranosyl)-succinimidyl)butyramide (MAD-Gal) (after Y. Chevolot [5]).

Pristine samples were washed (5 min., ultrasonic treatment in hexane then in ethanol) and dried for 2 hours at room temperature under vacuum (6 mbar). Then a MAD solution (0.25 mM in ethanol, 10 μl) was deposited as a droplet released by a syringe. The samples were dried for 2 hours at room temperature under vacuum (6 mbar). For photobonding, the samples were irradiated for 20 min. using the Stratalinker 350 nm light source with an irradiance of 0.95 mW/cm^2 and washed again.

Photo-immobilization provided a versatile tool with respect to the substrate (organic and inorganic) and allows one to easily create micro-domains of bio-recognition with addressable printing, mask-assisted lithography techniques. The photo-reagents most often used for photo-immobilisation of bio-molecules are arylazides, trifluoromethyl-aryl diazirines and benzophenones.[10] Utilizing diazirine as a photoactivatable function, a reactive carbene was generated by the light activation (350 nm). A covalent carbene bond was generated. Reaction of aryl diazirine with various substrates has previously been described. The carbene may insert into bonds like C-H, C-C, C=C, N-H, O-H, S-H.[10]

MAD was derivatized with 1-thio-β-D-galactopyranose to give *N*-[*m*-[3-(trifluoromethyl) diazirin-3-yl] phenyl]-4-(-3-thio(-1-D-galactopyranosyl)-succinimidyl) butyramide (Figure 2). The parent molecule MAD provides a maleimide function, allowing the reaction with thiogalactopyranose. Both MAD and its derivative MAD-Gal contain the diazirine function that yields carbenes upon light activation at 350 nm.

After chemical analysis with XPS at the various states of samples preparation, imaging ToF-SIMS analysis was applied to identify characteristic fragments of MAD-Gal at the surface. Figure 3 illustrates the negative F^- ion distribution on a masked-assisted patterned sample after photobonding.

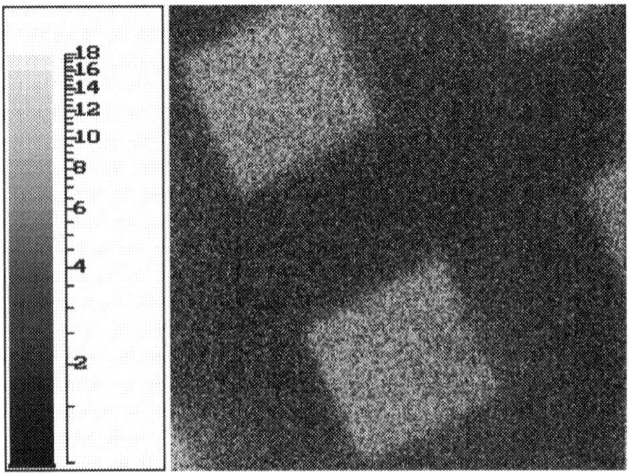

Figure 3. F^- ion distribution of MAD-Gal patterned sample after Y. Chevolot.[5]

After immobilization the grafted surfaces were probed by lection recognition and cell activity tests. Data were correlated with analytical data from XPS and ToF-SIMS. Allo A lectin binding was performed on γ-sterilized ELISA plates. The wells were modified with different concentration of MAD-Gal. Inhibition tests were performed on surfaces modified with 0.25 mM solution of MASD-Gal and asialofetuin (1 mg/ml) as an inhibitor. Figure 4 summarizes the results.

In summary, it has been shown that photo-immobilized MAD-Gal can be used as a linker and that the biological availability of the monosaccharide residue dishes was probed and ascertained chemically by XPS and ToF-SIMS.

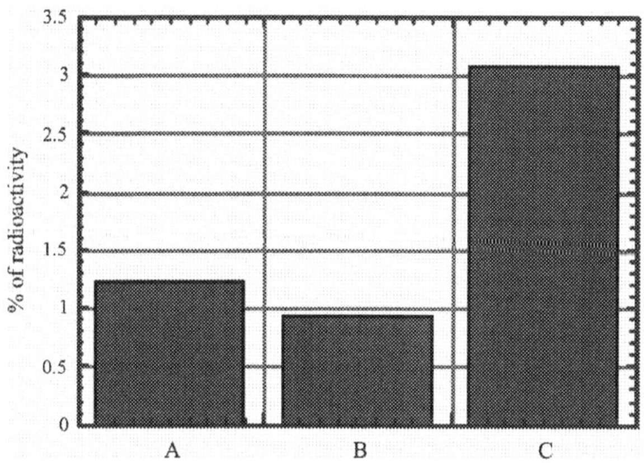

Figure 4. Binding of biotinylated Allo A lectin to aryl lactose derivatised PS surfaces. Derivatised surfaces were incubated with Allo A lectin and then extensively washed to remove physisorbed lectins. [^{35}S] streptavidin was incubated and the surfaces were again rinsed to remove excess streptavidin. (A) Pristine PS, (B) MAD-Gal desposited and washed, and (C) MAD-Gal light actived and washed. The radioactivity was measured by scintillation counting; after Chevolot.[5]

3. PS Functionalization for Biochip Fluorescence Immunoassays

Next, we report on a new diagnostic tool designed for use in applications point-of-care diagnostics, also known as near patient testing. The crucial part of this tool is the bio-sensing chip that is functionalized in order to bind bio-analytes. We will show the feasibility of preparing a chemically reactive yet specific surface for subsequent bio-immobilization. A Polystyrene (PS) surface has been activated by γ– and plasma activation rendering the biochip surface hydrophilic. The subsequent chemical modifications of the PS surface are monitored by surface analysis (X-ray Photoelectron Spectroscopy) identifying the various functional chemical groups at the PS surface.[11] Scheme 1 shows the 3 step surface modification procedure.

Scheme 1: 3-Step Surface Modification Procedure

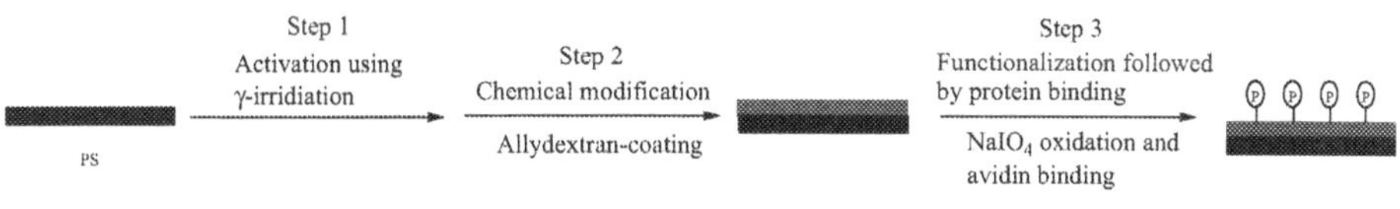

Coatings of allyldextran monolayers were carried out on PS chips activated with γ-irradiation. In a second step, sodium periodate chemistry was applied to functionalize the dextran layer, followed by the coupling with Streptavidin and/or Neutravidin. The modified PS surfaces are highly hydrophilic with low non-specific adsorption properties, and give a highly promising result for binding with biotin derivative bio-molecules. The surfaces are used in a heterogeneous immunoassay, in which Neutravidin is bound to the solid surface. In heterogeneous immunoassays, one reaction partner is immobilized on a solid PS surface, the second reaction partner being in the liquid phase. A very common reaction partner is Avidin immobilized and reacting with a biotin-modified antibody or antigen, because of the strong biotin-avidin interaction.[12,13]

XPS at two different take-off angles (0° and 70° with respect to the surface normal) provides an information depth of 8-10 nm or 3nm, respectively. After $NaIO_4$ oxidation Streptavidin or Neutravidin were subsequently bound to the surface, for further Avidin/biotin detection by the fluorescence reader.[14] XPS analysis of surfaces conjugated with Streptavidin or Neutravidin indicated the presence of 8.0% nitrogen. Figure 5 shows XPS high resolution C1s spectra. Functional groups like C-aromatic from PS substrate, C-O and C-OH from dextran moieties, and C-N from Streptavidin or Neutravidin molecules are observed. In addition, such surfaces turned out to be very hydrophilic, i.e. their contact-angles were too low (<5 degree) to be measurable.

We will now compare the γ-radiated sample to an NH_3 RF plasma activated surface. This part is focused on optimising and improving a method of surface modification of the PS biochip without the use of γ-radiation. The surface modification process is shown in Scheme 2. Firstly, a NH_3 RF plasma was used to graft amino groups onto PS surfaces. Secondly, the resulting PS surface carrying primary amino groups was chemically activated with glutaraldehyde ($OHC-CH_2-CH_2-CH_2-CHO$) and only then Neutravidin was covalently coupled. Thirdly, fluorescence immunoassay tests on the Avidin modified PS surface were performed on the basis of Neutravidin-Biotin binding.

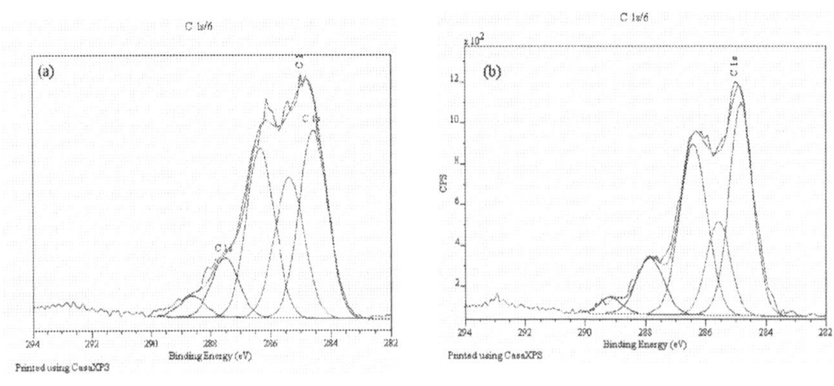

Figure 5. High resolution C1s spectra of the chip conjugated with (a) Streptavidin and (b) Neutravidin (take-off angle: 70 degree)- after X. Gao *et al.*[11]

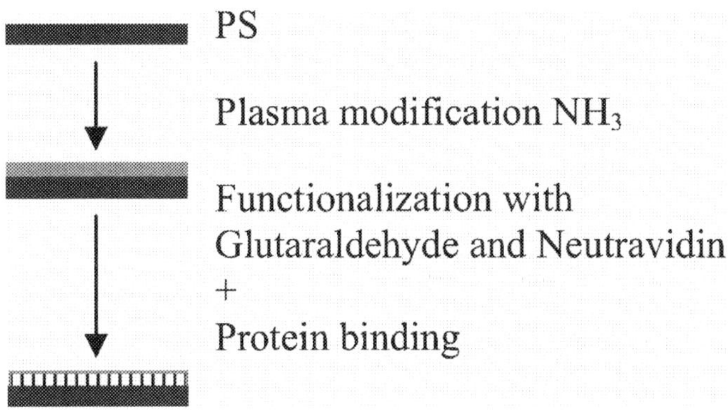

Scheme 2: 2-step NH_3 – Glutaraldehyde Neutravidin surface functionalization of PS

Aldehydes react with primary and secondary amines to form Schiff bases, which can be converted to an alkylamine linkage by using reducing reagents. Glutaraldehyde is used for protein conjugation. It reacts with amine groups to create cross-links by one of several routes. Under reducing conditions, the aldehydes on both ends of glutaraldehyde will couple with amines to form secondary amine linkages.[15] Using a pH of 6 a successful reaction between glutaraldehyde molecules having two oxygen atoms and five carbon atoms and amino groups on PS surfaces was observed.[16] As shown in Figures 6(a-b), a significant increase in C=O and C-aliphatic peaks and an increase of the C-N peak after Neutravidin coating are observed.

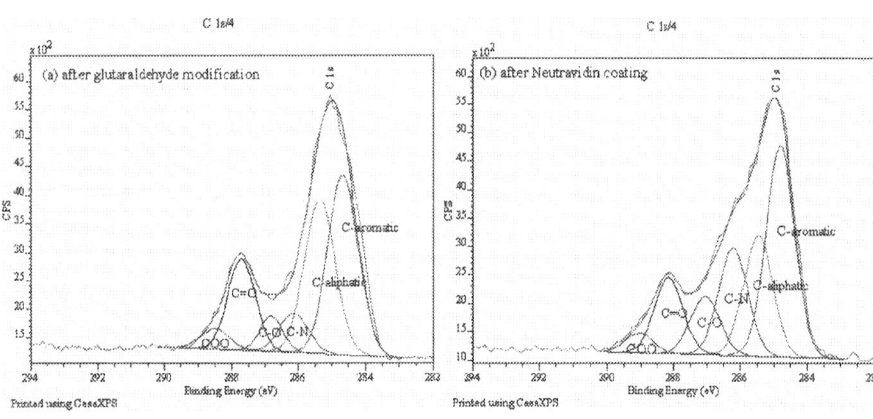

Figure 6. High energy resolution XPS spectra of C1s after (a) glutaraldehyde modification and (b) Neutravidin coating of the NH_3 RF plasma activated PS surface.[16]

PS surfaces, however, one observes different activities. In particular, the relative C-N concentration in Figure 5b is significantly lower after the plasma modification. Such a hydrophilic surface with biotin binding functionality is applied for the measurement of biological samples such as serum or other biological fluids because they reduce non-specific binding while presenting a high biotin binding activity to generate an increased signal intensity of the biosensor reported elsewhere.

4. Anti-Bacterial Treatment of Endotracheal Tubes

Pseudomonas aeruginosa is one of the most prevalent bacterial strains in a clinical environment, responsible for 30% of pneumonia cases occurring in intubated patients.[17]

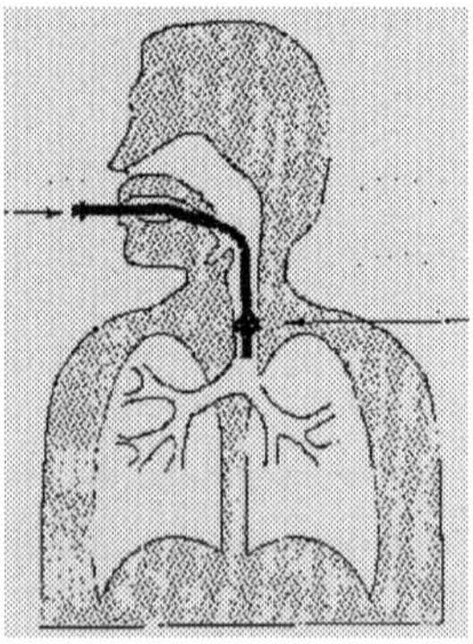

Figure 7. Endotracheal tubing (schematic).

We report here on the systematic study of plasma deposition to create a plasma fluoropolymer film on native Polyvinylchloride (PVC) based on a Ph.D. thesis of D. J. Balazs.[18] One approach is coating of Polyethylene oxide-like (PEO) of PVC serving as a platform for the preparation of non-fouling surfaces through the physisorption PEO polymer. XPS measurements will be presented to highlight the modifications induced.[18] Figure 7 illustrates schematically the endotracheal tubes, on which a biofilm is formed during extended intubation times. The substrates used for the investigation of PluronicsTM-PEO-like and Ag/PEO-like films were PVC, issued from Mallinckrodt Hi-LoTM endotracheal tubes, which were flattened to allow the microscopic counting of bacteria. Survey analysis data for the native PVC surface at 5 W power of the PEO-like films are presented in Figure 8. The silver-free PEO-like films serve as a point of comparison for the investigation of the operating parameters employed for the Ag/PEO-like film deposition.

The elemental composition data was determined from XPS data of the native PVC. The chlorine content was significantly lower than what is expected for standard PVC samples, which do not contain additives, or plasticizers. In an

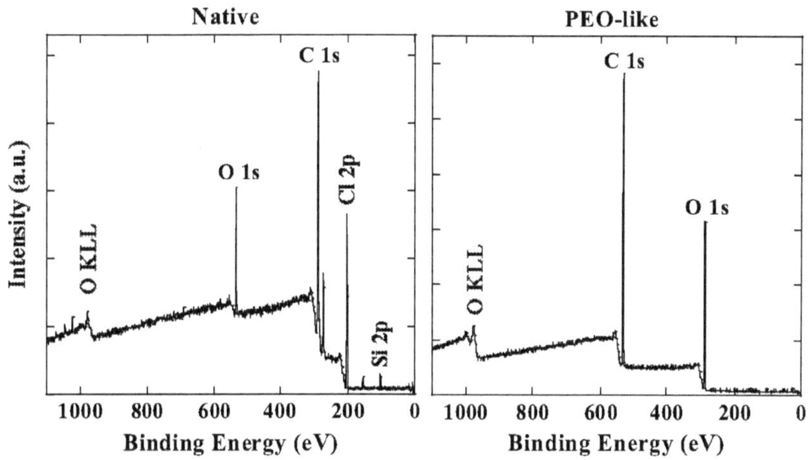

Figure 8. Wide scan spectra of native PVC (left) and a PEO-like film (right) deposited with a DEGDME flow rate of 0.4 sccm, an Ar flow rate of 5 sccm, a chamber pressure of 400 mTorr and an input power of 5 W.[18]

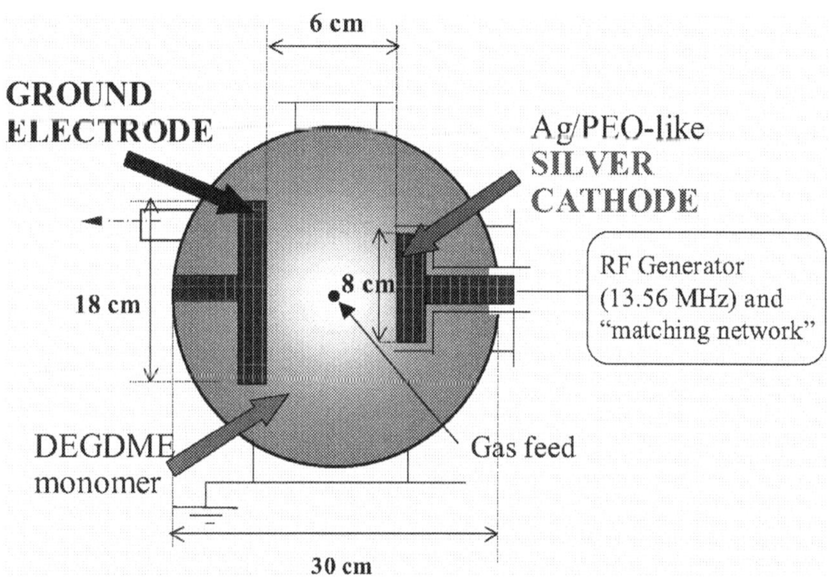

Figure 9. Schematic representation of the asymmetric plasma reactor used for the PEO-like and Ag/PEO-like coating deposition.[18]

additional step thin Ag/PEO-like coatings were deposited on PVC substrates through a mixed PE-CVD/sputtering process using an asymmetrical, parallel plate reactor with a silver cathode. Di-ethylene-glycol-dimethyl-ether (DEGDME) vapors were used as monomers and argon (Ar) as a buffer gas.[19] The plasma reactor is illustrated in Figure 9.

The effects of conditioning layers of two important blood serum proteins, albumin and fibrinogen were investigated. Protein adsorption was studied using bovine serum albumin (BSA) and fibrinogen (F) from Sigma. The samples were incubated for 3 h at 37°C in solutions of albumin (1 mg/mL) and fibrinogen (0.2 mg/mL) prepared in phosphate buffered saline (PBS, 0.01 M phosphate buffer, 0.0027 M KCl, 0.137 M NaCl, pH 7.4). After the incubation period, the samples were rinsed 3 times with PBS and analyzed by the various surface characterization techniques.

As determined by XPS analysis, the surface composition and O/C and N/C elemental ratios for 5 W and 15 W PEO-like and Ag/PEO-like films, following incubation in the protein solutions are illustrated in Table 1 for a power of 5 W. The data (Table 1) demonstrate that the PEO-like coatings are effective at reducing and/or preventing protein adsorption. The PEO-like films completely prevented the adsorption of albumin, as XPS did not detect a nitrogen signal. Moreover, the O/C ratio of the film is maintained after albumin exposure.

Table 1. Atomic percentages for the 5 W PEO-like and Ag/PEO-like films, following incubation in the protein solutions, bovine serum albumin and fibrinogen. In this figure BSA represents bovine serum albumin and F represents fibrinogen.[18]

	C 1s (%)	**O 1s (%)**	**Ag 3d (%)**	**N 1s (%)**	**O/C**	**N/C**
5 W PEO-like	71.3	28.7	-	-	0.40	-
5 W PEO-like + BSA	71.7	28.3	-	-	0.39	-
5 W PEO-like + F	70.6	28.9	-	0.4	0.41	0.01
Ag/ PEO-like	76.3	20.2	3.5	-	0.27	-
Ag/ PEO-like + BSA	73.8	18.9	1.6	5.7	0.26	0.08
Ag/ PEO-like + F	67.0	21.1	0.9	11.0	0.31	0.16

In order to determine the effects of combining the non-fouling and germicidal properties into one surface, the Ag/PEO-like coatings that were deposited on native PVC were tested in a bacterial adhesion assay, using native PVC and PEO-like coatings as reference substrates (Figure 9). Adhesion was evaluated for four different *P. aeruginosa* strains including, a wild-type PAO1, one mutant (AK44) lacking various surface structures such as Lipopolysaccharide (LPS), and two different clinical isolates (1.1.A1 and

19G12). The Ag/PEO-like films completely prevented the adhesion of the bacteria, for the four *P. aeruginosa* strains tested, yielding a 100% reduction in bacterial adhesion. On the contrary, the silver-free PEO-like films were not nearly as effective in preventing bacterial adhesion. Bacterial adhesion to the PEO-like films reached the same value as for native PVC. Although the PEO-like surfaces were determined to be non-fouling, as was demonstrated in D.J. Balazs' thesis[18], it was not sufficient to have an effect on bacterial adhesion. Therefore, it was concluded that it is the silver content of the Ag/PEO-like films that is essential to the anti-bacterial properties of these coatings.

In the following we discuss the combination of anti-adhesive and anti-colonization surfaces. Such a modification manifests itself after an oxygen glow discharge treatment, followed by a two step wet treatment in sodium hydroxide and silver nitrate solutions. The methodology employed is illustrated in Figure 10. In step 1, saponification with sodium hydroxide (NaOH) of esters, like those of the phthalic esters of the diiso-octyl phthalate additive of PVC[20] is performed, which produces a sodium phthalate salt[21] and iso-octyl alcohol.[21,22] Silver ions can be trapped when the sodium of sodium phthalate and sodium carboxylate are substituted in step 2 by silver after performing a treatment with a monovalent silver-containing solution, e.g. silver nitrate. The replacement reaction results in a surface containing silver ions, which are free to interact with bacteria.

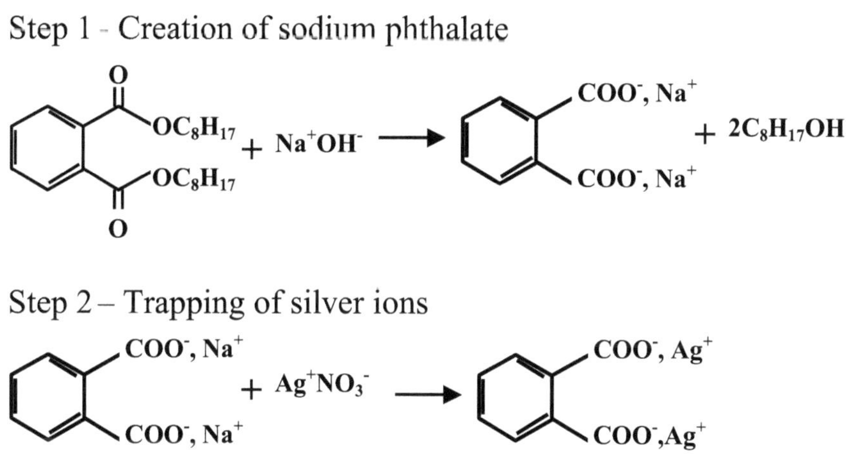

Figure 10. Schematic diagram illustrating the chemical modification by NaOH and $AgNO_3$.

Figure 10 suggests that the amount of sodium phthalate salt generated, will ultimately determine the quantity of Ag^+ on the surface. Therefore, XPS analysis and contact angle measurements were used (data not shown here) to investigate the chemical nature and surface wettability of the films following each step of the modification. Finally, bacterial adhesion studies of the $NaOH/AgNO_3$ modified surfaces were completed using the above-mentioned four different strains of *P. aeruginosa.* The anti-colonization effects of this two-step modification were also evaluated in a biofilm assay.

Biofilms, established on the surface of native and N + $NaOH/AgNO_3$, were processed and blotted over a succession of 25 pre-dried Luria agar (LA) plates. Calculations of total biofilm populations demonstrated the extremely high efficiency of this form of surface treatment in controlling bacterial biofilm development (Figure 11A). Indeed, total biofilm populations on native PVC were of 3×10^9 cfu/g of PVC at 24 h and 2×10^9 cfu/g of PVC at 72 h, whereas those on $NaOH/AgNO_3$ treated PVC were of the order 4.5×10^2 cfu/g of PVC at 24 h and 2.2×10^1 cfu/g of PVC at 72 h (CFU colony forming unit). This represents a 7-logarithmic drop in biofilm population for silver-treated surfaces at 24h and an 8-logarithmic reduction at 72 h.[18]

Figure 11B illustrates that the bacterial culture supernatants taken from the silver-treated surfaces, remained sterile throughout the 72 h test period. This indicates that silver was leaching from the surfaces and exerted an anti-bacterial effect in the surrounding medium. The diffusive leaching of silver from the substrates was not quantified in the scope of this work carried and is currently carried out at EPFL by M. Ramstedt.[23]

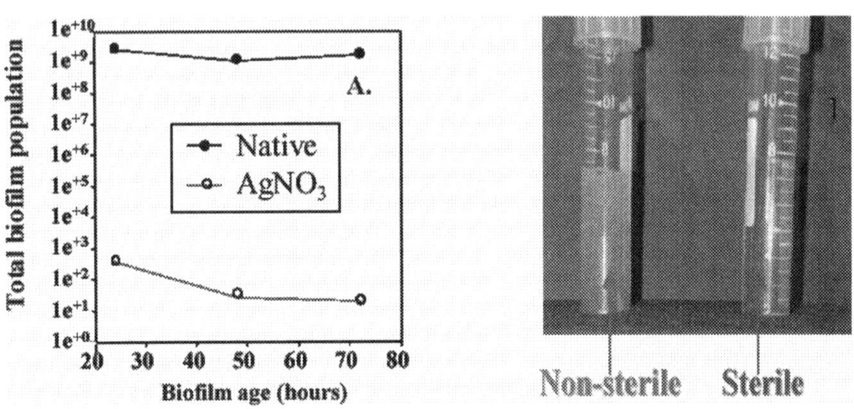

Figure 11. A-B illustrating the effect of the $NaOH/AgNO_3$ wet treatment of native PVC on survival of biofilm cells. Graph A. illustrates the evolution of the total biofilm population for both native and NaOH + $AgNO_3$ treated PVC. B illustrates the sterility of the supernatant following the $NaOH/AgNO_3$ modification (right), where the supernatant following biofilm development on the native PVC surface is illustrated on the left.[18]

5. Conclusions

Functionalization of biomedical materials controlled by high sensitivity of spectroscopical methods like ToF-SIMS and XPS allow one to build sensors and to control adhesion characteristics in a nicely controlled manner:

- - Photo-immobilization for glycoengineering of sensors can be achieved in a controlled manner.
- - Covalent Neutravidin immobilization on NH_3 rf plasma modified PS surfaces allows one to build PS biochips for fluorescence immunoassays.
- - *Pseudomonas aeruginosa* bacterial adhesion on PVC endotracheal tubes can be reduced either by rf plasma modification or wet chemical treatment using $AgNO_3$.

Acknowledgements

The authors like to thank. D. Léonard for discussion and help concerning Tof-SIMS, and N. Xanthopoulos for XPS measurements. Financial support is acknowledged from the commission for technology and innovation CTI, Bern (grants no. 5170.1 MTS and 6142.1 MTS).

References

1. H. J. Mathieu, Y. Chevolot, L. Ruiz-Taylor, and D. Léonard, Engineering and Characterization of Polymer Surfaces for Biomedical Applications, *Adv. Polymer Sci.* (Wiley) 162, 1-34 (2003).
2. H. J. Mathieu, E. Bergmann, and R. Gras, *Analyse et Technologie des Surfaces - Couches minces et tribologie, Traité des Matériaux* (Presses polytechniques et universitaires romandes, CH-1015 Lausanne, 1ère éd. Vol. 4, 2003, ISBN 2-88074-454-7).
3. H. J. Mathieu, Elemental Analysis by AES, XPS and SIMS, in: *Non-destructive Elemental Analysis*, edited by B. Z. Alfassi (Blackwell, Oxon, 2001 ISBN 0-632-05366-6), ch. 6, pp. 201-232.
4. H. J. Mathieu, Auger Electron Spectroscopy, in: *Surface Analysis - The Principal Techniques*, edited by J. C. Vickerman (John Wiley & Sons Ltd., New York, 1997) ch. 4, pp. 99-133.
5. Y. Chevolot, *Surface Photoimmobilization of Aryl Diazirine Containing Carbohydrates – Tools toward Surface Glycoengineering* (PhD Thesis EPFL, 1999), no. 1995.
6. D. Leonard, Y. Chevolot, O. Bucher, H. Sigrist, and H. J. Mathieu, Part I. *N*-[m-(3-(trifluoromethyl)diazirine-3-yl)phenyl]-4-maleimido-butyramide on Silicon, Silicon nitride and Diamond (MAD), *Surf. Interface Anal.* 26, 783-792 (1998).
7. D. Leonard, Y. Chevolot, O. Bucher, W. Haenni, H. Sigrist, and H. J. Mathieu, Part 2. *N*- [*m*-(3-(trifluoromethyl)diazirine-3-yl)phenyl]-4- (-3-thio(-1-D-galactopyrannosyl)-maleimidyl)butyramide (MAD-Gal) on Diamond, *Surf. Interface Anal.* 26, 793-799 (1998).

8. A. Collioud, J.-F. Clemence, M. Sänger, and H. Sigrist, Oriented and covalent immobilization of target molecules to solid supports: synthesis and application of a light-activatable and thiol-reactive cross-linking reagen, *Bioconjug. Chem.* 528-536 (1993).
9. C. Mueller, W. Haenni, M. Binggeli, and H. E. Hintermann, *Diamond Relat. Mater.* 2, 1211-1214 (1993).
10. H. Sigrist, A. Collioud, J.-F. Clemence, H. Gao, R. Luginbuehl, M. Saenger, and G. Sundarababu, *Opt. Eng.* 34, 2339-2348 (1995).
11. X. Gao, H. J. Mathieu, and M. Schawaller, Surface modification of polystyrene biochip for biotin labelled protein/StreptAvidin and NeutrAvidin coupling used in fluorescence assay, *Surf. Interface Analysis* 36, 1507-1512 (2004).
12. J. E. Pearson, J. W. Kane, I. Petraki-Kallioti, A. Gill, P. Vadgama, Surface plasmon resonance: A study of the effects of biotinylation on the selection of antibody, *J. Immunol. Methods* 221, 87-94 (1988).
13. J. H. Peterman, P. J. Tarcha, V. P. Chu, and J. E. Butler, The Immunochemistry of Sandwich-ELISAs. IV. The Antigen Capture Capacity of Antibody Covalently Attached to Bromoacetyl Surface-Functionalized Polystyrene, *J. Immunol. Methods* 111, 271-275 (1988).
14. G. Quapil, M. Schawaller, Vorrichtung zur Durchführung von Immunoassays, Europ. Patent WO 01/14859 A1, EP 1 079 226 A1.WO 01/14859 A1, EP 1 079 226 A1 (1999).
15. S. Y. Kim, Y-A. Jo, J. Choi, and M. J. Choi, Characterization of s-triazine antibodies and comparison of enzyme immunoassay and biotin-avidin enzyme immunoassay for the determination of s-triazine, *Microchemical J.* 68, 163-172 (2001).
16. X. Gao, M. Schawaller, and H. J. Mathieu, Covalent Neutravidin Immobilization on NH_3 RF Plasma Surface Modified Polystyrene Biochip and Its Application in Fluorescence Immunoassay, to be submitted to *Surf. Interface Anal.*
17. J. L. Vincent, D. J. Bihari, P. M. Suter, H. A. Bruining, J. White, M. H. Nicolas-Chanoin, M. Wolff, R.C. Spencer, and M. Hemmer, The prevalence of nosocomial infection units in Europe. Results… *J. Am. Med. Assoc.* 274, 639-644 (1995).
18. D. J. Balazs, *Surface Modification and Characterization of Endotracheal Tube Surfaces to Reduce Pseudomonas Aeruginosa Adhesion: Plasma treatment, Plasma Polymerization and Chemical Methods* (PhD Thesis EPFL no. 2748, 2003).
19. P. Favia, M. Vulpio, R. Marino, R. d'Agostino, R. P. Mota, and M. Catalano, Plasma-Deposition of Ag-Containing Polyethyleneoxide-Like Coatings, *Plasma. Polym.* 5, 1-14 (2000).
20. D. J. Balazs, K. Triandafillu, Y. Chevolot, B.-O Aronsson, H. Harms, P. Descouts, and H. J. Mathieu, Surface Modification of PVC Endotracheal Tubes by Oxygen Glow Discharge to Reduce Bacterial Adhesion, *Surf. Interface Anal.* 35, 301-309 (2003).
21. H. Hart, L. E. Craine, and D. Hart, *Organic Chemistry: a Short Course* (Houghton Mifflin, Co., New York, ed. Tenth, 1998).
22. S. M. Shin, H. S. Jeon, Y. H. Kim, T. Yoshioka, and A. Okuwaki, Polymer Degradation and Stability, *Polym. Degrad. Stabil.* 78, 511 (2002).
23. M. Ramstedt and H. J. Mathieu, Ways to decrease the adhesion of *Pseudomonas Aeruginosa* bacteria to the surfaces of endotracheal tubes, ECASIA'05 Book of Abstracts, Vienna, 2005, p. 76.

SURFACE MODIFICATIONS TO INFLUENCE ADHESION OF BIOLOGICAL CELLS AND ADSORPTION OF GLOBULAR PROTEINS

WILLEM NORDE
Laboratory of Physical Chemistry and Colloid Science, Wageningen University, Dreijenplein 6, 6703 HB Wageningen, and University Medical Centre Groningen, University of Groningen, Antonius Deusinglaan 1, Groningen, The Netherlands; e-mail: willem.norde@wur.nl

Abstract. It is desired to immobilize intact biological cells or protein molecules at surfaces for enzymes or cells in bioreactors and biosensors, immuno-proteins in solid-state diagnostics and proteinaceous farmacons in drug delivery systems. In other cases immobilization of cells and proteins should be avoided. A generic approach to influence the interaction between a particle (e.g. a cell or a globular protein molecule) and a sorbent surface is to manipulate both long- and short-range forces by grafting soluble polymers or oligomers onto the sorbent surface. Oligomers of ethylene oxide (EO) prevents the particle from making intimate contact with the surface so enzymes may retain their native structure and their enzymatic activity. Another interesting example is the steering effect of pre-adsorbed polymers of EO (PEO) on the orientation of subsequently adsorbing IgG molecules. Most recent research on modifying surfaces by grafting soluble polymers aims at the prevention of protein adsorption and/or adhesion of biological cells. The efficacy of grafted PEO layers to reduce protein adsorption and microbial adhesion will be illustrated for blood plasma proteins, saliva proteins and a number of bacterial and yeast cells.

Keywords: protein adsorption; cell adhesion; polymer brush; surface modification; biofouling; PEO; α-chymotrypsin; IgG

1. Introduction

Adhesion and adsorption occur widespread in nature, where it is often of vital importance. Perhaps the most well-known example is that of blood proteins and

J.P. Blitz and V.M. Gun'ko (eds.), Surface Chemistry in Biomedical and Environmental Science, 159–176.

platelets to injured blood vessels to stop bleeding and to defend the organism against wound infection. Blood proteins and platelets also adhere when they meet inanimate surfaces, but then they induce an adverse effect: a thrombus may develop. However, most cell adhesion concerns bacteria. In natural environments about 99% of bacterial mass exists at surfaces, which seems to imply that the adhered state is beneficial. Indeed, bacteria may adhere to survive. To mention a few examples: (a) in aquatic systems nutrients tend to accumulate at surfaces and this is a good reason for micro-organisms to adhere there as well; (b) bacteria may adhere to avoid transport by flow to a hostile environment, for instance from the oral cavity into the gastro-intestinal tract; (c) in the adhered state the bacteria are less susceptible to environmental attacks, such as by antibiotics.

In some technological and medical applications protein adsorption and/or cell adhesion is advantageous, but in others it is detrimental. In bioreactors it is stimulated to obtain favourable production conditions. In contrast, biofilm formation may cause contamination problems in water purification systems, in food processing equipment and on kitchen tools. Similarly, bacterial adhesion on synthetic materials used for e.g. artificial organs and prostheses, catheters, blood bags, etc., may cause severe infections. Furthermore, biofilms on heat exchangers, filters, separation membranes, and also on ship hulls oppose heat and mass transfer and increase frictional resistance. These consequences clearly result in decreased production rates and increased costs.

Thus, protein adsorption and cell adhesion occur for various reasons and in different appearances. When surfaces of living systems are involved, specific recognition mechanisms undoubtedly play crucial roles. Nevertheless, since we are dealing with a rather general phenomenon, it is likely that these specific interactions are superimposed on a generic interaction mechanism. Bioadhesion and adsorption is very complicated from a physical chemical point of view. Interfacial tensions, wetting and electrical properties of the surfaces are prominently involved.

When the (aqueous) medium contains both proteins and cells the smaller protein molecules (of which the number-concentration is usually higher than that of the cells) win the race for the surface and cover the surface before the cells arrive. Hence, as a rule, the cells adhere onto an adsorbed proteinaceous layer.

Whether protein adsorption or cell adhesion is intended or not, for each particular application the interaction of the bio-particle with the material's surface should be tuned to reach the optimal result. Protein adsorption and bacterial adhesion, further denoted as particle adhesion, may be manipulated in a controlled way if the various types of interaction determining the adsorption and adhesion processes are identified.

2. Physical-Chemical Interactions Determining Particle Adhesion

The interaction between particulate (bio)components and interfaces is influenced by many variables, including pH, ionic strength, temperature, properties of the particles and the inter-faces, the nature of the solvent and of other components present.

Whatever the mechanism is, particles adhere spontaneously if, at constant temperature and pressure, the Gibbs energy G of the system decreases. The main contributions to the Gibbs energy of particle adhesion ΔG_{ad} are from electrostatic, hydrophobic and dispersion forces,[1-5] and, furthermore, in case of protein adsorption, from rearrangements in the structure of the protein molecule.[6-9] When the sorbent surface is not smooth but "hairy", additional, mainly steric, interactions come into play.[4,10-12] Hairy surfaces are often encountered in nature as a result of adsorbed or grafted natural polymers, such as polysaccharides, that reach out in the surrounding medium with some flexibility. Interaction of particles with such hairy surfaces will be dealt with in section 3.

2.1. INTERACTION BETWEEN ELECTRICAL DOUBLE LAYERS

As depicted in Figure 1a both the particle and the sorbent surface are electrically charged. In an aqueous environment they are surrounded by counterions and co-ions, which together with the surface charge form the so-called electrical double layer. When the dispersed particle approaches the surface the electrical double layers overlap giving rise to a redistribution of charged groups, mainly low-molecular-weight ions of the electrolyte present in the solution. The contribution from electrical double layer overlap $\Delta G_{\mathrm{ad,cd}}$ may be assessed by comparing the Gibbs energies of the charge distributions before and after adhesion.[13] Obviously, electrical double layer overlap opposes deposition of the particle at the surface when the interacting species have the same charge sign and deposition is favoured in case of opposite charges.

The separation distance over which interaction between electrical charged bodies is effective depends primarily on the ionic strength of the surrounded medium, represented by the Debye length κ^{-1}.[14] In an aqueous medium of 0.01 M ionic strength, at ambient temperature, κ^{-1} is about 1 nm and it increases to 10 nm in a medium of 0.001 M ionic strength. Note that the dimensions of protein molecules are typically in the nm-range, whereas those of bacteria are up to a (few) thousand(s) of nm.

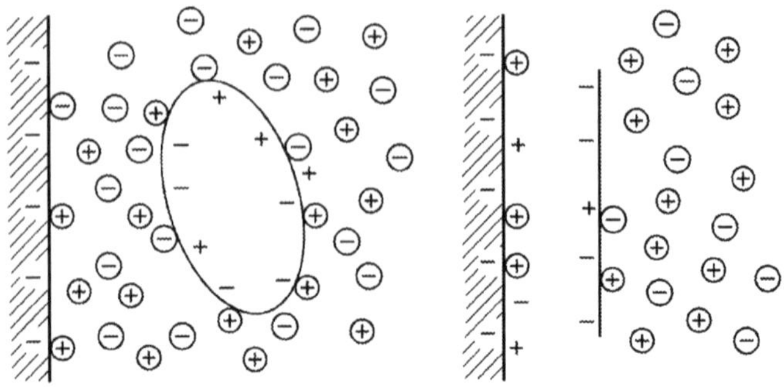

Figure 1. Schematic representation of charge distributions (a) before and (b) after particle deposition at a surface. The charged groups on the surface and the particles are indicated by + and – and the low molecular weight ions by ? and ?.

2.2. DISPERSION INTERACTION

Dispersion interaction between (macroscopic) bodies may be approximated using the Hamaker-De Boer theory.[15] In an aqueous environment the dispersion interaction between a bio-particle (globular protein, bacterial cell, …) and a (solid) material is usually attractive. The interaction increases with the size of the particle (proportionally with its radius) and it varies reciprocally with the separation distance. Typically dispersion interactions between a flat surface and a globular protein molecule of 5 nm radius operates over a distance of a few nm and for a bacterial cell of 500 nm that distance is in the range of tens of nm.[16,17]

2.3. CHANGES IN THE STATE OF HYDRATION

The composition of globular protein molecules and bacterial surfaces is usually complex comprising hydrophilic and hydrophobic groups. The surfaces of synthetic sorbent materials are in most cases less complex.

When the surfaces of the particle and the sorbent are predominantly hydrophilic it is probable that in the adhered state some hydration water is retained at their surfaces preventing intimate contact. However, if the surfaces are hydrophobic dehydration would stimulate adsorption and adhesion. The contribution from changes in hydration to ΔG_{ad} may be estimated from the partitioning of model compounds between water and a non-aqueous medium.[18] Hydration changes involve only a few layers of water molecules and hence are effective only over a separation distance of less than a nm.

2.4. REARRANGEMENTS IN THE PROTEIN STRUCTURE

The three-dimensional (3D) structure of a native protein (in aqueous solution) is only marginally thermodynamically stable and sensitive to changes in its environment. It is therefore not surprising that adsorption is often accompanied by rearrangements in the protein's 3D structure.

After adsorption one side of the protein molecule is oriented towards the sorbent surface, turned away from the aqueous solution. As a consequence, hydrophobic parts of the protein that are buried in the interior of the dissolved molecule may become exposed to the sorbent surface where they are still shielded from contact with water. Because hydrophobic interaction between apolar amino acid residues in the protein's interior support the formation of secondary structures as α-helices and β-sheets, a reduction of this interaction destabilizes such structures. Breakdown of the α-helices and/or β-sheets content is, indeed, expected to occur if peptide units released from these ordered structures can form hydrogen bonds with the sorbent surface. This is the case for polar surfaces such as oxides, e.g. silica and metal oxides, and with sorbent retaining residual water at their surfaces. Then the decrease in ordered secondary structures leads to an increased conformational entropy of the protein. This may favour the protein adsorption process considerably.[13] It may be understood that proteins having an intrinsically low structural stability are more prone to undergo adsorption-induced structural changes.

Based on the foregoing considerations the affinity between a particle and a surface may be semi-quantitatively predicted and the adsorption or adhesion tuned accordingly. Thus, in an aqueous environment, the tendency of a particle to adhere increases with increasing hydrophobicity and with increasing charge contrast between the interacting species. For the relatively large bacterial cells dispersive attraction may overcompensate electrostatic repulsion and (partial) hydrophilic dehydration, so that even hydrophilic bacteria adhere spontaneously on hydrophilic, like-charged surfaces.[12] For the much smaller proteins dispersive forces play only a minor role. Proteins are expected to adsorb at hydrophobic surfaces even when they are electrostatically repelled. With respect to their affinity for hydrophilic surfaces we have to distinguish between proteins having high and low structural stabilities (referred to as 'hard' and 'soft' proteins respectively). The hard proteins adsorb at hydrophilic surfaces only if they are electrostatically attracted, whereas structural rearrangements, i.e. decrease in ordered structure, in the soft proteins may result in a sufficiently large conformational entropy increase to make them adsorb at a hydrophilic electrostatically repelling surface.[13]

3. Tuning Particle Adhesion for Applications

As outlined in the Introduction, in various applications it is desired that proteins and/or biological cells are immobilized at surfaces, e.g. in bioreactors and biosensors, in solid-state diagnostics and in drug delivery systems. To retain biological activity the structural integrity of the immobilized component should not be perturbed too much. In other cases where surfaces are brought in contact with biological fluids particle adhesion should be avoided as much as possible, for instance to suppress fouling of biomedical materials, food processing equipment, etc.

Knowledge of the mechanisms of protein adsorption and cell adhesion provides clues to influence these processes, for example by adapting the charge and the hydrophobicity of the interacting species and by selecting environmental conditions such as pH, ionic strength and temperature. However, practice often does not allow much freedom of choice. The composition of a biological fluid is a pre-set condition and only the surface properties of the (synthetic) material can be chosen to some extent. Moreover, most biological fluids contain a mixture of proteins and/or cells. Selection of a combination of surface properties with respect to charge and hydrophobicity may result in a low affinity for the one component but a high affinity for another.

A generic approach to influence particle deposition at a surface is to manipulate the long- and short-range interactions by grafting soluble polymers or oligomers onto the surface. By way of illustration we will briefly present a few cases.

3.1. POLYMER BRUSHES TO PREVENT PARTICLE ADHESION

Water-soluble polymers densely grafted on a surface render that surface less accessible for particles approaching from the adjoining (aqueous) phase. The efficacy of such a polymer coating in reducing particle deposition depends primarily on two characteristics of the polymer layer: (a) the grafting density, i.e. the average distance between two neighbouring polymer chains at the surface, and (b) the extension into the solution, i.e. the thickness of the polymer layer.[19] As expected, deposition of indwelling particles at the sorbent surface decreases with increasing grafting density and increasing thickness of the layer. More specifically, it appears that particle repellency increases sharply when the polymer grafting density reaches a value corresponding to the onset of brush formation, that is when the polymer chains are forced to stretch out from the surface.[19] See Figure 2. In a brush conformation the separation distance d between neighbouring grafting points is less than the characteristic dimensions (e.g. end-to-end distance $< r^2 >^{0.5}$) of the coiled polymer molecule in solution.

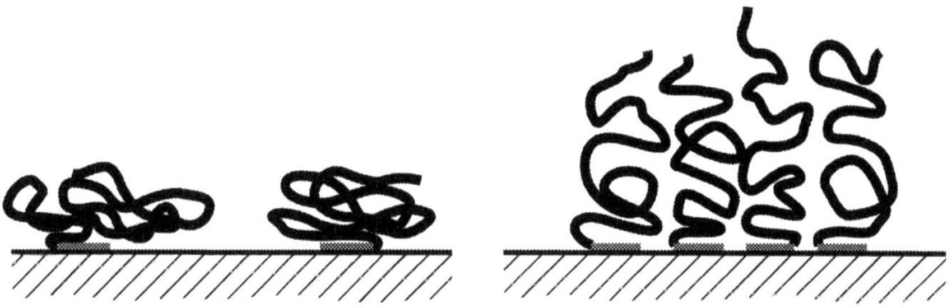

Figure 2. Conformational states of polymer chains grafted at a surface: mushroom (left) and brush (right).

The presence of a polymer brush at a surface influences particle deposition kinetically and thermodynamically. The principles are well-described in Halperin's model.[20,21] In this model the interaction between a particle (which is assumed to be inert to the polymer chains) and a brush-coated surface comprises four contributions: (1) short-range particle-surface contact (e.g. hydrogen bonding, ion pairing, hydrophobic interaction), (2) long-range dispersion interaction, (3) long-range electrical double layer interaction, and (4) repulsive osmotic and steric interaction when the particle enters the brush. The resulting interaction profile for a particle at a brushed surface is qualitatively depicted in Figure 3. The model predicts two possible modes of particle deposition: in the primary minimum at the surface and in the secondary minimum at the outer periphery of the brush. Primary minimum deposition represents equilibrium (Gibbs energy minimum), whereas in the secondary minimum the particles are kinetically trapped but they will, by an activated process, eventually reach the deeper primary minimum. Hence, to suppress particle adhesion altogether it is required to reduce the depths of the primary and secondary minima and to increase the activation energy for reaching the primary minimum. The depth of the primary minimum depends strongly on the grafting density of the brush. For high grafting densities a dense compressed layer of polymer chains will prevent the particle from making intimate contact with the underlying surface. Also a high grafting density increases the osmotic repulsion when a particle enters the brush implying a higher activation energy for reaching the primary minimum. Decreased deposition in the secondary minimum can only be achieved in a thermodynamic way by reducing the depth of that minimum. Note that this minimum is essentially determined by the thickness of the polymer brush and the magnitude of the attractive long-range interactions.

The polymer most commonly used to prepare brushes is poly (ethylene oxide) (PEO), which is alternatively called poly (ethylene glycol)(PEG). This is primarily because in an aqueous environment PEO molecules are highly mobile[22] and strongly hydrated, attaining large exclusion volumes.[23] Furthermore, PEO is biocompatible.[24] Sometimes the role of polysaccharides to prevent bioadhesion is reported.[25]

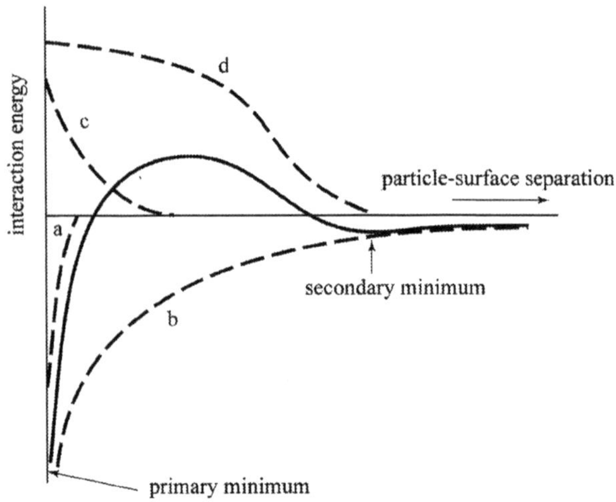

Figure 3. Particle-polymer brushed surface interaction profile. (a) particle-surface contact; (b) dispersion interaction; (c) electrical double layer interaction; osmotic polymer-brush particle interaction.

3.1.1. *Blood plasma proteins*

PEO molecules consisting of 700 ethylene oxide monomers, $(PEO)_{700}$, were in a controlled way grafted on a surface of polystyrene, PS. This allows a systematic study of the effects of grafting density, σ, and brush thickness, L, on particle-brush interaction.[10] Characteristics of the $(PEO)_{700}$ brushes are given in Table 1. The parameter $2<r^2>^{0.5}/d$ compares two times the end-to-end distance $<r^2>^{0.5}$ of the $(PEO)_{700}$ molecule in aqueous solution with the separation distance d between the grafting points of two neighbouring $(PEO)_{700}$ molecules at the surface. For the $(PEO)_{700}$-tethered surfaces investigated the values of $2<r^2>^{0.5}/d$ exceed unity by far, indicating a brush conformation. These $(PEO)_{700}$-brushed surfaces were exposed to human blood plasma (dilution 1:100 in 0.15 M PBS, pH 7.4). Figure 4 shows adsorption saturation Γ^{sat} of blood plasma components (= proteins) as a function of the reciprocal grafting density σ^{-1}. For densely packed brushes, $\sigma^{-1} \leq 8$ nm^2, plasma protein adsorption is reduced below the detectable level. At $\sigma^{-1} \geq 10$ nm^2 the protein repelling efficacy of the brush decreases steeply with decreasing grafting density. This is ascribed to penetration of the brush and adsorption in the primary minimum. For $\sigma^{-1} \geq 15$ nm^2 the adsorbed mass exceeds that at the bare PS surface. A similar trend has been observed by others.[26] It may be due to entrapment of protein in a tenuous brush having a thickness that largely exceeds the size of the protein molecules. It confirms the hypothesis that the protein resilience of dense PEO brushes results from an osmotic activation energy barrier rather than from unfavourable contacts between the proteins and the PEO. This is in accordance with force-distance experiments using similar systems.[27,28] The data in Figure 4 furthermore reveal that the grafting density range over which the brush suppresses adsorption of the blood proteins is not significantly shifted when changing the temperature from 25°C to 38°C.

Table 1. Characteristics of $(PEO)_{700}$-tethered surfaces.

σ^{-1} (nm^2)	$2\langle r^2\rangle^{0.5}/d$	L (nm)
3.24	15.1	35.1
3.92	13.8	32.9
4.41	13.0	31.6
5.70	11.4	29.1
6.17	11.0	28.1
7.07	10.3	26.9
8.20	9.5	25.6
9.23	9.0	24.7
10.00	8.6	24.0
12.00	7.9	22.8
13.20	7.5	21.8
14.00	7.3	21.5
14.90	7.1	21.2
15.80	6.9	20.6
17.20	6.5	20.2
20.00	6.1	19.3
23.00	5.7	18.3
24.50	5.5	18.0
33.00	4.8	16.1

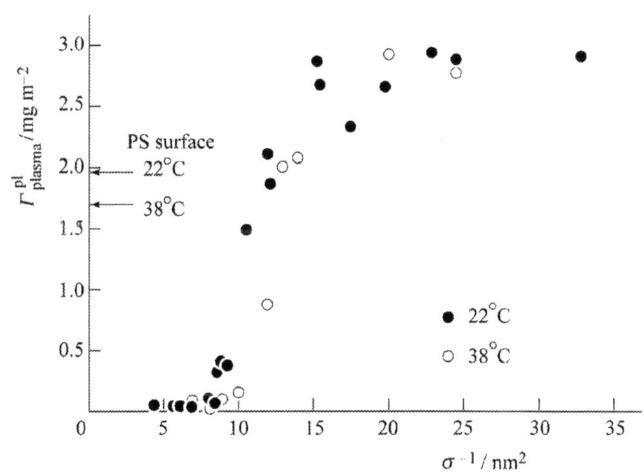

Figure 4. Adsorption of plasma components on $(PEO)_{700}$ tethered surface. Explanation is given in the text.

3.1.2. *Saliva proteins*

The interaction of saliva proteins with $(PEO)_{772}$-brushed surfaces has been investigated as a function of σ and compared with solutions containing a single model protein.[29] Some relevant properties of these proteins are listed in Table 2. The saliva protein mixture was prepared from whole human saliva. Each of these protein samples were dissolved (1 mg/mL) in 10 mM phosphate buffer pH 7. The $(PEO)_{772}$ brushed surfaces of which characteristics are given in Table 3, were exposed to the protein solutions. After 24 hours the adsorbed protein mass Γ as well as the adsorption-induced change in the zetapotential $\Delta\zeta_{ad}$ were determined. Results are presented in Figure 5. All model proteins, HSA, βLG and LSZ, adsorb at the bare negatively charged ($\zeta = -28$ mV) SiO_2/PS surface, even under electrostatically adverse conditions (as is the case for the negatively charged HSA and βLG). This is to be expected for a hydrophobic sorbent surface. The adsorption of the positively charged LSZ is also electrostatically favoured; this shows up in the relatively high adsorbed amount (1-2 mg m^{-2}). The different adsorbed amounts of HSA (1.0 mg m^{-2}) and βLG (0.5 mg m^{-2}) may be explained by the larger dimensions of HSA resulting in a thicker adsorbed layer.

Table 2. Characteristics of some model proteins, relevant for their adsorption behaviour.

	LSZ	HSA	βLG
Molar mass (g/mol)	14,500	68,000	18,300
Dimensions (nm^3)	4.5 × 3.0 × 3.0	14.0 × 3.8 × 3.8	6.5 × 3.6 × 3.6
Diffusion coefficient (m^2/s)	1.2×10^{-10}	0.70×10^{-10}	1.0×10^{-10}
Isoelectric points (pH units)	10.7	4.7	5.1

LSZ: lysozyme; HAS: human serum albumin; βLG: β-lactoglobulin.

Table 3. Characteristics of $(PEO)_{772}$-tethered surfaces.

σ^{-1} (nm^2)	$2\langle r^2\rangle^{0.5}/d$	L (nm)	ζ (mV)
12.0	15.4	24	−18
5.5	22.5	31	−13
4.0	27.0	24	−8

The presence of the brush suppresses the adsorption of these model-proteins. At $\sigma^{-1} = 12$ nm^2 adsorption of HSA is more strongly reduced than for the smaller LSZ and βLG. Furthermore, at higher grafting densities, LSZ adsorption is relatively poorly suppressed. Apparently, the positively charged LSZ molecules are pulled into the brush by the negative electric field. Thus, there seems to be a trade-off between repulsive steric and osmotic interactions on the one hand and an attractive electrostatic interaction on the other.

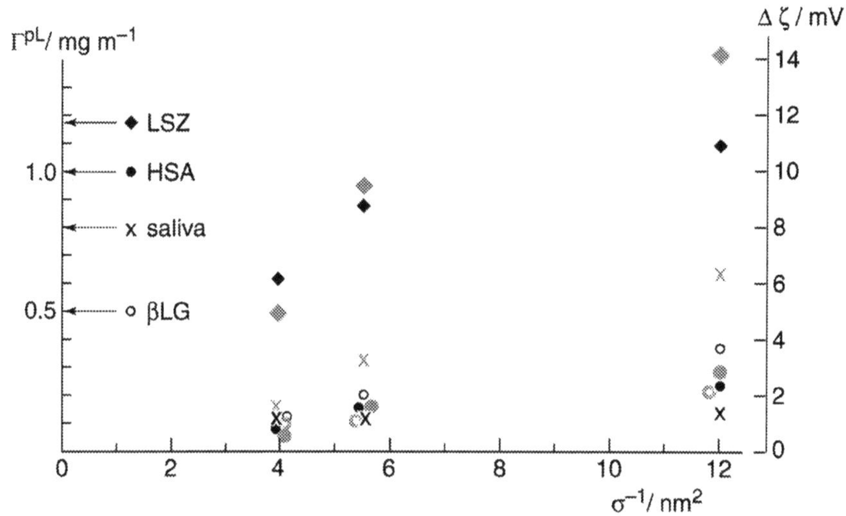

Figure 5. Interaction of saliva proteins with a $(PEO)_{772}$ brush. Comparison with some model proteins. Explanation is given in the text.

According to expectation, the positively charged LSZ causes a strong reduction of the zeta potential of the negatively charged (brushed) surfaces, whereas the effects due to the negatively HSA and βLG are much smaller.

Exposing the saliva protein mixture to the brushed surfaces reveals two remarkable features: (a) the brush, even at $\sigma^{-1} = 12$ nm^2, strongly suppresses adsorption, whereas (2) the shift in the zeta potential is still relatively large. These observations suggest that the small amounts of saliva protein deposited at the brushed surface are positively charged and that small proteins (i.e. of the size of that of LSZ) are hardly involved.

3.1.3. *Micro-organisms*

The interaction of two bacterial strains (*Staphylococcus epidermidis* and *Pseudomonas aeruginosa*) and two yeasts (*Candida albicans* and *Candida tropicana*) to PEO-brushed glass surfaces was studied at 20°C and 37°C.[12] The effective radius *r*, and the hydrophobicity (probed by the contact angle θ of a sessile drop of water) of the micro-organisms are given in Table 4 and properties of the PEO-brush at the glass surface in Table 5. The micro-organisms and the surfaces are negatively charged but in the 0.15 M PBS solution their zeta potential is rather low, not more than a few tens of milliVolts.

Adhesion data are presented in Figure 6. The adhesion of *S. epidermidis* is suppressed to essentially zero by each of the PEO brushes investigated. However, for *P. aeruginosa* adhesion is only slightly reduced at the short brush.

Table 4. Properties of micro-organisms relevant for their adhesion behavior.

	θ_{water}	r (nm)
bacteria:		
S. epidermidis	29°	500
P. aeruginosa	106°	500
yeasts:		
C. albicans	51°	1500
C. tropicalis	119°	1500

Table 5. Characteristics of $(PEO)_n$-tethered surfaces.

n	σ^{-1} (nm^2)	$2<r^2>^{0.5}/d$	L (nm)
12	0.4	22	1.6
46	1.0	10	3.5
223	5.0	11	9.3

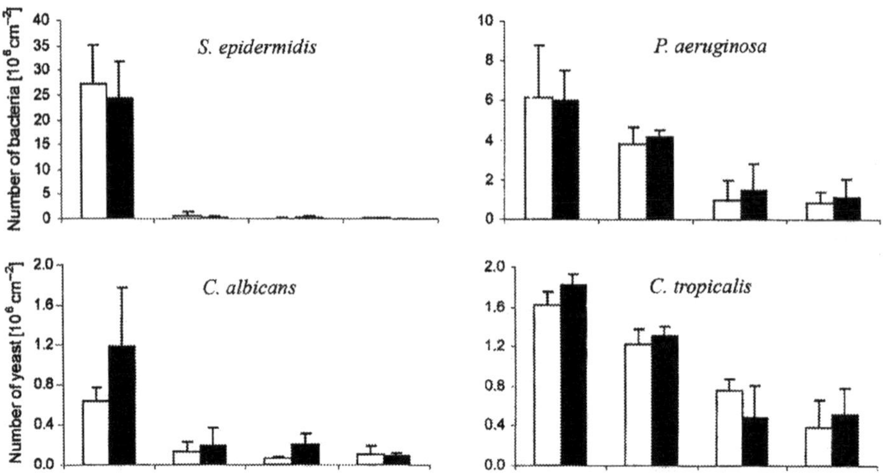

Figure 6. Effect of PEO brushes on the adhesion of micro-organisms at glass surfaces at 20°C (white bars) and 37°C (black bars). For further details refer to the text.

The more extended brushes cause a further reduction but a significant number of cells still adhere.

The adhesion of *C. albicans* is strongly lowered by the brushes, but not completely. The adhesion of *C. tropicalis* is less effectively suppressed, especially in case of the short brush.

Microscopic pictures shown in Figure 6 illustrate the influence of the brush on the adhesion of *S. epidermidis* and *C. albicans*.

Three main trends emerge from the data presented in Figure 6: (1). For both types of microbes, the bacteria and the yeasts, the adhesion of the hydrophobic cells is less strongly reduced. Apparently, hydrophobic interaction favours PEO-cell interaction. (2) The PEO brushes are less effective in repelling the larger yeast cells compared to the smaller bacteria. This may be due to the stronger dispersion interaction between the glass and the larger cells such that yeast cells are captured in a secondary minimum at the outer periphery of the brush. (3) Variation of the temperature from 20°C to 37°C does not significantly influence the adhesion pattern of both the bacteria and the yeasts.

It is furthermore noteworthy that a passing air bubble removes the bacteria and the yeasts much more easily from the brush than from the bare glass, indicating weak interaction between the cells and the brush.

3.2. SURFACE-TETHERED POLYMERS OR OLIGOMERS TO INFLUENCE THE CONFORMATION AND ORIENTATION OF ADSORBED PROTEIN MOLECULES

Polymers grafted at the surface at a density below the brush regime (see Figure 1) do not frustrate subsequent particle deposition. Still, the surface will dynamically respond to the indwelling particles. For instance, the conformation and orientation and, hence, the biological activity of adsorbed protein molecules may be manipulated. Two cases are discussed below.

3.2.1. Enzymatic activity of adsorbed α-chymotrypsin

The influence of adsorption on the biological activity of a proteolytic enzyme, α-chymotrypsin was investigated. The enzyme was adsorbed from 0.01 M phosphate buffer at pH 7.0 and at 22°C onto solid surfaces of different hydrophobicities and morphologies.[30,31]

The sorbents were hydrophobic Teflon and hydrophobic polystyrene (PS). These sorbents were supplied as negatively charged colloidal particles having smooth hydrophobic surfaces. In addition, PS particles at the surface of which oligomers (8-mers) of ethylene oxide ($(EO)_8$) were grafted at a density of one $(EO)_8$-moiety per 2.5 nm^2, were used. Because of the water-solubility of EO, these flexible $(EO)_8$ oligomers reach out from the surface into the aqueous solution causing a hairy sorbent surface. A more detailed description of these sorbent materials is described elsewhere.[30,31]

The molecular shape of α-chymotrypsin is an ellipsoid of 5.1 nm × 4.0 nm × 4.0 nm. Its molar mass is 25200 Da and its isoelectric point is 8.1.

Figure 7 shows adsorption isotherms for this protein on the different sorbents. The adsorption plateau-values at PS-$(EO)_8$, approximately 2.5 mg m^{-2}, is compatible with a complete monolayer of side-on adsorbed α-chymotrypsin molecules. Adsorption saturation at the PS and, even more so, the Teflon surfaces, is beyond monolayer coverage suggesting that on these hydrophobic surfaces the protein molecules are severely perturbed as to accommodate more protein mass in the adsorbed layers and/or adsorption of a second layer of protein molecules (possibly triggered by structurally altered molecules in the first layer). Modifying the PS-surface with $(EO)_8$ moieties lowers the adsorption affinity. In view of the dimensions of the α-chymotrypsin molecules and the $(EO)_8$ grafting density it is inferred that the surface area per adsorbed α-chymotrypsin molecule comprises about 8 $(EO)_8$-oligomers.

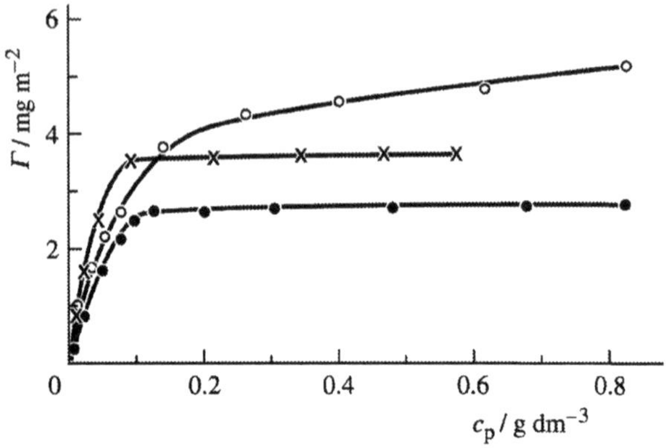

Figure 7. Adsorption of α-chymotrypsin on Teflon (o), polystyrene (x) and $(EO)_9$-tethered polystyrene (•). For details refer to the text.

The enzymatic activities of α-chymotrypsin in solution and adsorbed at the different surfaces are presented in Figure 8.

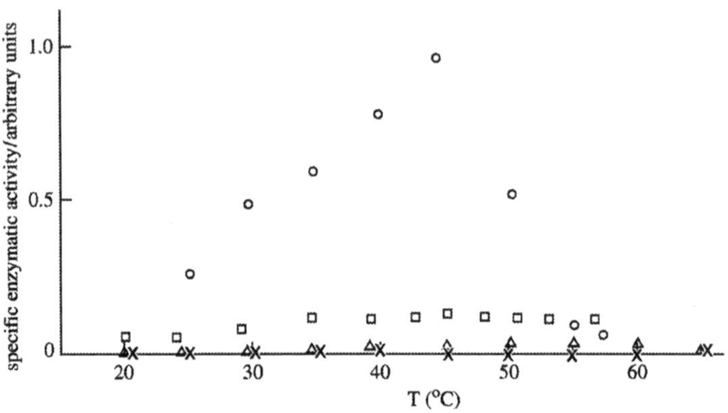

Figure 8. Temperature dependency of the specific activity of α-chymotrypsin in solution (o), adsorbed on Teflon (x), polystyrene (Δ) and $(EO)_8$ tethered (□).

In this figure the specific enzymatic activity (defined as activity per unit mass of protein) is plotted as a function of temperature. The data show that the enzyme loses activity due to adsorption. On the hydrophobic Teflon and PS surfaces the activity is completely vanished. The application of the $(EO)_8$-oligomers on the PS surface leads to retention of some of the enzymatic activity of adsorbed α-chymotrypsin. The short $(EO)_8$ chains trapped between the adsorbed protein molecules and the PS surface, frustrate intimate contact. As a result, the stress exerted by the sorbent surface on the protein molecule is diminished so that the protein's structural integrity and, hence, its biological functioning is less perturbed. It is furthermore remarkable that in the adsorbed state α-chymotrypsin is less sensitive to heat-induced inactivation. Beyond, say 55°C the specific activity of the adsorbed enzyme is higher than in solution. This is advantageous when applying the enzyme in a bioreactor that, because of various reasons (e.g. fluidity of the medium, microbial contamination, etc.) is preferably operated at elevated temperatures.

3.2.2. Immunological activity of adsorbed antibodies

Antibodies (immunoglobulin molecules) bind target molecules, antigens, with high specificity and affinity. Therefore, antibodies are widely used as sensing elements in a range of technical and medical applications. It is advantageous to immobilize the antibodies onto a (solid) surface. Then, the locally high concentration of the antibodies at the surface (e.g. a dip stick, a microtiter plate, the surface of a screen chip) exposed to an antigen containing sample, yields strong response. Immobilization is often realized by physical adsorption of the immunoglobulin molecules. Maximal immunological activity is achieved when the antibodies adsorb in an orientation where the binding sites are exposed to the test liquid and when the structural integrity of the protein is retained (see Figure 9).

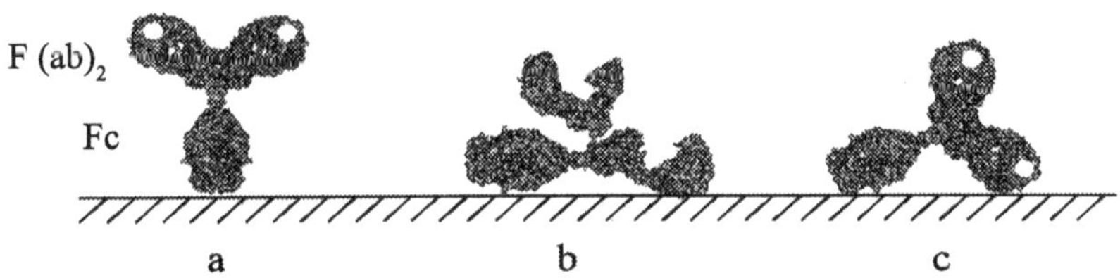

Figure 9. The effect of physical adsorption on the biological activity of IgG (Schematic). (a) Optimal orientation and conformation and, hence, maximal biological activity. (b) Reduced activity caused by conformational changes or (c) reduced accessibility of antigen bonding sites. The antigen bonding sites are represented by the white dots at the top of the F(ab) parts.

Controlling the orientation and conformation of adsorbed antibodies has been achieved by grafting soluble polymers at the surface with a density such that interstitial spaces allow sufficient room for the Fc parts of the immunoglobulin to adsorb whereas the more bulky Fab parts are excluded to reach the surface.[32] This is schematically illustrated in Figure 10a. Moreover, the antibodies thus adsorbed will be frustrated to undergo structural rearrangements by the neighbouring polymer chains (Figure 10b). The success of this 'sieving' principle is presented below for immunoglobulin G (IgG) directed against the human pregnancy hormone (hCG). The dimensions of that IgG molecule are given in Figure 11.

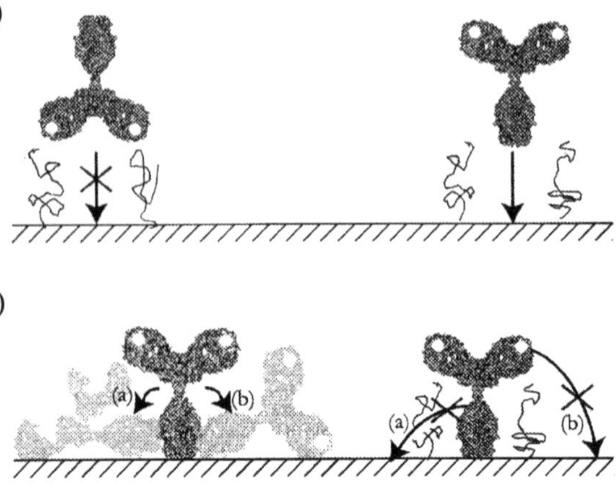

Figure 10. (a) Oriented adsorption of IgG on a pre-coated surface, leading to a biologically more active layer. (b) Prevention of (a) spreading and (b) tilting of adsorbed IgG by pre-adsorbed molecules. The antigen binding sites are represented by the white dots at the top of the F(ab) parts.

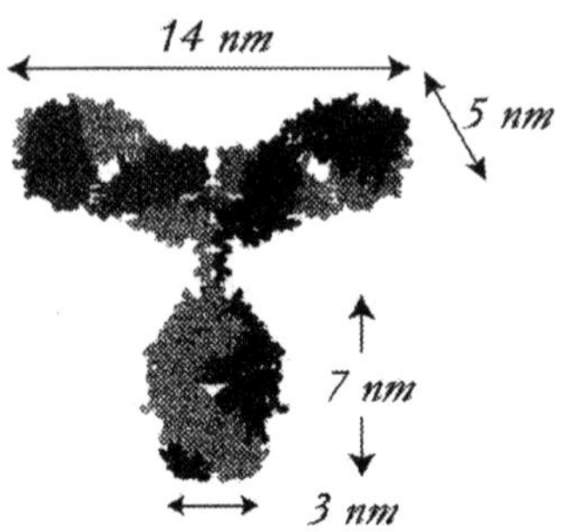

Figure 11. Dimensions of IgG molecule.

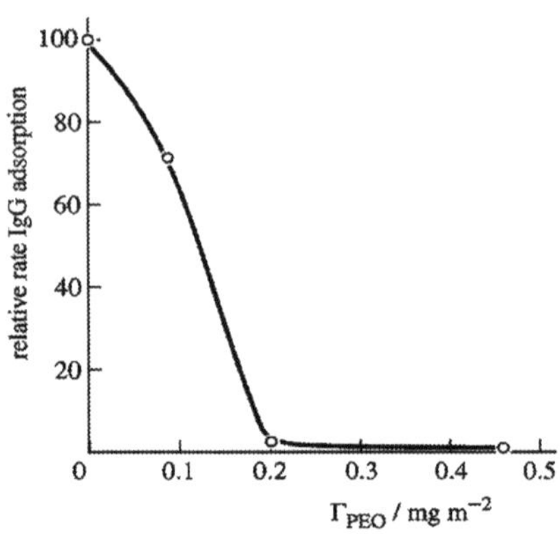

Figure 12. Relative IgG adsorption rate on pre-adsorbed $(PEO)_{24}$ surfaces.

Figure 12 shows how IgG adsorption on a silica surface is affected by pre-grafting $(PEO)_{24}$ (24 ethylene oxide monomers results in a PEO molecule of which the size matches the length of the Fc part of the IgG molecule). The effect of the $(PEO)_{24}$ coating on the hCG binding capacity may be judged from the data given in Table 6. To make a fair comparison equal amounts of adsorbed IgG are considered. It is clear that the sieving by the PEO chains forces the IgG molecules in the desired orientation resulting in almost doubling of the antigen binding capacity. (Further optimization may be possible since the theoretically maximum binding capacity is 2 hCG molecules at each IgG molecule). Comparing the binding capacities at $t = 0$ and 90 min reveals that the

PEO-moieties prevent the adsorbed IgG molecules from conformational changes that would cause a substantial reduction of their immunological activity.

Table 6. Biological activity of IgG adsorbed on surfaces pre-coated with $(PEO)_{24}$. For further explanation is referred to the text.

Adsorbed amount (mg m^2)		Antigen binding (mol hCG/mol IgG)	
PEO	IgG	$t = 0$	$t = 90$ min
0	0.70	0.66	0.47
0.09	0.70	1.11	1.06
0.13	0.70	1.23	1.17

4. Conclusions

In various biotechnological and biomedical applications control of cell adhesion and protein adsorption is required. Control may be facilitated by decorating the sorbent material with (water-)soluble polymers that yields a malleable, responsive surface. Thus, undesired conformational and orientational rearrangements of adsorbed protein molecules can be minimized resulting in a biologically active protein layer at the surface. Higher grafting densities, where the polymer layer attains a brush conformation, may yield a non-fouling surface that resists deposition of indwelling particles such as globular protein molecules and biological cells.

References

1. M. C. M. van Loosdrecht, J. Lyklema, W. Norde, and A. J. B. Zehnder, Bacterial adhesion: a physicochemical approach, *Microb. Ecol.* 17, 1-15 (1989).
2. M. C. M. van Loosdrecht, J. Lyklema, W. Norde, and A. J. B. Zehnder, Use of hydrophobic and electrostatic parameters in the interpretation of bacterial adhesion, *Aquatic Sci.* 52, 103-114 (1990).
3. H. Rijnaarts, W. Norde, J. Lyklema, and A. J. B. Zehnder, DLVO and steric contributions to bacterial depositions in media of different ionic strengths, *Colloids Surf. B: Biointerfaces*, 14, 179-195 (1999).
4. C. A. Haynes and W. Norde, Globular proteins at solid-liquid interfaces, *Colloids Surf. B: Biointerfaces*, 2, 517-566 (1994).
5. W. Norde and C. A. Haynes, in: *Interfacial Phenomena and Bioproducts,* edited by J. L. Brash and P. W. Wojciechowski (Marcel Dekker, 1996), pp. 123-144.
6. W. Norde and J. P. Favier, Structure of adsorbed and desorbed proteins, *Colloids Surf.* 64, 87-93 (1992).
7. C. E. Giacomelli and W. Norde, The adsorption-desorption cycle. Reversibility of the BSA-silica system, *J. Colloid Interface Sci.* 233, 234-240 (2001).
8. C. E. Giacomelli, M. G. E. G. Bremer, and W. Norde, ATR-FTIR study of IgG adsorbed on different silica surfaces, *J. Colloid Interface Sci.* 220, 13-23 (1999).
9. T. Arai and W. Norde, The behaviour of some model proteins at solid-liquid interfaces. 1. Adsorption from single protein solutions, *Colloids Surf.* 51, 1-15 (1990).

10. W. Norde and R. A. Gage, Interaction of bovine serum albumin and human blood plasma with PEO-tethered surfaces: influence of PEO-chain length, grafting density and temperature, *Langmuir* 20, 4162-4167 (2004).
11. W. T. E. Bosker, P. A. Iakovlev, W. Norde, and M. A. Cohen Stuart, BSA adsorption on bimodal PEO brushes, *J. Colloid Interface Sci.* 286, 496-503 (2005).
12. A. Roosjen, H. C. van der Mei, H. J. Busscher, and W. Norde, Microbial adhesion to poly(ethylene oxide) brushes: influence of polymer chain length and temperature, *Langmuir* 20, 10949-10955 (2004).
13. W. Norde, in *Biopolymers at Interfaces,* edited by M. Malmsten (Marcel Dekker, 2003), pp. 21-43.
14. J. Lyklema, in: *Fundamentals of Interface and Colloid Science*, Vol. I: *Fundamentals* (Academic Press, 1991) pp. 5.15-5.23.
15. J. Lyklema, in: *Fundamentals of Interface and Colloid Science*, Vol. I: *Fundamentals* (Academic Press, 1991) pp. 4.57-4.87.
16. W. Norde and J. Lyklema, Protein adsorption and bacterial adhesion to solid surfaces: a physicochemical approach, *Colloids Surf.* 38, 1-13 (1989).
17. W. Norde, in: *Colloids and Interfaces in Life Sciences* (Marcel Dekker, 2003) pp. 313-341.
18. G. Némethy and H. A. Scheraga, Structure of water and hydrophobic bonding in proteins, *J. Chem. Phys.* 36, 3401-3417 (1962).
19. E. P. K. Currie, W. Norde, and M. A. Cohen Stuart, Tethered polymer chains: surface chemistry and their impact on colloidal and surface properties, *Adv. Colloid Interface Sci.* 100-102, 205-265 (2003).
20. D. Leckband, S. Sheth, and A. Halperin, Grafted poly(ethylene oxide) brushes as non-fouling surface coatings, *J. Biomater. Sci. Polymer Ed.* 10 1125-1147 (1999).
21. A. Halperin, Polymer brushes that resist adsorption of model proteins: Design parameters, *Langmuir* 15, 2525-2533 (1999).
22. J. M. Harris, in: *Poly(ethylene glycol) Chemistry: Biotechnical and Biomedical Applications*, (Plenum Press, 1992) pp. 1-13.
23. A. P. Ryle, Behavior of polyethylene glycol on dialysis and gel filtration, *Nature*, 206, 1256 (1965).
24. P. A. Albertsson, in: *Partition of cell particles and macromolecules*, 3rd edition (1986), Wiley.
25. N. B. Holland, Y. Qiu, M. Ruegsegger, and R. E. Marchant, Biomimetic engineering of non-adhesive glycocalyx-like surfaces using oligosaccharide surfactant molecules, *Nature* 392, 799-801 (1998).
26. E. P. K. Currie, J. van der Gucht, O. V. Borisov, and M. A. Cohen Stuart, Stuffed brushes: theory and experiment, *Pure Appl. Chem.* 71, 1227-1241 (1999).
27. N. V. Efremova, S. R. Sheth, and D. E. Leckband, Protein-induced changes in poly (ethylene glycol) brushes: molecular weight and temperature dependence, *Langmuir* 17, 7628-7636 (2001).
28. S. R. Sheth, N. V. Efremova, and D. E. Leckband, Interactions of poly(ethylene oxide) brushes with chemically selective surfaces, *J. Phys. Chem. B.* 104, 7652-7662 (2000).
29. K. Kawasaki, M. Kambara, H. Matsumura, and W. Norde, Protein adsorption at polymer-grafted hydroxyapatite surfaces: comparison between a mixture of saliva proteins and some well-defined model proteins, *Biofouling* 19, 355-363 (2003).
30. T. Zoungrana and W. Norde, Thermal stability and enzymatic activity of α-chymotrypsin adsorbed on polystyrene surfaces, *Colloids Surf. B: Biointerfaces* 9, 157-167 (1997).
31. W. Norde and T. Zoungrana, Surface-induced changes in the structure and activity of enzymes physically immobilized at solid/liquid interfaces, *Biotechnol. Appl. Biochem.* 28, 133-143 (1998).
32. M. G. E. G. Bremer, *Immunoglobulin adsorption on modified surfaces*, Ph.D. Thesis, Wageningen University, The Netherlands (2001).

FUNDAMENTALS OF NANOSILICA APPLICATIONS FOR HUMAN PROTECTION

V. M. GUN'KO,* V. V. TUROV, A. A. CHUIKO†
Institute of Surface Chemistry, 17 General Naumov Street, 03164 Kiev, Ukraine

Abstract. Structural and adsorptive characteristics of nanosilica and its interaction with organics, biomacromolecules, and microorganisms are analyzed to elucidate fundamentals of its application as a medicinal preparation for human protection. Nanoscaled primary particles (5-50 nm), forming aggregates (50-1000 nm), and agglomerates (>1 μm), are responsible for a high specific surface area (up to 500 m^2/g), a high adsorption capacity for proteins (>300 mg/g) and microorganisms ($>10^8/g$), and fast adsorption of proteins or toxins (~50% for 1 min and ~90% for 10 min).

Keywords: nanosilicas; structural and adsorption characteristics; surface charge density; aqueous suspension; particle mobility; protein adsorption; *Proteus mirabilis*

1. Introduction

Fumed nanosilica Silics (A-300 at $S_{BET} \approx 300$ m^2/g), as a medicinal preparation with high sorption capability, possesses certain unique properties which make it possible to use it effectively for the treatment of different diseases.[1,2] These features of Silics result from at least 6 factors: (i) the chemical nature of amorphous nonporous primary nanoparticles which are passive in redox reactions, possess weak reactivity in acid-base reactions and low surface charge density at pH <8; (ii) a small primary particle size (average ~9.1 nm) which form aggregates by hydrogen bonding and partially by siloxane bonds (density of aggregates is ~30% of the true density ρ_0), and more fragile agglomerates of aggregates (~0.05-0.07ρ_0) which can be easily destroyed; (iii) a local interaction of nanoparticles with certain membrane structures of cells and microorganisms (e.g. integrated proteins); (iv) good transport properties of nanoparticles

* To whom correspondence should be addressed. V. M. Gun'ko, Institute of Surface Chemistry, 17 General Naumov Street, 03164 Kiev, Ukraine; e-mail: gun@voliacable.com

J.P. Blitz and V.M. Gun'ko (eds.), Surface Chemistry in Biomedical and Environmental Science, 177–190.

in aqueous medium because of fast diffusion; (v) adsorption of substances on the external surface of nonporous primary particles; and (vi) changes in the interfacial water structure several nanometers thick because of the chaotropic properties of the silica surface.[1-14] These unique properties of Silics suggest that its application can change treatment techniques of many diseases.[1,2] The aim of this work is to various reasons for the effectiveness of applications of nanosilica as a medicinal preparation, or as a component of complex drugs.

2. Experimental

Various nanosilicas (pilot plant at the Institute of Surface Chemistry, Kalush, Ukraine) were used as the initial materials (Table 1).

Table 1. Structural characteristics of nanosilicas.

Samples	S_{BET}, M^2/g	S_{nano}, m^2/g	S_{mes}, m^2/g	S_{mac}, m^2/g	V_p, cm^3/g	V_{nano}, cm^3/g	V_{mes}, cm^3/g	V_{mac}, cm^3/g	Δw_{mix}
A-50	52	30	20	2	0.13	0.01	0.06	0.06	-0.09
A-100	143	78	53	13	0.31	0.03	0.25	0.04	0.19
A-200	226	128	77	21	0.50	0.05	0.42	0.03	0.08
A-300	337	161	139	37	0.71	0.05	0.62	0.04	0.33
A-380	378	130	201	47	0.94	0.04	0.81	0.09	0.40
A-400	409	156	209	44	0.86	0.05	0.73	0.09	0.46
A-500	492	185	258	49	0.87	0.05	0.77	0.06	0.39

Note. The values of S_{nano} and V_{nano} ($R_p \leq 1$ nm), S_{mes} and V_{mes} ($1 < R_p \leq 25$ nm), and S_{mac} and V_{mac} ($R_p > 25$ nm) were calculated using a model of complex pores including gaps between spherical nonporous particles and cylindrical pores.

Nitrogen adsorption isotherms were recorded at 77.4 K using a Micromeritics ASAP 2405N adsorption analyzer. Calculations of the structural and adsorption characteristics are described in detail elsewhere,[6-10] as well as the adsorption of drugs, polymers, proteins, and cells.[1,2,10-16] Electrophoretic and particle size distribution (PSD) investigations were performed using a Zetasizer 3000 (Malvern Instruments) apparatus ($\lambda = 633$ nm, $\Theta = 90^\circ$).[3,4] The ^{1}H NMR spectra were recorded using a Varian 400 Mercury spectrometer or a Bruker WP-100 SY spectrometer of high resolution with layer-by-layer freezing-out of bulk and interfacial water at $200 < T < 273$ K.[5-7,10,12,13,15]

3. Results and Discussion

The specific surface area (S_{BET}) of silicas produced by burning of $SiCl_4$ in an $O_2/H_2/N_2$ flame can be varied over a large range from 50-500 m^2/g (Table 1).[6,7] Features of the flame synthesis and the nature of amorphous nanosilicas cause certain generic characteristics: (i) a roughly spherical shape of nonporous primary nanoparticles (d = 5-50 nm ~ $1/S_{BET}$); (ii) stepwise structural hierarchy of protoparticles (1-2 nm), primary particles, their aggregates (<1 μm), agglomerates (>1 μm) of aggregates, and visible flocks; (iii) low bulk density ρ_b = 0.02-0.14 g/cm^3 of powders; (iv) broadening of the primary particle size distribution with decreasing S_{BET} value; (v) surface hydrophilicity caused by silanols; (vi) lower amounts of water adsorbed under standard conditions in comparison with porous silicas; (vii) possibility of decomposition of secondary to primary particles under specific treatments; and (viii) formation of stable concentrated suspensions characterized by slow sedimentation.[3-7]

The structure of secondary particles of nanosilicas is random and loose with an empty volume $V_{em} = \rho_b^{-1} - \rho_0^{-1} > 10$ cm^3/g.[6-9] Changes in synthesis conditions allow one to vary the structure of contacts between adjacent primary particles in aggregates[6,7], which affect the properties of powders and dispersions. Different treatments of the powders and suspensions result in changes in particle-particle interactions in aggregates, that leads to variation of the adsorption capacity for various adsorbates.[1-16] From the textural characteristics (Table 1 and Figure 1) one can surmise that the structures of

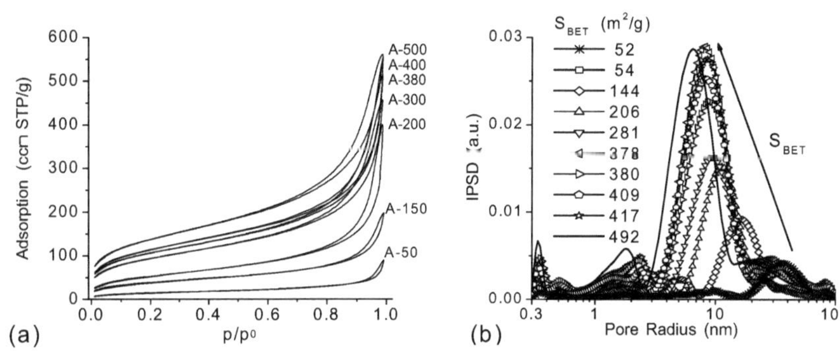

Figure 1. (a) Nitrogen adsorption-desorption isotherms and (b) incremental pore size distribution (calculated using a mixture of cylindrical pores and gaps between spherical particles) of nanosilicas.

secondary particles of fumed silicas are closely related for samples of different S_{BET} values. This is in agreement with microscopic images of nanosilicas.[2,3,9] The shape (type II) of all nitrogen isotherms in Figure 1a suggests that there is similarity in the texture of nanosilicas of different S_{BET} values. However, hysteresis loops become shorter and narrower with decreasing S_{BET} values, suggesting a reduction in the aggregation of primary particles. If the aggregate size becomes smaller, then the bulk density of secondary particles increases, and A-50 has the largest bulk density ($\rho_b \approx 0.14$ g/cm^3) among the studied nanosilicas. The incremental pore size distributions (IPSDs) (Figure 1b) demonstrates a certain similarity in the shape, but the intensity of the ~10 nm mesopores peak decreases and shifts toward larger pore radius R_p with decreasing S_{BET} value. This result suggests that compacting of primary particles in aggregates decreases with decreasing S_{BET} value. However, changes in the surface tension measured by nitrogen adsorption onto A-500 and A-50 are nearly the same ($-\Delta\sigma = 38$ and 37 mJ/m^2).[6] This suggests a similarity in certain properties of both primary and secondary particles of fumed silicas with very different S_{BET} and V_p values. There is the tendency of an increase in the V_p value with increasing S_{BET} value (Table 1). Enhanced compacting of secondary particles up to $\rho_b \approx 0.25$ g/cm^3 from ball-milling of the A-300 powder for 1-24 h[6,8] leads to a very different relationship between S_{BET} and V_p, which exhibits a sharp maximum. Heating of nanosilicas at different temperatures for different times leads to changes in the structural parameters because of desorption of water and condensation of silanols to form ≡Si-O-Si≡ bridges.[8] This treatment can affect the reactivity and adsorption capacity of nanosilicas. The amounts of water adsorbed as intact water and surface hydroxyls, and changes in the surface tension ($\Delta\sigma$) and interfacial free energy (γ_S) calculated per surface unit (or changes in the chemical potential $\Delta\mu_s$) upon water adsorption (Table 2), poorly correlate to changes in the specific surface area because of different synthesis conditions.[7] However, there is a good correlation between γ_S, $\Delta\sigma$ and C_{OH} values (Table 2) that shows the dependence of the adsorption

Table 2. Parameters of interface layer upon water adsorption onto fumed silicas synthesized under different conditions.

S_{BET} m^2/g	γ_S[a], mJ/m^2	$-\Delta\sigma$, mJ/m^2	$\Delta\mu_s$, kJ/mol	$C_{OH,378}$, wt%	$C_{OH,1173}$, wt%	C_{OH}, mg/m^2
300	304	56	11.1	1.8	1.7	0.117
267	315	63	8.3	1.6	1.4	0.112
290	245	47	7.3	1.0	0.8	0.062
144	124	26	4.1	0.4	0.4	0.055

Note. [a]γ_S is total changes in Gibbs free energy of interfacial water calculated from the ^{1}H NMR data;[6] $C_{OH,378}$ and $C_{OH,1173}$ are the amounts of water desorbed on heating at $293 < T < 378$ K and $378 < T < 1173$ K respectively, $C_{OH} = (C_{OH,378} + C_{OH,1173})/S_{BET}$.[7]

characteristics of silicas on silanol content. For instance, an increase in the C_{OH} value leads to an increase in the $\Delta\mu_s$, γ_S, and $-\Delta\sigma$ values, which causes an increase in water adsorption (Table 2, $C_{OH,378}$), as well as other compounds like proteins.[10-15] Thus changes in synthesis conditions allows the production of fumed silicas with different hydrophilicity (Table 2, C_{OH}).[6-10] It should be noted that the morphological characteristics of fumed oxides based on silica depend on the chemical composition of primary particles much less than on synthesis conditions.[6-9] This allows one to control the adsorption of compounds such as water, or polymers and proteins.[10-16] There are weakly acidic Brönsted sites of two types on a silica surface: silanols and adsorbed water molecules. Si atoms in SiOH or $Si(OH)_2$ are considered weak electron-acceptors.[1,2] The absorbance spectrum of DMAAB (Figure 2) shows that complexes of hydrogen-bonded DMAAB on silica are weakest compared to other oxides. Surface sites of nanosilica are not strong enough to disrupt cell membranes. For instance, amorphous nanosilica does not cause silicosis, unlike crystalline silica.

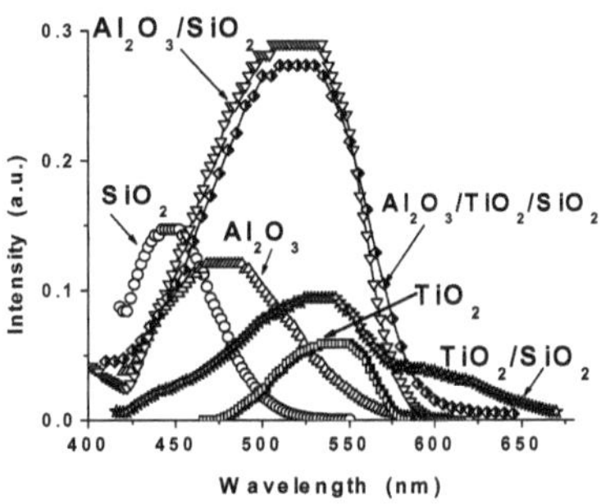

Figure 2. Optical spectra of (dimethylamino) azobenzene, DMAAB, adsorbed onto individual fumed alumina, silica A-300, and titania, and mixed oxides SiO_2/TiO_2 (37 wt% TiO_2), Al_2O_3/SiO_2 (30 wt% Al_2O_3) and $Al_2O_3/SiO_2/TiO_2$ (22, 28, and 50 wt% respecttively).

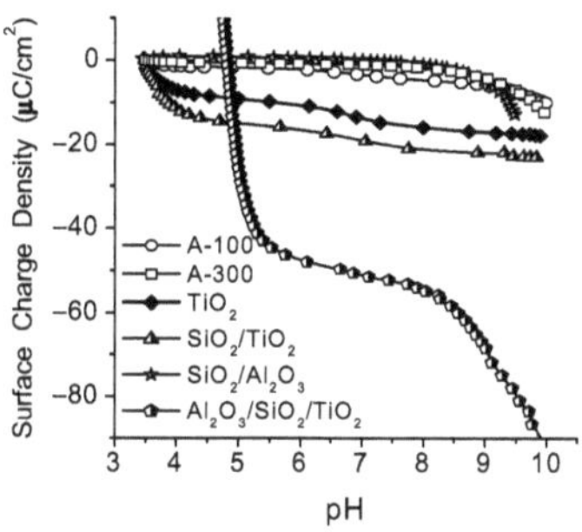

Figure 3. Surface charge density of A-100 and A-300, titania, alumina/silica (23 wt.% Al_2O_3), titania/silica (35 wt% TiO_2), and alumina/silica/titania (21, 8, and 71 wt% respectively) as a function of pH.

The presence of weakly acidic SiOH sites leads to a negative surface charge at pH >7 (Figure 3) from the transfer of protons from silanols into solution, or because of OH^- adsorption. The surface charge density of nanosilica is less than other nanooxides. Silica particles can strongly interact with positively charged structures, e.g. protonated amino groups of proteins; however, these interactions are weaker than those of mixed oxides[11] which have a higher surface charge density. Silanols can interact as either a proton-donor or an electron-donor (proton-acceptor) with protein molecules, water and other polar substances. As a result Silics sorbs proteins strongly (because of multi-centered adsorption complexes), and in large amounts (300-800 mg per gram of silica).[10-16]

Comparison of equilibrium adsorption (Figures 4 and 5) and minute protein adsorption/flocculation as a function of protein concentration, C_p, demonstrates strong but variable effects of pH and salinity.[4] The equilibrium adsorption of proteins is as large as 1 mg/m^2 (or ≈ 300 mg/g) at pH 3.5 (i.e. between pH(IEP) of silica and proteins) for bovine serum albumin (BSA) with 0.9 wt.% NaCl and gelatin without NaCl, or at pH(IEP) of protein for ovalbumin without NaCl. The lowest equilibrium adsorption (0.1-0.2 mg/m^2) is typically observed at pH = 2, which is close to pH(IEP_{SiO2}) ≈ 2.2, and without NaCl (Figure 4). It should be noted that the maximal adsorption values for BSA preparations with marked amounts of lipids and fatty acids can be up to 600 mg/g (at C_{SiO2} = 3-5 wt.%), and significantly lower (280-300 mg/g) for pure BSA (BSA(S)). This may be caused by the effect of hydrophobic admixtures on the structure of adsorbed complexes, as polar and charged proteins shield hydrophobic compounds from

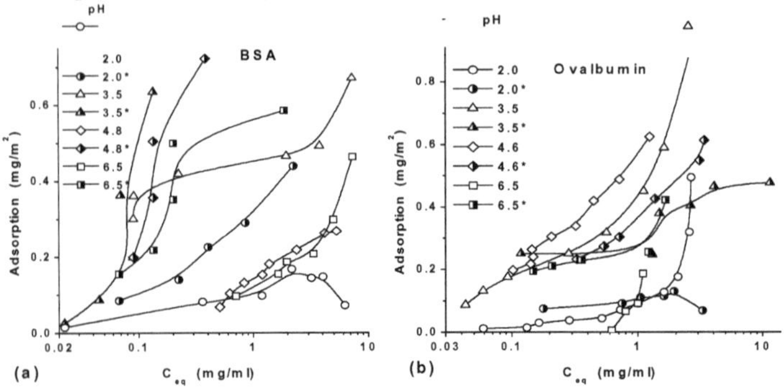

Figure 4. Adsorption isotherms of different proteins: (a) BSA and (b) ovalbumin on nanosilica A-300 (3.6 wt.%) at different pH values, which with asterisk show the systems with addition of 0.9 wt.% NaCl.

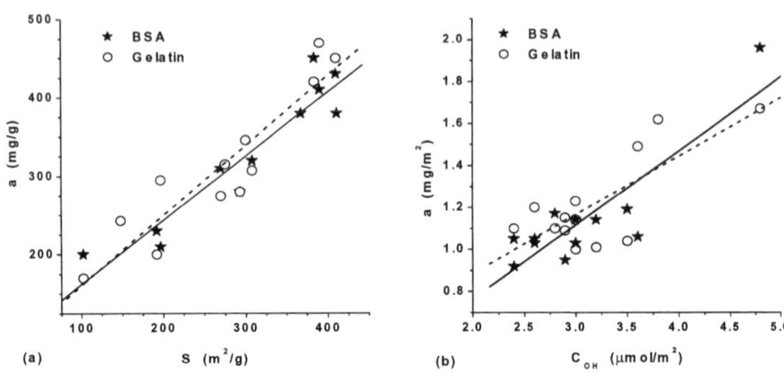

Figure 5. Protein adsorption in (a) mg/g and (b) mg/m^2 as a function of (a) S_{BET} and (b) content of silanols.

water. In dilute suspensions (C_{SiO2} = 0.1-0.2 wt.%), adsorption increases up to 1.5-4.5 g/g due to extensive flocculation of proteins with silica aggregates by formation of protein bridges between them.[4] The adsorption of proteins increases with increasing S_{BET} and C_{OH} values (Figure 5).

To elucidate some of these effects, PCS (photon correlations spectroscopy) measurements of the protein/A-300 systems were performed.[4] The D_{ef}(pH) graphs have a maximum close to pH(IEP) of the proteins (Figure 6). However, at pH far from pH(IEP) of the proteins, the PSDs of protein/A-300 are similar to those for pure silica suspensions, but D_{ef} is smaller pure silica due to decomposition of silica agglomerates and aggregates from the action of proteins. This effect is independent of protein type, since the interaction of protein molecules with silica particles can be stronger than between silica particles, especially in loose agglomerates.[4] The PSDs of A-300 in aqueous suspension at C_{SiO2} = 0.1-3.0 wt.% (without NaCl), and far from IEP_{SiO2}, are typically bimodal. This distribution corresponds to small aggregates of 20-60 nm (including from several to dozens of primary particles with the size of 5-12 nm), and larger agglomerates with sizes up to 500 nm (up to dozens of thousands of primary particles) (Figure 7).

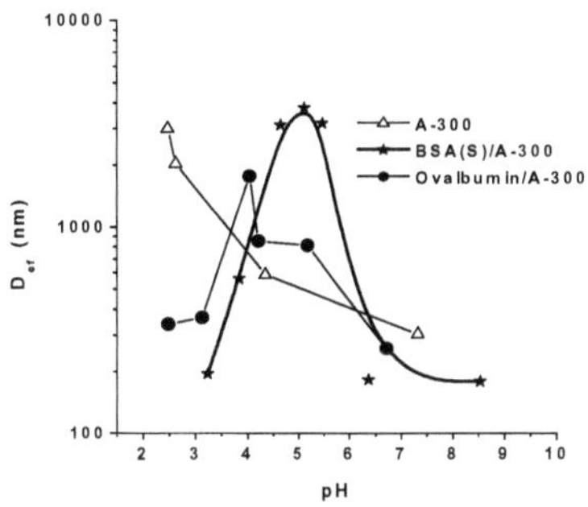

Figure 6. Effective diameter of particles as a function of pH in aqueous suspensions of nanosilica A-300 and protein/A-300.

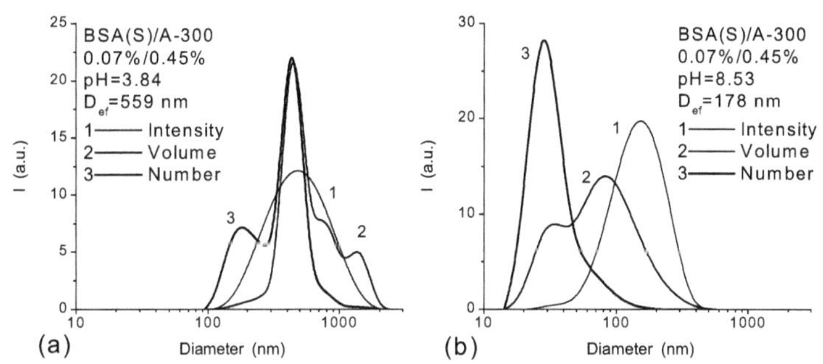

Figure 7. Particle size distributions in aqueous suspensions of A-300/BSA(Sigma) at C_{SiO2} = 0.45 wt.% and C_{BSA} = 0.07 wt.% and pH: (a) 3.84 and (b) 8.53 with respect to the light scattering intensity, particle volume and number.

Agglomerates (which are typical for untreated or weakly treated suspensions[3,4,6]), with a size d_{PCS} >1 μm, are not observed in sonicated suspensions at pH >5. Addition of 0.15 M NaCl to a 3% suspension results in the disappearance of the first PSD peak at d_{PCS} = 10-30 nm, observed without NaCl. As a whole the bimodal PSD shifts slightly toward larger d_{PCS} values. The interaction of BSA with nanosilica after pre-adsorption of (poly(ethylene glycol), PEG (~2 kDa) or poly(vinyl alcohol) (PVA), (~43 kDa), depends on time (Figure 8) because of the rearrangement of secondary particles including silica and the polymers. This effect can play an important role in the use of Silics as a component of complex preparations which include polymers. The second and perhaps most important and unique factor, which causes a high efficiency of Silics as a medicinal preparation, is the nano-dimensionality of primary particles which form stable concentrated suspensions with a major contribution from individual primary particles (Figure 9). This nano-dimensionality provides effective local interaction of silica particles with small and positively charged structures of cell membranes which have a total negative charge. These local structures can be fragments of integrated transport proteins, products of

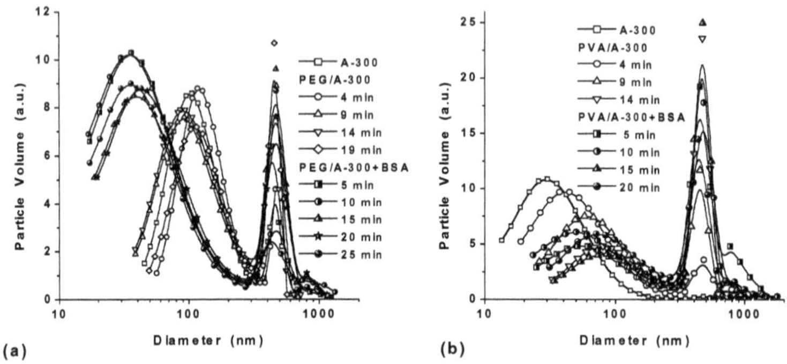

Figure 8. Particle size distributions with respect to particle volume in the aqueous suspension of A-300, (a) PEG/A-300, BSA(II)/PEG(I)/A-300 and (b) PVA/A-300, BSA(II)/PVA(I)/A-300; I and II is the first and second adsorbates respectively.

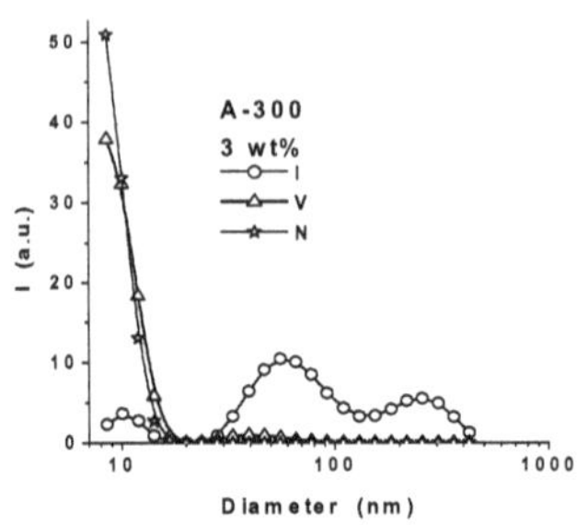

Figure 9. Particle size distributions with respect to light scattering (I), particle volume (V) and particle number (N) in an aqueous suspension A-300 (sonication time 25 min).

metabolism or cell decay, and others.[1,2] If the size of negatively charged colloidal particles considerably exceeds the size of positively charged structures (fragments) of the membranes, then negatively charged cells can repulse these particles due to long-range electrostatic forces. Therefore direct contact between large sorbent particles and cell membranes can be absent. This can decrease the effectiveness of sorbents consisting of particles of micron or millimetre size. Furthermore, low-mobility large particles of a roughly disperse sorbent are less effective as a carrier of drugs, in contrast to silica nanoparticles which can be excellent carriers for medicinal substances immobilized on their surface because they are very mobile (Figures 10 and 11). These nanoparticles can also come into close contact with cell membranes. Nanoparticles and their aggregates adsorbing macromolecules near cell membranes can change the state of interfacial water, thus altering the diffusion of molecules both to the membrane and from it. Silics particles can adsorb substances (e.g. toxins) which should be removed from the organism, and can simultaneously fulfil a carrier function for drugs transported to the cell membranes. Desorption of medicines occurs more easily from nonporous primary particles than from porous particles. Desorption can also be governed by changes in the morphology of silica, and surface modification. Furthermore, as previously discussed, Silics particles diffuse rapidly (Figure 11), and their mobility ensures the fast

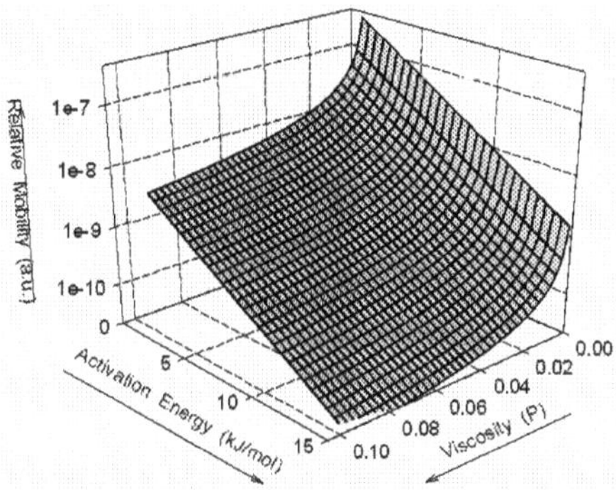

Figure 10. Dependence of mobility of particles versus viscosity and activation energy of diffusion.

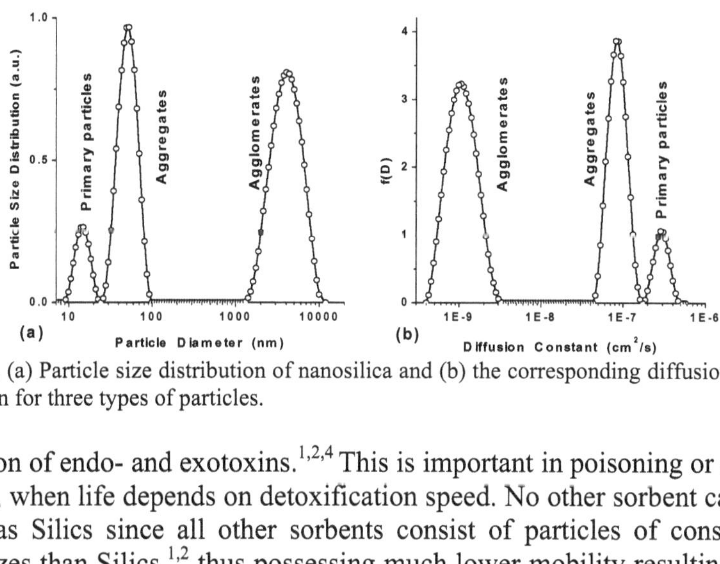

Figure 11. (a) Particle size distribution of nanosilica and (b) the corresponding diffusion constant distribution for three types of particles.

absorption of endo- and exotoxins.[1,2,4] This is important in poisoning or elapsing diseases, when life depends on detoxification speed. No other sorbent can act as rapidly as Silics since all other sorbents consist of particles of considerably larger sizes than Silics,[1,2] thus possessing much lower mobility resulting in less rapid absorption. The adsorption time of macromolecules (toxins) in pores of porous sorbents is quite long. This reduces the quantity of sorbed substance in comparison with nonporous Silics which sorbs proteins much faster and effectively (~50% for 1 min and ~90 % for 10 min) than porous adsorbents.[1-4] Biomacromolecules are rapidly sorbed only on the external surface of porous adsorbents. However, they practically do not penetrate into nanopores (<2 nm) and the adsorption in mesopores (2-50 nm) and macropores (>50 nm) requires a certain amount of time. This is crucial, since a substantial portion of harmful substances which must be moved away from the organism, wounds, and damaged organs are proteinaceous origin and of high molecular weight. This again emphasizes the uniqueness of Silics. Notice that the interaction of polymers (polyvinylpyrrolidone, PVP, PEG, and others) or proteins with nanosilica leads to the partial destruction of secondary structures (sometimes all the way to primary particles), but new secondary structures (50-1000 nm) form with oxide particles and macromolecules. This causes some reduction in the diffusion of modified Silics particles; however, this can also provide for greater biocompatibility because of the formation of an outer adsorption layer with biomacromolecules. Therefore, it is possible to expect that the rapid adsorption of proteins can reduce the deleterious effects of Silics on the membranes of erythrocytes and other cells. For example, hemolysis is observed when nanosilica interacts with erythrocytes in the absence of plasma proteins.[16] However, PVA immobilized on A-300 causes a reduction in hemolysis (Figure 12).

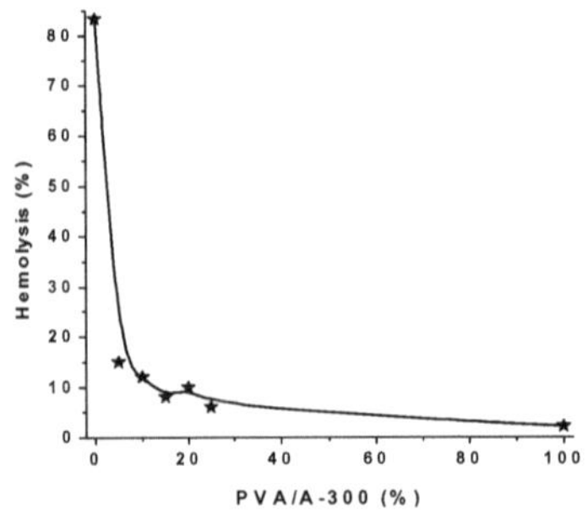

Figure 12. Influence of PVA immobilized onto nanosilica A-300 on RBC hemolysis as a function of PVA amount (last point corresponds to pure PVA solution without silica).

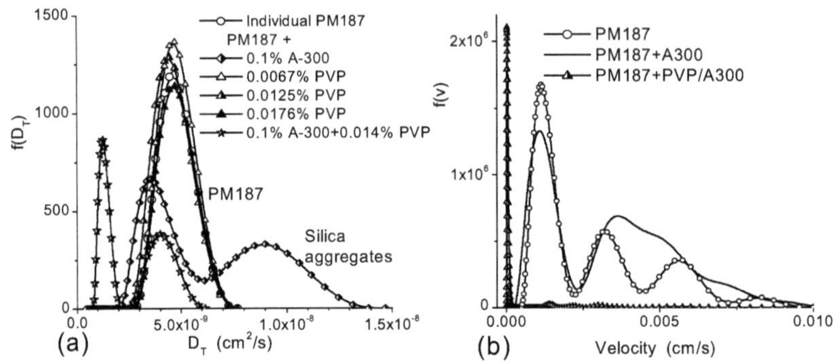

Figure 13. (a) Distribution function of the diffusion coefficient for individual *Proteus mirabilis 187* (PM 187) $10^6/cm^3$, and with addition of fumed silica A-300 (0.1 wt.%), PVP (12 kDa), and PVP/A-300; (b) velocity distribution function for individual PM 187, and with addition of A-300 (0.1 wt.%) and 0.014 wt.% PVP/0.1 wt.% A-300.

Changes in particle surface structure by chemical modification provide increased selectivity in the interaction with certain biostructures and affinity to specific substances, cells or microorganisms, e.g. *Proteus mirabilis* (Figure 13). However, Silics particles affect the mobility of these microorganisms more strongly in the presence of immobilized polymer (PVP),[4,10] which enhances the agglutination of complex formation with microorganisms. Silics can be modified to allow one to regulate the desorption rate of medicinal substances, or to increase the selectivity of the absorption of many harmful substances. Furthermore, particles of a sorbent or the carrier of medicinal substances can be magnetic, i.e., their transport will depend not only on random diffusion but also may have a local target in the organism, e.g., tumor or another local stricken area. On the other hand, partial hydrophobization of the nanosilica surface ensures reduction of the adsorption potential (surface becomes partially kosmotropic). In this case, the adsorption of small molecules decreases more substantially than macromolecules. This is an additional factor in the regulation of adsorption properties of Silics. Thus, Silics bioactivity can be caused by adsorption of toxins, metal ions, and low and high molecular weight compounds. Action may be on the mucous membranes, the medium as a whole in the gastrointestinal tract, and by structuring of water near cell membranes.

An important problem related to the stability of dilute and concentrated aqueous suspensions of nanosilica (because these suspensions can be used as medicinal preparations), dependent on particle-particle (or EDL-EDL) interactions.[3-6] Some aspects of this problem are linked to the properties of interfacial water.[5] Changes in these properties were analyzed for suspensions with A-300 and A-50 at C_{SiO2} = 1.0-20.0 wt.% sonicated for 5 min (power 500 W and frequency 22 kHz) or ball-milled (MCA) for 3 h (Figure 14). Suspensions at C_{SiO2} >8 wt.% can transform to a gel-like bound state over the course of several days (C_{SiO2} = 5-8 wt.% corresponds to the bulk density of the fumed silica powder). An increase in C_{SiO2} is accompanied by rearrangement of the silica particle network in the suspension, and the γ_S value increases (Figure 14f). The main changes in the Gibbs free energy of interfacial water occur in one-two statistical monolayers ($h \approx$ 0.3-0.6 nm) (Figure 14b). This thickness can also be estimated considering the structure of secondary particles (i.e., gaps in aggregates) (Figure 14c). This shows that strongly disturbed water is located in narrow gaps R_p <1 nm between adjacent primary particles in aggregates (Figure 14c) (if these aggregates remain after treatment of the suspension), and the portion of the occupied pore volume is less than 0.5 cm^3/g (Figure 14d). The long-range forces (of electrostatic and hydrogen-binding nature, gradually decreasing with increasing distance from the solid surface), result in distortion of a thicker layer of interfacial water up to $h \approx$ 10 nm (Figure 14c) corresponding to a distance of 20 nm between particles. This value corresponds to a cubic packing of primary particles of A-300 at C_{SiO2} = 17-18 wt.% and to the main IPSD peak (Figure 1b). However, the corresponding changes in the ΔG value are observed at smaller C_{SiO2} values (Figure 14). This may be caused by different types of particle packing, depending on the C_{SiO2} value, and retention of some aggregates after sonication (Figure 9). The distances between adjacent primary particles can be smaller with increasing C_{SiO2} value, despite changes in the type of packing due to variation in the coordination number (n_c).

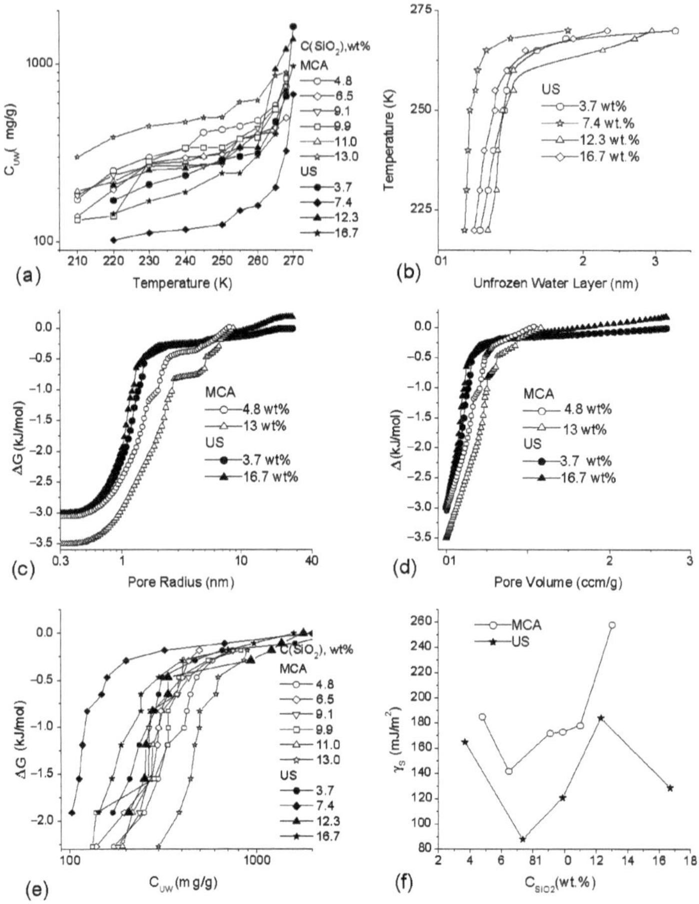

Figure 14. Characteristics of interfacial water in aqueous suspensions of A-300 sonicated (US) or treated in a ball-mill (MCA) at different concentration of silica: (a) amounts of unfrozen water as a function of temperature at T < 273 K; (b) relationship between the thickness of unfrozen water layer and temperature; and changes in Gibbs free energy of interfacial water versus (c) pore radius, (d) pore volume, and (e) amounts of water unfrozen in these pores; (f) interfacial Gibbs free energy as a function of silica concentration in suspensions differently treated.

This value can be relatively low, n_c = 2-4 in the initial aggregates.[3] An increase in n_c in the concentrated suspensions, and the shortening of distances between particles with increasing C_{SiO2}, affect the dispersion structure with time. Therefore, gelation of the concentrated dispersion (at 293 K) occurs faster, e.g. at C_{SiO2} = 16.7 wt.% in 2 days, than at C_{SiO2} <8 wt.%. Sonication leads to a lower turbidity of the suspensions compared to suspensions made from ball-milled silicas because of a different size of residual aggregates possessing substantially higher light-scattering ability compared to individual primary particles. This difference is reflected in the $\gamma_S(C_{SiO2})$ graphs (Figure 14f) as the MCA suspension demonstrates higher γ_S values. An increase in the S_{BET} value leads to a decrease in the transverse relaxation time ($T2$) for aqueous suspensions at the same C_{SiO2} values.[6] The $T2$ value decreases with increasing C_{SiO2}. Observed diminution of $T2$ corresponds to a decrease in the diffusion rate of water molecules. Thus, the smaller the S_{BET} and C_{SiO2} values, the higher the mobility of water molecules in aqueous suspensions of nanosilicas. These results can be explained by two reasons: (i) enhancement of aggregation with increasing S_{BET} value (decreasing primary particle size) and (ii) increase in the amount of bound water with an increase in S_{BET} at the same C_{SiO2} value. Both factors reduce the mobility of interfacial water molecules. An increase in the C_{SiO2} value affects the structure of unfrozen water layers filling gaps between primary particles in aggregates and agglomerates.[5,6] By comparison of IPSDs calculated from the nitrogen desorption data for the powder (Figure 1) and the ^{1}H NMR data for the aqueous suspension,[5,6] one can conclude that the type of aggregation of particles in air and water differ. For instance, narrow gaps at $1.3 < R_p < 4$ nm are less characteristic for the suspension than for the powder. This result can be caused by rearrangement of aggregates destroyed due to sonication, and repulsive forces between negatively charged silica particles, confirmed by the PCS data (Figure 7). In other words, the packing density of primary particles in concentrated suspensions is lower than in dense aggregates (about 30% of the true density of silica) but higher than in loose agglomerates. The concentrated suspensions at C_{SiO2} >8 wt.% (when strong changes in primary particle packing occur in the suspension) stored for several days transform their state from free (relatively low viscosity and low turbidity) to a bound (very high viscosity and higher turbidity) state. Thus, changes in the concentration of silica in the suspension and the kind of treatment allow one to vary the surface free energy, dispersion structure, as well as its stability, turbidity and other characteristics.

4. Conclusion

The specific surface area, i.e. primary particle size, and the corresponding order of aggregation of primary particles, as well as the content of silanols, are key-parameters affecting many physicochemical properties of powders and aqueous suspensions of nanosilica. These key characteristics are responsible for the properties of Silics as a medicinal preparation acting as a sorbent.

References

1. *Silicas in Medicine and Biology*, edited by A. A. Chuiko (SMI, Stavropol, 1993).
2. *Medical Chemistry and Clinical Application of Silica*, edited by A. A. Chuiko (Naukova Dumka, Kiev, 2003).

3. V. M. Gun'ko, V. I. Zarko, R. Leboda, and E. Chibowski, Aqueous Suspensions of Fumed Oxides: Particle Size Distribution and Zeta Potential, *Adv. Colloid Interface Sci.* 91, 1-112 (2001).
4. V. M. Gun'ko, A. V. Klyueva, Y. N. Levchuk, and R. Leboda, Photon correlation spectroscopy investigations of proteins, *Adv. Colloid Interface Sci.* 105, 201-328 (2003).
5. V. M. Gun'ko, V.V. Turov, V. M. Bogatyrev, V. I. Zarko, R. Leboda, E. V. Goncharuk, A. A. Novza, A. V. Turov, and A. A. Chuiko, Unusual properties of water at hydrophilic/hydrophobic interfaces, *Adv. Colloid Interface Sci.* 118, 125-172 (2005).
6. V. M. Gun'ko, I. F. Mironyuk, V. I. Zarko, E. F. Voronin, V. V. Turov, E. M. Pakhlov, E. V. Goncharuk, Yu. M. Nicheporuk, T. V. Kulik, B. B. Palyanytsya, S. V. Pakhovchishin, N. N. Vlasova, P. P. Gorbik, O. A. Mishchuk, A. A. Chuiko, J. Skubiszewska-Zięba, W. Janusz, A. V. Turov, and R. Leboda, Morphology and surface properties of fumed silicas, *J. Colloid Interface Sci.* 289, 427-445 (2005).
7. V. M. Gun'ko, I. F. Mironyuk, V. I. Zarko, V. V. Turov, E. F. Voronin, E. M. Pakhlov, E. V. Goncharuk, R. Leboda, J. Skubiszewska-Zięba, W. Janusz, S. Chibowski, Yu. N. Levchuk, and A. V. Klyueva, Fumed Silicas Possessing Different Morphology and Hydrophilicity, *J. Colloid Interface Sci.* 242, 90-103 (2001).
8. V. M. Gun'ko, E. F. Voronin, I. F. Mironyuk, R. Leboda, J. Skubiszewska-Zięba, E. M. Pakhlov, N. V. Guzenko, and A. A. Chuiko, The effect of heat, adsorption and mechanochemical treatments on stuck structure and adsorption properties of fumed silicas, *Colloids Surf. A* 218, 125-135 (2003).
9. V. M. Gun'ko, J. Skubiszewska-Zięba, R. Leboda, K. N. Khomenko, O. A. Kazakova, M. O. Povazhnyak, and I. F. Mironyuk, Influence of Morphology and Composition of Fumed Oxides on Changes in Their Structural and Adsorptive Characteristics on Hydrothermal Treatment at Different Temperatures, *J. Colloid Interface Sci.* 269, 403-424 (2004).
10. V. M. Gun'ko, V. I. Zarko, E. F. Voronin, V. V. Turov, I. F. Mironyuk, I. I. Gerashchenko, E. V. Goncharuk, E. M. Pakhlov, N. V. Guzenko, R. Leboda, J. Skubiszewska-Zięba, W. Janusz, S. Chibowski, Yu. N. Levchuk, and A. V. Klyueva, Impact of Some Organics on Structural and Adsorptive Characteristics of Fumed Silica in Different Media, *Langmuir* 18, 581-596 (2002).
11. V. M. Gun'ko, N. N. Vlasova, L. P. Golovkova, N. G. Stukalina, I. I. Gerashchenko, V. I. Zarko, V. A. Tischenko, E. V. Goncharuk, and A. A. Chuiko, Interaction of Proteins and Substituted Aromatic Drugs with Highly Disperse Oxides in Aqueous Suspensions, *Colloid. Surf. A* 167, 229-243 (2000).
12. I. F. Mironyuk, V. M. Gun'ko, V. V. Turov, V. I. Zarko, R. Leboda, and J. Skubiszewska-Zięba, Characterization of Fumed Silicas and their Interaction with Water and Dissolved Proteins, *Colloid. Surf. A* 180, 87-101 (2001).
13. V. M. Gun'ko, E. F. Voronin, V. I. Zarko, E. V. Goncharuk, V. V. Turov, S. V. Pakhovchishin, E. M. Pakhlov, N. V. Guzenko, R. Leboda, J. Skubiszewska-Zięba, W. Janusz, S. Chibowski, E. Chibowski, and A. A. Chuiko, Interaction of poly(vinyl pyrrolidone) with fumed silica in dry and wet powders and aqueous suspensions, *Colloid. Surf. A* 233, 63-78 (2004).
14. E. F. Voronin, V. M. Gun'ko, N. V. Guzenko, E. M. Pakhlov, L. V. Nosach, M. L. Malysheva, J. Skubiszewska-Zięba, R. Leboda, M. V. Borysenko, and A. A. Chuiko, Interaction of poly(ethylene oxide) with fumed silica, *J. Colloid Interface Sci.* 279, 326-340 (2004).
15. V. V. Turov, V. M. Gun'ko, V. M. Bogatyrev, V. I. Zarko, S. P. Gorbik, E. M. Pakhlov, R. Leboda, O. V. Shulga, and A. A. Chuiko, Structured Water in Partially Dehydrated Yeast Cells and at Partially Hydrophobized Fumed Silica Surface, *J. Colloid Interface Sci.* 283, 329-343 (2005).
16. B. I. Gerashchenko, V. M. Gun'ko, I. I. Gerashchenko, R. Leboda, H. Hosoya, and I. F. Mironyuk, Probing the Silica Surfaces by Red Blood Cells, *Cytometry* 49(2), 56-61 (2002).

MEDICAL ASPECTS OF APPLICATION OF HIGHLY DISPERSE AMORPHOUS SILICA

ALEXEY CHUIKO,[a†] ALEKSANDR PENTYUK,[b] ELENA SHTAT'KO,[b,*] NATAL'YA CHUIKO[c]
[a] *Institute of Surface Chemistry, 17 Generala Naumova Str., Kyiv 03164, Ukraine*
[b] *Vinnitsa State Medical University, 56 Pirogova Str., Vinnitsa 21018, Ukraine*
[c] *Institute for Occupational Health of the Academy of Medical Sciences of Ukraine, 75 Saksaganskogo Str., Kyiv 01033, Ukraine*

Abstract. A novel preparation Silics for efferent therapy has been created, distinguished for the practically complete absence of toxic properties. Silics possesses a wide variety of valuable characteristics such as an optimum balance of osmotic and hydrophilic indices; high affinity for proteins, microorganisms, and a number of low-molecular weight biologically active substances. These characteristics give rise to successful applications of Silics in all fields of medicine, but first and foremost the management of toxoinfectious and pyoinflammatory processes.

Keywords: silicon dioxide, medicinal chemistry, surface chemistry, sorbent, efferent therapy, enteosorption, medicinal preparation, infectious diseases, detoxication, bioactive silica, purulent wound, pyoinflammatory diseases, diarrhea

1. Introduction

In the last twenty years, research of the medicinal chemistry of surfaces has shown increasing progress.[1–3] Studies have resulted in the development of methods for synthesizing numerous sorbents; revealing how their structure, adsorptive, and biomedical properties affect their efficacy in the treatment of various diseases. The sorbents find much use for removal of toxic substances and microorganisms from the gastrointestinal tract (enterosorption), and for management of such external lesions of as wounds, burns, frostbite, trophic

* To whom correspondence should be addressed. E. I. Shtat'ko, Vinnitsa State Medical University, 56 Pirogova Str., Vinnitsa 21018, Ukraine; e-mail: igorgera@vsmu.vinnitsa

J.P. Blitz and V.M. Gun'ko (eds.), Surface Chemistry in Biomedical and Environmental Science, 191–204.

ulcers, etc. (applicative sorption). At present, the efferent methods (which are known to provide removal of toxic substances from an organism, and which can be employed individually or in conjunction with adjuvant therapy) find wide use in many branches of medicine. The data on their experimental and clinical applications have been covered at scientific conferences and consultative meetings,[4–7] elucidated in periodical editions, and expounded in monographs.[1–3,8-11]

2. Biologically Active Silica

In collaboration with various medical institutions of Ukraine and Russia, the Institute of Surface Chemistry of the NAS of Ukraine has developed a novel medicinal preparation Silics (the term *silics* is derived from *sil*ica and the *I*nstitute of *C*hemistry of *S*urface), and designed a nanotechnology for its production.[2,3] This bioactive silica (BAS) is distinguished for its extended surface and high adsorption properties.

On the assumption that a particular silica possesses a specific surface area of 400 $m^2\ g^{-1}$, and consists of particles of the same size and regular spherical shape, simple calculations make it possible to show that each particle should have a diameter of about 7 nm and the ratio of surface silicon atoms to their total number in a particle should be equal to 0.31.

BAS is a silicon oxide polymer whose structural cells are silicon–oxygen tetrahedra bonded by disiloxane (Si–O–Si) bridges. On the surface of a particle there are O–H groups chemically bonded with silicon atoms (silanol groups Si–OH) and molecules of coordinated water at these silicon atoms. Hydroxyl groups and coordinated water molecules on the BAS surface bring about its high hydrophilicity and, as a consequence, possibility for sorbing polar molecules (especially water molecules).

The bioactivity of silica is attributed to the nanodimensionality of its primary particles, to the presence of surface coordination compounds of water molecules, hydroxylated silicon atoms and strongly sorbed/highly structured bioactive water. Biologically active water (BAW) is a special form of water whose molecules are strongly bound to the BAS surface but at the same time are weakly associated. BAW is formed within gaps among biologic objects and particles of amorphous nanodimensional silica at a certain water content, spatial structure, and hydrophobic–hydrophilic balance at interfaces.[10]

The above-mentioned physico-chemical properties of the BAS give rise to its major characteristic, namely its high protein-sorbing capacity. This characteristic formed the foundation for employing the BAS to remove exo- and endotoxins, pathogenic immunocomplexes, products of degradation of necrotic tissues, and other harmful substances of protein origin as well as to immobilize microorganisms. Our pharmacological trials and toxicity tests of Silics have provided strong evidence for its reduced toxicity for enteral and applicative administration at higher doses than those approved for use of silica as an adjuvant. Thus, it has been proved that in the case of its peroral introduction the preparation does not practically exhibit any toxic properties. As shown earlier[3] long-term administration (for 6–8 months) of Silics at a dose of 100, 330, 1000 mg kg^{-1} to white mice, rats, rabbits, dogs, and piglets did not bring about any significant pathologic changes to these animals (Table 1).

Table 1. Toxicity of Silics in the case of its peroral administration to animals.

Toxic Effect	Presence and Degree of Manifestation
Cytotoxicity	moderate
Acute toxicity	absent
Chronic toxicity	absent
Influence on gastrointestinal tract	within limits of adaptative changes
Cumulation in tissues and lymph nodes	absent
Locally irritant action	insignificant
Teratogenicity and gonadotoxicity	absent
Carcinogenicity	absent

The biologic activity of the silica surface depends on (1) high hydrophilicity of the surface, (2) high protein-sorbing activity, (3) binding of great numbers of microorganisms and microbial toxins, and (4) adsorption of low-molecular substances.

1. <u>High hydrophilicity of the BAS surface</u> is attributed to the presence of hydroxyl and hydrate coatings, and electron-accepting silicon atoms of silanol groups. Therefore, the BAS surface is able to sorb polar molecules. In particular, it is well wetted by water, which leads to formation of suspensions or gels (depending on the amount of water). Comparative studies carried out to evaluate osmohydrophilic properties of various agents for applicative therapy have shown that with respect to its osmotic activity and water absorption capacity Silics is considerably superior to activated charcoal sorbents and Debrisan®. Although with reference to its water absorption Silics is at a slight disadvantage in relation to such agents as Gelevin® and Celosorb® (Table 2), it exhibits the most balanced state of osmotic and hydrophilic properties. The above-mentioned hydrophilic properties of Silics are employed to reduce edematization and exudation in topical treatment of wounds during inflammation, to bind and structure water in the bowels in the case of diarrhea, and to effect dehumidification in dermatological practice.

Table 2. Comparison of characteristics of agents for applicative therapy.

Applicative Sorbent	Sorption of Protein mg g^{-1}	Sorption of Micro-organisms milliard g^{-1}	Osmotic Activity %	Water Absorption %	Osmohydrophilic Coefficient
Silics	up to 600	up to 10.0	503	944	0.53
Debrisan®	60–70	up to 0.8	52	281	0.19
Celosorb®	7–10	up to 5.0	459	1900	0.24
Gelevin®	5–7	up to 5.0	317	1826	0.17

2. <u>High protein-sorbing capacity</u> of the BAS is due to the combined actions exerted on protein molecules by the silica surface from electrostatic surface charges, reaction sites including hydroxyl groups and silicon atoms of silanol groups, water coordinated with silicon atoms of Si–OH groups, and electron-donating oxygen atoms of siloxane bonds. From Tables 2 and 3 it is evident that in its capacity to sorb proteins and microorganisms Silics is substantially superior to the comparison preparations. Our research into the adsorption of proteins [bovine serum albumin (BSA), egg albumin, dried blood plasma, horse hemoglobin, gelatin] on the BAS surface have shown that this silica possesses a substantial sorptive capacity with respect to proteins (up to 600 mg g^{-1}) and a high rate of adsorption(≥90% of protein is adsorbed within the first 10 min of contact). Moreover, the protein-sorbing capacity of the BAS considerably exceeds (by a factor of 20–30) the corresponding capacity of known medicinal sorbents (e.g., Debrisan®). As seen from Table 3, with increasing ionic strength the adsorption value of the BAS increases. An analogous phenomenon is observed as the pH value of a solution approaches the isoelectric point of a given albumin. The results achieved when studying the specificity of interaction between human blood plasma and the BAS give evidence for its increased affinity to lipoproteins in comparison with proteins that are not saturated with lipids. Therefore one can, to some extent, explain hypolipidemic activity of the BAS in the case of its peroral administration, because lipids in a gastrointestinal tract are in the form of micelles consisting of proteins, phospholipids, and bile acids. Silics provides fixation of food fats and other lipid components, which decreases their absorption and, therefore, produces a desired treatment response. It has been found that ingestion of a medicinal agent developed on the basis of Silics by patients who suffer from disturbances of lipid metabolism within 3–4 weeks brings about a substantial decrease in the level of cholesterol in blood serum, and this decrease is greater, the higher the initial level of cholesterol at the beginning of treatment.

Table 3. Degree of extraction (in %) of proteins from an aqueous medium by Silics and other medicinal sorbents.

Sorbent	Protein Preparation	Degree of Removal, %		
		Distilled Water, pH 6.5	0.9% NaCl Sol-ution, pH 6.5	0.1 M Phosphate Buffer, pH 5.7
Silics	BSA	27	60	62
	plasma	64	90	95
Debrisan®	BSA	0	0	0
	plasma	0	1.4	1.5

As is known, sorbents of medicinal specification must meet certain requirements, with one of the most important being their physico-chemical stability. First of all, such a sorbent should not be dissolved when in contact with biologic fluids (exudate of a wound, contents of a digestive tract). From the results achieved during experiments with test animals (rats and rabbits) it follows that after a single intragastric administration of even a large dose of BAS (1 g kg^{-1}) any statistically reliable variations in concentration of silicon in blood are not revealed. A similar result has also been obtained in the case of a long-term administration (for 30 days) of BAS at the above dose to rats. Thus, it has been established that Silics satisfies the requirement with reference to its physico-chemical stability, which is one of the most stringent requirements for sorbents of medicinal specification.

3. The protein-sorbing properties characteristic of Silics impart the ability to fix microorganisms. The interaction between silica and microorganisms is not noted for any specific nature. It is attributed to the affinity of silica particles to glycoprotein structures and to phospholipids in the membranes of microbe cells. This characteristic of Silics is of great importance for medical practice because its application as a medicinal agent does not bring about selection of stable strains, but provides proportional decontamination of bowels or wound contents. Research has also been done into the interaction of Silics with enteropathogenic colon bacilly (*Escherichia coli*, *Staphylococcus aureus*, *Proteus vulgaris*, *Bacillus pynocyaneus*). The results of the research show that even at low concentrations of Silics (0.33–1.33 wt%) it provides fixation of practically all the microorganisms, with the fixation value being virtually independent of type of microorganisms. The interaction of microorganisms with BAS is different than what might be expected as discussed below.

First, particle sizes (4–40 nm) are considerably smaller than those of microorganisms (1000–10000 nm) so that the aforementioned interaction brings about the effect of agglutination. Thus, it is BAS particles that are sorbed on microbial cells and not vice versa. Even at low concentrations BAS particles are able to agglutinate microorganisms, that is, to act in the capacity of a glue that unites microorganisms into an agglomerate. Evidently, it is this phenomenon of agglutination of microorganisms by particles of BAS that explains its unique ability to bind enormous amounts of microorganisms in comparison with other sorbents.

Second, binding of highly disperse amorphous silica particles and microbial cells is affected by the nature of a given silica surface. Thus, introduction of aluminium oxide (2.8 wt%) into highly disperse silica composition leads to enhancement of its ability to bind *Escherichia coli* by about 33% in comparison with the initial preparation. One of the substantial advantages of Silics over other agents for management of pyoinflammatory diseases is its ability to decrease resistance of wound microflora to antibiotics. For instance, a seven-day course of vulnerosorption by Silics results in an enhancement of wound microflora sensitivity to penicillins, aminoglycosides, tetracyclines, and other antibiotics (especially for erythromycine, gentamycin, and streptomycin).

Third, after contact with BAS the microorganisms become sensitive to the action of proteolytic enzymes and cationic and anionic SAS, such as bile acids and phospholipids; that is to natural components of intestinal and gastric juices. The high affinity of BAS to microorganisms, and its influence on processes of vital activity of microorganisms, provides an explanation for mechanisms of its curative effect. The appearance of infectious diseases and their progress are directly dependent on the contagious dose and number of bacteria that are accumulated in an intestine in the course of colonization.

4. Regarding the adsorption of low-molecular weight substances, of significance are regularities of adsorption of medicinal substances on BAS, because the data form a scientific basis for the development of medicinal preparations with modulated pharmacokinetics. A study has been made of adsorption of orthophen, quinidine, scopolamine, amphotericin, and some vitamins on BAS from aqueous solutions. For example, it has been shown that for the slightly soluble antibiotic amphotericin B, it is possible to increase the rate of absorption of the curative substance by simultaneous peroral introduction of the substance and BAS into intestines of rats. The results of the experiments on rats show that introduction of the antibiotic, simultaneously with an aqueous suspension of BAS, leads to an increase in the maximum concentration of the preparation in blood from 2 to 21 μg mL^{-1}. The sharp increase in the absorption of the curative substance provides an enhancement of its biologic availability. Analogous effects were observed for substances of other classes, such as alkaloids (quinidine), carbohydrates (xylose), and organic acids (voltarene). The studies of absorption showed that the maximum concentration of quinidine (antiarrhythmic preparation) in blood increased from 2.6 to 4.6 μg mL^{-1} and that of voltarene (antiinflammatory agent) from 16 to 26 μg mL^{-1}. A study has also been conducted of the release of quinidine from various medicinal forms produced by immobilization of quinidine and its complexes with surface active substances or proteins on BAS. It has been

inferred from this work that the requirements of medicinal agents of prolonged action are most fully met by a preparation produced by coprecipitation of complexes of quinidine with molecules of serum albumin on the surface of BAS. Administration of this preparation does not give rise to a sharp peak of concentration of quinidine in the blood; after the administration over a long period of time a gradual decrease in the antiarrhythmic preparation concentration is observed. Besides, administration of this medicinal form makes it possible to provide the maximum bioavailability of quinidine in comparison to other medicinal forms.

Results of the above-outlined research into physico-chemical and medico-biological properties of BAS, conducted at the Institute of Surface Chemistry of the NAS of Ukraine in collaboration with Vinnitsa State Medical University, named after N.I. Pirogov, and other medical institutions of Ukraine and Russia give reasons to infer that Silics is an active medicinal substance which on its own can function as a therapeutic agent. Sorptive detoxication with the aid of the bioactive silica brings about a profound effect in the case of acute intestinal infections, diarrheas of various origins, viral hepatitis, as well as for local treatment for pyoinflammatory diseases and purulent wounds.

3. Clinical Applications of Silics

The data about fields of application of Silics in clinics for treatment for infectious diseases are presented in Table 4. From Table 4 it is evident that the field of application of Silics is rather large and covers both intestinal infections and toxicoses which victimize infants, as well as viral hepatitis, and botulism. It is appropriate to mention here that inclusion of Silics into the complex treatment of patients suffering from salmonellosis, dysentery, and intestinal toxicoses accelerates normalization of clinic manifestations of these diseases by a factor of two and more. In the case of botulism the normalization of symptoms characteristic of lesions of the nervous system is shortened by almost 4 days. If intestinal infections are not severe, Silics can be recommended as a single therapeutic agent. In the case of a considerable diarrheal syndrome it is more expedient to use it together with rehydration substances. Inclusion of Silics into a complex of therapeutic agents for patients suffering from viral hepatitis substantially accelerates recovery rates of patients, so that their normal level of bilirubin and activity of alanine aminotranspherase are recovered within shorter periods of time.

Table 4. Clinical applications of Silics for treatment of infectious diseases.

Fields of Application	Pathologic Syndromes	Pharmacologic Effect	Particularities of Application
Intestinal infections (toxi-coinfections, salmonellosis, shigellosis, cholera, etc.)	Diarrhea, intoxication, dyspepsia	Fixation of microorganisms and their toxins, normalization of absorption and secretion in GIT	Monotherapy; combination with antibacterial agents and agents for rehydration
Intestinal toxicoses in children	Dehydration, diarrhea, intoxication, dyspepsia	Sorption of toxins, normalization of absorption and secretion in GIT	Monotherapy; combination with antibacterial agents and agents for rehydration
Viral hepatitises	Cholestasis, cytolysis of hepatocytes, intoxication	Sorption of viruses, bile acids, and bilirubin	Monotherapy; combination with hepatoprotectors
Botulism	Neurotoxicosis, dyspepsia	Sorption of toxins, enhancement of action of immunopreparations	Combination with specific serums

The therapeutic effects produced by Silics are not confined only to binding of pathogenic microorganisms. We have shown that Silics decreases the ability of pathogenic *Escherichia coli* and *Salmonella typhimurium* to adhere to mucosa of bowels and to form intraintestinal colonies. Moreover, the sorbent prevents their translocation and generalization.[3] Inclusion of Silics into a complex of therapeutic agents for patients suffering from viral hepatitis substantially accelerates recovery rates of patients, so that their normal level of bilirubin and activity of alanine aminotranspherase are recovered within shorter periods of time. The therapeutic effects produced by Silics are not confined only to binding of pathogenic microorganisms. We have shown that Silics decreases the ability of pathogenic *Escherichia coli* and *Salmonella typhimurium* to adhere to mucosa of bowels and to form intraintestinal colonies. Moreover, the sorbent prevents their translocation and generalization.[3]

Besides, it has been found[3] that Silics possesses some antidiarrheal properties that are not related to sorption of microorganisms or their toxins. A Silics medicinal preparation retards secretion of water in isolated intestinal loops of rats in the case of its combined introduction with diarrhea mediators such as sodium desoxycholate, *cyclo*-adenosine 3',5'-monophosphate (*c*-AMP), and serotonin (Table 5). The decrease of water release into intestinal segment lumens seems to be related to the modifying action of the preparation on mucosa of bowels, because sorption of the stimulators themselves on Silics does not take place. The modifying action of the preparation on the small intestine mucosa is also attributed to the stimulation exerted by Silics on transport of glucose and electrolytes through bowel walls.[3]

Table 5. Influence of HDS on accumulation of fluid (mL) in isolated intestinal loops under the action of secretion stimulators.

Group	Sodium Desoxycholate	*c*-AMP	Serotonin
Control (intact)	0.10 ± 0.032	0.10 ± 0.032	0.10 ± 0.032
Secretion stimulators	0.55 ± 0.039'	0.69 ± 0.057'	0.49 ± 0.039'
Secretion stimulators + 0.5 wt % HDS	0.36 ± 0.039'*	0.11 ± 0.028*	0.08 ± 0.018*
Secretion stimulators + 2.5 wt % HDS	0.02 ± 0.036*	0.06 ± 0.016*	0.03 ± 0.005*

Notes: 'significant differences in comparison with control; * significant differences in comparison with secretion stimulators.

Binding of causative agents of enteric infections and their protein toxins by Silics, as well as its antisecretory properties and ability to enhance absorption processes in bowels, give rise to its high efficacy in the treatment for infectious diarrheas. Thus, addition of Silics (at a daily dose of 100–250 mg kg^{-1}) to the basic therapeutic complex makes it possible to accelerate (by 2–3 days) the positive dynamics of symptoms in patients and, as a result, to attain a shortening of their hospital stay. Besides, administration of Silics leads to a decrease in the necessary amounts of agents for dehydration therapy and prevents development of bacteria carrying states (Table 6).

According to the results of our experiments *in vitro*, the highest adsorptive capacity with reference to proteins and protein-associated phospholipids and triglycerides of blood serum was characteristic of the highly disperse silica modified with aluminium oxide. The mechanism of the therapeutic effect of HDS on treatment for intestinal infections seems to involve the following major aspects.

Table 6. Effect of medication on the dynamics of clinical symptoms in patients suffering from salmonellosis and shigellosis (M± m)

Symptoms	Duration of Symptoms (h)			
	Salmonellosis		Shigellosis	
	Traditional Therapy	Traditional Therapy+Silics	Traditional Therapy	Traditional Therapy+Silics
Nausea	61.7 ± 14.5	40.0 ± 4.2	54.0 ± 6.7	20.0 ± 5.0*
Diarrhea	133.7 ± 7.3	47.7 ± 10.9*	172.8 ± 36.1	56.0 ± 10.0*
Anorexia	78.9 ± 14.5	54.9 ± 7.3	72.0 ± 20.6	48.0 ± 20.1
Pains	72.0 ± 7.3	51.4 ± 7.3	139.2 ± 36.1	56.0 ± 10.0
Intestinal spasm	136.0 ± 29.7	60.0 ± 6.7	206.4 ± 41.2	72.0 ± 20.1*
Coprograms normalization	133.7 ± 25.4	62.0 ± 8.5*	168.0 ± 36.1	96.0 ± 30.1

Note: * Significant differences in comparison with the group 'traditional therapy' ($p < 0.05$).

1. *Direct influence on intestinal microflora:*

– fixation of bacteria and their removal from an organism with stools; sorption of proteinaceous microbial toxins and other pathogenic proteins (neuraminidase, hyaluronidase, contact hemolysins) that promote the pathogenic action of microorganisms; a bactericidal effect is also possible (in the presence of bile acids and proteolytic enzymes).

2. *Indirect influence on intestinal microflora:*

– creation of conditions unfavourable for the vital activity of pathogenic microorganisms (concentration of microorganisms on the sorbent results in a local deficit of nutrients, fixation of hemoglobin sets a limit on iron that is necessary for microorganisms, etc.);

– adjuvant action (concentration of microbial cells and their toxins on the sorbent enhances the antigenic action and immune response of an organism).

3. *Interaction of Silics with intestinal walls and intestine contents:*

– blocking of receptors of the mucous membrane of the stomach responsible for adhesion of microorganisms and fixation of toxins; intensification of transport of water, electrolytes, and other substances from the intestine into internal medium; modelling of baroreceptors and chemoreceptors of intestinal walls responsible for motility;

– clearance of intestinal juices from toxic substances (products of vital activity of microorganisms and of microbial putrefaction of proteins), toxic metabolites of endogenous origin (bilirubin, bile acids, micelle complexes, medium-molecular peptides, etc.);

– fixation of cholesterol and other nonpolar lipids that are constituents of complexes with proteins and phospholipids;

– particles of Silics perform the role of sites of concentration and transport of ingredients of the intestine contents so that the sorbent acts as a coenzyme, thereby favouring the interaction between metabolites and accelerating the natural course of the process of their transformation, which leads to a decrease in the amount of intermediate products with toxic properties;

– the presence of Silics in a gastrointestinal tract induces immobilization of digestive enzymes and, thereby, intensifies digestion.

In the course of the above-mentioned mechanisms Silics remains unchanged and thus preserves its activity during its residence in the intestine. Silics may also have promise in the clinical treatment for some internal diseases. The major lines of research into applications of this medicinal preparation in the relevant therapies are presented in Table 7. Of significance is the ability of Silics to lower cholesterol levels and triglycerides, as well as to retard aggregation of thrombocytes. Thus, with the help of Silics it becomes possible to treat the main pathogenic factors of atherosclerosis, namely hyperlipidemia and hypercoagulation (Table 7).

Table 7. Major lines of application of SILICS in therapy.

Fields of Application	Pathologic State	Pharmacologic Effect	Particularities of Application
Disturbances of lipoid metabolism	Hypercholesterinemia	Reduction of absorption and synthesis of cholesterol	Monotherapy. Combination with statines
	Hypertriglyceridemia	Reduction of absorption and synthesis of triglycerides	Combination with nicotinic acid and fibrates
Thrombocytarno-coagulative homeostasis	Hypercoagulation	Retardation of aggregation of thrombocytes	Monotherapy. Combination with aspirin
Allergology	Bronchial asthma, food allergy, psoriasis, eczema	Detoxication, sorption of immune complexes	Monotherapy. Combination with antihistaminic preparations, steroids
Correction for effect of toxicity of medicinal agents	Insufficient efficacy, side effects	Acceleration of absorption, decrease in dosage, detoxication	Amiodarone, nicotinate, symvastatine, orthophen, quinidine

Administration of Silics to patients with dislipoproteinemias facilitates a noticeable decrease in total cholesterol, and cholesterol of low and very low density, as well as a decrease in blood coagulation potential (Table 8). Of significance here is the fact that the Silics therapy does not interfere with the action of other medicinal agents, does not have side effects, and is the most efficacious medication for patients with high initial levels of cholesterol in blood serum. In the case of profound disturbances of lipoid metabolism it proved advantageous to employ Silics together with other hypolipodemic agents. The use of the complexes Silics–symvastatine makes it possible to decrease the dose of the latter by a factor of two without any lowering of the intensity of hypolipodemic action. Positive results have also been achieved when using Silics as a therapeutic agent in allergology. This medicinal preparation has been successfully employed for treatment of patients suffering

Table 8. Effect of Silics on the dynamics of the total cholesterol concentration in blood serum of patients suffering from atherosclerosis (M ± m)

Initial Level of Total Cholesterol' mmol L^{-1}	Before Treatment	After Treatment for 1 Month
5.5–6.5 ($n = 26$)	5.7 ± 0.02	5.0 ± 0.02*
6.5–7.5 ($n = 11$)	7.0 ± 0.15	5.8 ± 0.22*
7.5 and more ($n = 10$)	8.6 ± 0.14	6.8 ± 0.18*

Notes: Total cholesterol level in healthy persons: 5.0 ± 0.06 mmol L^{-1}; * Significant differences in comparison with control ($p < 0.001$)

Table 9. Application of Silics in surgery, oncology, stomatology, obstetrics, and gynecology.

Fields of Application	Pathologic State	Pharmacologic Effect	Particularities of Application
Surgery	Hemorrhages	Hemostasis	Applications
Purulent wounds, destructive peritonitis, ileus, purulent pleurisy	Localized aerobic and anaerobic infections	Sorption of microorganisms, endo- and exotoxins, dehydration	Applications, drainage of cavities
Oncology	Intoxication syndrome	Sorption of endo- and exotoxins, chemopreparations	Enterosorption
Stomatology	Pyoinflammatory processes	Sorption of microorganisms, endo- and exotoxins, dehydration, abrasive action	Applications, drainage of cavities
Gynecology	Endometritis, vaginitis, pelvioperitonitis	Sorption of microorganisms, endo- and exotoxins	Applications, lavage
Operative obstetrics	Hemorrhages, pyoinflammatory processes	Hemostasis, sorption of microorganisms, endo- and exotoxins, localized dehydration	Applications, lavage, combination with antibacterial agents
Pregnancy	Gestoses of pregnancy	Detoxication	Monotherapy. Combination with antioxidants

from bronchial asthma, chronic obstructive lesions of lungs, food allergies, psoriasis, and eczema. The profound detoxication effect of Silics and its high affinity to proteins and medium-molecular weight peptides form the foundation for its application for treatment of patients of this profile. The results of such a treatment have shown that the use of Silics for treatment of patients suffering from chronic obstructive lesions of the lungs makes it possible to produce a more complete and more prolonged curative effect in comparison with the traditional treatment.

Important fields of application of Silics are surgery, stomatology, oncology, obstetrics, and gynecology (Table 9). As far as these fields are concerned, there are two directions of application of Silics, namely as an enterosorption agent and as an agent for local (applicative) use.

Local applicative use of Silics may have much promise in the treatment of purulent wounds, destructive pancreatitis, peritonitis, purulent pleurisy, and odontogenous phlegmons. The mechanism of the curative action resides in the sorption of pathogenic microorganisms and microbial toxins, and in the dehydration of wound tissues. In particular, administration of Silics for sanation of postinjective suppurative foci (abscesses, phlegmons, empyemas) in patients with opiomania makes it possible to attain a rapid debridement of wounds from pathogenic microorganisms (Table 10).

Table 10. Effect of Silics on the dynamics of microbial seeding in postinjective supperative foci in treatment for patients with narcomania.

	Microbial Seeding (CFU g^{-1})	
Day of Observation	Traditional Treatment	Traditional Treatment + Silics
1st day	10^7–10^8	10^6–10^7*
3rd day	10^4–10^5	10^3–10^4*
5th day	10^2–10^3	0–10^2
7th day	10–10^2	solitary bacteria

Note: * Significant differences ($p < 0.05$).

Table 11. Effect of Silics on the dynamics of a wound process in treatment for patients with purulent wounds.

	Day of Observation	
Symptoms of a Wound Process	Traditional Treatment	Traditional Treatment + Silics
Debridement of a wound from devitalized tissues	8.94 ± 0.55	4.90 ± 0.16*
Appearance of granulations	6.70 ± 0.55	3.45 ± 0.18*
Beginning of epithelization	9.57 ± 1.21	6.80 ± 0.43*

When Silics is used for the management of purulent wounds or odontogenous phlegmons, it exerts a marked effect on all the phases of a disease process, which manifests itself in the acceleration of healing of wounds, with the time interval necessary for restoration of the function of an injured organ being shortened by 3–4 days (Table 11). In such a case it is possible to perform a more effective debridement of a wound from necrotic tissues, to attain more rapid granulations and epithelisation.

The local applicative use of Silics reveals one more salient feature of the preparation, namely its hemostatic effect. It has been proved that this BAS activates the first phase of coagulation of blood, so that it can be employed in the case of moderate external or internal hemorrhages. Depending on localization of a hemorrhage, Silics is used for applications or for drainage and insufflation. Hemostasis is usually attained after 1–2 treatment procedures involving the sorbent. It is also possible to employ Silics for treatment in the case of puerperal infections occurring after cesarean sections. Lavage and drainage of the cavity of the uterus shortened the duration of fever, which, in turn, decreased the time span of antibacterial therapy. Silics as an agent for applications or as an ingredient for drainage is also successfully employed for treatment for gynecologic diseases (such as vaginitis, endometritis, pelvioperitonitis). The general detoxicating action of Silics manifests itself in the case of its use for treatment of oncologic patients. Silics seems to decrease levels of endogeneous intoxication that was a side effect of application of chemotherapeutic preparations and radiation therapy, with the efficiency of enterosorption being comparable to that of plasmapheresis.

In summary, our research has corroborated and substantiated the idea that Silics is an efficacious medicinal agent, and has much potential as a nonselective enteral and applicative therapeutic preparation. Also, its use as a component of multimodality therapy for a broad spectrum of diseases is pathogenically validated and innocuous.

References

1. N. A. Lopatin and Yu. M. Lopukhin, *Efferent Methods in Medicine* (Meditsina, Moscow, 1989).
2. *Silicas in Medicine and Biology,* edited by A. A. Chuiko (IPF 'Stavropol'ye', Kyiv–Stavropol, 1993).
3. *Medicinal Chemistry and Clinical Application of Silicon Dioxide,* edited by A. A. Chuiko (Naukova Dumka: Kyiv, 2003).
4. O. O. Chuiko and O. O. Pentyuk, Scientific Principles of Development of Medicinal Preparations on the Basis of Highly Disperse Silica, in: *Scientific Principles of Development of Medicinal Preparations* (Osnova, Kharkiv, 1998), pp. 35-51.
5. V. V. Strelko and N. T. Kartel, Activated Carbons of Medicinal Duty, in: *Scientific Principles of Development of Medicinal Preparations* (Osnova, Kharkiv, 1998), pp. 490-516.
6. Book of Abstracts, NATO ARW *"Pure and Applied Surface Chemistry of Nanomaterials for Human Life and Environment Protection,* September 14-17, Kyiv, Ukraine (Tov. INFODRUK, Kyiv, 2005).
7. Book of Abstracts, NATO ARW *"Combined and Hybrid Adsorbents: Fundamentals and Applications*, International Conference, September 15-17, Kyiv, Ukraine (Institute of Sorption and Problems of Endoecology, 2005).
8. *Enterosorption,* edited by N. A. Belyakov, (Centre of Sorption Technologies, Leningrad, 1991).
9. *Sorbents and Their Clinical Application*; edited by C. Giordano, (Vyshcha Shkola, Kyiv, 1989).
10. V. M. Gun'ko, V. V. Turov, V. M. Bogatyrev, V. I. Zarko, R. Leboda, E. V. Goncharuk, A. A. Novza, A. V. Turov, and A. A. Chuiko, Unusual properties of water at hydrophilic/ hydrophobic interfaces, *Adv. Colloid Interface Sci.* 118, 125-172 (2005).
11. B. M. Dazenko, S. G. Belov, and T. I. Tamm, *Festering Wound* (Zdorov'ja, Kyiv, 1985).

THE INTERACTION OF NANOSTRUCTURED BIOMATERIALS WITH HUMAN CELL CULTURES. THE CHOICE OF CELL CULTURES FOR USE AS BIOCOMPATABILITY PROBES

S. L. JAMES,* M. ILLSLEY, S. E. JAMES, E. MENDOZA, S.R.P. SILVA, P. VADGAMA,[5] P. TOMLINS,[4] S.V. MIKHALOVSKY[1]

[1]*School of Pharmacy and Biomolecular Science, University of Brighton, Moulsecoomb, Brighton BN2 4GJ, UK*

[2]*Blond McIndoe Centre, Queen Victoria Hospital, East Grinstead RH19 3DZ, UK*

[3]*Nano-electronics centre, Advanced Technology Institute, University of Surrey, Guildford GU2 7XH, UK*

[4]*Materials Centre, National Physical Laboratory, Queens Road, Teddington, Middlesex TW11 0LW, UK*

[5]*IRC in Biomedical Materials, Queen Mary, University of London, Mile End Road, London E1 4SN, UK*

Abstract. The study of the interaction of cells with nanoparticles or nanosurfaces is increasingly being performed using *in vitro* methods, largely with immortal cell lines. Immortal cell lines may behave differently from cells initially isolated from living tissue in these assays. Thus we have established a cell bank of primary human cell cultures stored at different stages from initial isolation through to senescence. We have examined the effects of culture age on proliferation upon carbon nanotube arrays, and adhesion to polystyrene surfaces. These data have been compared with an equivalent telomerase transformed line, and a commercially available line. Initial results suggest very large errors can be introduced into nanotoxicity and adhesion assays by inappropriate use of cell strains and lines.

Keywords: biocompatibility, cell culture, nanotubes, telomerase

*To whom correspondence should be addressed. S. L. James, School of Pharmacy and Biomolecular Science, University of Brighton, Moulsecoomb, Brighton, BN2 4GJ, UK, e-mail: S.L.James@bton.ac.uk

J. P. Blitz and V. M. Gun'ko (eds.), Surface Chemistry in Biomedical and Environmental Science, 205–214.

1. Introduction

Carbon nanotubes are a logical structural extension of the fullerene (C-60) allotrope of carbon, yielding carbon tubes of micrometer dimensions in length, but only a few nanometres in diameter. These may be manufactured in a number of forms, viz. single walled (SWCNT) double walled (DWCNT) and multiwalled (MWCNT), as well as being produced as individual nanotubes, or attached to an original catalytic base (array or brush form). CNTs can be modified in a number of ways, such as filling them with DNA, or attaching biomolecules to functionalised nanotubes.

During the last decade, advances in understanding of the physics and physical chemistry of carbon nanotubes (CNT) have led to an expanding field of applications, including important areas in the biological and medical sciences. Bioelectrodes, vaccine carriers, drug delivery systems, tissue engineering constructs, gene therapy delivery systems and biosensors have all been suggested as *in vivo* applications.[1-6] *In vitro* applications of biological importance would include the manufacture and sensing of DNA microarrays, and guidance of cultured cell growth. Furthermore, a range of non-biological applications have been discussed, including their use as nanowires and nanoswitches, which would imply relatively large scale manufacture of these materials. However, the biological safety of these nanotubes is an area of investigation still in its infancy. The literature is starting to show studies on nanotube-cell interactions only in the last two to three years, and two European research programmes investigating some aspects of this problem (Nanosafe and Nanoderm) are current. It is clear that materials, for which there is good safety data in bulk form, behave quite differently in nanoparticulate form in biological environments.[7,8]

The problem of the toxicity of nanoparticles has to be addressed at a number of levels, because such materials may not only enter the body by deliberate action, as in the applications above, but accidentally, during manufacture, or use in an *ex vivo* application. Adverse biological reactions therefore need to be studied at the histological level after exposure of a living animal model; at the tissue explant level, giving a greater degree of experimental control; at the cell culture level, giving the advantage of using an homogenous cell type; and at the molecular level, yielding more sophisticated information about patterns of gene up and down regulation after exposure to these nanotubes.

Of these approaches, currently the most widely used are the cell culture models widely employed elsewhere in toxicological studies. While it is common to prepare biomaterials for testing for biocompatibility with great care, cleaning surfaces and then analysing them with a variety of techniques yielding information on surface rugosity, surface energy and surface composition, for instance, less emphasis has been placed on the preparation of the cells to be used in such assays. The literature contains reports of work using whole animals, excised organ culture, and more relevantly, immortal, immortalised and primary cell types of various lineages. Frequently the cell type used is not clearly related to the system of relevance to the biomaterial, and no account has been taken of potential differences between primary and immortal cells during interactions with biomaterials.

A number of recent publications have considered the effects of CNTs on the lungs of rodents in *in vivo* situations (for example 7 and 8), and on a variety of cell cultures, for example uses HaCaT cells[9] and alveolar macrophages.[10] Of particular interest is the work of Cui et al.[11] and Monteiro-Riviere et al.[12] Both groups use cells confusingly referred to as HEK cells. In ref 11 these are human embryo kidney cells, an immortal line, while in Ref. 12 these are human epidermal keratinocytes, a primary culture. Cui et al. have performed an excellent, wide ranging study, including morphological analysis, Western blot protein analysis, flow cytometry, immunofluorescence and biochip analysis of cells after exposure to CNTs. However, this study cannot avoid the criticism that the cells employed are abnormal, as they are a transformed line. For instance, the authors show marked changes in gene expression associated with both cell cycle control, and apoptosis. However, the cells are genetically very abnormal, being hypotriploid human cells with a modal chromosome number of 64. Similarly, they perform adhesion studies with these cells, but this particular cell line is known to possess an unusual vitronectin receptor (ATCC Data sheet CRL-1573). We believe, therefore, that considerable attention must be paid to the cell type and status used in biocompatibility studies so that results from a range of workers are comparable.

2. Types of Models Available

2.1 IN VIVO MODELS

To study the physiological behaviour of particulates in mammals, clearly whole organisms are needed. This is particularly true when distribution studies are carried out, when translocation of particles from the lung or gut epithelium, in particular, are studied. However, such studies are complex to analyse at the cellular level as such a variety of cell types are involved in any tissue or organ. Similarly, tissue-tissue interactions via hormonal or cytokine signalling may confound an already complex situation. Furthermore, it might be argued that data derived from, for instance, rodents, may not be wholly applicable to humans where differences in metabolism may lead to inappropriateness of the model. There are also ethical considerations concerning animal welfare.

2.2 TISSUE PREPARATIONS

To retain some information furnished by animal models, but reduce the variability caused by whole animal models, many workers have resorted to excised tissue preparations. Classical preparations include excised rat gut models and various muscle models using organ baths. To further reduce variation, only parts of tissue may be used. For instance, we routinely use a sheep tracheal model in which the tracheal epithelium is stripped from the tracheal surface after a subepithelial injection of dispase.[13] This allows study of the epithelial reaction to insult, in particular mucus secretion by goblet cells, without the confounding involvement of mucus secretion by submucosal gland cells. Similarly, such preparations can be used to study ciliary beat frequency in response to insult without the involvement of innervation from the surrounding tissue. More recently, it has become possible to grow so-called organotypic tissues from primary cell cultures with appropriate culture conditions and growth factors, mechanical stimulii etc. For instance, we are able to obtain primary skin fibroblasts and produce a structure similar enough to skin epidermis to be able to replace epidermis on patients suffering epidermal destruction due to burns.[14] Such organotypic models may be valuable for testing nanoparticulate behaviour in the near future.

2.3 CELL CULTURES

Cell cultures commonly used to test biomaterials fall into three categories:

1. Immortal lines are readily available from commercial suppliers and are generally very easy to grow and maintain. However, these cells are almost always derived from a tumour, either in an animal or human, and display properties which might make them inappropriate for biomaterials testing. For instance, such cells frequently have abnormal and/or unstable chromosome numbers, they do not (by definition) show a natural Hayflick limit, ie they do not show natural senescence.[15] They also almost always do not display contact inhibition, ie the cessation of growth once they have become a confluent layer, and they tend to pile up on one another. This is particularly importance because this represents an abnormal interaction with a material surface, which is the crucial interaction in biomaterials testing. The appropriateness of immortal lines for biomaterials testing must be questioned, and such applications scrutinised carefully.
2. Immortalised cell lines. It is possible to transfect primary cells with constructs which overcome the natural limit to proliferation observed in most cells. A widely used transforming element is the large-T antigen derived from SV40, which, if stably incorporated into a host cell genome such as a human or rodent cell, will allow the cell to maintain its ability to proliferate theoretically indefinitely. Furthermore, such constructs can be used where the promoter for the large-T is under exogenous control, for instance temperature or presence of metals. The gene can then be virtually switched off after growth of cultures but prior to testing, so the cells revert to a near-normal phenotype. An alternative,

and perhaps even better understood immortalising construct is the gene encoding the enzyme telomerase. Senescence in many cell types is known to be associated with shortening of the telomeres of the chromosome ends. Such telomeres shorten by a specific amount at each cell division as an inevitable result of the mechanism of DNA replication. Rebuilding these telomeres via telomerase appears to delay senescence indefinitely.[16] Such immortalised cells may provide a suitable probe for the rational testing of cell-material interactions since they tend to retain near-normal phenotype.

3. Primary cell strains. Cells removed from living tissue can now be cultured successfully using rather more sophisticated culture media and conditions than used for immortal lines, but still well within the capabilities of any cell culture lab. These cell strains typically show contact inhibition, and a normal morphology and genetic compliment. The success of such cultures is demonstrated by the wide use of such cells as allografts to repair human tissues in living patients. The cell strains would seem ideal for testing biomaterials, but naturally exhibit a Hayflick limit, ie a finite number of population doublings as a result of increasing numbers of senescent cells in the culture. However, samples of such cells can be successfully frozen over liquid nitrogen at stages through their growth for later use. Clearly, if no cells are discarded, even a small number of cells from an initial biopsy will become a workably large number after ten to fifteen population doublings.

3. Potential Variabilities in Primary Cell Strains

It is conventional wisdom in cell culture laboratories that the same primary cell types may grow rather differently in two laboratories due to subtle variations in culture conditions. Variations which we have considered are:

1. Total number of population doublings (PDs). The more population doublings a culture has experienced, the greater number of senescent, or possibly near-senescent cells there will be in that culture. The problem will be that a culture at a low PD number may superficially look the same as a population with a higher PD number, and be growing at the same rate, but might behave quite differently in a biocompatibility assay.
2. Number of PDs per passage. Clearly if cells are grown in small flasks, they will have to be passaged more frequently than a culture growing at the same PD rate, but in a larger flask. This may be important as passaging includes treating the adherant cells with trypsin to release them from the flask plastic, and so rapidly passaged cells will have been trypsinised more frequently than the cells going through the same number of PDs at a lower passaging rate.
3. Culture pH. This is controlled by the proportion of the atmosphere in the incubator which is carbon dioxide. So cells will usually grow at 5% or 10%

carbon dioxide, but the culture medium will display a different pH at these two concentrations.

4. Age of, and genetic polymorphisms in the original human donor. Little can be done to control this variation, particularly as most polymorphisms are not expressed, and their effect on assays unknown.

Briefly, human skin samples are taken by surgical biopsy under local anasthesia and after treatment with dispase, the epidermis and dermis seperated manually. Both samples are trypsinized, and epithelial keratinocytes grown in Rheinwold and Green medium,[17] while skin fibroblasts are grown in DMEM + FCS. Similarly, corneal limbal biopsies are rescued if not appropriate for allografting, and treated as above, except that corneal keratocytes are recovered from tissue outgrowths. Chondrocytes are obtained from human cartiledge shards, digested overnight in collagenase type II, and cells grown in a standard chondrocyte medium.

The storage of primary strains must not lead to variation between cell deposits. In order to overcome this problem we use a standard freezing and thawing protocol to store primaries over liquid nitrogen. Our cell bank currently holds about 700 deposits of the following human cell types; skin keratinocytes, skin epithelial cells, corneal limbal epithelial cells, corneal fibroblasts, and chondrocytes.

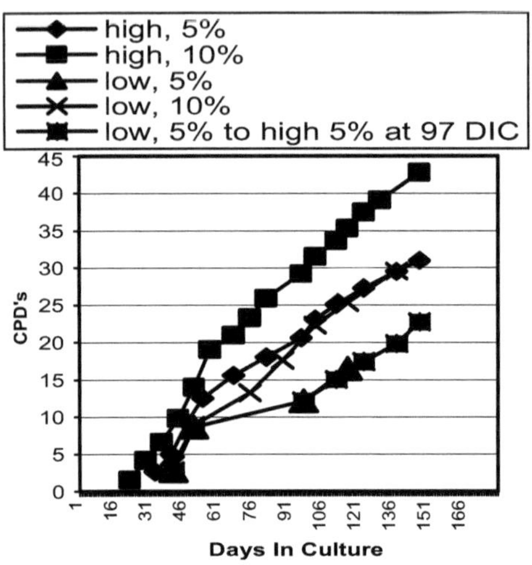

Figure 1. Growth rates of human primary skin fibroblasts at high and low PD/passage rate, and 5% and 10% carbon dioxide.

These cells have all been grown under conditions which vary the number of PDs per passage, and at 5% or 10% CO_2 in the incubator. and samples frozen at each passage up to senescence. Standard protocols are also used for the initial isolation of the cells from human skin, cartiledge and limbal biopsies.

High and low PD/passage ratios and high and low carbon dioxide concentrations clearly affect initial growth rates, but with time, growth rates become similar in all cultures of one cell type (Figure 1).

4. Exposure of Cells to Nanotextured Surfaces

To produce a surface with an observable texture at the nanometer range, we use a carbon nanotube array for toxicology testing. Nanotubes are produced by catalytic growth from a nickel plate, yielding multiwalled nanotubes of 2-3 microns in length, by about 4-5 nanometers in diameter. These nanotubes are produced in an AC field, so they are lying on the nickel surface, rather than standing up in a brush-like form. Controls used are an uncoated nickel plate, and a polished polystyrene surface to monitor natural growth rates.

Human skin fibroblasts were generated from our cell bank, and vigorous growth ensured. Cells of 12 PDs and 31 PDs after isolation were used, together with fibroblasts transfected with a telomerase construct at 12 PDs, and shown to have extended lifespan after 44 PDs. Cells were applied to the surfaces, test and control, in known quantities (40 cells per surface), and allowed to grow under optimum conditions for 5 days. Surfaces were gently rinsed to remove non-adherant cells, and stained with propidium iodide (nuclear flourescent stain), and counted under flourescence microscopy.

The results can be expressed as the percentage of cells grown on the arrays, relative to the number of cells which would have grown under optimal conditions as derived from the growth on the control plates.

Fibroblasts at 12 PDs-	$97 \pm 10\%$ of control value
Fibroblasts at 31 PDs-	$5.1 \pm 5\%$ of control value
Telomerase transfected fibroblasts-	$84 \pm 9\%$ of control value

Within experimental error the fibroblasts having undergone 12 PDs, and the telomerase transfected cells at 12 PDs but having undergone 44 PDs, show similar results. However, the cells having undergone 31 PDs in primary culture show no growth, the figure representing cells placed on the array at the start of the experiment.

The critical implication is that, as can be seen from Figure 1, the two cell cultures were growing at comparable rates, and morphologically looked the same: they are the same culture at two different growth stages. However, if these were used in a nanotoxicology assay, without attention to their culture proliferative age, completely contradictory results would be obtained.

5. Adhesion of Cells of Various Ages to a Polystyrene Surface

The same cell cultures used in the previous experiments were employed, except the low PD culture had been grown through two more PDs to 14 rather than 12. Human primary fibroblasts were seeded onto a polystyrene surface at a predetermined density (10^5 cells per surface), and allowed to attach over a period of 100 mins. At six time points the flasks were gently agitated, and the supernatent drawn off and the cell content counted. This simple procedure yielded a characteristic and highly reproducable attachment/time curve (Figure 2).

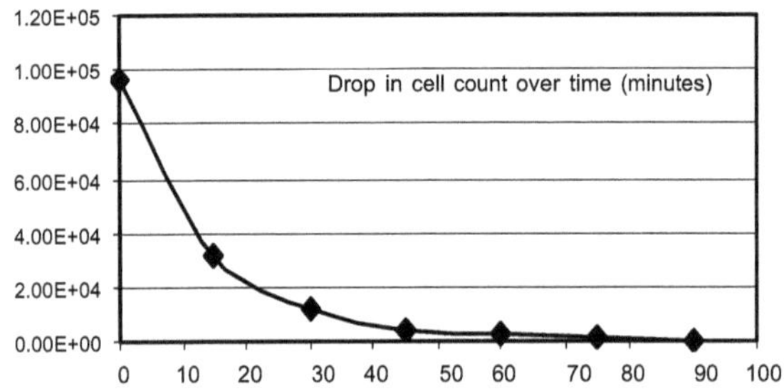

Figure 2. Unattached human primary fibroblasts vs time at 37 degrees centigrade.

For ease of comparison, the percentage attachment at 15 mins was taken, and a comparison drawn between Human primary skin fibroblasts at 14 and 31 PDs, the telomerase transfected human fibroblasts at 44 PDs, and a commercially available mouse immortal fibroblast line, 3T3 cells (Figure 3).

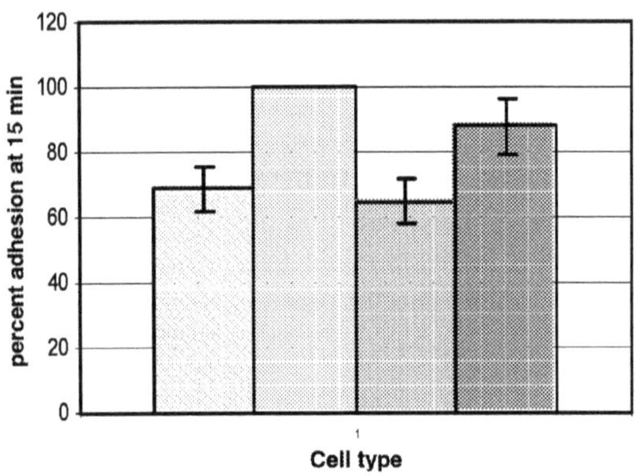

Figure 3. Cell attachment to a polystyrene surface at 15 mins. Comparison of four different cell types at 37 degrees centigrade.

It is interesting to note that when the adhesion of 3T3 cells to the surface was re-tested at 20 degrees centigrade (room temperature) there was zero adhesion at 15 mins and only 52% adhesion at 30 mins. Therefore, the human primary fibroblast adhesion experiment was repeated at both 37 and 20 degrees centigrade (Figure 4).

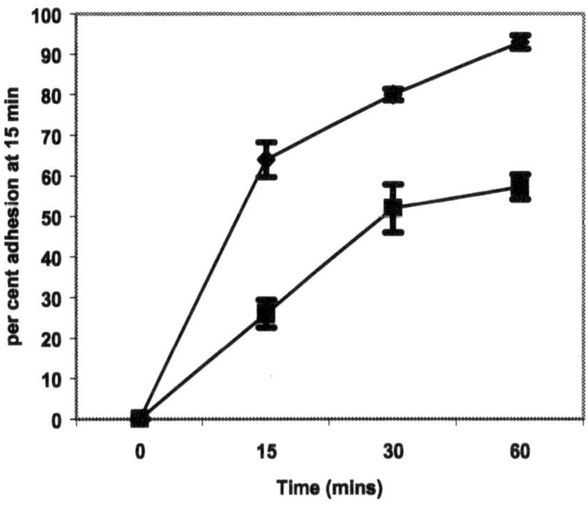

Figure 4. Primary human skin fibroblasts adherent to a polysyrene surface at 20 and 37 degrees.

6. Conclusion

A variety of assays are being reported in the literature which measure the interaction of cells with biomaterials, be they surfaces of various textures or micro or nano-particulates. While considerable attention has been paid to proper preparation of the materials and their surfaces, less has been paid to the cells used as probes in such assays. We suggest that where possible, primary human cell cultures at relatively low PD numbers from isolation are used, of a cell type directly relevant to the material application, or in toxicity studies, the exposed tissue. We suggest very significantly erroneous conclusions may be drawn from assays using inappropriate cells. If sufficient quantities of low PD primaries are not available, it may be possible to use telomerase transfected cells with extended lifespan, but these must first be verified against the non-transfected cells in the assay to be employed. Furthermore, care must be taken to conduct such assays at a controlled temperature, preferably at the temperature the material will be subjected to in the body.

References

1. J. Wang and M. Musameh, Carbon nanotube screen-printed electrochemical sensors, *Analyst* 129, 1-2 (2004).
2. Y. H. Lin, F. Lu, Y. Tu, and Z. F. Ren, Glucose biosensors based on carbon nanotube nanoelectrode ensembles, *Nano Lett.* 4, 191-195 (2004).
3. D. Pantarotto, C. D. Partidos, J. Hoebeke, F. Brown, E. Kramer, J. P. Briand, S. Muller, M. Prato, and A. Bianco, Immunization with peptide-functionalized carbon nanotubes enhances virus-specific neutralizing antibody responses, *Chem. Biol.* 10, 961-966 (2003).

4. Y. Lin, S. Taylor, H. P. Li, K. A. S. Fernando, L. W. Qu, W. Wang, L. R. Gu, B. Zhou, and Y. P. Sun, Advances toward bioapplications of carbon nanotubes, *J. Mat. Chem.* 14, 527-541 (2004).
5. E. Katz and I. Willner, Integrated nanoparticle-biomolecule hybrid systems: Synthesis, properties, and applications, *Angew Chem. Int. Edit.* 43, 6042-6108 (2004).
6. R. Singh, D. Pantarotto, D. McCarthy, O. Chaloin, J. Hoebeke, C. D. Partidos, J. P. Briand, M. Prato, A. Bianco, and K. Kostarelos, Binding and condensation of plasmid DNA onto functionalized carbon nanotubes: Toward the construction of nanotube-based gene delivery vectors, *J. Am. Chem. Soc.* 127, 4388-4396 (2005).
7. C. W. Lam, J. T. James, R. McCluskey, and R. L. Hunter, Pulmonary toxicity of single-wall carbon nanotubes in mice 7 and 90 days after intratracheal instillation, *Toxicol. Sci.* 77, 126-134 (2004).
8. D. B. Warheit, B. R. Laurence, K. L. Reed, D. H. Roach, G. A. M. Reynolds, and T. R. Webb, Comparative pulmonary toxicity assessment of single-wall carbon nanotubes in rats, *Toxicol. Sci.* 77 , 117-125 (2004).
9. A. A. Shvedova, V. Castranova, E. R. Kisin, D. Schwegler-Berry, A. R. Murray, V. Z. Gandelsman, A. Maynard, and P. Baron, *J. Toxicol. Environment. Health.* A 66(20), 1909-1926 (2003).
10. G. Jia, H. F. Wang, L. Yan, X. Wang, R. J. Pei, T. Yan, Y. Zhao, and X. Guo, *Env. Sci. Technol.* 39(5), 1378-1383 (2005).
11. D. Cui, F. Tian, T. Ozkan , M. Wang, and H. Gao, *Toxicol. Lett.* 155, 73-85 (2005).
12. N. A. Monteiro-Riviere, R. J. Nemanich, A. O. Inman, Y. Y. Wang, and J. E. Riviere, *Toxicol. Lett.* 155(3), 377-384 (2005).
13. S. M. Clancy, M. Yeadon, J. Parry, M. Yeoman, and M. I. Lethem, Endothelin-1 inhibits mucin secretion from ovine airway epithelial goblet cells, *Am. J. Respir. Cell. Mol. Biol.* 31, 663-671 (2004).
14. R. Martin, S. Bevan, J. Boorman, I. Grant, S. E. James, I. Jones, N. Parkhouse, R. Ng, P. Rubin, and B. Woodward, Cultured keratinocytes; experimental and clinical directions in the quest for tissue engineered new skin, in: *Cultured Human Keratinocytes and Tissue Engineering Skin Substitutes* (Pubs. Thieme Verlag, Stuttgart, 2001), pp. 107-116.
15. J. C. Houck and V. K. Sharma, Functional failures of cultured human diploid fibroblasts after continued population doublings, *Proc. Soc. Experim. Biol. Med.* 137(1), 331 (1971)
16. A. G. Bodnar, M. Ouelette, M. Frolkis, A. E. Holt, C.-P. Chiu, G. B. Morin, C. B. Harley, J. W. Shay, S. Lichsteiner, and W. E. Wright, Extension of lifespan by introduction of telomerase into normal human cells, *Science* 279, 349-352 (1998).
17. J. G. Rheinwald and H. Green, Formation of a keratinizing epithelium in culture by a cloned cell line derived from a teratoma, *Cell* 6, 331-343 (1975).

PHYSICAL CHARACTERISATION OF A POLYCAPROLACTONE TISSUE SCAFFOLD

PAUL V. GRANT, CLAUDIA M. VAZ, PAUL E. TOMLINS*,
Division of Engineering and Process Control, National Physical Laboratory, Hampton Road, TW11 0LW
United Kingdom
LYUBA MIKHALOVSKA, SERGEY MIKHALOVSKY, STUART JAMES
Department of Pharmacy and Biomolecular Sciences, University of Brighton, Cockcroft Road, Brighton, BN2 4GJ
United Kingdom
PANKAJ VADGAMA
IRC in Biomedical Materials, Queen Mary, University of London, Mile End Road, London, E1 4NS
United Kingdom

Abstract. The suitability of tissue scaffolds to act as temporary housings for culturing cells depends on their size, shape and structural characteristics. Factors such as porosity, the type of pores present and their size distribution play a key role in ensuring appropriate levels of metabolite transport as well as influencing scaffold degradation behavior and any associated changes in mechanical resilience. This paper presents the findings of a preliminary study of the structure of a range of porous polycaprolactone tissue scaffolds manufactured using the particulate leaching technique.

Keywords: tissue scaffolds, mercury porosimetry, capillary flow porometry, scanning electron microscopy, image analysis

* To whom correspondence should be addressed. Paul Tomlins, Division of Engineering and Process Control, National Physical Laboratory, Hampton Road, Teddington, Middlesex, TW110 LW, United Kingdom; email: paul.tomlins@npl.co.uk

J.P. Blitz and V.M. Gun'ko (eds.), Surface Chemistry in Biomedical and Environmental Science, 215–228.

1. Introduction

Porous foams manufactured from a range of polymers including mixtures of polylactic acid, polycaprolactone and hyaluronic acid are being used a temporary housings in which to culture cells.[1-3] These tissue scaffolds are typically relatively stiff materials that degrade over a period of time either *in vitro* or *in vivo*. Many papers have been published on this generic group of tissue scaffolds[1-3] in the search for materials and structures, which can be used with specific cell types to produce new functional tissue. Ideally this new tissue should be identical, both phenotypically and genotypically, to that of the native material from which the cells have been harvested. There are many well recognised issues that remain to be addressed in achieving these objectives, including the practical problems of providing a suitable culture environment within the scaffold and detecting minor changes in both the short and long-term behaviour of the cells. Exactly what a suitable culture environment within a scaffold means is open to debate; it is clear that cells in the core of a developing tissue construct will require an appropriate level of oxygen to prevent hypoxia, a supply of nutrients, removal of waste products and some stability in the local pH. The increase in cell density that occurs over time will increase demand for nutrients and oxygen, and the production of waste. Increased cell density, the build up of extra-cellular matrix and protein accretion will inevitably lead to a reduction in the diameter or even blockage of pores that will impact on the diffusivity of molecules through the scaffold. Time dependent degradation of the scaffold, that may or may not be enhanced by enzymes produced by the growing cells, adds to the complexity of this situation by altering the local pH, local chemistry and pore size distribution. The kinetics of this process is also affected by the presence of adsorbed proteins and attached cells.

The majority of publications in tissue engineering do not dwell on the structural characteristics of the scaffolds that are currently being used in this developing field either in the short- or long-term. Indeed most investigators[1,2] report some average pore size dimensions based on images derived from electron microscopy. These data are usually in the micrometre range and may include some measure of the range of pore sizes, i.e. upper and lower bounds. This lack of detailed data makes it very difficult to optimise the structure of scaffolds, manufacturing processes or indeed to begin to understand the complex perfusion characteristics of a cell-seeded scaffold over time. In this paper we compare structural information obtained from experimental data generated by a range of techniques including imaging, intrusive methods and density measurements for a scaffold manufactured from polycaprolactone.

2. The Relevance of Structural Information

The structure of tissue scaffolds is very dependent on the manufacturing route used to produce them: more controllable methodologies such as direct deposition tend to generate 'simpler' structures[4] than, for example, particulate leaching where the final structure depends on factors such as the rate of solvent evaporation and ambient humidity.[5] Porous structures usually consist of a range of different pore sizes and pore types. The current IUPAC definitions[6] of pore sizes cover a much smaller length scale than encountered in tissue engineering and are more appropriate for the structure of catalysts: here we use the term 'pore' regardless of diameter. Pores can also be categorized as closed, open or blind-end, terms that reflect their connectivity with external surfaces. Closed pores are voids within a material, blind-end pores are voids that are open on one side to an external surface and open pores are in contact with at least two external surfaces. All of these pore types serve a useful function within a tissue scaffold; closed and blind end pores can facilitate gas diffusion through walls by reducing the path length. However the bulk of molecular transport and all cell movement takes place through the network of interconnected pores that permeate the scaffold. The path length through open pores can be very tortuous depending on the pore interconnectivity.

The pore size distribution is often neglected in tissue engineering publications or determined over a narrow length scale ranging from 10's to 100's of micrometers despite its key role in scaffold performance. There is evidence to suggest that the pore sizes in a tissue scaffold can extend from nanometers to several hundred micrometers.[7] The detailed distribution of pore sizes over this extended range will impact on the rate at which the scaffold degrades, molecular transport through the structure and on mechanical performance.

A range of techniques are being assessed for their potential to characterize tissue scaffolds that include some of the methods used in this study.[8]

3. Experimental

Polycaprolactone (PCL, density 1.145 gcm^{-3}, Aldrich; product code 440744, M_n = 86.8 kDa, M_w = 136 kDa) is a biodegradable polyester extensively used in medical applications. PCL is a semicrystalline polymer that has a melting point close to room temperature and a T_g of -60°C. Tissue scaffolds can be made from PCL using the particulate leaching method: a solution of PCL in chloroform (weight ratio 1:9) is poured over previously sieved sodium chloride particles (NaCl) in a petri-dish. Evaporation of the solvent results in the formation of a polymer matrix containing a high proportion of NaCl particles that can be removed by soaking the polymer-salt composite in deionised water for at least 48 hours. Adjusting the grain size distribution of the porogen and the proportion used in the manufacturing process can vary the level of porosity and the pore size distribution in the PCL matrix. This particulate leaching method can easily be used to manufacture highly porous scaffolds that have a high degree of interconnectivity as shown in Figure 1. The grain size distributions of the NaCl porogen used to produce the scaffolds in this study are shown in Figure 2 and details of the samples produced are given in Table 1.

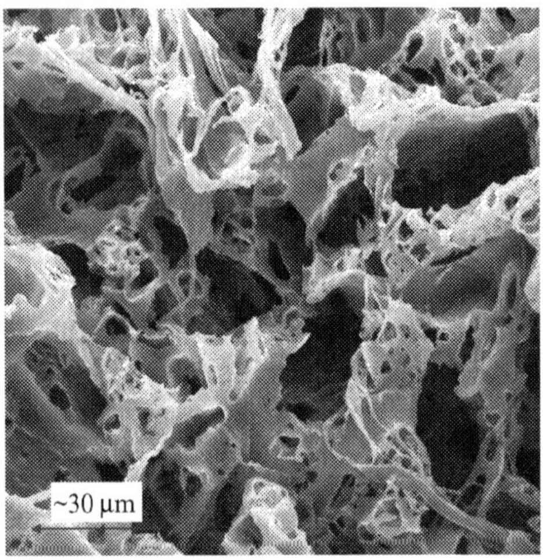

Figure 1. The highly porous, interconnected structure of a particulate leached PCL tissue scaffold.

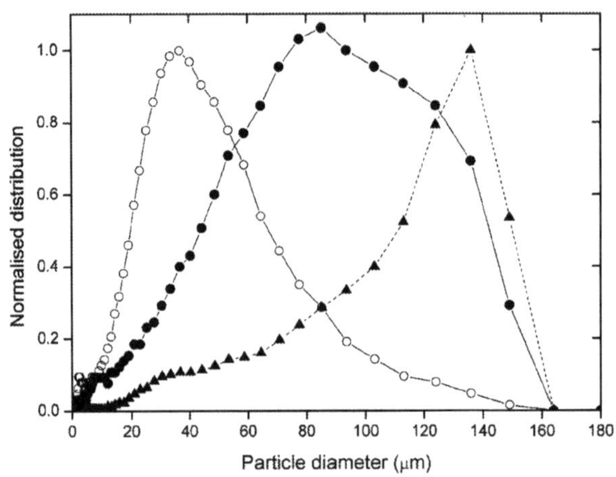

Figure 2. The normalized particle size distribution of the salt grains used to produce tissue scaffolds can be obtained by light scattering (Malvern Mastersizer series 2600c).

Table 1. In theory the porosity of a scaffold and its pore size distribution is determined by the proportion of salt to PCL and the size of the salt grains used in its manufacture.

Scaffold sample	Theoretical scaffold porosity (%)	Salt fraction	Salt grain sizes based on sieving (μm)
A	62	I	$0 < d < 53$
B	62	II	$53 < d < 106$
C	62	III	$106 < d < 125$
D	54	II	$53 < d < 106$
E	66	II	$53 < d < 106$

4. Structural Characterisation

4.1. DENSITY DETERMINATIONS

A simple method for estimating scaffold porosity is to weigh a scaffold and measure its volume; its porosity can then be derived if the density of the solid material is known. In practice measuring the volume of an object directly can be experimentally challenging, however this issue can be overcome by employing Archimedes Principle, which states that the volume of liquid displaced by an object completely immersed in it is equal to the volume of the object. This relationship is also used to determine the density of an unknown material either by measuring the decrease in weight of a body due to buoyancy (buoyancy method) or by measuring the weight of the fluid or gas displaced (e.g. helium pycnometry).

The buoyancy method relies on weighing the sample in air, w_a and in a wettable liquid, w_{fl}, e.g. water of density ρ_{fl}. The weight of liquid displaced due to the volume of the sample is given by $w_{(fl)} = w_a - w_{fl}$ and the apparent solid density[†], ρ_s, is then defined by equation 1:[9]

$$\rho_S = \frac{W_a}{W_a - W_{fl}} \rho_{fl} \,. \tag{1}$$

The apparent porosity[‡], π_a is given by

$$\pi_a = \frac{W_{soaked} - W_a}{W_{soaked} - W_{fl}} 100 \tag{2}$$

† The apparent solid density is the ratio of the mass of the dry material to its apparent solid volume. Apparent solid volume is the sum of the volume of closed pores and true volume where the true volume is that of a body occupied by solid material excluding all forms of porosity (BS EN 623-2, 1993).

‡ Apparent porosity is the ratio of total volume of open pores in a porous body to its bulk volume (BS EN 623-2, 1993).

Where w_{soaked} is the weight of the sample soaked in fluid in air.

Table 2 lists the apparent densities and porosities for samples A to E. From this table it is apparent that there is an excellent agreement between the expected porosity values (based on the known proportions of porogen and PCL) and those determined from weighing. There is a less satisfactory agreement between the measured densities of the tissue scaffolds and that determined for an as received granule. This disparity could be due to incomplete wetting of the sample or the presence of closed pores, both of which increase the buoyancy of the sample reducing w_{fl}. Helium pycnometry can be used to distinguish between these two effects.

The basic technique relies on measuring the material volume within the scaffold from a change in helium pressure due to the presence of the scaffold in a known volume of gas. Helium is able to fully penetrate the porous structure of the scaffold without any encountering any problems with surface tension. The true densities[§] obtained using this method are also shown in Table 2. The measured values for the true density of the PCL that forms the walls of the scaffold are within 10% of the as received non-porous granules which implies that the differences obtained using the buoyancy method are due to surface tension effects and incomplete wetting. The densities of the as received granules are in very good agreement between the two techniques.

Helium pycnometry can also be used to obtain an estimate of porosity providing that the porous sample is sealed to prevent gas penetration. There are experimental problems that need to be addressed to ensure that the seal is effective and that it doesn't significantly alter the volume or weight of the sample which are beyond the scope of this study.

Table 2. Material density and scaffold porosity comparisons.

Sample	Sample code	Apparent density (Buoyancy method) (gcm^{-3})	Apparent porosity (%)	Theoretical porosity (%)	True density (Helium pycnometry) (gcm^{-3})
A	B18504/2	0.796 ± 0.003	63	62	1.148 ± 0.009
B	B18504/3	0.777 ± 0.003	62	62	1.172 ± 0.005
C	B18504/4	0.808 ± 0.007	62	62	1.175 ± 0.031
D	B18504/6	0.911 ± 0.002	53	54	1.173 ± 0.013
E	B18504/5	0.735 ± 0.003	68	66	1.081 ± 0.005
PCL granules		1.116 ± 0.022			1.162 ± 0.025

[§] True density is the ratio of the mass of dry material to its true volume.

4.2. IMAGE ANALYSIS

Figure 1 is a scanning electron micrograph of a PCL scaffold. From this figure it is apparent that the structure is complex comprising of a network of highly interconnected pores, the walls of which have windows or fenestrations. Figure 3A is a lower magnification image of the same structure. Both images can be used to provide estimates of porosity based on the proportion of 'wall' to 'pore' area. The fundamental step in this analysis is to define those parts of the images that form the walls and those parts that correspond to the pores. This is achieved by plotting a histogram of the greyscale intensities of each pixel within the image as shown in Figure 3B. Monochrome images are subdivided into 256 divisions spanning a range from 0 (pure black) to 256 (pure white). The histogram in Figure 3B shows a typical bimodal distribution ranging from dark grey to light grey pixels. The challenge of thresholding is to define a point along the x axis of this histogram that splits the overlapping distributions into contributions from the walls and pores.

Many approaches to defining the threshold have been published.[10] In this study we have used visual inspection to define this point as well as an automated routine within the software package image.[11] This routine assumes a starting value for the threshold and averages all the pixels that fall on the greyscale above and below this value. The two averages are then further averaged to produce an 'actual' threshold. If the actual threshold is less than the initial threshold then the initial value is adjusted. This process is repeated until the two threshold values coincide. An alternative approach, proposed by Otsu[12], utilises the variance as a measure of homogenity and minimizes the variance in the pixels associated with the pores and walls.

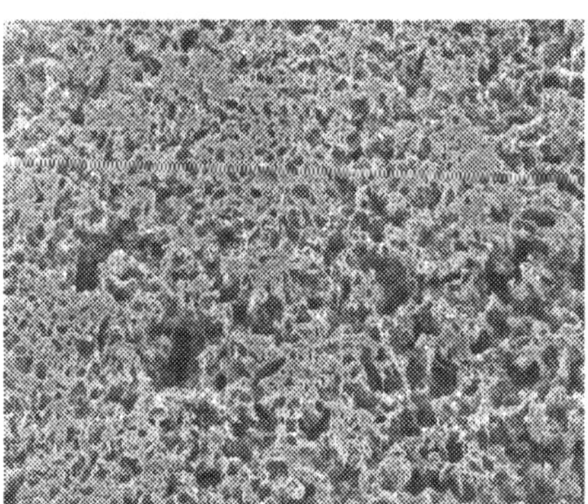

(A)

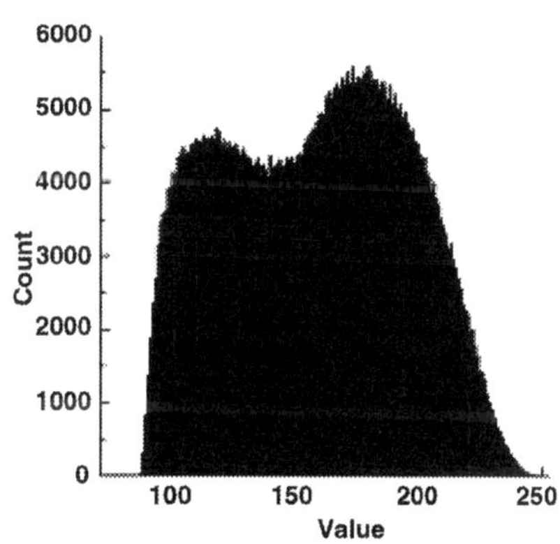

(B)

Figure 3. (A) A typical low contrast micrograph of a PCL scaffold (sample B18304/1, image 10.tif) and (B) its corresponding greyscale histogram.

Thresholding produces a black and white binary image that represents contributions from noise, pores and walls. The noise contribution can be removed from the porosity calculation by assigning a lower pixel limit below which information will be ignored. The choice of cut-off can have a significant impact on the calculated porosity for fine structures i.e. those where the wall thicknesses are of the order of a few pixels. A common practical problem encountered in analysing 2-D micrographs of 3-D structures is that the greyscale intensity of wall material located at the bottom of a pore is often comparable to that located at the rim. Consequently the image analysis software is unable to detect the actual boundaries of the pore. This issue is more of a problem in interpreting high magnification images than low magnification images which can also produce more statitically robust data by analysing a greater number of pores.

4.3. INTRUSION TECHNIQUES

Scaffold porosity and information on the pore size distribution can be obtained from intrusion techniques. The most commonly used methods are mercury porosimetry and capillary flow porometry. In mercury porosimetry the pressure required to fill a tissue scaffold with non-wetting mercury is monitored over a set period of time. Higher pressures are required to fill small pores than large pores; a fact that can be exploited using the Washburn equation[13] to extract structural information where D is the diameter of the pore at a particular differential

$$D = -(4\gamma \cos\theta)/p \qquad (3)$$

pressure p, θ is the surface tension of mercury, and γ is the contact angle of mercury with the material. This method assumes that the pores are interconnected and have a regular geometry with a circular cross-section.[14,15] The porosities and pore size distributions derived from mercury intrusion porosimetry are listed in Table 4.

Table 4. Porosities and pore size distributions derived from mercury intrusion porosimetry.

Sample	Sample identity	Theoretical NaCl particle distribution (μm)	Pore size distribution (μm)	Theoretical Porosity (%)	Measured Porosity (%)
A	B18504/1	$0 < d < 53$	$7 < d < 60$	62	83
B	B18504/4	$53 < d < 106$	$10 < d < 100$	62	67
C	B18504/3	$106 < d < 125$	$40 < d < 120$	62	43
D	B18504/6	$53 < d < 106$	Doesn't reach unity*	54	-
E	B18504/5	$53 < d < 106$	$10 < d < 100$	66	72

Note, the porosity values are calculated using a measured density for the solid material
*data ignored due to low permeability (see Table 5)

Capillary flow porometry is essentially mercury porosimetry in reverse where the increasing gas pressure required to displace a fluid (not mercury) from a fluid filled scaffold is monitored as a function of time. Higher pressures equate to smaller pore sizes again following Washburn's equation i.e.

$$D = (4\gamma \cos\theta)/p \tag{4}$$

where D is the pore diameter at a particular location, p is the pressure required to displace fluid at that point. γ and θ are the surface tension and contact angle of the wetting fluid respectively. The pressure required to initiate the formation of the first bubble is referred to as the bubble point and equates to the largest pore diameter and lowest pressure. Measurement of the volume of displaced fluid provides a direct measure of pore volume as a function of gas pressure. In practice the pressure and flow rates through wet and dry samples are determined to give 'wet' and 'dry' curves.[16,17] The pore size distribution is calculated from the difference between the wet and dry curves. The median flow pore diameter corresponds to 50% cumulative flow. The median pore diameter, and range of pore diameters for each of the scaffold samples are listed in Table 5. The rate, Q at which air flows through a sample of length, L and cross sectional area, A for a range of differential pressures, ΔP provides a measure of permeability, k, according to Darcy's Law

$$Q = (kA\Delta P)/|L \tag{5}$$

Values for the Darcy permeability constant for the PCL scaffolds are given in Table 5.

Table 5. Pore diameters obtained for PCL scaffolds from capillary flow porometry.

Sample	Sample identity	Theoretical porosity (%).	Darcy permeability constant	Minimum pore diameter (µm)	Median pore diameter (µm)	Largest pore diameter (µm)	NaCl particle size range (µm)
A	B18504/1	62	0.3382	0.2	4.1	25.6	$0 < d < 53$
B	B18504/2	62	0.5509	0.3	6.3	50.9	$53 < d < 106$
C	B18504/3	62	0.3338	0.3	5.9	54.9	$106 < d < 125$
D	B18504/6	54	0.0116	Cumulative flow not reached			$53 < d < 106$
E	B18504/5	66	1.344	0.3	6.3	54	$53 < d < 106$

5. Discussion

5.1 DETERMINATION OF POROSITY

The porosity of a scaffold is an often quoted characteristic of tissue scaffolds, *per se* porosity is a measure of the void volume contained within a porous material and as such conveys no information about the type of pore present, their size, shape or how interconnected they are. Measures of porosity are susceptible to the presence of defects, i.e. voids produced during manufacture that are commonly found in samples produced by particulate leaching.

Three methods have been used in this investigation to estimate the porosity of PCL tissue scaffolds, namely weight determination, mercury porosimetry and analysis of scanning electron micrographs. The results are shown in Figure 4.

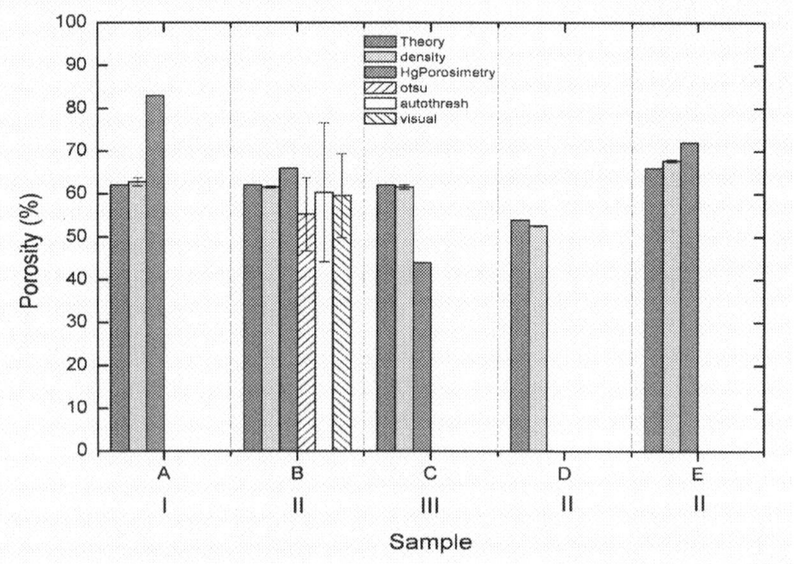

Figure 4. Comparison between scaffolds porosity obtained via different measurement techniques for samples A to E containing different salt fractions.

From this figure it is evident that estimates of porosity based on weight determination provides the best agreement with that expected according to the ratio of porogen to PCL used to manufacture the scaffolds. The measurements that contribute to the porosity rely on fluid penetration into the scaffold, which can be hampered by surface tension and trapped air. The influence of these factors can be significantly reduced by soaking the sample in the wetting fluid for >24 hours with repeated agitation to remove air bubbles, by adding a very small amount of wetting agent or by partially evacuating the sample. However, the calculation of scaffold porosity from buoyancy method measurements is much less sensitive to incomplete wetting than density determinations as can be seen from equations (1) and (2).

Estimates of scaffold porosity can also be obtained from image analysis. The success of this approach is highly dependent on the contrast between the walls of the structure and the voids. In analyzing 2-D scanning electron micrographs, e.g. figure 1, similarities between the greyscale intensity of material clearly at the base of a pore and that near to the top surface makes it difficult to automatically define regions that correspond to pores. This could be overcome by analyzing reconstructed 3-D images based on the analysis of, for example, stereo pairs of 2-D images. However this approach is rarely, if at all used to examine tissue scaffolds. Despite this limitation a reasonable agreement between different thresholding methods can be obtained (Figure 4) for samples that have been manufactured to have the same level of porosity. The uncertainty in these data is probably too high to be of much use as an approach to optimizing scaffold structures or as a quality control tool. The variance in the estimated porosity increases with increasing image magnification for this particular complex scaffold geometry in which the pores are irregularly shaped and the walls have apertures.

Table 4 lists porosity values determined from mercury porosimetry. These preliminary values agree with those expected based on the proportion of porogen used to make the scaffolds for samples B and E. However there are significant differences between the porosity estimates for samples A and C and those expected based on sample composition. Intrusion methods were not suitable to characterize sample D due to its low permeability.

5.2 DETERMINATION OF PORE SIZE DISTRIBUTIONS

The pore size distribution is key to its performance influencing solute transport, cell migration and scaffold degradation[18] but not easy to determine, especially for complex structures where the definition of a pore is open to debate. The analysis of 2-D images obtained from scanning electron microscopy relies on accurate thresholding, i.e. on being able to differentiate between a cavity and a wall. This procedure is open to ambiguity especially for high magnification images, as the greyscale intensity of material located at the base of a pore can be comparable to that at its rim. This situation improves significantly using lower magnification images that also improves the statistical significance of the data. However resolving the pore size distribution still depends on being able to define a clear continuous edge around a pore in order to map its size. This process for the complex scaffolds studied here is open to substantial error. On this basis, generating a pore size distribution based on the analysis of images of complex structures is not to be recommended.

Figure 5 compares the pore size distributions of the scaffold computed from the intrusive techniques of capillary flow porosimetry and mercury porometry. From this figure it is apparent that the range of pore sizes derived from capillary flow porometry occurs over a smaller length scale than those based on mercury porometry data. This difference is expected since underlying physics of the former technique is sensitive to the bottleneck regions of the pores as shown in Figure 6A. The pore diameters based on mercury porosimetry are based on a range of diameters that all contribute to the pressure –time curve from which the pore size distribution is based as shown in Figure 6B.

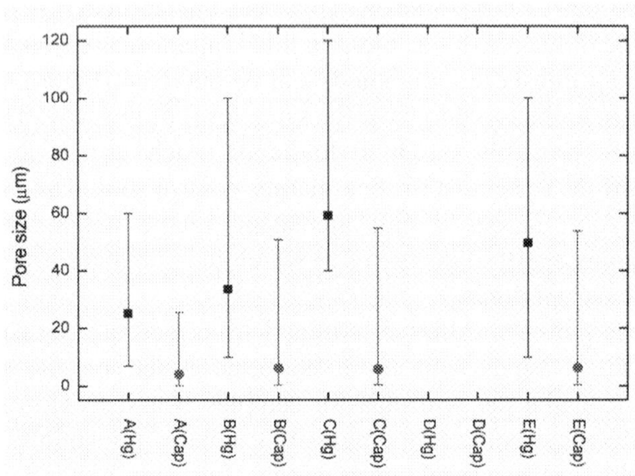

Figure 5. Comparison between scaffolds pore distributions obtained via different measurement techniques (cap = capillary flow porometry, Hg = mercury porosimetry, ■ = median value, ● = median value). The error bars represent the span of the distribution.

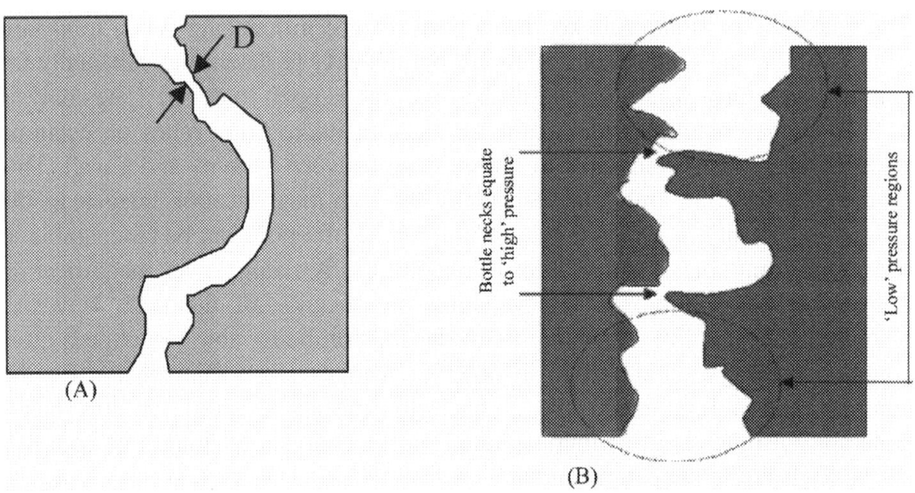

Figure 6. (A) Capillary flow porosimetry is sensitive to the bottlenecks within a porous structure (B) All regions of the pores contribute to the pressure – time curve that is obtained from mercury porometry.

6. Conclusion

Various techniques, such as intrusion methods, image analysis and weight determinations are used to provide estimates of the porosity and the distribution of pore sizes within a tissue scaffold. This preliminary study of the structural features of particulate-leached PCL scaffolds has shown that reliable measures of porosity can be obtained from simple weighing measurements (Buoyancy method).

The distribution of pore sizes can be obtained from both mercury porosimetry and capillary flow porometry. These distributions are only representations of the actual scaffold structure reflecting the limitations of the underlying physics behind each technique. For this reason it is very difficult to compare pore size distributions for complex structures, such as particulate-leached tissue scaffolds.

Acknowledgements

The authors would like to thank Sam Gnaniah for determining the level of crystallinity within the polycaprolactone and Adam Calver for performing many of the density measurements. This work was funded by the United Kingdom Department of Trade and Industry as part of its programme of reseach on Materials for Processing and Performance (Project MPP 4.2: Physical Characterisation of Tissue Scaffolds).

References

1. A. S. Goldstein, G. Zhu, G. E. Morris, R. K. Meszlenyi, and A. G. Mikos, Effect of osteoblastic culture conditions on the structure of poly(DL-lactic-co-glycolic acid) foam scaffolds, *Tissue Eng.* 5, 421-433 (1999).
2. X. M. Mo, C. Y. Xu, M. Kotaki, and S. Ramakrishna, Electrospun P(LLA-CL) nanofiber: a biomimetic extracellular matrix for smooth muscle cell and endothelial cell proliferation, *Biomater.* 25, 1883-1890 (2004).
3. L. A. Solchaga, J. E. Dennis, V. M. Goldberg, and A. I. Caplan, Hyaluronic acid-based polymers as cell carriers for tissue-engineered repair of bone and cartilage. *J. Orthop. Res.* 17, 205-123 (1999).
4. E. Sachlos and J. T. Czernuska, Making tissue scaffolds work. Review on the application of solid freeform fabrication technology to the production of tissue engineering scaffolds, *Europ. Cells Mater.* 5, 29-40 (2003).
5. A. G. Mikos and J. S. Temenoff, Formation of highly porous biodegradable scaffolds for tissue engineering, Elect. *J. Biotechnol.* 3(2), 1-6 (2000).
6. IUPAC Compendium of Chemical Technology, The Gold Book, 2nd Edition, A. D. McNaught and A. Wilkinson' Blackwell Science.
7. This reference was not provided in the original manuscript (Karageorgiu and Kaplan, 2005).

8. F2450-04 ASTM Standard guide for assessing microstructure of polymeric scaffolds for use in tissue engineered medical products.
9. BS EN 623-2 (1993), Advanced technical ceramics – monolithic ceramics – general and textural properties. Part 2: Determination of density and porosity.
10. M. Sezgin and B. Sankur, Survey over thresholding techniques and quantitative performance evaluation, *J. Electronic Imaging*, 13(1), 146-167 (2004).
11. Image J. The image processing and analysis package, Image J is freely downloadable from http://rsb.info.nih.gov/ij/.
12. N. Otsu, A threshold selection method from gray-level histograms, *IEEE Trans. Syst. Man. Cybern.* 9(1), 62-66 (1979).
13. E. W. Washburn, The dynamics of capillary flow, *Phys Rev.* 17, 273-283 (1921).
14. V. Maquet, S. Blacher, R. Pirard, J.-P. Pirard, M. N. Vyakarnum, and R. Jerome, Preparation of macroporous biodegradable poly(L-lactic-co-ε-caprolactone) foams and characterization by mercury intrusion porosimetry, image analysis and impedance spectroscopy, *J. Biomed. Mater. Res.* 66A, 199-213 (2003).
15. A. B. Abell, K. L. Willis, and D. A. Lange, Mercury intrusion porosimetry and image analysis of cement-based materials, *J. Colloid Interf. Anal.* 211, 39-44 (1999).
16. ASTM E 1294-89 Standard test method for pore size characteristics of membrane filters using automated liquid porosimeter, ASTM International.
17. ASTM F316-03 Standard test methods for pore size characteristics of membrane filters by bubble point and mean flow pore test.
18. H. Lo, S. Kadiylala, S. E. Guggino, and K. W. Leong, Poly(L-lactic acid) foams with cell seeding and controlled-release capacity, *J. Biomed. Mater. Res.* 30, 475-484 (1996).

NANOSTRUCTURED AND NANOCOMPOSITE HYDROGELS FOR BIOMEDICAL APPLICATIONS

ANNA SPANOUDAKI, DANIEL FRAGIADAKIS, KALLIOPI VARTZELI-NIKAKI, POLYCARPOS PISSIS*, JOSE C. RODRIGUEZ HERNANDEZ†, MANUEL MONLEON PRADAS†

Department of Physics, National Technical University of Athens, 15780 Athens, Greece

†Center for Biomaterials, Universidad Politecnica de Valencia, E-46071 Valencia, Spain

Abstract. Polymer hydrogels are interesting for biomedical applications due to their biocompatibility and good water sorption/diffusion properties. In previous work their mechanical stability could be improved by combination with a hydrophobic polymer. Instead, here we prepare poly(hydroxyethyl acrylate), PHEA/silica nanocomposites by sol-gel techniques as potential scaffold materials for tissue engineering. Their morphology and properties were investigated by scanning electron microscopy, thermogravimetry, dynamic mechanical analysis, water sorption/desorption, dielectric spectroscopy, and thermally stimulated depolarization currents techniques at various levels of relative humidity/water content. The results show that the nanoparticles are homogeneously distributed, form a continuous phase, and significantly reinforce the matrix, without affecting its hydrophilicity and, thus, biocompatibility. Water is more homogeneously distributed, as compared to pure PHEA matrix, forming smaller clusters. The glass transition temperature is controlled by water content, and less affected by the concentration of silica.

Keywords: hydrogels; poly(hydroxyethyl acrylate); nanocomposites; silica nanoparticles; water clustering

*To whom correspondence should be addressed. P. Pissis, Department of Physics, National Technical University of Athens, 15780 Athens, Greece, e-mail: ppissis@central.ntua.gr

J.P. Blitz and V.M. Gun'ko (eds.), Surface Chemistry in Biomedical and Environmental Science, 229–240.

1. Introduction

Polymer hydrogels are hydrophilic crosslinked polymers, which may absorb large amounts of water preserving their integrity. Of special interest are hydrogels based on biocompatible polymers for applications as biomaterials in drug delivery systems, implantation, preparation of scaffolds for tissue engineering, etc.[1,2] To optimize composition, preparation conditions and processing of such materials tailored to specific end-use requirements, a better understanding of structure - property relationships is essential. Issues related to the organization (state) of water, and the influence of water on the final properties of the hydrogel, are of fundamental interest in such considerations.

In previous work we employed broadband dielectric relaxation spectroscopy (DRS) and thermally stimulated depolarization currents (TSDC) techniques, in combination with other complementary techniques, to study water effects in biocompatible hydrogels based on poly(hydroxyethyl acrylate), PHEA.[3-7] In several cases PHEA was combined with a second, hydrophobic polymer in the form of blends, copolymers, and more successfully, interpenetrating polymer networks (IPNs), to improve mechanical stability.[6,7] Special attention has been paid to the investigation of the effects of water on polymer (chain) dynamics, and on the final properties of the hydrogel, including those of biotechnological significance. This work has been recently extended to include porous hydrogels, either homopolymers[8] or IPNS,[9] prepared by solution polymerization using water, methanol or ethanol as diluents, to improve water capacity. Pores are formed due to segregation of the solvent from the polymer during the polymerization process. They collapse during the drying process after polymerization, but reopen when the xerogel is immersed in water. The results show that volume fraction, size and morphology (connectivity) of the pores can be controlled by solvent type, composition of the mixture, and polymerization conditions.[8] Thus, it is possible to predict the hydrogel properties in terms of composition and preparation conditions.

An interesting result with respect to applications obtained with the IPN hydrogels is that these are two- phase systems (two glass transition temperatures), with the hydrophilic domains behaving essentially like the pure hydrophilic component.[6,7,9] Thus, the two basic functions of these IPN hydrogels with respect to applications, namely hydrophilicity and mechanical stability, are separately taken over by the two IPN components, the hydrophilic and hydrophobic domains, respectively. Figure 1 shows TSDC and DMA results for the water content dependence of the α relaxation (dynamic glass transition) of PHEA in sequential IPNS prepared from PHEA and poly(ethyl methacrylate) (PEMA) as the hydrophobic component.[9] In these IPNs a porous PEMA network was prepared first, and PHEA was then polymerized in the pores. In addition to the strong plasticization action of water, we observe in Figure 1 that practically a single reduced curve is obtained, demonstrating that all hydrophilic phases of the IPNs are equally plasticized. Water sorption isotherms show the same effect when the water content is referred to the PHEA weight fraction. Another interesting observation in Figure 1, is that water in excess of about 0.3 g of water per gram of PHEA, does not further contribute to plasticization of the PHEA chains. This indicates that water forms a separate phase at higher water contents, in agreement with the results of DSC measurements.[9]

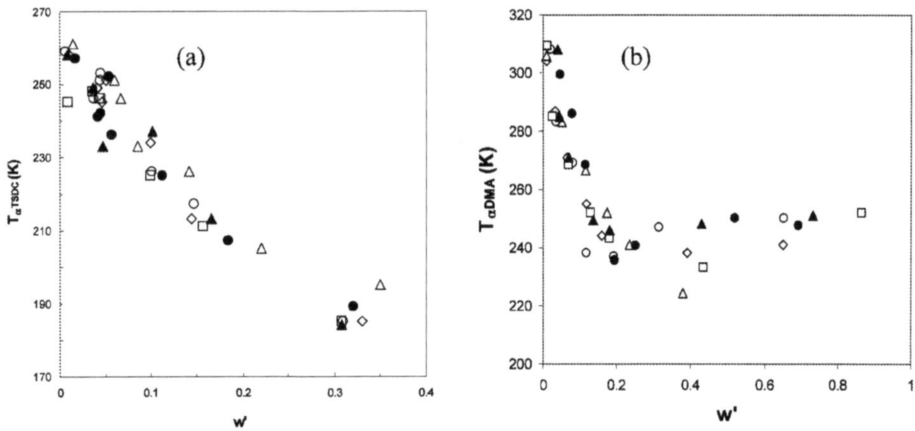

Figure 1. Temperature of the TSDC peak (a) and DMA loss tangent peak (b) corresponding to the main relaxation of PHEA, α, as a function of water content of the hydrogel referred to PHEA weight in the sample, w'. The samples are identified by their PHEA mass fraction equal to: 0.81 (□), 0.71 (◊), 0.64 (▲), 0.45 (O), 0.34 (•), 0.15 (Δ).

To further improve hydrogel mechanical stability, in particular for the preparation of scaffolds for tissue engineering, the concept of nanocomposites may be used. Polymeric nanocomposites, consisting of a polymeric matrix and an inorganic filler of typical size 1-100 nm, have been shown by several investigators to exhibit, for the same filler fraction, far better mechanical properties than conventional composites (micro- or macrocomposites).[10] The reasons for this improvement have not been yet fully understood. It is commonly accepted, however, that the large surface to volume ratio of the nanofiller plays an important role. Several other aspects of polymer nanocomposites are currently being intensively investigated, such as the effects of nanofiller on chain dynamics, and the dependence of the final properties on the degree of dispersion.[11]

In this paper we report results of our studies on PHEA hydrogels reinforced with silica nanoparticles. These nanocomposite hydrogels were prepared and investigated as potential materials for the development of scaffolds for tissue engineering. Other applications may also be envisaged, such as drug delivery systems and implantation materials. Sol-gel techniques in the presence of the polymeric matrix were employed for preparing the nanocomposite hydrogels. Owing to the combination of nanocomposites and hydrogel in one material several interesting fundamental questions arise. Issues such as the organization of water in relation with the hydrophilic/hydrophobic nature of the filler, the changes in free volume due to loosened molecular packing of the polymeric chains, interface and interphase effects etc. Several experimental techniques were employed to investigate the organization of water, and the effects of water and of the silica nanoparticles on the final properties of the nanocomposite hydrogels in relation to structure and morphology. These include, next to DRS and TSDC, electron microscopy, thermogravimetric analysis (TGA), water sorption/diffusion techniques, differential scanning calorimetry (DSC) and dynamic mechanical analysis (DMA). The results are discussed in terms of effects of the nanoparticles on chain dynamics, confinement of water in a small volume, specific polymer-water interactions, interfacial effects and filler-water interactions, all in relation to recently obtains results with other nanohydrogels.[12-14]

2. Experimental

2.1. MATERIALS

The nanocomposites were prepared in two steps. In the first step PHEA hydrogels were prepared following procedures described in previous work.[3-5] In the second step standard sol-gel techniques were employed to prepare silica nanoparticles in the presence of the PHEA hydrogel. Details of the preparation will be given elsewhere. The silica concentration in the dry nanocomposites was varied between 0 and 30 wt%.

2.2. EXPERIMENTAL TECHNIQUES

Scanning electron micrographs were taken with an ISIDS-130 microscope. Thermogravimetric analysis (TGA) was performed with an SDT Q600 (TA Instruments). A Seiko DMS 210 dynamic-mechanical analyzer was used for DMA measurements. For water content determination samples were allowed to equilibrate over saturated salt solutions in sealed jars at controlled relative humidities. The water content h, defined as the ratio of the weight of water in the hydrogel to the weight of the dry sample (dry basis), was determined by weighing. Dry weights were determined by drying in vacuum (5×10^{-2} Torr) for 24 h at 60°C. DRS (10^{-2}-10^{6} Hz) and TSDC techniques (−180 to 30°C) have been described elsewhere.[3,4] The less familiar TSDC technique corresponds to measuring dielectric losses against temperature at low frequencies of 10^{-2}-10^{-4} Hz, i.e. in a range not easily available for DRS measurements. Moreover, the technique is characterized by high sensitivity and high peak resolving power[3]. A Novo control Alpha Analyzer was used for DRS measurements. The temperature was controlled to better than 0.1°C with a Novo control Quarto system. TSDC measurements were carried out using a Deathly 617 electrometer in combination with the Novocontrol Quatro cryosystem and a Novocontrol sample cell.

3. Results and Discussion

3.1. MORPHOLOGICAL, THERMAL AND MECHANICAL CHARACTERIZATION

SEM micrographs showed an excellent dispersion of nanoparticles in the matrix. Pyrolysis in TGA measurements revealed, in agreement with results of other measurements to be reported elsewhere, that the nanoparticles form a not fully-interconnected silica network in the samples with filler content below about 15%, and a fully interconnected network in the samples with higher filler content. Thus, the nanocomposites under investigation may be considered, at least at higher filler contents, as IPNs consisting of two networks, an organic (PHEA) and an inorganic network (silica).

Figure 2 shows results of DMA measurements: storage modulus as a function of temperature measured at 1 Hz for pure PHEA, and two nanocomposites equilibrated in air. We observe that the glass transition temperature T_g of the hydrogels is in the range of 10-20°C, slightly shifting to higher temperatures and becoming broader on addition of silica nanoparticles. The main result in Figure 2, however, is the significant and systematic increase of the storage modulus of the nanocomposites in the rubbery phase up to about three orders of magnitude on addition of 30 wt% silica, as compared to pure PHEA. This significant improvement, by far more than typically reported for silica nanocomposites, as well as for nanocomposite hydrogels,[12,13] is essential for scaffold applications. It may be suggested that the network structure of the silica phase, revealed by pyrolysis, is responsible for that. In that respect, it is interesting to note that the significant improvement of the storage modulus in the rubbery phase in Figure 2 is achieved for filler contents higher than about 15%.

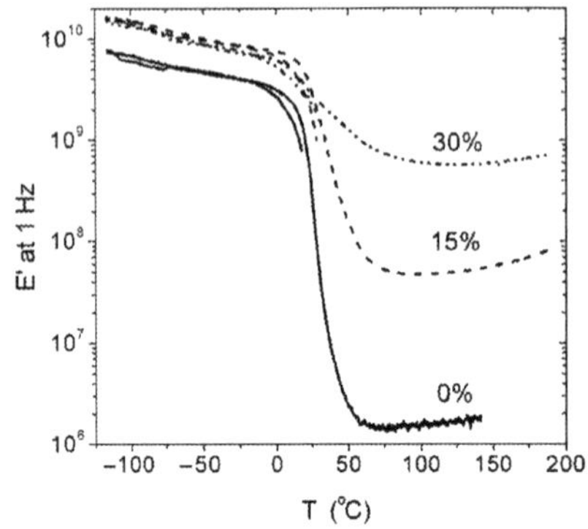

Figure 2. Temperature dependence of the storage modulus *E*' of the samples indicated on the plot at 1 Hz.

3.2. WATER SORPTION

Figure 3 shows preliminary results of water sorption measurements of several nanocomposites at a few values of relative humidity at room temperature. The water content has been normalized to PHEA content. Several interesting results are observed in Figure 3. Similar to the IPNs in Figure 1, the normalized water content does not depend on composition, i.e. hydrophilicity and thus biocompatibility of the nanocomposite hydrogels are not affected by the presence of the nanoparticles. This result suggests that water sorption on the surface of silica nanoparticles is not significant. In addition, no hysteresis effects were observed in sorption/desorption measurements, indicating good performance of the hydrogels under investigation. Closer inspection of the data in Figure 3 suggests a higher water uptake for the samples with 10% silica. This is an interesting observation, since preliminary water uptake measurements by immersion in water indicate higher water uptakes for that composition, probably related to the connectivity of the silica network. Further immersion experiments are in progress to explore these results, and also for differences in the behavior of the hydrogels in sorption from the vapor phase and by immersion.[8,9] For comparison, the water uptake of poly(vinyl alcohol)/silica hydrogels immersed in water was found to be reduced with respect to the pure matrix , in particular at higher pH values, probably due to rigidity of the silica network restricting the extensibility of the composite.[12]

A second interesting result in Figure 3 is that the sorption isotherms can be classified as type III according to Brunauer,[15] indicating that the results should not be described in terms of free and bound water and of hydration layers. Instead, the water molecules are distributed randomly at the hydrophilic sites up to a critical water content h_c, above which clusters begin to form. The data were

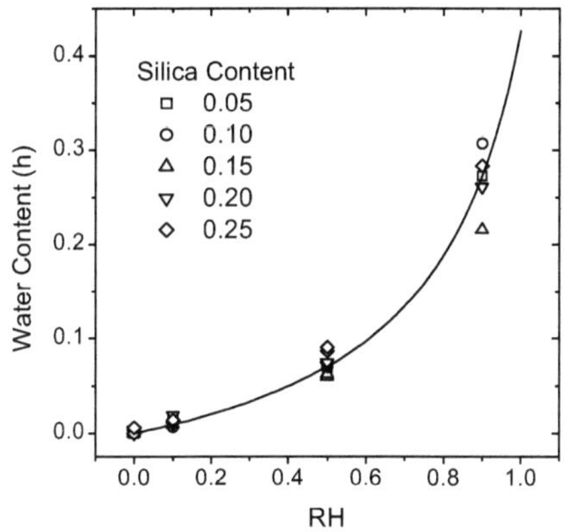

Figure 3. Water content h, normalized to PHEA content, against relative humidity RH of the samples indicated on the plot at 25°C. The line is a fit of Eq. (1) to the experimental data.

quantitatively analyzed by following the analysis proposed by Brown.[16] This analysis combines conventional solution theory (Flory-Huggins) and cluster theory (Zimm and Lundberg) and provides an interpretation for the sorption of water molecules in polar materials.[5] The data in Figure 3 were fitted to the equation

$$h = \frac{P_1 a_w}{P_2 + a_w} \tag{1}$$

where α_w is the water activity (equal here to relative humidity RH) and P a fit parameter, equal to the critical hydration layer h_c above which water clusters are formed, $h_c = P$. The inset to Figure 4 shows results for h_c of the nanocomposites. The mean number of water molecules in a cluster N_c at water content h can be calculated.[16]

Figure 4 shows results for N_c at $h = 0.9$. For comparison h_c was estimated to be 0.05 in pure PHEA hydrogels, and N_c about 4.5 at RH = 0.9 (Ref. 5). We observe that at higher silica contents h_c increases and N_c decreases. Thus, water is more homogeneously distributed in the presence of the silica nanoparticles. Water clusters are formed at higher water contents and are smaller, compared to pure PHEA, i.e. clustering is suppressed in the nanocomposites. Although quantitative analysis of the results suggests that water is sorbed only in the hydrophilic PHEA phase, it may be that the silica nanoparticles interact with and contribute to the good dispersion of water in the nanocomposites. This point, which is also of interest for silica nanocomposites prepared for other applications using various polymer matrices and preparation methods, in particular mixing of the silica nanoparticles with the polymer melt or solution, will be explored in future work by tailoring the hydrophilic/hydrophobic nature of silica used, and the type of chemical treatment.[10,11]

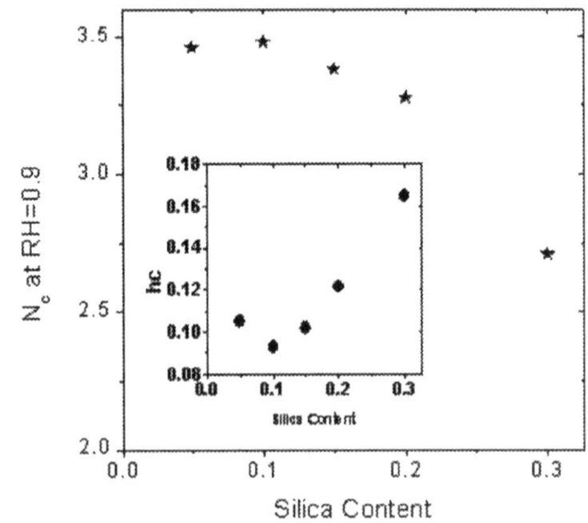

Figure 4. Number of water molecules in a cluster N_c at RH = 0.9 against silica content. The inset shows the critical water content h_c against silica content.

3.3. MOLECULAR MOBILITY

The effects of silica nanoparticles and sorbed water molecules on polymer (chain) dynamics investigated by dielectric techniques are now discussed. Figure 5 shows results of DRS measurements: real, ε'(a), and imaginary dielectric permittivity (dielectric loss), ε''(b), at room temperature for samples equilibrated in air (RH around 0.5, i.e. with water contents between 0.05 and 0.10 (Figure 3)). At the measurement temperature the PHEA polymer is in the rubbery phase.[3-5] This fact is reflected in the high values, and the frequency dependence of dielectric loss at low frequencies, related with dc conductivity. A shoulder observed in $\varepsilon''(f)$ and, less clear, a drop in $\varepsilon'(f)$ in the kHz frequency region are related with the α relaxation, associated with the glass transition (dynamic glass transition[17]) of the PHEA matrix. The increase of ε''at higher frequencies in Figure 5 arises from the secondary γ and β_{sw} relaxations, which are intimately related with polymer-water interactions.[4] The secondary relaxations will be studied in detail by both DRS and TSDC in future work. ε' and ε'' in Figure 5 do not change systematically with composition, even at high frequencies where space charge and conductivity effects are significantly suppressed. Two explanations may account for that behavior: (1) the water content at ambient conditions is different for different samples, in correlation with the well established strong dependence of dielectric properties on water content;[4] and (2) real differences exist between the various compositions with respect to the degree of dispersion of the nanoparticles, related with details of preparation which should be better controlled in the future. We will come back to this point later.

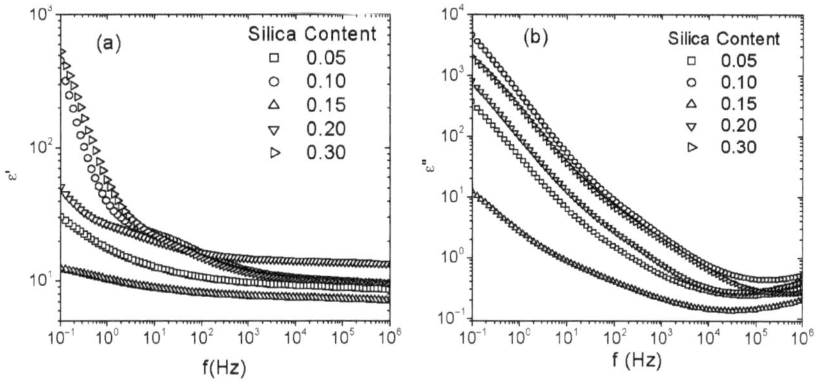

Figure 5. Real ε′(a) and imaginary part ε″(b) of complex dielectric permittivity against frequency *f* of the nanocomposites indicated on the plot at ambient temperature and relative humidity conditions.

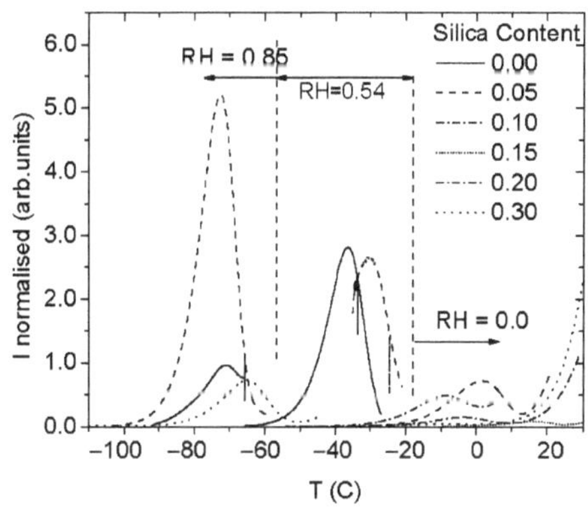

Figure 6. TSDC thermograms (normalized depolarization current *I* against temperature *T*) in the glass transition region of the samples and the relative humidity conditions indicated on the plot.

The α relaxation in Figure 5 is masked by conductivity, in particular at higher water contents, similar to many other polymers.[17] TSDC, which is less influenced by dc conductivity, offers an attractive alternative to investigate the α relaxation. Figure 6 shows results of TSDC measurements on the nanocomposites in the region of the α relaxation at three values of relative humidity. The depolarization current *I*, normalized to the same polarizing field, sample surface area and heating rate, is a measure of $\varepsilon''(T)$ at a frequency range of 10^{-2}-10^{-4} Hz (equivalent to TSDC measurements[3]). The TSDC peak temperature is a good measure of the calorimetric glass transition temperature T_g. Detailed TSDC studies of PHEA hydrogels have indicated that dc conductivity makes a significant contribution to the TSDC α peak without significantly affecting the temperature position of the dipolar contribution.[3] We observe in Figure 6 that the α peak (and thus T_g) shifts significantly to lower temperatures on addition of water. This shift reflects the strong plasticizing action of water on the glass transition, and the corresponding α relaxation studied in detail in previous work.[3-8]

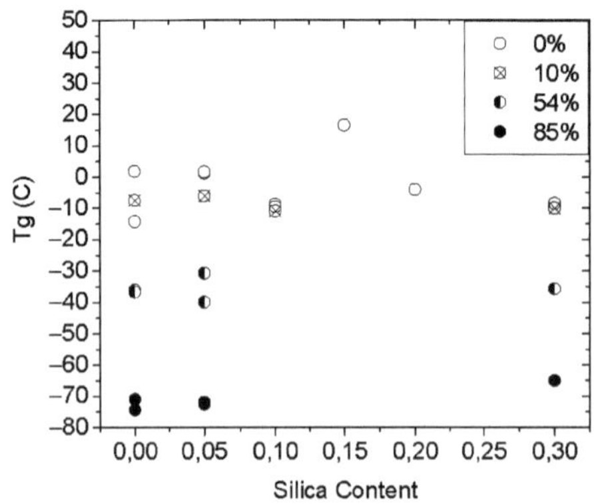

Figure 7. Glass transition temperature T_g against silica content at four levels of relative humidity indicated on the plot

The composition and relative humidity dependence of T_g determined by TSDC is shown in Figure 7. The silica content dependence is rather weak compared to the relative humidity dependence, in particular at higher levels of relative humidity/water content. A possible explanation is that any differences between the dry nanocomposites, associated with the presence of various amounts of silica nanoparticles and their effects on chain dynamics, are masked by the strong plasticization induced by water at higher water contents. It may also be that drying of the samples was incomplete, resulting in different amounts of residual water in the samples considered completely dry in Figure 7. Detailed studies of the water content dependence of T_g in pure PHEA by DSC and TSDC have revealed that this dependence is particularly pronounced at the initial levels of hydration, in agreement with phenomenological descriptions by Fox and Couchman-Karasz.[5]

TSDC measurements in pure PHEA hydrogels have shown that, under the conditions adopted for drying in this work, T_g is around –10°C, (Ref. 7), i.e. in the same range as in the nanocomposites in Figure 7. Compared to DMA, the values of T_g by TSDC are shifted to lower temperatures. The higher frequency of DMA measurements can explain this shift. Much work has been devoted in recent years to the investigation of the composition dependence of chain dynamics and glass transition in organic-inorganic nanocomposites.[18-20] The shift to higher temperatures due to constraints imposed on the motion of polymeric chains by rigid nanoparticles, in particular in the case of covalent bonds between the two components,[18] is often cancelled out, or even overcompensated, by an increase in molecular mobility. A shift of T_g to lower temperatures, due to loosened molecular packing of thc polymeric chains, results in an increase in free volume. This point will be studied in future work by systematic DSC, DMA and TSDC studies on the nanocomposites under investigation here, at defined levels of relative humidity / water content.

4. Conclusion

PHEA/silica nanocomposites were prepared by sol-gel techniques. Morphology, water sorption, and chain dynamics were investigated by various techniques. Results indicate that the PHEA/silica nanocomposites exhibit enhanced properties compared to pure PHEA regarding their application as scaffold materials for tissue engineering. DMA measurements show that the polymer matrix is significantly reinforced by the nanoparticles. SEM and pyrolysis studies show that the nanoparticles are homogeneously distributed, forming a not fully-interconnected silica network for filler contents below about 15% and a fully interconnected network at higher filler contents. Water sorption/desorption measurements from the vapor phase show that the hydrophilicity, and with that the biocompatibility of PHEA, is preserved in the nanocomposites. No hysteresis effects exist. Water molecules are more homogeneously distributed in the nanocomposites, forming smaller water clusters at the same relative humidity. Water sorption/desorption measurements by immersion in water to check the effects of the silica network on swelling are in progress. The dynamic properties, investigated by dielectric techniques, are governed by water content, similar to the pure PHEA matrix, and to a lesser extent by the concentration of nanoparticles.

Acknowledgments

The project is co-founded by the European Social Fund (75%) and National Resources (25%) – (EPEAEK II) – PYTHAGORAS.

References

1. *Hydrogels in medicine and pharmacy,* edited by N. A. Peppas (CRC Press, Boca Raton, FL, 1986), vol I.
2. V. Stoy, and C. Kliment, *Hydrogels: Speciality plastics for biomedical and pharmaceutical applications*, (Technomic Pub., Basel, 1996).

3. A. Kyritsis, P. Pissis, J. L. Gomez Ribelles, and M. Monleon Pradas, Depolarization thermocurent studies in poly(hydroxyethyl acrylate)/water hydrogels, *J. Polym. Sci. B* 32, 1001-1008 (1994).
4. A. Kyritsis, P. Pissis, and J. Grammatikakis, Dielectric relaxation spectroscopy in poly(hydroxyethy acrylate)/water hydrogels, *J. Poly. Sci. B* 33, 1737-1750 (1995).
5. A. Kyritsis, P. Pissis, J. L. Gomez Ribelles, and M. Monleon Pradas, Polymer-water interactions in poly(hydroxyethyl acrylate) hydrogels studied by dielectric, calorimetric and sorption isotherm measurements, *Polym. Gels Networks* 3, 445-469 (1995).
6. G. Gallego Ferrer, M. Monleon Pradas, J. L. Gomez Ribelles, and P. Pissis, Swelling and thermally stimulated depolarization currents in hydrogels formed by interpenetrating polymer networks, *J. Non-Cryst. Solids* 235-237, 692-696 (1998).
7. J. L. Gomez Ribelles, M. Monleon Pradas, G. Gallego Ferrer, N. Peidro Torres, V. Perez Gimenez, P. Pissis, and A. Kyritsis, Poly(methyl acrylate)/poly(hydroxyethyl acrylate) sequential interpenetrating polymer networks. Miscibility and water sorption behavior, *J. Poly. Sci. B* 37, 1587-1599 (1999).
8. M. Monleon Pradas, J. L. Gomez Ribelles, A. Serrano Aroca, G. Gallego Ferrer, A. Suay Anton, and P. Pissis,, Interaction between water and polymer chains in poly(hydroxyethyl acrylate) hydrogels, *Colloid Polym. Sci.* 279, 323-330 (2001).
9. G. Gallego Ferrer, J. M. Soria Melia, J. Hernandez Canales, J. M. Meseguer Duenas, F. Romero Colomer, M. Monleon Pradas, J. L. Gomez Ribelles, P. Pissis, and G. Polizos, Poly(2-hydroxyethyl acrylate) hydrogel confined in a hydrophobous porous matrix. *Colloid Polym. Sci.* 283, 681-690 (2005).
10. L. Bokobza, Elastomeric composites. I. silicone composites. *J. Appl. Polym. Sci.* 205, 61-69 (2004).
11. D. Fragiadakis, P. Pissis, and L. Bokobza, Glass transition and molecular dynamics in poly(dimethylsiloxane)/silica nanocomposites, *Polymer* 46, 6001-6008 (2005).
12. J.-P. Boisvert, J. Persello, and A. Guyard, Influence of the surface chemistry on the structural and mechanical properties of silica-polymer composites, J. Polym. Sci. Part B Polym. Phys., 41, 3127-3138 (2003).
13. F. Bignotti, L. Sartore, M. Penco, G. Ramorino, and I. Peroni, Effect of montmorillonite on the properties of thermosensitive poly(N-isopropylacrylamide) composite hydrogels, J. Appl. Polym. Sci., 93, 1964-1971 (2004).
14. Ye. P. Mamunya, V. I. Shtompel, E. V. Lebedev, P. Pissis, A. Kanapitsas, and G. Boiteux, Structure and water sorption of polyurethane nanocomposites based on organic and inorganic components, *Eur. Polym. J.* 40, 2323-2331 (2004).
15. R. Pethig, *Dielectric and electronic processes of biological materials* (Wiley, Chichester, 1979).
16. G. L. Brown, Clustering of water in polymers, in: *Water in polymers*, edited by S. P. Rowland (ACS Symposium Series 127, American Chemical Society, Washington, DC, 1980), pp. 441-450.
17. *Dielectric spectroscopy of polymeric materials*, edited by J. P. Runt and J. J. Fitzgerald (American Chemical Society, Washington, DC, 1997).
18. V. A. Bershtein, L. M. Egorova, P. N. Yakushev, P. Pissis, P. Sysel, and L. Brozova, Molecular dynamics in nanostructured polyimide-silica hybrid materials and their thermal stability, *J. Polym. Sci. B* 40, 1056-1069 (2002).
19. L. Matejka, O. Dukh, and J. Kolarik, Reinforcement of crosslinked rubbery epoxies by in-situ formed silica, *Polymer* 1449-1459 (2000).
20. R. Kotsilkova, D. Fragiadakis, and P. Pissis, Reinforcement effect of carbon nanofillers in an epoxy resin system: rheology, molecular dynamics, and mechanical studies, *J. Polym. Sci. B* 43, 522-533 (2005).

AN OVERVIEW ON THE TOXICITY OF INHALED NANOPARTICLES

BICE FUBINI,*[,1] IVANA FENOGLIO,[1] GIANMARIO MARTRA,[1] RAFFAELLA CESCHINO,[1] MAURA TOMATIS,[1] ROBERTA CAVALLI,[2] MICHELE TROTTA[2]

[1] Dip. di Chimica Inorganica, Chimica Fisica e Chimica dei Materiali and Interdepartmental Centre "G. Scansetti" for Studies on Asbestos and Other Toxic Particulates, University of Torino, via Pietro Giuria 7, 10125 Torino, Italy

[2] Dip. di Scienza e Tecnologia del Farmaco, University of Torino, via Pietro Giuria 9, 10125 Torino, Italy

Abstract. The general alarm raised concerning the potential toxicity of inhaled nano-sized particles has moved several particle toxicology groups towards thorough investigations on the peculiar toxicity of ultrafine (nano) dusts. The new term Nanotoxicology has been proposed for these kinds of studies. An overview of the factors which may affect the toxicity of inhaled nano and micro-sized inhaled particles is herein presented.

Keywords: nanoparticles, ultrafine particles, pulmonary toxicology, toxicity, liposomes, carbon nanotubes, iron oxide, titania, silica, asbestos

1. Nanotoxicology: A New Branch of Science?

Much emphasis has been given in the past two years to the possible toxicity of nanoparticles, by both media and scientific journals, with a general fear that they might turn up in an occupational and environmental disaster, as with asbestos, or become a serious and debated issue, like OGM. Because of their peculiar physical and chemical features, the study of nanoparticles as potential toxic agents requires a multidisciplinary approach involving disciplines from physics & chemistry to biology and medicine.

*To whom correspondence should be addressed. Bice Fubini, Dip. di Chimica Inorganica, University of Torino, via Pietro Giuria 7, 10125 Torino, Italy; e-mail: bice.fubini@unito.it

J.P. Blitz and V.M. Gun'ko (eds.), Surface Chemistry in Biomedical and Environmental Science, 241–252.

In the past it was assumed that ultrafine particles would be easily exhaled, unlike micron-sized particles, which would be retained in the lung. Conversely an enhanced toxicity to experimental animals and cell cultures, when passing from micron to nano-sized particles of the same composition (e.g. titania, carbon particles etc.), has raised alarms.[1,2,3] The recent experimental evidence that in exposed rats some nanoparticles reached the brain through the olfactory system, thus overcoming the blood brain barrier, suggests for nanoparticle toxicity quite a new scenario.[4] These observations have led some scientists to organize meetings entirely devoted to the toxicity of nanoparticles, and the term "nanotoxicology" has been proposed as a discipline.[1] The new Journal of Nanotoxicology is duly expected.

Nanosize particles constitute a double edged sword for human health. On the one hand, as previously outlined, nanoparticles may turn out to be a serious hazard. On the other hand they may also be useful as drug delivery systems to well defined targets in the body.

2. Mechanisms of Toxicity of Inhaled Particles

Humans are often exposed to particulates of different chemical composition at the workplace and in daily life. Some of these particles, when inhaled, may be very harmful to the respiratory system. A few diseases that originate in this way are silicosis from quartz dust, pleural mesothelioma and bronchogenic carcinoma from asbestos fibres and coal miner's pneumoconiosis - have been long recognized, and their association with exposure to a particular kind of particulate is well established. Several physicochemical properties, which differ for each pathogenic dust, are implied in the mechanisms of toxicity which are still partially obscure at the molecular level even for these well known occupational diseases. However the scientific community agrees that **form** and **micromorphology** of the particle, **chemical composition** and **biopersistence** contribute altogether to particle toxicity.

2.1. INTERACTIONS WITH CELLS AND TISSUES

Particle dimension is crucial for the distribution of particles in the various regions of the human respiratory tract. The various mechanisms of clearance act differently for nano and micro-sized particles.[1]

Figure 1 shows the various steps of cell and tissue interaction with particles in the lung, the major effects seen with micron and nano-sized particles.

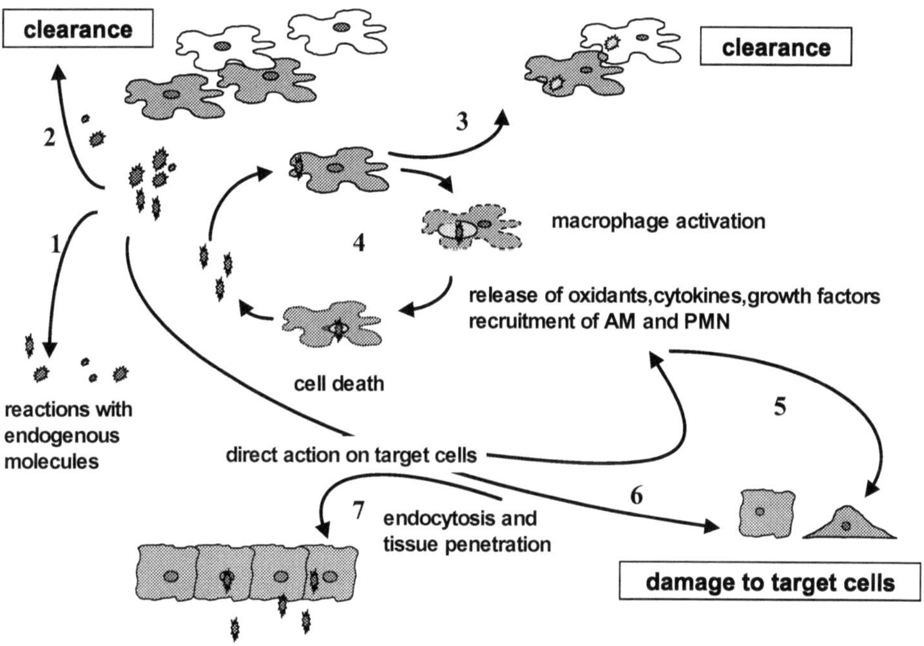

Figure 1. Steps of cell and tissue interaction with nano and micro-sized particles in the lung.

Once in the alveolar space the particle may react with extracellular matter (step 1) and be engulfed by alveolar macrophages (AMs), which clear the particles out of the lungs (step 2). Depending upon the surface characteristics of the particle itself this clearance process may either succeed (step 3) or fail (step 4). In the latter case macrophages will become activated at the cellular and molecular level with the activation of transcription factors, and the release of ROS (Reactive Oxygen Species) and RNS (Reactive Nitrogen Speices), chemotactic factors, lytic enzymes, cytokines, growth factors, with eventual cell death (necrosis/apoptosis), releasing the particle. Subsequent ingestion-reingestion cycles accompanied by a continuous recruitment of AM, neutrophils (PMN) and lymphocytes are the cause of sustained and chronic inflammation. Target cells such as bronchiolar and alveolar epithelial cells will then be affected by both AM products (step 5) and the extracellular particles (step 6), again resulting in activation and/or cell death. Particle derived ROS may also react with cell derived ROS and RNS yielding new toxic moieties, e.g. peroxynitrite ($ONOO^-$) from nitric oxide (NO) and superoxide anion ($O_2^{-\bullet}$) (step 7).[5] Free radicals and ROS play a key role in steps 1, 5, 6 and 7, whereas the distribution of hydrogen bonding sites at the surface, (e.g. silanols, SiOH, in silica-based matrials), which govern hydrophilicity and adsorption processes[5] are mostly related to steps 2, 3 and 4.

Some evidence suggests that inhaled nanoparticles, after deposition in the lung, largely escape alveolar macrophage surveillance and clearance, and gain access through translocation to the pulmonary interstitium (step 7). Furthermore, the reaction with endogenous molecules may affect the fate of nanoparticles in a different manner than micron-sized particles since they may have not the same reactivity. In the case of proteins, the adsorption processes with particles having similar mean diameter may be very different.

2.2. PHYSICO-CHEMICAL FEATURES RELEVANT TO PARTICLE TOXICOLOGY

A variety of experimental studies performed so far indicate clear-cut links between some physico-chemical features and the response elicited in cellular models relevant to lung toxicity. We describe here briefly the most relevant ones, reported in most review papers in the field.[6-10]

- Surface area which is related to particle size and morphology.
- Particle shape such as isometric vs fibrous habit in asbestos and asbestos substitutes. Among non fibrous materials, e.g. silicas, the particles may exhibit different micromorphologies. Those obtained by grinding have acute points and spiky edges, while those generated at high temperature or via hydrothermal synthesis have a smooth and roundish aspect. Furthermore, crystalline oxidic or metal particles can expose highly rough surfaces when in a nanosized form, that are converted to more regular terminations after sintering at high temperature.
- The hydrophilic character of the surface which is related to the distribution of different hydrophilic and hydrophobic patches (defect of the bulk composition, presence of impurities).
- The reactivity of the surface such as the generation of radical species and reaction with organic molecules.

3. Nanosized Particles

3.1. HOW CAN WE DEFINE A NANOPARTICLE

Nanoparticles maybe defined by size, which is common for toxicologists, who usually define them as particles with a **mean diameter > 100 nm.** However some scientists, particularly after the outburst of interest in the field, claim that any particle under study less than one micron should fall in the nanoscale particle group.

Physical-chemistry offers a different approach to the definition of nanoparticles based on the fact that several properties – typically melting point, phase stability, electronic states and defects (usually resulting in peculiar colours) – change from those typical of larger particles when approaching nano-dimensions:

"…an upper limit for nanoparticles is the size at which one of the properties deviates from the value for the equivalent bulk material by an amount that is significantly larger than the error of the method used to make the measurement" (from Banfield & Zhang[11])

Thus the nanoparticle range may differ from one material to the next.

3.2. ORIGIN, SIZE & FORM

Nanoparticles are not only a product of new nano-technologies, but are also present in the environment, and nanoscale phenomena permeate and often control natural processes.[12] Humans have always experienced exposure to nanosized particles, but with the advent of the industrial revolution, this exposure from "anthropogenic" sources (e.g. internal combustion engines, power plant etc.) has increased dramatically.

Under the general term of nanoparticles we find different kinds of materials which can be classified on the basis of their origin, shape, or chemical composition.

3.2.1. Origin

As reported in Table 1, from G. Oberdoster and coauthors,[1] NPs may be distinguished as natural and anthropogenic, the latter being divided into Unintentional or Intentional. Although their toxicity is still under debate, many intentionally prepared NPs are used in medical devices, such as for the slow release of drugs or anticancer agents.

Table 1. Classification of NP following their origin, (from Oberdoster et al.[1]).

	Anthropogenic	
Natural	Unintentional	Intentional (NPs)
Gas-to-particle conversions	Internal combustion engines	Controlled size and shape, designed for functionality
Forest fires	Power plants	
Volcanoes (hot lava)	Incinerators	Metals, semiconductors, metal oxides, carbon, polymers
Viruses	Jet engines	
Biogenic magnetite: magnetotactic bacteria protoctists, mollusks, arthropods, fish, birds human brain, meteorite (?)	Metal fumes (smelting, welding, etc.)	Nanospheres, -wires, -needles, -tubes, -shells, -rings, -platelets
	Polymer fumes	
	Other fumes	Untreated, coated (nanotechnology applied to many products: cosmetics, medical, fabrics, electronics, optics, displays, etc.)
Ferritin (12.5 nm)	Heated surfaces	
Microparticles (< 100 nm; activated cells)	Frying, broiling, grilling	
	Electric motors	

3.2.2. Size and form

NPs have physical dimensions close to cell membrane receptors and other biomolecules. This opens a new scenario to be explored. For example, the small dimensions are responsible for an enhanced penetrability of the cell membrane, since endocytosis is favored. The interaction with proteins is also affected by dimensions: differences in surface curvature influence the ability of proteins to interact with surface functionalities, and consequently possible differences in conformational modification may occur.[13] This is particularly relevant in the case of enzymes, where activity may depend on conformation.

Several nanoparticles having a non-spherical shape have been synthesized, e.g., carbon nanotubes, nanofibers, nanorods, and nanowires which exhibit, similar to asbestos, a fibrous shape.

3.3. PECULIARITIES OF NANOPARTICLES

Three major facts are new and typical of nanoparticles.

- Aggregation: because of strong inter-particle forces, originated by an excess of surface free energy, nanoparticles may easily aggregate. When hydrophobic they will clump together in polar/aqueous media such as any body fluid. This may be inconvenient with animal tests as such clumped material will not disperse,[14] and in some cases even cause immediate animal death.[15]
- Enhanced surface reactivity: because of the high surface to bulk atomic ratio, many atoms at the surface will be poorly coordinated hence particularly reactive.
- Migration: because of their small size, opposite to micron sized particles, nanoparticles may readily move between biological compartments.

3.4. DOSE AND BIOLOGICAL EFFECTS

Several pulmonary toxicology studies in rats have demonstrated that sometimes nanoscale particles, when administered to the lung of experimental animals, cause a greater inflammatory response when compared to micron-scale particles of identical composition, at equivalent mass concentrations.

From the standpoint of surface reactivity we may distinguish three separate cases.

- A higher effect than larger particles per unit mass, but same per unit surface:[1,16] this mainly shows that the effect elicited is driven by the exposed surface, thus bound to increase in intensity when particle size decreases and consequently the specific surface increases.

- An effect similar or lower than with larger particles: this indicates that other factors, besides exposed surface play a role, which might be more effective in larger particles than for nano-sized particles. For example the effects of nanoscale TiO_2 rods, as well as nano TiO_2 dots (20 nm), were not significantly different from larger TiO_2 particles (300 nm) at equivalent doses.[17]
- A higher effect even when compared per unit surface. Only in such cases a truly adverse surface reactivity appears specifically at the nano-scale.
- Recent experimental studies on the mechanism of toxicity of nanoparticles comparing micron vs nano-sized particles

3.5. QUARTZ

A specimen of quartz at the nano-scale was prepared by V. Colvin and coauthors[18] in hydrothermal conditions, to study whether a well known toxic particulate such as quartz would show an enhanced toxicity as a nanoparticle. The particles, with a size of 50 nm, were crystalline as evidenced by XRD and tested *in vivo* (instillation on rats) by D. Warheit with the same amount in weight of the largely studied Min-U-Sil 5 micron-sized quartz. This is one of the most often employed commercial quartz specimen for *in vitro* and *in vivo* experimental studies on silicosis, reported to be a highly inflammatory, fibrogenic and cancerogenic kind of quartz.[19,20] Unexpectedly the nano-sized quartz was less toxic than Min-U-Sil 5 quartz, even when the effects were compared on a unit weight base.[17]

Results from several studies on the toxicity of various forms of quartz[19,21-23] suggest that such differences in toxicity may stem from the fact that the crystal examined originates from different sources. Thus the differences in behavior between Min-U-Sil 5 and the nano-quartz may also be ascribed to the well known variability of quartz hazard[24,25] and not size. It must be considered that the hydrothermal conditions in which the sample was prepared usually provide smooth particles whose surface is equilibrated with moisture, thus much less reactive.

3.6. IRON OXIDE: NANO VS. MICRON SIZE

Iron oxides are not toxic per se, and iron ions in the oxide framework do not yield ROS as do most asbestos fibers[26] or quartz particles, particularly when doped by iron. To investigate whether this also takes place when the oxides are prepared in nano-size form we have compared the surface reactivity of one fine, micron-sized specimen of Fe_2O_3 (mean diameter 0.1 μm, Fe_2O_3-F) to that of an ultrafine nano-size particle (mean diameter 3 nm, Fe_2O_3-UF). The surface generation of carbon centered radicals by homolytic rupture of the carbon-hydrogen bond in formate ions was measured on both specimens by means of the spin trapping technique (EPR/spin trapping with DMPO). This reaction may be assumed as a model for C-H cleavage in endogenous biomolecules. The dependence of radical activity on the presence of ascorbic acid, one of the most abundant antioxidant defences, was also considered. In the original form, both Fe_2O_3-UF and Fe_2O_3-F were inactive in the release of radical species (Figure 2), by interaction with formate ions, because for this kind of reaction the presence of Fe^{2+} ions is required. By contrast, when ascorbic acid was added as a reductant to the reaction medium, Fe_2O_3-UF exhibited a substantial radical release activity, comparable with what observed for asbestos (amosite), whereas Fe_2O_3-F was still inactive. Such different behavior between the two hematite powders can be related to differences in the structure of surface iron sites, as suggested by TEM and IR studies. The activation of Fe_2O_3-UF by ascorbic acid should result from an effective reduction of surface Fe^{3+} ions in a low coordination state, that are present in this sample in a significantly larger amount. Furthermore, ascorbic acid extracted iron ions from Fe_2O_3-UF (ca 1%) and not from Fe_2O_3-F, generating ferrous complexes in the supernatant, which are also active in the generation of radicals.

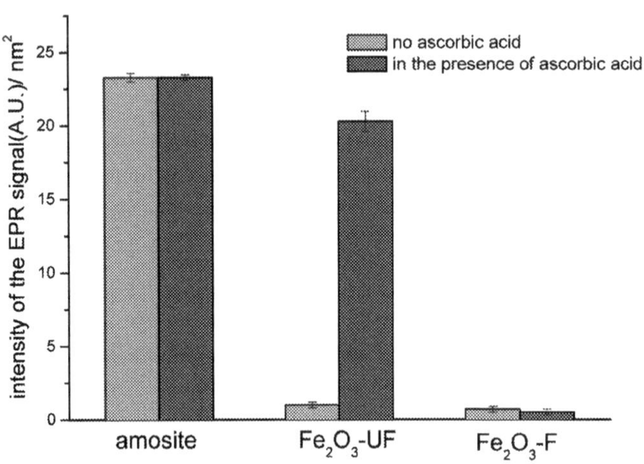

Figure 2. Free radical release, measured by means of ESR/spin trapping technique, from micron-sized and nano-sized Fe_2O_3 particles compared with amosite (asbestos).[27]

4. The Peculiar Case of Carbon Nanotubes

Carbon nanotubes (CNTs) currently attract intense interest because of their unique properties which make them suitable for many industrial applications.[28] Carbon nanotubes exhibit some of the properties implied in asbestos toxicity. Carbon nanotubes share with asbestos the fibrous habit – long fibers with a diameter of a few nanometers -and a very high biopersistence. On this basis they are suspected to be hazardous and indeed the first studies in vivo[14,29,30] have shown an inflammatory response followed by some evolution towards fibrosis. When inhaled, CNTs may thus constitute a possible hazard to human health. The inflammatory and fibrotic responses elicited by CNTs is similar to that caused by other toxic particles which might be the result of oxidative stress caused by particle- and/or cell-derived free radicals. There is no direct experimental evidence of a capacity of carbon nanotubes to generate free radicals similar to silica asbestos and nano sized iron oxide particles.

Recent reports show unexpected information on the role of free radicals in the health effects of nanotubes currently employed in many industries.[31] Unlike asbestos and most toxic particles, nanotubes do not release but blunt free radicals, which are considered one of the features imparting toxicity to particulates. Multi-wall carbon nanotubes (MWCN) in aqueous suspension do not generate oxygen or carbon centered free radicals detectable with the spin-trapping technique. Conversely, when in contact with an external source of hydroxyl (HO·) or superoxide radicals ($O_2 \cdot^-$), MWCN exhibit a remarkable radical scavenging capacity (Figure 3). It is therefore possible that the inflammatory reaction reported in vivo should be ascribed to MWCN features other than particle derived free radical generation.

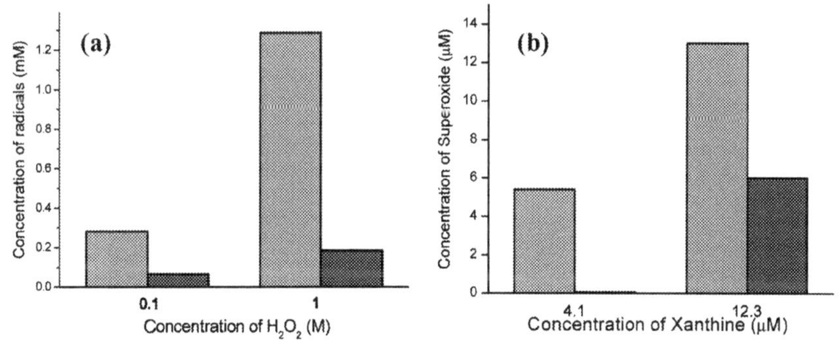

Figure 3. Effect of MWCNT on a) amount of hydroxyl radicals generated by photolysis of hydrogen peroxide, measured by ESR/spin trapping technique; b) amount of superoxide radicals generated by a xantine/xantine oxidase system, measured by UV spectroscopy. Light gray bars: no NT, dark gray bars: in the presence of NT.[31]

5. Effect of the Presence of Titania and Amorphous Silica Nanoparticles on the Gel-Liquid Transition of Liposomes

Titania and silica nanoparticles are widely used in cosmetics and pharmaceutical technology as additives. To mimic the major effects taking place when a nanoparticle gets in contact with the cellular membranes, we have started a systematic investigation on the interaction of nanoparticles of different chemical composition with liposomes, usually employed in drug delivery systems. When heated, liposomes undergo a phase transition, which may be monitored by DSC, due to a gel-liquid crystal transition. We have studied the influence of titania nanoparticles on the transition temperature and on the enthalpy change associated with the transition, since it would measure the ability of the nanoparticles to perturb the order of the double phospholipidic layer. Solid lipid nanoparticles were employed as blanks. Table 2 contains the results obtained with and without titania and silica nanoparticles. The transition temperature is virtually unaffected by the presence of the particles, but a remarkable variation on the enthalpic effect associated with the transition, reveals that the mineral, but not the lipidic particles, disrupt the organization of the liposome. Thus the organized phase which undergoes the phase transition is markedly reduced by the presence of the TiO_2 and SiO_2 particles. Preliminary results on the effect of nanoparticles on the rate of release of a lipophilic marker introduced into the liposome (betamethasone) reveal that the marker was released more rapidly when in presence of the titania nanoparticles (Figure 4).

Table 2. Effect of TiO_2 and SiO_2 nanoparticles on the gel-liquid transition temperature (T_m) of liposomes.

	ΔH (kJ/mol)	T_m (°C)
Liposomes	24.0	50.3
Liposomes+SiO_2		
Liposomes+TiO_2		
Liposomes+SLN	21.2	50.1

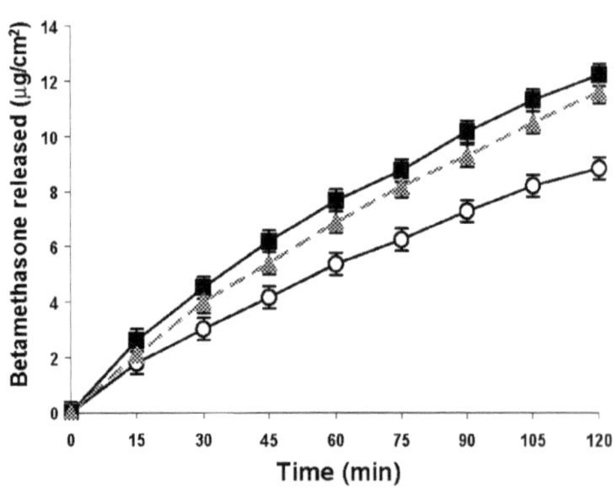

Figure 4. Effect of TiO_2 and SiO_2 nanoparticles on the release of betamethasone from liposomes: (○) no particles and in the presence of (▲) TiO_2 and (■) SiO_2 particles.

Acknowledgments

The research has been carried out with financial support by the Italian Ministry of Education, University and Research (MIUR) no. 2003032158-002 project title: "Correlation between surface properties and cellular response of fibres substituting asbestos and other silica-based materials".

References

1. G. Oberdörster, E. Oberdörster, and J. Oberdörster, Nanotoxicology: An emerging discipline evolving from studies of ultrafine particles, *Environ. Health Persp.* 113, 823-839 (2005).
2. V. L. Colvin, The potential environmental impact of engineered nanomaterials, *Nature Biotech.* 21, 1166-1170, (2003).
3. P. H. M. Hoet, Health impact of nanomaterials? *Nature Biotech.* 22, 19 (2004).
4. G. Oberdörster, Z. Sharp, V. Atudorei, A. Elder, R. Gelein, W. Kreyling, and C. Cox, Translocation of inhaled ultrafine particles to the brain, *Inhal. Toxicol.* 16, 437-445 (2004).
5. B. Fubini, Health effect of silica, in: *The surface properties of silicas,* edited by J. P. Legrand (J. Wiley and Sons, Chichester, 1998), chap 5, pp. 415-464.
6. J. A. Hardy and A. E. Aust, Iron in asbestos chemistry and carcinogenicity, *Chem. Rev.* 118, 95-97 (1995).
7. D. W. Kamp and S. A. Weitzman, The molecular basis of asbestos induced lung injury, *Thorax* 54, 638-652 (1999).
8. B. Fubini and W. E. Wallace, Modulation of silica pathogenicity by surface processes. In *Adsorption Silica Surfaces*, edited by E. Papirer (M. Dekker, Mulhouse, France, 1999) chap. 20, pp. 645-664.
9. B. Fubini and C. Otero-Aréan, Chemical aspects of the toxicity of inhaled mineral dusts, *Chem. Soc. Rev.* 28, 373-381,(1999).
10. B. Fubini and A. Hubbard, ROS, and RNS Generation by Silica in Inflammation and Fibrosis, *Free Rad. Biol. Med.* 34 (12), 1507-1516 (2003).
11. J. F. Banfield and H. Zhang, Nanoparticles in the Environment, in: *Nanoparticles and the Environment*, edited by J. F. Banfield and A. Navrotsky, Reviews in Mineralogy & Geochemistry, Vol. 44 (Mineralogical Society of America, USA, 2001).
12. M. F. Hochella, There's plenty of room at the bottom: Nanoscience in geochemistry *Geochim. Cosmochim. Acta* 66, 735-743 (2002)
13. H. Larsericsdotter, S. Oscarsson, and J. Buijs, Structure, stability, and orientation of BSA adsorbed to silica *J. Colloid Interface Sci.* 289, 26-35 (2005).
14. J. Muller, F. Huaux, J. F. Heilier, M. Arras, M. Delos, J. B. Nagy, and D. Lison, Respiratory toxicity of carbon nanotubes, *Toxicol. Appl. Pharmacol.* 197, 305-305 (2004).
15. D. Warheit, B. Laurence, K. Reed, D. Roach, G. Reynold, and T. Webb, Comparative Pulmonary Toxicity Assessment of Single Wall Carbon Nanotubes in Rats, *Toxicol. Sci.* 77, 117-125 (2004).
16. D. Hohr, Y. Steinfartz, R. P. F. Schins, A. M. Knaapen, G. Martra, B. Fubini, and P. J. A. Borm, The surface area rather than the surface coating determines the acute inflammatory response after instillation of fine and ultrafine TiO_2 in the rat, *Int. J. Hyg. Environ. Health* 205, 239-244 (2002).
17. D. B. Warheit, T. R. Webb, K. L. Reed, C. M Sayes, and V. L. Colvin, Impact of titania and ouartz nanoparticulate exposures on respiratory health: Role of particle size, *Abstracts of Papers of the American Chemical Society* 229, U911-U911 015-IEC, Part 1 (2005).

18. J. F. Bertone, J. Cizeron, R. K. Wahi, J. K. Bosworth, and V. L. Colvin, Hydrothermal synthesis of quartz nanocrystals, *Nano Letters* 3, 655-659 (2003).
19. L. N. Daniel, Y. Mao, T.-C. L. Wang, C. J. Markey, S. P. Markey, X. Shi, and U. Saffiotti, DNA strand breakage, thymine glycol production, and hydroxyl radical generation induced by different samples of crystalline silica in vitro, *Environ. Res.* 71, 60-73 (1995).
20. IARC Monographs on the Evaluation of the carcinogenic risk of chemicals to humans, *Silica, some silicates, coal dusts para-aramid fibrils,* vol. 68 (Lyon, France, 1997).
21. K. Donaldson and P. J. A. Borm, The quartz hazard: a variable entity, *Ann. Occup. Hyg.* 42, 287-294 (1998).
22. A. Clouter, D. Brown, P. Borm, and K. Donaldson, Inflammatory Effects of Respirable Quartz Collected in Workplaces versus Standard DQ12 Quartz: Particle Surface Correlates, *Toxicol. Sci.* 63, 90-98 (2001).
23. B. Fubini, I. Fenoglio, Z. Elias, and O. Poirot, On the Variability of the Biological Responses to Silicas: Effect of Origin, Crystallinity and State of the Surface on the Generation of Reactive Oxygen Species and Consequent Morphological Transformations in Cells, *J. Environ. Pathol. Toxicol. Oncol.* 20, 87-100 (2001).
24. B. Fubini, Surface chemistry and quartz hazard, *Ann. Occup. Hyg.* 42, 521-530 (1998).
25. B. Fubini, I. Fenoglio, R. Ceschino, M. Ghiazza, G. Martra, M. Tomatis, P. Borm, R. Schins, and J. Bruch, Relationships between the state of the surface of four commercial quartz flours and their biological activity in vitro and in vivo, *Int. J. Hyg Environ Health* 207, 89-104 (2004).
26. B. Fubini, Use of physico-chemical and cell free assays to evaluate the potential carcinogenicity of fibres, in: *Mechanisms of Fibre Carcinogenesis*, edited by A. B. Kane, P. Boffetta, R. Saracci, and J. D. Wilbourn (IARC Scientific Publication no 140, International Agency for Research on Cancer, Lyon 1996), pp. 35-54.
27. R. Ceschino, I. Fenoglio, M. Tomatis, D. Ghigo, B. Fubini, and G. Martra, Ultrafine versus fine iron oxide particles: size effects on the surface reactivity towards model biological systems, *Chem. Res. Toxicol*, submitted for publication.
28. C. R. Martin and P. Kohli, The emerging field of nanotube biotechnology, *Nat. Rev. Drug. Discov.* 2, 29-37 (2003).
29. A. A. Shvedova, E. R. Kisin, A. R. Murray, D. Schwegler-Berry, V. Z. Gandelsman, P. Baron, A. Maynard, M. R. Gunther, and V. Castranova, Exposure of human bronchial epithelial cells to carbon nanotubes causes oxidative stress and cytotoxicity, *Proc. Soc. Free Rad Research Meeting*, European Section (June 26-29, 2003, Ioannina, Greece, 2004), pp. 91-103.
30. A. A. Shvedova, E. R. Kisin, R. Mercer, A. R. Murray, V. J. Johnson, A.I. Potapovich, Y. Y. Tyurina, O. Gorelik, S. Arepalli, D. Schwegler-Berry, A. F. Hubbs, J. Antonini, D. E. Evans, B.-K. Ku, D. Ramsey, A. Maynard, V. E. Kagan, V. Castranova, and P. Baron, Unusual inflammatory and fibrogenic pulmonary responses to single walled carbon nanotubes in mice, *Am. J. Physiol. Lung Cell Mol. Physiol.* 289, L698-L708 (2005).
31. I. Fenoglio, M. Tomatis, D. Lison, J. Muller, A. Fonseca, J. B. Nagy, and B. Fubini, Reactivity of carbon nanotubes: free radicals generation or scavenging activity? *Free Rad. Biol. Med.* 2005, in press.

DETECTION OF V-TYPE NERVE AGENT DEGRADATION PRODUCTS USING A POLYPYRROLE / PYRROLOQUINOLINE QUINONE-MODIFIED ELECTRODE

OLGA V. SHULGA, CHRISTOPHER PALMER*
Department of Chemistry, University of Montana, Missoula, MT 59812, USA

Abstract. Electropolymerization of pyrrole was used to entrap the biocatalyst Pyrroloquinoline quinone (PQQ) to produce a sensor for electrochemical detection of thiol-containing degradation products of V-type nerve agents. Cyclic voltammetry, linear sweep voltammetry and amperometry were used to characterize the electrode performance. Various parameters that affect the detection process and the performance of the sensor, such as the supporting electrolyte used during electrodeposition and the thickness of the polypyrrole (PPy) film, were examined and optimized. Electocatalytic oxidation of thiols at the PPy/PQQ electrode was strongly affected by the presence Ca^{2+} cations during the electrodeposition process. Amperometric detection of 2-(dimethylamino) ethanethiol (DMAET) and 2-(diethylamino) ethanethiol (DEAET) was performed at 0.25 V, compared with much higher oxidation potentials at unmodified electrodes. Linear calibration plots were observed for both DEAET and DMAET, and detection limits of 50 μM were obtained for both compounds with sensitivities of 45 and 60 nA/mM, respectively.

Keywords: pyrroloquinoline quinone, polypyrrole, electrodeposition, thiols, amperometry

1. Introduction

During the last several years the detection of chemical warfare agents (CWA) and their degradation products have received substantial attention in response to

*To whom correspondence should be addressed. Christopher Palmer, Associate Professor, Department of Chemistry, University of Montana, Missoula, MT 59812-1656. e-mail: christopher.palmer@umontana.edu

J.P. Blitz and V.M. Gun'ko (eds.), Surface Chemistry in Biomedical and Environmental Science, 253–264.

recent terrorist activity. Organophosphate (OP) nerve agents are among the most toxic CWAs, and compounds of similar structure are also used as pesticides. The high toxicity of OP compounds is due to their irreversible binding to the serine residue at the active site of acetylcholine esterase (AChE). This hinders the hydrolysis of the neurotransmitter acetylcholine, resulting in poisoning effects or even death due to dysfunction of the transmission of nerve impulses.[1,2]

Organophosphorus nerve agents are divided into three major types: "G", "V", and "GV".[3] The distinction is based on the differences in structure and toxicity. For example, the LC_{50} for inhalation of G-type nerve agent is as low as 1 ppm after 10 minutes exposure, whereas the LC_{50} of V-type nerve agents is only 0.3 ppm for the same time period. V-type nerve agents, which are more toxic and persistent than the G-type compounds, were developed during the 1950s.[4] The two major V-type agents are VX [*O*-ethyl *S*-(2-diisopropylaminoethyl) methylthiophosphonate] and its Russian analogue R-VX [*O*-isobutyl *S*-(2-diethylaminoethyl) methylthiophosphonate].

Organophoshorous CWAs are hydrolyzed in water at rates that strongly depend on the pH and temperature. As illustrated in Figure 1, the hydrolysis of V-type nerve agents produces a mixture of thiol-containing species. These hydrolysis products are less toxic and can be safely analyzed to provide information regarding the presence and concentration of the parent compound. At pH < 6.5 and > 10, the predominant hydrolysis pathway of VX and other "V" agents proceeds through cleavage of the P-S bond (see Figure 1 (a)), while from pH 7-10 another degradation pathway is dominant (Figure 1 (b)).[5] Table 1 contains the structure of known V-type nerve agents and their thiol hydrolysis products.

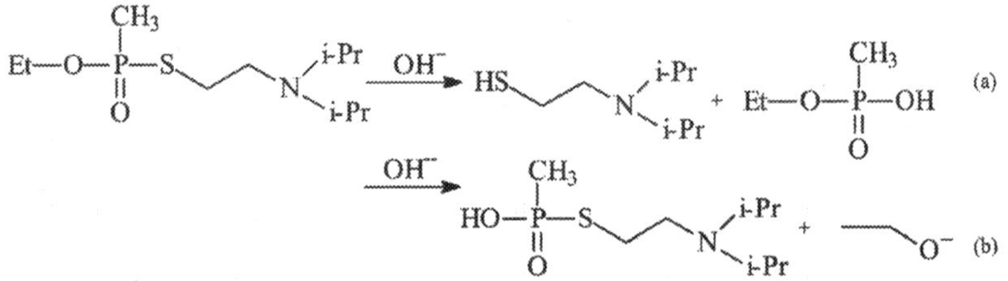

Figure 1. Hydrolysis pathway for VX.

A large number of analytical methods for detection of CWA degradation products are available. Gas chromatography/mass spectrometry (GC/MS) and high performance liquid chromatography (HPLC) with fluorescence detection are the most widely used.[6,7] However, limitations of these methods include complicated derivatization procedures, long analysis time, and difficulty in miniaturization.

Table 1. Structure of V-type nerve agents and the corresponding hydrolysis products.

CWA structure	Name	Actual Hydrolysis product	Hydrolysis product examined
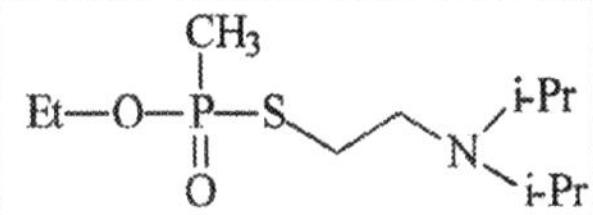	VX		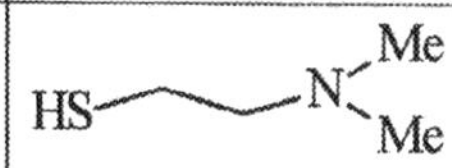
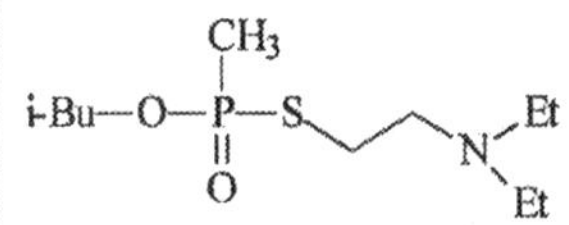	R-VX	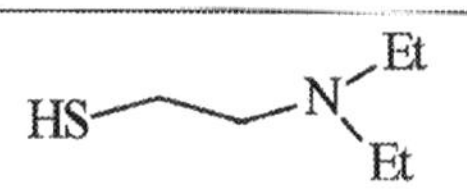	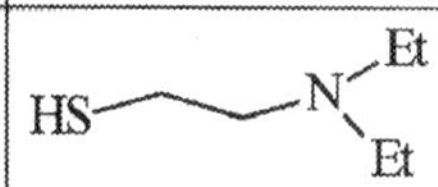
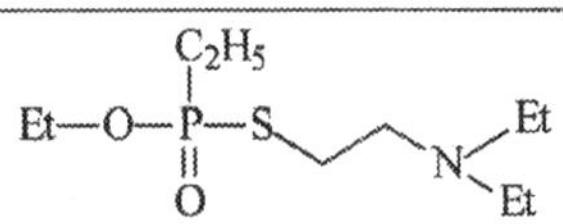	VE		same
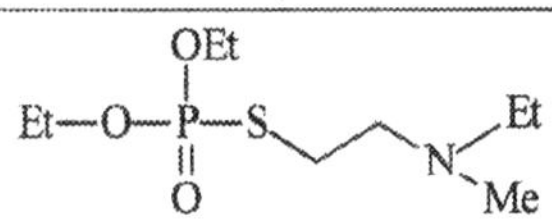	VG	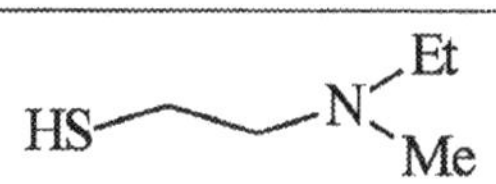	not examined

The determination of V-type nerve agent thiol degradation products has been a significant challenge for analytical chemists. Since thiols do not have significant UV absorption, UV or fluorescent detection cannot be employed unless a derivatization procedure is used. A number of other analytical methods including colorimetry,[8] gas chromatography[9] and radioimmunoassay[10] have been developed. Conventional CE was employed for measuring V-type nerve agent hydrolysis products in connection with UV absorption,[11] conductivity,[12] fluorescence[13] and electrochemical[14] detection schemes.

Electrochemical (EC) techniques provide an alternative way to detect sulfur containing molecules. Earlier methods of EC detection involve the application of a gold/mercury electrode.[15] Platinum and gold electrodes have also been used for anodic detection of thiols,[16] but this requires high oxidation potentials, which complicates analytical applications. Thus, chemically modified electrodes with inorganic or organic mediators have been employed to facilitate electron-transfer between the electrode and the analyte, and therefore reduce the oxidation potential. Recently, pyrroloquinoline quinone (PQQ) modified electrodes have been developed for detection of endo- and exogenous thiols.[17]

PQQ has been extensively studied for the last two decades. It was discovered as a redox co-factor in bacterial alcohol dehydrogenases,[18] and more recently has been found widely distributed in eukariotes. PQQ has been reported to function as an antioxidant, a tissue protective agent, and an essential nutrient. As a part of its structure, PQQ contains an *o*-quinone moiety that exhibits an efficient, pH-dependent and reversible electron transfer between its oxidized and reduced forms (Figure 2). PQQ has high reactivity towards nucleophiles, with its 5-position being the most electrophilic center. It is attacked by water, alcohols, alkyl and aryl amines, diamines, amino acids, acetone, thiols, urea, cyanide and others.[19] Here we report an application of PQQ as a redox mediator for electrochemical detection of the hydrolysis products of V-type CWAs.

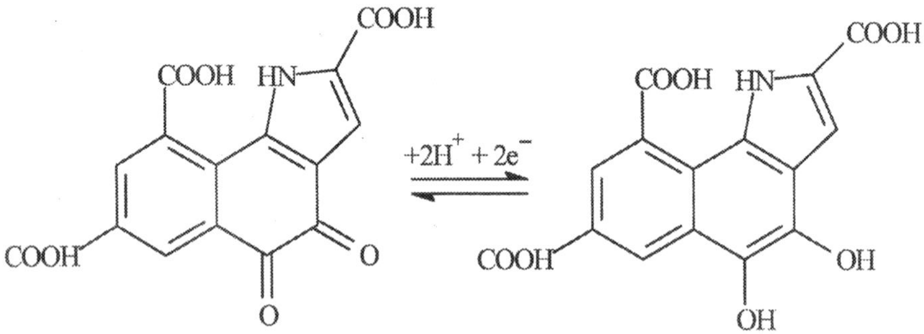

Figure 2. Redox reaction of PQQ.

2. Experimental

2.1. REAGENTS

PQQ (methoxatin; 4,5-dihydro-4,5-dioxo-1H-pyrrolo [2,3-f] quinoline-2,7, 9-tricarboxylic acid), 2-(dimethylamino) ethanethiol (DMAET), 2-(diethylamino) ethanethiol (DEAET), and pyrrole were purchased from Sigma (St. Louis, MO). Potassium hydrogen phthalate was purchased from Acros (Pittsburgh, PA). Pyrrole was distilled every time immediately before use. All other regents were used without further purification.

2.2. INSTRUMENTATION

A CH Instruments 800B electrochemical analyzer (Austin, TX) was used for all electrochemical measurements and electrode preparation. Glassy carbon (GC) electrodes (3 mm diameter, geometrical area 0.07 cm^2) were purchased from CH Instruments (Austin, TX) and polished with polishing alumina solution (BAS, West Lafayette, IN). All electrochemical measurements were conducted in a conventional three-electrode cell using Ag/AgCl (3M KCl) (CH Instruments,

Austin, TX) and Pt wire (BAS, West Lafayette, IN) as reference and auxiliary electrodes, respectively.

2.2.1. *Methods*

Solutions of DMAET and DEAET were prepared in 0.1 M pH 4 phthalate buffer. PQQ solution was prepared by dissolving 1 mg of PQQ in 5 drops of 0.33 M KOH (EMD Chemicals, Gibbstown, NJ), followed by addition of 1 mL of 0.1 M KCl (Mallinckrodt, Paris, KY) or 0.1M $CaCl_2$ (EMD Chemicals, Gibbstown, NJ) resulting an olive-green (KCl) or bright green-yellow ($CaCl_2$) solution. A 0.1 M solution of pyrrole was prepared by adding 70 μL of freshly distilled pyrrole to 10 mL of 0.1 M KCl or $CaCl_2$.

PPy/PQQ modified GC electrodes were constructed by entrapment of PQQ into the polypyrrole membrane during electropolymerization of pyrrole. This method was first developed by Shinohara,[20] and was later used by Inoue[17] for detection of endo- and exogenous thiols. Several modifications to this procedure were necessary for the current studies. PQQ solution was degassed for at least 30 minutes prior to addition of 100 μL of 0.1 M pyrrole. GC, Ag/AgCl and Pt electrodes were immersed in the electrochemical cell, electropolymerization was initiated by the application of a potential of 1.2V versus Ag/AgCl. The thickness of the polypyrrole film was controlled by the quantity of charge that passed across the solution/electrode interface. It was assumed that a charge of 45 mC/cm^2 produced a film 100 nm thick.[21] After deposition the electrodes were washed with distilled deionized (DDI) water, and were stored in DDI water when not in use.

The effectiveness of PQQ entrapment and the dependence of the electrode response on the PPy film thickness were investigated by the application of cyclic voltammetry. The PPy/PQQ modified electrode was immersed in 10 or 20 mM solution of the thiol and potential was scanned from –0.5V to 0.5V monitoring the resulting current.

Amperometric detection of DMAET and DEAET was performed at the PQQ/PPy-modified GC electrode by measuring the current response at a fixed potential of +0.250 V as a function of time. Stock solutions of DMAET and DEAET were prepared in phthalate buffer. An electrochemical cell initially contained 3.5 mL of phthalate buffer to which an aliquot of thiol stock solution was injected. The current response (Δi) was measured in triplicate for each concentration and was plotted against the concentration of the thiol to obtain calibration plots for each thiol.

3. Results and Discussion

3.1. ELECTRODE DESIGN

Cyclic voltammetry experiments of 20 mM DMAET in phthalate buffer were performed to evaluate the effectiveness of PQQ entrapment. Figure 3 shows CV responses of PPy/PQQ-modified electrodes to phthalate buffer (A) and 20 mM DMAET (B). It is seen that in the absence of thiol no current is observed (A), and when thiol is present an irreversible oxidation peak appears at 0.25 V (B).

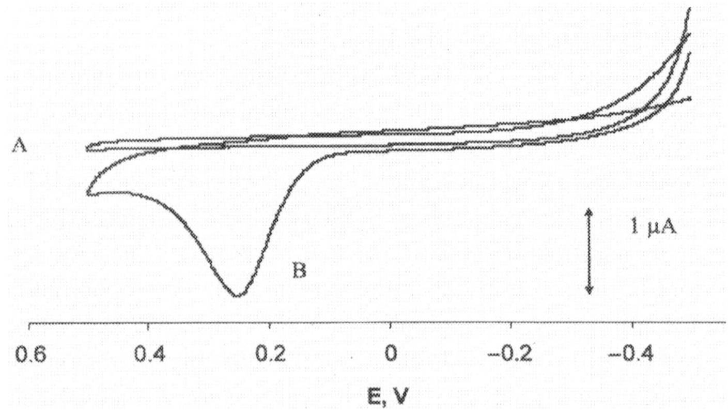

Figure 3. V response of PPy/PQQ modified GC electrode to (A) phthalate buffer, and (B) 20 mM DMAET. Supporting electrolyte during electrodeposition KCl. Film thickness 132 nm.

Catalytic oxidation of thiols by PQQ has been studied by Itoh and co-workers.[22] The reaction between PQQ and thiol is shown below:

$$PQQ_{ox} + RSH \rightarrow PQQH_{2\,red.} + RSSR \qquad (1)$$

The electrochemical process at the modified electrode can thus be separated into two steps. First is a chemical reaction between PQQ and thiol to produce the reduced form of PQQ, $PQQH_2$. This is followed by electrochemical oxidation of $PQQH_2$ to produce PQQ, which completes the cycle. We observed an oxidation potential of $PQQH_2$ much lower than the 0.5 V vs Ag/AgCl reported earlier for the detection of endo- and exogenous thiols such as cysteine and glutathione[17]. This difference could be due to changes in the procedure (smaller amount of pyrrole used) as well as different thiol structures.

The supporting electrolyte used during PPy/PQQ deposition plays an important role in the electrode performance. It has been reported that the physico-chemical properties and elecrocatalytic behavior of PQQ are strongly influenced by the presence of metal cations,[22-24] particularly Ca^{2+}.[22] Earlier, Katz and co-workers showed enhanced electrocatalytic oxidation of NADH by cysteamine/PQQ modified electrode in the presence of Ca ions.[25] In order to investigate the effect of Ca^{2+} ions $CaCl_2$ was used in place of KCl in the supporting electrolyte during PPy/PQQ deposition. Linear sweep voltammetry (LSV) was used to characterize this new electrode (Figure 4).

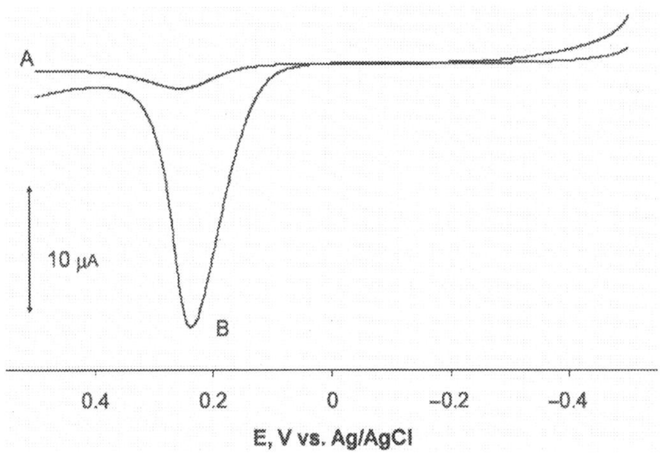

Figure 4. Linear sweep voltametric response of PPy/PQQ modified electrode to 20 mM DMAET as a function of supporting electrolyte used during electrodeposition: (A) KCl , (B) CaCl2. Film thickness 90 nm.

A significant increase in anodic current from 1.8 µA to 21 µA was observed when $CaCl_2$ was used as the supporting electrolyte in place of KCl. A similar effect was observed by Katz and co-workers for the oxidation of NADH at a cysteamine/PQQ modified electrode[25]. A slight shift of the oxidation potential from 0.25 to 0.21 V was also observed. LSV also showed similar results for DEAET.

Such an enhancing effect most likely arises from metal ion coordination. PQQ behaves as a strong tridentate complexing agent[26] involving the carboxylic group at the C(7) position, N(6) and the C(5) carbonyl oxygen (Figure 5). Introduction of Ca^{2+} ion and formation of a complex with PQQ pulls the electron density from the conjugated system making the carbonyl oxygen more susceptible for nucleophilic attack.

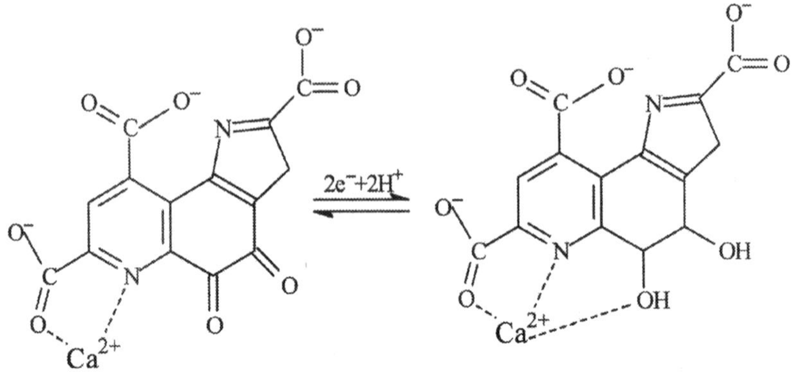

Figure 5. Coordination of Ca^{2+} ion with oxidized and reduced forms of PQQ.

It has been shown that the thickness of the polypyrrole (PPy) film has a significant effect on the electrode performance.[17] Figure 6 shows the dependence of the response of PPy/PQQ modified electrode to 10 mM DMAET (A) and 10 mM DEAET (B) as a function of PPy film thickness. As film thickness increases the oxidation current increases for both DMAET and DEAET, presumably due to increases in the amount of PQQ loaded in the PPy film. The maximum current for the oxidation of $PQQH_2$ is observed when 200 nm films are used. When the PPy film thickness was larger then 200 nm a decrease in the sensor response was observed, which could be due to increased resistance (R) of the thicker film. The optimum 200 nm PPy film thickness was used to characterize the performance of the electrode for amperometric detection of thiols.

3.2. AMPEROMETRIC DETECTION OF DMAET AND DEAET

Electrochemical detection of DMAET and DEAET was performed by PPy/PQQ-modified GC electrode at a constant potential of 0.25 V. The electrodeposition procedure using KCl as a supporting electrolyte resulted in sensors which had poor sensitivity and high detection limit. For example, the detection limit and sensitivity for DMAET using a 136nm PPy film thickness were 0.11 mM and 1.6 nA/mM, respectively. However, using $CaCl_2$ as the supporting electrolyte during electrodeposition significantly improves the electrode characteristics. Figure 7 shows representative calibration plots for

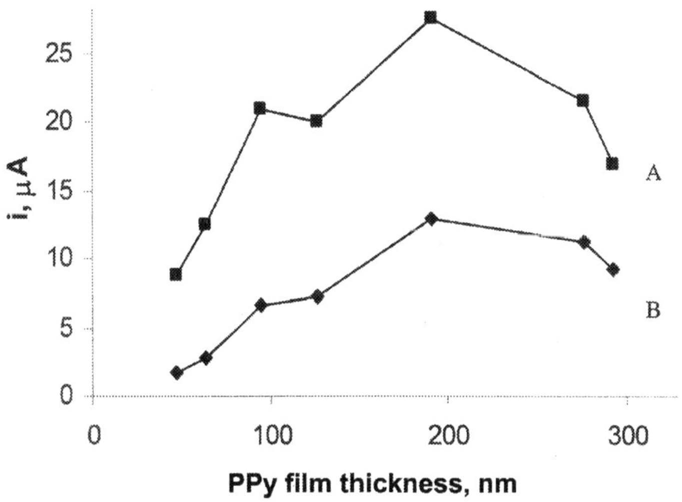

Figure 6. The dependence of the oxidation current generated by PQQH2 as a function of PPy film thickness in presence of (A) 10 mM DMAET and (B) 10 mM DEAET.

DMAET and DEAET using a 200 nm film deposited with the $CaCl_2$ supporting electrolyte. The detection limit of DEAET was 50 μM and the sensitivity was 45 nA/mM. For DMAET these characteristics were 57 μM and 60 nA/mM, respectively. Longer dynamic range (0.050-0.5 mM) was observed for DMAET compare to 0.050-0.26 mM for DEAET.

4. Conclusion

A PPy/PQQ modified GC electrode was used for amperometric detection of V-type nerve agent decomposition products. The electropolymerization of pyrrole was efficiently used for immobilization of the biocatalyst, PQQ. The introduction of $CaCl_2$ as a supporting electrolyte during electrodeposition significantly improves the response of the sensor to DMAET and DEAET. Amperometric studies targeted to detection of DMAET and DEAET by PPy/PQQ electrode were performed at a constant potential set at 0.25 V, and the electrode characteristics such as sensitivity and the analyte detection limit were determined.

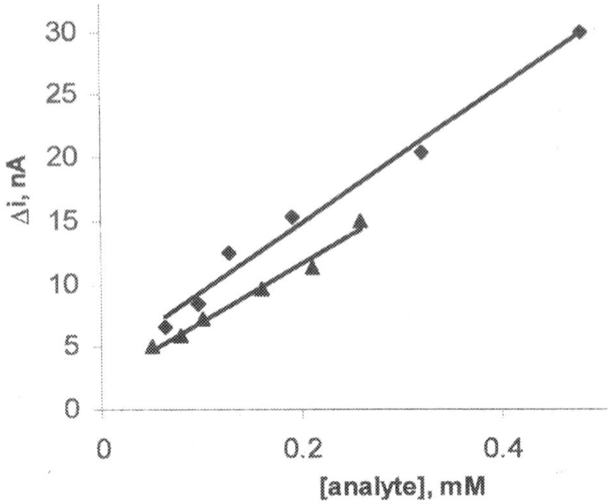

Figure 7. Calibration plot for DMAET (A) and DEAET (B). Film thickness 200 nm.

Acknowledgements

This research was sponsored by the Army Research Laboratory and was accomplished under Cooperative Agreement Number W911NF-04-2-0043. The views and conclusions contained in this document are those of the authors and should not be interpreted as representing the official policies, either expressed or implied, of the Army Research Laboratory or the U.S. Government. The U.S. Government is authorized to reproduce and distribute reprints for Government purposes notwithstanding any copyright notation hereon.

References

1. D. M. Quinn, Enzyme structure, reaction, dynamics, and virtual transition states, *Chem. Rev.* 87, 955-979 (1987).
2. D. Voet and J.G. Voet, *Biochemistry* (John Wiley: New York, 1990).
3. H. Ellison, *Handbook of Chemical and Biological Warfare Agents* (CRC Press, Boca Raton, Florida, 2000).
4. Y. Yang, Chemical detoxification of nerve agent VX, *Acc. Chem. Res.*, 32, 109 (1999).
5. C. Cooper and G. Collins, Separation of thiol and cyanide hydrolysis products of chemical warfare agents by capillary electrophoresis, *Electrophoresis* 25, 897-902 (2004).
6. J. A. Tornes and B. A. Johnsen, Gas chromatographic determination of methylphosphonic acids by methylation with trimethylphenylammonium hydroxide, *J. Chromatogr.* 467, 129-138 (1989).

7. J. G. Purdon, J. P. Pagotto, and R. K. Miller, Preparation, stability, and quantitative analysis by gas chromatography and gas chromatography-electron impact mass spectrometry of tert-butyldimethylsilyl derivatives of some alkylphosphonic and alkyl methylphosphonic acids, *J. Chromatogr.* 475, 261-272 (1989).
8. P. R. Pal, Colorimetric Estimation of Penicillamine, *J. Biol. Chem.* 234, 618-619 (1959).
9. E. Jelleum, V. A. Bacon, and W. Patton, W., Quantitative determination of biologically important thiols and disulfides by gas-liquid chromatography, *Anal. Biochem.* 31, 339-347, (1969).
10. J. Vrbíková, J. Tallová, M. Bičíková, K. Dvořáková, M. Hill, and L. Stárka, Plasma Thiols and Androgen Levels in Polycystic Ovary Syndrome, *Clin Chem Lab Med* 41(2):216-221 (2003).
11. J. E. Melanson, B. L. Wong, C. Boulet, and C. Lucy, High-sensitivity determination of the degradation products of chemical warfare agents by capillary electrophoresis-indirect UV absorption. *J. Chromatogr. A* 920:359-365, (2001).
12. A. E. Nassar, S. Lucas, W. Jones, and L. Holland, Separation of chemical warfare agent degradation products by the reversal of electroosmotic flow in capillary electrophoresis, *Anal. Chem.* 70, 1085-1091 (1998).
13. C. L. Copper and G. E. Collins, Separation of thiol and cyanide hydrolysis products of chemical warfare agents by capillary electrophoresis, *Electrophoresis* 25, 897-902 (2004).
14. J. Wang, J. Zima, N. S. Lawrence, M. P. Chatrathi, A. Mulchandani, and G. E. Collins, Microchip Capillary Electrophoresis with Electrochemical Detection of Thiol-Containing Degradation Products of V-Type Nerve Agents, *Anal. Chem.* 76, 4721-4726 (2004).
15. L. A. Allison and R. E. Shoup, Dual-electrode liquid chromatography detector for thiols and disulfides, *Anal. Chem.* 55, 8-12 (1983)
16. J. C. Hoekstra and D. C. Johnson, Comparison of Potential-Time Waveforms for the Detection of Biogenic Amines in Complex Mixtures Following Their Separation by Liquid Chromatography, *Anal. Chem.* 70, 83-88, (1998).
17. T. Inoue and J. R. Kirchhoff, Electrochemical Detection of Thiols with a Coenzyme Pyrroloquinoline Quinone Modified Electrode. *Anal. Chem.* 72, 5755-5760 (2000).
18. S. A. Salisbury, H. S. Forrest, W. B. Cruse, and O. Kennard, A novel coenzyme from bacterial primary alcohol dehydrogenases, *Nature* 280, 843-844 (1979).
19. McIntire, W. *Essays in Biochem.* 27, 119-134 (1997).
20. H. Shinohara, G. F. Khan, and M. Aizawa, Electrochemical oxidation and reduction of PQQ using a conducting polypyrrole-coated electrode, *J. Electroanal. Chem.* 304, 74-84 (1991).
21. S. Holdkroft and B. L. Funt, Preparation and electrocatalytic properties of conducting films of polypyrrole containing platinum miroparticulates, *J. Electroanal. Chem.* 240, 89-103 (1998).
22. S. Itoh, N. Kato, Y. Ohshiro, and T. Agawa, Catalytic oxidation of thiols by coenzyme PQQ, *Chem. Lett.* 135-136 (1985).
23. J. Frebortova, K. Matsushita, H. Arata, and O. Adachi, Intramolecular electron transport in quinoprotein alcohol dehydrogenase of Acetobacter methanolicus: a redox-titration study, *Biochim. Biophys. Acta* 1363, 24-34 (1998).
24. Y. Oshiro and S. Itoh, *PQQ and Quinoproteins* (Kluwer, Dordrech, 1988) p. 195.

25. E. Katz, T. Lotzbeyer, D. Schlereth, W. Schuhmann, and H.-L. Schmidt, Electrocatalytic oxidation of reduced nicotinamide coenzymes at gold and platinum electrode surfaces modified with a monolayer of pyrroloquinoline quinone. Effect of Ca^{2+} cations, *J. Electroanal. Chem.* 373, 189-200 (1994).
26. J. B. Noar, E. J. Rodriguez, T. C. Bruice, Synthesis of 9-decarboxymethoxatin. Metal complexation of methoxatin as a possible requirement for its biological activity, *J. Am. Chem. Soc.* 107, 7198 (1985).

BIOLOGICAL APPLICATIONS OF FUNCTIONALIZED CARBON NANOPARTICLES

JURGIS BARKAUSKAS[1]
Vilnius University, Naugarduko 24, 03225 Vilnius, Lithuania

Abstract. Biological applications of carbon nanotubes have been the focus of significant interest since the discovery of these one dimensional carbon allotropes. Indeed, carbon nanotubes have many interesting and unique properties potentially useful in a variety of biological and biomedical systems and devices. Significant progress has been made in the effort to overcome fundamental and technical barriers toward bio-applications, especially on issues concerning the aqueous solubility and biocompatibility of carbon nanotubes and other carbon nano-structures. Potential bio-applications of carbon nanotubes are wide ranging, yet the use of these structures in biosensors has received special attention. One of the key issues in biosensor design is the establishment of fast, unrestricted electron transfer between the enzyme active site and the electrochemical transducer. The latest advances in production of functionalized carbon nanoparticles have been utilized to design third generation of biosensors, where direct electron transfer between the enzyme and the nanoparticle occurs.

Keywords: bio-application, biocompatability, biosensor, carbon nanotube, direct electron transfer, electroconductivity, enzyme, functionalization, immobilization, quantum dot

1. Introduction

Carbon nanotubes and other nano-sized carbon structures are of special interest due to their many unique properties. These nano-materials are very promising for the development of novel technological applications. They exhibit extreme toughness and high elastic moduli, their electronic behaviour can vary from semiconducting to metallic, depending on structure. Promising optical and magnetic properties of these structures have also been recently reported. The nano-sized structures used for all potential applications must be modified to

[1] To whom correspondence should be addressed: Dept. General & Inorganic Chemistry, Vilnius University, Naugarduko 24, 03225 Vilnius, Lithuania; e-mail: jurgis.barkauskas@chf.vu.lt

J. P. Blitz and V. M. Gun'ko (eds.), Surface Chemistry in Biomedical and Environmental Science, 265–276.

make them more processable. To achieve this objective, the nano-particles can be functionalized by tailoring a wide range of chemical groups in order to impart to the surface desirable properties.

2. Carbonaceous Materials

Carbonaceous materials are defined as structures composed from elemental carbon. This includes a great variety of materials with novel properties, which attract considerable interest among engineers and scientists. Carbon exists in a number of forms and allotropes. The best known examples found in nature are diamond (the hardest naturally-occurring substance ever known) and graphite (the most stable electroconductive carbon allotrope). The numerous group of amorphous and polycrystalline carbons includes activated carbons, diamond-like carbons, glassy carbons, pregraphitic carbons, soots, pitches and many others. The structure of amorphous and polycrystalline carbons is highly heterogeneous, different domains containing either diamond or graphite phases. Many members of this group contain considerable quantities of impurities, mostly hydrogen. Review articles on amorphous and polycrystalline carbons are presented elsewhere.[1,2] The solids, which are already known, have been prepared as films, fibres, foams, aerogels, carbon blacks, as required by their ultimate end use application. These solids may potentially bridge the gap between polymers and ceramics combining surface and bulk properties, which lead to specific properties useful for many applications. In principle, these carbons may be intensively applied for bio-medical applications for the same reason as other carbonaceous materials.

Vigorous development in the field of carbon science has began in the mid 1980s, when Kroto, Curl, and Smalley[3] at Rice University reported the existence of fullerenes. A few years later carbon nanotubes (CNTs) were discovered.[4] The interest in nanotubes has not yet peaked: the number of publications increasing at a yearly rate of nearly 80%.

Mechanical properties, electrical properties, thermodynamic stability, surface chemical activity, and other important parameters can all be discussed relative to the structure of the carbon network, composed of both aromatic layers and 3D-arranged (diamond-like) phases.

The nature of the chemical bonding between neighbouring carbon atoms is unique in several respects. Electronic hybridization (sp, sp^2 and sp^3) leads to several types of covalent bonding, which are quite different in geometry. The most surprising and important discovery concerning carbonaceous materials is that the carbon atom is able to undergo an intermediate hybridization, leading to the formation of carbon allotropes with surface curvature (fullerenes and nanotubes).

At the heart of carbon reactions, and the key to the usefulness of many carbonaceous materials, is the interfacial chemistry of carbon surfaces. The field of carbon surface reactions has been frequently reviewed. The structural arrangements in carbons are responsible for some remarkable chemical, electronic, and electrical properties of basic carbon polymorphs: diamond and graphite, which clearly set them apart from other materials. This is readily apparent when compared with the surface chemical properties of inorganic solids. Carbon surfaces can contain markedly different type of surface groups that can be used for surface modification.[5] The surface of carbon is unique due to the combination of a number of properties of this material: strength and covalency of the chemical bond, inertness at room temperature, electro-conductivity, and structural diversity. For the same reasons, carbons (especially recently discovered carbon forms) hold considerable interest for applications in emerging technologies, including nanobiotechnology.

3. Carbon Nanotubes and Other Nano-Sized Carbon Particles

Carbon nanotubes are cylindrical carbon molecules with properties that make them potentially useful in extremely small scale electronic and mechanical applications. They exhibit unusual strength and unique electrical properties, and are efficient thermal conductors.

A graphene sheet can be rolled up in different ways to form a certain type of CNT. Similarly, various defects in the graphene plane (flaws in the regular hexagonal structure, substitution of carbon atoms by other atoms, etc.) may form new types of CNTs. Multi-walled, functionalized, capped and opened CNTs broaden this list.

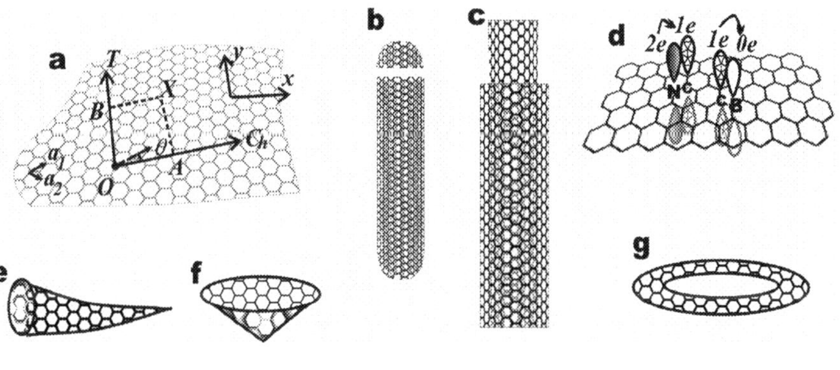

Figure 1. CNTs and other nano-sized carbon particles: (a) scrolling a graphene sheet; (b) closed and opened single-walled CNT; (c) double-walled CNT; (d) CNT sites doped with boron and nitrogen; (e) nanopipettes; (f) nanocones; and (g) nanorings.

One way to define the structure of single-walled CNT is to think of each CNT as a result of rolling a graphene sheet; specifying the direction of rolling and the circumference cross-section (Figure 1a). The unit vectors of the planar hexagonal lattice are defined by a_1 and a_2. A single walled CNT is constructed by rolling the graphene sheet along a certain direction $C_h = n\ a_1 + m\ a_2$ (with n and m being integers) making OB and AX coincide. The chiral indices (n, m) are commonly used to label CNTs. The chiral angle θ between a_1 and C_h ranges within 30° due to the symmetry of the hexagonal lattice. CNTs with $\theta = 30°$ or $\theta = 0°$ have the highest symmetry, known as armchair (n, n) or zigzag nanotubes $(n, 0)$. For a given (n, m) nanotube, if $2n + m = 3q$ (where q is an integer), then the nanotube is metallic, otherwise the nanotube is a semiconductor. Thus all armchair nanotubes are metallic.

The ends of CNTs may be either closed or open (Figure 1b). Based on Euler's theorem on polyhedra, along with additional knowledge of the minimum energy structure of fullerenes, Fujita et al.[6] have shown that any cap at the end of small-radius CNT must contain 6 pentagons isolated from each other. For CNTs with large radius, there are different possibilities of forming caps that satisfy this requirement.

A multi-walled CNT is a coaxial assembly of single-walled CNTs, normally encased by 10 to 30 consecutive layers of carbon atoms. Multi-walled CNTs are more synthesized than single-walled CNTs. In most cases multi-walled CNT contain many structural defects. Recently the synthesis of double-walled CNTs has been reported[7] (Figure 1c).

Electronic, mechanical and chemical properties of CNTs may be altered significantly by doping (in most cases with B or/and N). The presence of holes (B-doped tubes) or donors (N-doped tubules) make the surface of CNTs more reactive[8] (Figure 1d).

The synthesis of other nano-sized carbon structures (nanocones,[9] nanopipettes[10] and nanorings,[11] presented in Figure 1e-g) has been recently reported. These structures have promising mechanical, electronic and surface properties, and are of interest for applications in nano-scale devices.

Economically feasible large-scale production and purification techniques must still be developed. In arc discharge, a vapour is created between two carbon electrodes with or without catalyst. In the laser ablation technique, a high-powered laser beam impinges on a volume of carbon–containing feedstock gas. Flame synthesis is used in a controlled flame environment.[8]

Currently, chemical vapour deposition is most commonly used for CNT production (Figure 2a). In this method Fe, Ni, Co or an alloy of the three catalytic metals is initially deposited on a substrate. The precise mechanism is not yet exactly known. The leading theory suggests that first a radical organic precursor is dissolved in the metal catalyst particle. From this metastable carbide particle, a rodlike carbon is rapidly formed (Figure 2b). Formation of tubular structures with a growth axis perpendicular to the catalyst surface is thermodynamically favorable due to the absence of dangling bonds at the end of the graphene layer. Plasma enhanced chemical vapour deposition, sol-gel catalyst preparation, template synthesis, flying catalyst and other protocols have also been used to achieve the controlled synthesis of CNTs.[8]

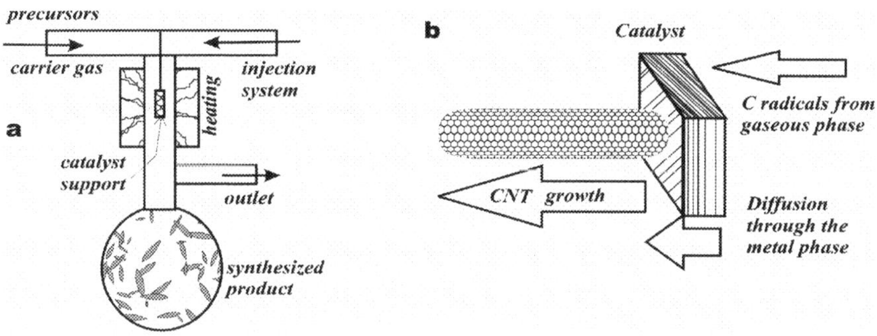

Figure 2. CNT formation in the catalytic growth process from the vapor phase: (a) equipment used for the synthesis; (b) formation of the CNT.

Purification and separation of CNTs after syntheis is a difficult problem. Purification by oxidation was the first successful method.[12] However, > 99% of the original sample is lost. Bandow et al.[13] separated coexisting carbon nanospheres, metal nanoparticles, polyaromatic carbons, and fullerenes from a single-walled CNT fraction by microfiltration. This technique has recently been widely employed in spite of its low efficiency. Field-flow fractionation is a chromatography-like separation and sizing technique based on elution through a thin, empty channel. Currently, in association with ultrasonically assisted filtration[14] it is one of the most commonly used techniques for nanotube purification. Among the other separation techniques, methods based on CNT functionalization[15] should also be mentioned.

CNTs are valued for their novel properties: mechanical strength, chemical inertness, electronic and thermal conductivity. Mechanical properties of CNTs, especially their extreme flexibility and strength equal to steel with one-sixth the mass, is most impressive. An excellent resistance of carbon nanotubes to bending has been observed experimentally and studied theoretically. CNTs are not nearly as strong under compression. Due to their hollow structure, they undergo buckling when placed under compressive, torsional or bending stress.[16]

As mentioned above, electronic properties of CNTs depend on the chirality and presence of defects in the scrolled graphene layer. Metallic nanotubes can have an electric current density more than 1,000 times greater than metals such as silver and copper. All nanotubes are expected to be very good thermal conductors along the tube, but good insulators laterally to the tube axis.[8]

Quantum dots have been widely explored during the last few years for their interesting optical properties and for their possible applications in diode lasers, amplifiers, and detectors. Quantum dot behaviour, observed both in single- and in multi-walled CNTs,[17,18] is determined by the unique electronic structure of these particles. Moreover, a CNT can act as a quantum wire capable of tranferring quantum energy in a desired direction.[19] CNTs have excellent material properties which impart attractive field emission characteristics.[20]

4. Functionalization of Carbon Nanoparticles

The question about functionalization of CNTs in nature has a close connection with their bio-application. Functionalization is closely associated with the ability to disperse and perhaps dissolve the nanotubes, which would greatly improve processability.[21,22] Due to the fact that the majority of popular synthetic methods produce samples yielding a mixture of many different diameters and chiralities of nanotubes, post-synthesis chemical processing protocols[23] are the most popular among the methods of chemical modification.

One aspect of research in this field involves non-covalent interactions with the surfaces of CNTs[24] (Figure 3a). An example of non-covalent functionalization may be a dispersion of single walled CNTs solubilized in water by wrapping them with linear polymers.[25] Wrapping of single walled CNTs with DNA oligonucleotide sequences has been aimed at improved separation of metallic versus semiconducting nanotubes.[26] In addition, a wide range of biological molecules have been immobilized onto CNT sidewalls through π-stacking interactions, this involves the mediation of 1-pyrenebutanoic acid and succinimidyl ester linkers. This approach tends to retain biological activity and functionality.[27] Mechanistically, the non-specific adsorption of proteins onto the nanotube surface may be more complicated than the widely attributed hydrophobic interactions.[28] For the nonspecific adsorption of proteins on other surfaces, electrostatic interactions, hydrogen bonding, and other mechanisms are generally important.[29] It is established that the adsorption of proteins on single-walled CNTs is insensitive to the protein isoelectric point, with both positively and negatively charged proteins showing strong adsorption.[30] This is inconsistent with an electrostatic interaction mechanism. It is possible that the observed protein adsorption is at least in part associated with the amino affinity of carbon nanotubes,[30] since there are abundant surface amino groups in the proteins under consideration. A comprehensive understanding of the non-specific protein – CNT interactions requires further investigation.

Covalent surface chemical modification strategies are also of interest. CNT covalent functionalization can be considered analogous to fullerene chemistry, with definite differences resulting from the difference in structure of these two nano-sized particles[31]. Covalent chemical functionalization of nanotubes has the potential to provide a high degree of tunability of CNT properties, however this potential has not yet been realized. This type of functionalization can be divided into two sub-sections based on the spatial target site for chemical modification involved: end and defect site functionalization, as well as sidewall-functionalization (Figure 3b). A review article concerning covalent functional modification of single walled CNTs is presented by Wong and co-workers.[32]

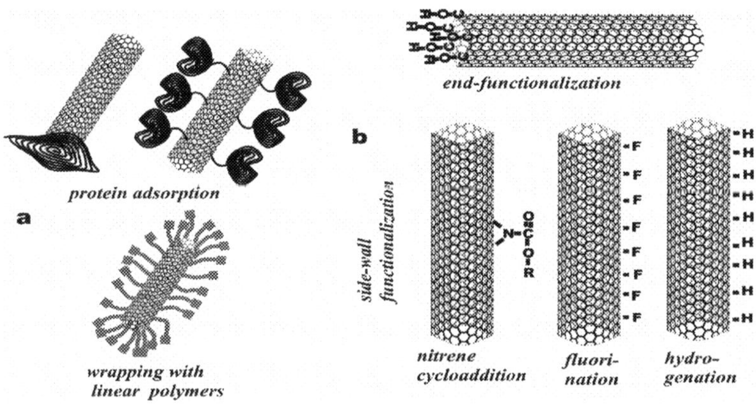

Figure 3. Functionalization of CNTs: (a) non-covalent interacions with polymers and biomolecules; (b) covalent surface chemical modification (end-functionalization and side-wall functionalization).

5. Potential of Carbon Nano-Structures for Bio-Application

Bio-applications of carbon nanotubes have been predicted and explored ever since their discovery. Significant progress has been made in the effort to overcome many fundamental and technical barriers toward bio-applications. Much of this research is in the field of biosensors; to be discussed in the next section.

CNTs can be applied in bioimaging and drug delivery, e.g. as transporters for delivering small molecules or proteins into cells.[33] Thus, CNTs may be considered as a new generation of biocompatible materials for drug, protein, and gene delivery applications. With the same goal of promoting information delivery into cells, CNTs have been utilized as 'electric needles' for localized perforation of cell membranes.[34]

CNTs are especially valued as implant materials thanks to their novel mechanical properties and surface functionability.[35] They have been found to make an ideal scaffold for the growth of bone tissue.[36] Moreover, many tissues and organs require bio-compatible substrates to facilitate tissue growth and implantation. The fabric made fom CNTs serves as an efficient tissue scaffold.[36] Several publications demonstrate that CNTs can be used as a substrate for neuronal growth, and that modifications of the CNTs can be employed to modulate the development of neurons. This suggests that it may be possible to employ suitably functionalized CNTs as neural prostheses in neurite regeneration.[35] Lipid bilayers have been developed using a nanotube template.

The nanotubes have been wrapped with hydrophilic polymer layers, which act as assembly templates for common phospholipids and provide a support surface to maintain the bilayer structure.[37] The bilayers were stable enough to survive multiple bleaching-recovery cycles.

6. Application of Carbon Nanoparticles in Biosensors

Recently, hopes have been raised for the use of CNTs as superior biosensor materials. Successful fabrication of various analytical nanotube devices, especially those modified with biomolecules, has made this a possibility. These prototype devices, sometimes prepared as ordered arrays or single-nanotube transistors, have shown efficient electrical communications and promising sensitivities required for such applications as antigen recognition,[38] enzyme-catalyzed reactions[39] and DNA hybridizations.[40] Publications considering a quantum dot bahaviour of CNTs show their promise for biorecognition devices with optical indication.[41]

Optical properties of CNTs have been recently employed in biosensors. Single semiconducting CNTs can emit strong luminescence in the near infrared region,[42] where human tissue and biological fluids are particularly transparent. Barone and coworkers[43] have made optical biosensors based on CNTs where absorbed electroactive species react selectively with a target analyte to modulate the fluorescence of CNTs. The assembled system was tested for the reaction of β-D-glucose to D-glucono-1,5-lactone with a H_2O_2 coproduct catalyzed by glucose oxidase.

Compared with the usual quantum dots based on semiconductors, quantum wires with large surface area offer a unique opportunity to improve the sensitivity of biosensors.[44] Heterojunctions of semiconductor quantum dots and CNTs may provide better alternatives for the synthesis of nanoscale devices which preserve the electronic properties of CNTs.[18] Covalent conjugation between an organic virus capsid and functionalized inorganic nanoparticles such as single-walled CNT has been reported.[45] By taking advantage of the receptor recognition to its virus counterpart, it has been shown that antibody functionalized CNTs attached only to where the virus counterpart was. Another application of CNTs functioning as quantum dots is as a potential artificial fluorophore for intra-operative detection of tumors using fluorescence spectroscopy.[35]

Electrical conductance of CNTs is highly sensitive to surface adsorbed molecules.[46] Only semiconducting single-walled CNTs, acting as field-effect transistors, exhibit a large conductance change in response to the electrostatic and chemical gating effects. Research on biomodified CNTs has demonstrated the selective detection of proteins in solution via specific antigen and antibody interactions.[28] The CNTs can be assembled in contact with Au electrodes via a three-strand homologous recombination reaction between a double-stranded DNA molecule serving as a scaffold and an auxiliary single-stranded DNA[47] with subsequent formation of Ag wires between Au electodes. This strategy offers a way to scale up one-dimensional nanomaterials into conventional sensor devices.

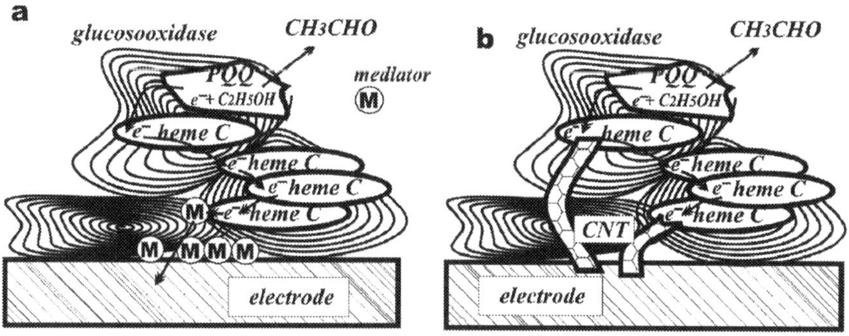

Figure 4. Enzyme (PQQ glucosooxidase) adsorbed on the surface of electrode: (a) electron transfer by means of mediator; (b) direct electron transfer in the case of enzyme adsorbed on the array of CNTs.

At present, amperometric biosensors are the most successful application area for CNTs.[35] A number of different nanotube based biosensing concepts developed thus far have resulted in significant contributions.[39] Mass-produced CNT based amperometric biosensors exhibit good analytical characteristics.[39] Nevertheless, challenges such as a low and superimposed analytical signal, and a limited number of suitable enzymes remain. These problems result from enzyme structure. A major barrier lies in the electrical communication between the biological recognition element, which is situated in the enzyme's active center, and the signal transducer (the electrode material). The most promising way to overcome this problem is to facilitate electron transfer between an enzyme and electrode surface. For this purpose, specific mediators (e.g. ferrocene) are used to enhance a mediated electron transfer by taking part in the enzymatic reaction[39] (Figure 4a). The main practical disadvantage of mediators is the necessity to use additional reagents. In the case of direct electron transfer, where the reaction itself causes the response and no product or mediator diffusion is directly involved, a reagentless (so called "third generation") biosensor is realized.[35]

In the range of nano-sized structures, CNTs are considered the best candidates for third generation biosensors (Figure 4b). The key advantage of a CNT as a transducer is its small diameter and long length which allows this structure to be plugged into proteins.[39] Promising features of carbon nanostructures are their good electro-conductivity, controllable surface electrochemistry and surface functionalization. Other carbon nanoparticles have also been reported suitable for third generation biosensors.[48]

7. Conclusion

CNTs and other nano-sized carbon structures are promising materials for bio-applications, which was predicted even previous to their discovery. These nanoparticles have been applied in bioimaging and drug delivery, as implant materials and scaffolds for tissue growth, to modulate neuronal development and for lipid bilayer membranes. Considerable research has been done in the field of biosensors. Novel optical properties of CNTs have made them potential quantum dot sensors, as well as light emitters. Electrical conductance of CNTs has been exploited for field transistor based biosensors. CNTs and other nano-sized carbon structures are considered third generation amperometric biosensors, where direct electron transfer between the enzyme active center and the transducer takes place. Nanoparticle functionalization is required to achieve their full potential in many fields, including bio-applications.

References

1. J. Robertson, Diamond-like amorphous carbon, *Mater. Sci. & Eng. R* 37(4-6), 129-281 (2002).
2. V. V. Brazhkin, A. G. Lyapin, S. V. Popova, S. K. Beiliss, T. D. Varfolomeeva, R. N. Voloshin, A. G. Gavrilyuk, M. V. Kondrin, V. V. Mukhamad'yarov, I. A. Troyan, S. V. Demishev, A. A. Pronin, N. E. Sluchanko, Interplay between the structure and properties of new metastable carbon phases obtained under high pressures from fullerite C-60 and carbyne, *JETP Lett.* 76(11), 681-692 (2002).
3. H. W. Kroto, J. R. Heath, S. C. O'Brien, R. F. Curl, and R. E. Smalley, Buckminsterfullerene, *Nature* 318, 162-163 (1985).
4. S. Iijima, Helical Microtubules of Graphitic Carbon, *Nature* 354, 56-58 (1991).
5. R. J. Cava, F. J. DiSalvo, L. E. Brus, K. R. Dunbar, C. B. Gorman, S. M. Haile, L. V. Interrante, J. L. Musfeldt, A. Navrotsky, R. G. Nuzzo, W. E. Pickett, A. P. Wilkinson, C. Ahn, J. W. Allen, P. C. Burns, G. Ceder, C. E. D. Chidsey, W. Clegg, E. Coronado, H. J. Dai, M. W. Deem, B. S. Dunn, G. Galli, A. J. Jacobson, M. Kanatzidis, W. B. Lin, A. Manthiram, M. Mrksich, D. J. Norris, A. J. Nozik, X. G. Peng, C. Rawn, D. Rolison, D. J. Singh, B. H. Toby, S. Tolbert, U. B. Wiesner, P. M. Woodward, P. D. Yang, Future directions in solid state chemistry: report of the NSF-sponsored workshop, *Progr. Solid State Chem.* 30(1-2), 1-101(2002).
6. M. Fujita, R. Saito, G. Dresselhaus, and M. S. Dresselhaus, Formation of general fullerenes by their projection on a honeycomb lattice, *Phys. Rev. B.* 45(23), 13834-13836 (1992).
7. G. Chen, S. Bandow, E. R. Margine, C. Nisoli, A. N. Kolmogorov, V. H. Crespi, R. Gupta, G. U. Sumanasekera, S. Iijima, and P. C. Eklund, Chemically doped double-walled carbon nanotubes: cylindrical molecular capacitors, *Phys. Rev. Lett.* 90(25), 257403-1-4 (2003).
8. M. Terrones, Carbon nanotubes: Synthesis and properties, electronic devices and other emerging applications, *Int. Mater. Rev.* 49(6), 325-377 (2004).
9. S. P. Jordan and V. H. Crespi, Theory of carbon nanocones: mechanical chiral inversion of a micron-scale three-dimensional object, *Phys. Rev. Lett.* 93, 255504-1-4 (2004).
10. B.M. Kim, T. Murray, and H. H. Bau, The fabrication of integrated carbon pipes with sub-micron diameters, *Nanotechnology* 16(8), 1317-1320 (2005).

11. Z. L. Wang, Semiconducting and piezoelectric nanobelts, nanosprings and nanorings, *JOM* 56(11), 110-119 (2004).
12. T. W. Ebbesen, P. M. Ajayan, P. Hiura, and K. Tanigaki, Purification of carbon nanotubes, *Nature* 367, 519-520 (1994).
13. S. Bandow, A M. Rao, K. A. Williams, A. Thess, R. E. Smalley, and P. C. Eklund,. Purification of single-wall carbon nanotubes by microfiltration, *J. Phys. Chem. B* 101(44), 8839-8842 (1997).
14. N. Sinha and J. T.-W. Yeow, Carbon nanotubes for biomedical applications, *IEEE Trans. Nanobiosci.* 4(2), 180-195 (2005).
15. D. Chattopadhyay, L. Galeska, and F. Papadimitrakopoulos, A route for bulk separation of semiconducting from metallic single-wall carbon nanotubes, *J. Amer. Chem. Soc.* 125(11), 3370-3375 (2003).
16. D. Qian, G. J. Wagner, W. K. Liu, M.-F. Yu and R. S. Ruoff, Mechanics of carbon nanotubes, *Appl. Mech. Rev.* 55(6), 495-553, (2002).
17. S. Moriyama, T. Fuse, M. Suzuki, Y. Aoyagi, and K. Ishibashi, Selecting single quantum dots from a bundle of single-wall carbon nanotubes using the large current flow process, *Sci. Technol. Adv. Mater.* 5(5-6), 613-615 (2004).
18. S. Ravindran, K. N. Bozhilov, and C. S. Ozkan, Self assembly of ordered artificial solids of semiconducting ZnS capped CdSe nanoparticles at carbon nanotube ends, *Carbon* 42(8-9), 1537-1542 (2004).
19. L. Balents and R. Egger, Spin transport in interacting quantum wires and carbon nanotubes, *Phys. Rev. Lett.* 85(16), 3464-3467 (2000).
20. G. Chai, L. Chow, D. Zhou, and S. R. Byahut, Focused-ion-beam assisted fabrication of individual multiwall carbon nanotube field emitter, *Carbon* 43(10), 2083-2087 (2005).
21. S. Niyogi, M. A. Hamon, H. Hu, B. Zhao, P. Bhowmik, R. Sen, M. E. Itkis, and R. C. Haddon, Chemistry of single-walled carbon nanotubes, *Acc. Chem. Res.* 35(12), 1105-1113 (2002).
22. C. A. Dyke and J. M. Tour, Overcoming the insolubility of carbon nanotubes through high degrees of sidewall functionalization, *Chem. Eur. J.* 10(4), 813-817 (2004).
23. M. S. Strano, C. A. Dyke, M. L. Usrey, P. W. Barone, M. J. Allen, H. W. Shan, C. Kittrell, R. H. Hauge, J. M. Tour, and R. E. Smalley, Electronic structure control of single-walled carbon nanotube functionalization, *Science* 301(5639), 1519-1522 (2003).
24. D. J. Hornbaker, S. J. Kahng, S. Misra, B. W. Smith, A. T. Johnson, E. J. Mele, D. E. Luzzi, and A. Yazdani, Mapping the one-dimensional electronic states of nanotube peapod structures, *Science* 295(5556), 828-831 (2002).
25. Y. Lin, S. Taylor, H. Li, K. A. S. Fernando, L. Qu, W. Wang, L. Gu, B. Zhou and Y. P. Sun Advances toward bioapplications of carbon nanotubes, *J. Mater. Chem.* 14(5), 527-541 (2004).
26. M. Zheng, A. Jagota, M. S. Strano, A. P. Santos, P. Barone, S. G. Chou, B. A. Diner, M. S. Dresselhaus, R. S. McLean, G. B. Onoa, G. G. Samsonidze, E. D. Semke, M. Usrey, D. J. Walls, Structure-based carbon nanotube sorting by sequence-dependent DNA assembly, *Science* 302(5650), 1545-1548 (2003).
27. J. Zhu, M. Yudasaka, M. F. Zhang, D. Kasuya, and S. Iijima, Surface modification approach to the patterned assembly of single-walled carbon nanomaterials, *Nano Lett.* 3(9), 1239-1243 (2003).
28. R. J. Chen, S. Bangsaruntip, K. A. Drouvalakis, N. W. S. Kam, M. Shim, Y. M. Li, W. Kim, P. J. Utz, and H. J. Dai, Noncovalent functionalization of carbon nanotubes for highly specific electronic biosensors, *Proc. Natnl. Acad. Sci. USA* 100(9), 4984-4989 (2003).
29. B. R. Azamian, J. J. Davis, K. S. Coleman, C. B. Bagshaw, and M. L. H. Green, Bioelectrochemical single-walled carbon nanotubes, *J. Am. Chem. Soc.* 124(43), 12664-12665 (2002).

30. J. Kong and H. J. Dai, Full and modulated chemical gating of individual carbon nanotubes by organic amine compounds, *J. Phys. Chem. B* 105(15), 2890-2893 (2001).
31. Y. Chen, R. C. Haddon, S. Fang, A. M. Rao, P. C. Eklund, W. H. Lee, E. C. Dickey, E. A. Grulke, J. C. Pendergrass, A. Chavan, B. E. Haley, and R. E. Smalley, Chemical attachment of organic functional groups to single-walled carbon nanotube material, *J. Mater. Res.* 13(9), 2423-2431 (1998).
32. S. Banerjee, T. Hemraj-Benny, and S. S. Wong, Covalent surface chemistry of single-walled carbon nanotubes, *Adv. Mater.* 77(1), 17-29 (2005).
33. N. W. S. Kam and H. J. Dai, Carbon nanotubes as intracellular protein transporters: Generality and biological functionality, *J. Amer. Chem. Soc.* 127(16), 6021-6026 (2005).
34. J. A. Rojas-Chapana, M. A. Correa-Duarte, Z. F. Ren, K. Kempa, and M. Giersig, Enhanced introduction of gold nanoparticles into vital Acidothiobacillus ferrooxidans by carbon nanotube-based microwave electroporation, *Nano Lett.* 4(5), 985-988 (2004).
35. E. Bekyarova, Y. Ni, E. B. Malarkey, V. Montana, J. L. McWilliams, R. C. Haddon, and V. Parpura, Applications of carbon nanotubes in biotechnology and biomedicine, *J. Biomed. Nanotechnol.* 1(1), 3-17 (2005).
36. R. A. Freitas Jr., Current status of nanomedicine and medical nanorobotics, *J. Comput. Theor. Nanosci.* 2, 1-25 (2005).
37. A. B. Artyukhin, A. Shestakov, J. Harper, O. Bakajin, P. Stroeve, and A. Noy, Functional one-dimensional lipid bilayers on carbon nanotube templates, *J. Amer. Chem. Soc.* 127(20), 7538-7542 (2005).
38. R. J. Chen, H. C. Choi, S. Bangsaruntip, E. Yenilmez, X. W. Tang, Q. Wang, Y. L. Chang, and H. J. Dai, An investigation of the mechanisms of electronic sensing of protein adsorption on carbon nanotube devices *J. Amer. Chem. Soc.* 126(5), 1563-1568 (2004).
39. J. J. Gooding, Nanostructuring electrodes with carbon nanotubes: A review on electrochemistry and applications for sensing, *Electrochim. Acta* 50 (15), 3049-3060 (2005).
40. J. Li, H. T. Ng, A. Cassell, W. Fan, H. Chen, Q. Ye, J. Koehne, J. Han, and M. Meyyappan, Carbon nanotube nanoelectrode array for ultrasensitive DNA detection, *Nano Lett.* 3(5), 597-602 (2003).
41. Y. Wang, Z. Tang, and N. A. Kotov, Bioapplication of nanosemiconductors, *Mater. Today* 8(5), 20-31 (2005).
42. M. J. O'Connell, S. M. Bachilo, C. B. Huffman, V. C. Moore, M. S. Strano, E. H. Haroz, K. L. Rialon, P. J. Boul, W. H. Noon, C. Kittrell, J. P. Ma, R. H. Hauge, R. B. Weisman, R. E. Smalley, Band gap fluorescence from individual single-walled carbon nanotubes, *Science* 297(5581), 593-596 (2002).
43. P. W. Barone, S. Baik, D. A. Heller, and M. S. Strano, Near-infrared optical sensors based on single-walled carbon nanotubes, *Nature Mater.* 4(1), 86 (2005).
44. Z. Y. Tang and N. A. Kotov, One-dimensional assemblies of nanoparticles: Preparation, properties and promise, *Adv. Mater.* 17(8), 951-962 (2005).
45. C. L. Tsai and C. S. Ozkan, Virus based self assembly of carbon nanotubes and quantum dots, *Abstr. Amer. Chem. Soc.* 229(1), U914- U914 (2005).
46. P. G. Collins, K. Bradley, M. Ishigami, A. Zettl, Extreme oxygen sensitivity of electronic properties of carbon nanotubes, *Science* 287(5459), 1801-1804 (2000).
47. K. Keren, R. S. Berman, E. Buchstab, U. Sivan, E. Braun, DNA-templated carbon nanotube field-effect transistor, *Science* 302(5649), 1380-1382 (2003).
48. J. Razumiene, J. Barkauskas, V. Kubilius, R. Meskys, and V. Laurinavicius, Modified graphitized carbon black as transducing material for reagentless H_2O_2 and enzyme sensors, *Talanta* 67(4), 783-790 (2005).

INTERACTION OF CELLS WITH NANOPARTICLES

V. A. POKROVSKIY*, N. P. GALAGAN, A. A. CHUIKO†
Institute of Surface Chemistry, 17 General Naumov Street, 03164 Kiev, Ukraine

Abstract. Bio-active nanocomposites were prepared by adsorption of fructose and glucose on fumed silica in parallel and by co-adsorption with bovine serum albumin (BSA). The adsorption data show a significant difference in the structure of surface complexes for the two monosaccharides under investigation. The bio-activity of synthesized nanocomposites used as admixtures to aqueous suspensions of native cells was estimated with respect to bovine spermatozoa using Dynamic Light Scattering, and to human erythrocytes using phase-contrast microscopy. Glucose-containing nanocomposites demonstrated a higher bio-activity than fructose-containing nanocomposites. Changes in the bio-activity were compared to results of thermal decomposition of an adsorbed layer at fumed silica using temperature-programmed desorption mass spectrometry.

Keywords: cell surface, nanoparticle, MALDI, TPD, mass spectrometry

1. Introduction

Interaction of nanomaterials with native cells is an important problem in modern life science. Recent progress in mass spectrometry provides a vital tool to study this problem. Advances in applications of mass spectrometry for investigating the interaction of nanoparticles with cell membranes and biomacromolecules are based on at least two methods. The first is matrix-assisted laser desorption ionization (MALDI),[1] and the second is temperature-programmed desorption mass spectrometry (TPD MS), newly developed to study the interaction of nanoparticles with a cell surface.[2-4]

Previously we have studied BSA and carbohydrate decomposition using TPD MS for individual organics in a condensed state, and those adsorbed to a

*To whom correspondence should be addressed. V. A. Pokrovskiy, Institute of Surface Chemistry, 17 General Naumov Street, 03164 Kiev, Ukraine, e-mail: ucs@ln.ua

J.P. Blitz and V.M. Gun'ko (eds.), Surface Chemistry in Biomedical and Environmental Science, 277–286.

surface of nanosilica.[5-7] It is expected that MS data obtained for synthesized nanocomposites, and analyzed in comparison with the above results, can provide an adequate explanation of the adsorption mechanism of carbohydrates and BSA, as well as the bio-activity of nanocomposites from the other.

The aim of this contribution is to present experimental data concerning the synthesis, physicochemical properties and bio-activity of several nanosilica-based composites tested by interaction with native red blood cells (RBCs). The investigations include (i) synthesis of nanocomposites by adsorption of BSA, fructose and glucose on a fumed silica surface, and (ii) analysis of the adsorption properties of composites and their bio-activity. It is known that both glucose and fructose do not individually adsorb on fumed silica; therefore, bovine serum albumin was used as a mediator to bind the monosaccharides to the fumed silica surface.

2. Experimental

2.1. SYNTHESIS OF NANOBIOCOMPOSITES

Fructose is a ketose, i.e. it possesses a 5-membered ring; while glucose is an aldose possessing a 6-membered ring (Figure 1). Comparison of the bio-activity of these monosaccharides embedded into the structure of nanocomposites together with BSA as an intermediate agent, in the framework of similar experimental procedures, was expected to give information about the specificity of interaction of these carbohydrates both with the silica and with cell surfaces.

Synthesis of nanocomposites using adsorption of BSA, fructose and glucose on ultra-fine silica was conducted using several combinations: BSA adsorption on the surface of ultra-fine silica (control); (1) adsorption of fructose on silica with pre-adsorbed BSA (subsequent adsorption); (2) simultaneous adsorption of fructose and BSA from aqueous solution; (3) subsequent adsorption of BSA and glucose on silica; and (4) simultaneous adsorption of glucose and BSA. In all the cases S-type isotherms were obtained. Values of the maximal adsorption determined from the isotherms are summarized in Table 1. It is interesting to note that the maximal adsorption of BSA/fructose is much greater than that of BSA/glucose.

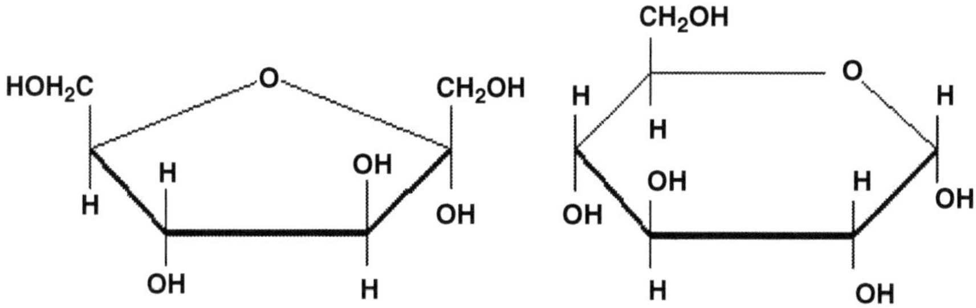

Figure 1. Chemical formulae of fructose (left) and glucose (right).

Table 1. Maximal adsorption of BSA and mono-sugars.

Nano-composite	BSA (control)	BSA+fructose subsequent adsorption (1)	BSA+fructose simultaneous adsorption (2)	BSA+glucose subsequent adsorption (3)	BSA+glucose simultaneous adsorption (4)
Adsorption capacity (g\g)	0.56	1.24	4.5	0.70	0.51

2.2. ESTIMATE OF BIOLOGICAL ACTIVITY BY DLS

The bio-activity of synthesized samples was estimated from mobility parameters of unicellular microorganisms measured by Dynamic Light Scattering (DLS). The following characteristics of cell population were measured: cell concentration (units/ml); % of mobile cells; mean translation velocity (μm/s), cell rotational frequency of (Hz); flagella beat frequency (Hz), distribution of beat frequencies, and kinetic energy of cell motion in a viscous environment (arbitrary units). Details of experimental equipment and measurement procedures may be found elsewhere.[7,8] All parameters were measured with an accuracy of about ±2%.[9]

The DLS experiments were conducted with populations of microorganisms corresponding to different stages of evolutionary development from protozoa (*Tetrhymena Pyriformis*), micro-algae (*Dunaliella Salina, Pedinomonas Masuik*, and *Euglena gracilis*) to mammal gametes (boar, bull, and human spermatozoa), to estimate the effects of external factors on cell mobility and other characteristics of cell suspensions.

One parameter, the kinetic energy of bovine spermatozoa motion (arbitrary units) in a viscous medium (standard cryo-protecting medium after a freeze-thaw cycle) was chosen to estimate bio-activity (Figure 2). Data corresponding to the admixture of nanocomposites (1)-(4) from Table 1 were measured using DLS in compared with the control cell suspension, and averaged over a wide concentration range of nanocomposite admixture.

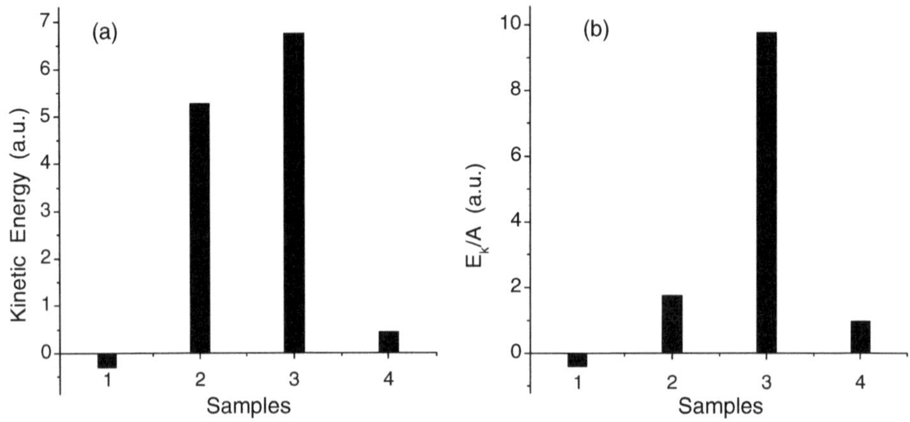

Figure 2. Kinetic energy of bovine spermatozoa motion in viscous medium (arbitrary units) for samples 1-4 (Table 1) measured in comparison with the control cell suspension; (a) corresponds to direct experimental measurements, and (b) is the specific activity with respect to the amounts of adsorbed sugar.

The picture becomes more informative if kinetic energy is referenced to the amount of sugar adsorbed on the nanocomposite surface.

2.3. ESTIMATE OF BIO-ACTIVITY BY PHASE CONTRAST MICROSCOPY

Phase-contrast microscopy allows the calculation of the number of erythrocytes in various stages of damage; namely discocytes, echinocytes, spherocytes and ghosts. This damage results from external factors causing injurious effects, and by the interaction with nanocomposites.

The protecting action of nanocomposites was characterized by a ratio of the number of discocytes and ghosts after admixture of the nanocomposites to donor blood under similar experimental conditions. The ratio of discocyte/ghost numbers presented in Figure 4a was normalized to the ratio observed for intact donor blood. Comparison of fructose with glucose containing nanocomposites has shown that the glucose containing nanocomposites exhibit a much lower level of cell damage. The data is more convincing when the protecting activity is normalized to the amounts of sugar adsorbed on the surface (Figure 3b). This analysis is similar to the right part of Figure 2. The difference in bio-activity of fructose and glucose with respect to erythrocytes is not as dramatic as for reproductive cells, but still exceeds experimental error.

To summarize, maximal adsorption of glucose in combination with BSA on the surface of ultra-fine silica was appreciably less than that of fructose under the same conditions. On the other hand, the positive bio-effect of glucose-containing composites is much higher for both reproductive cells and erythrocytes.

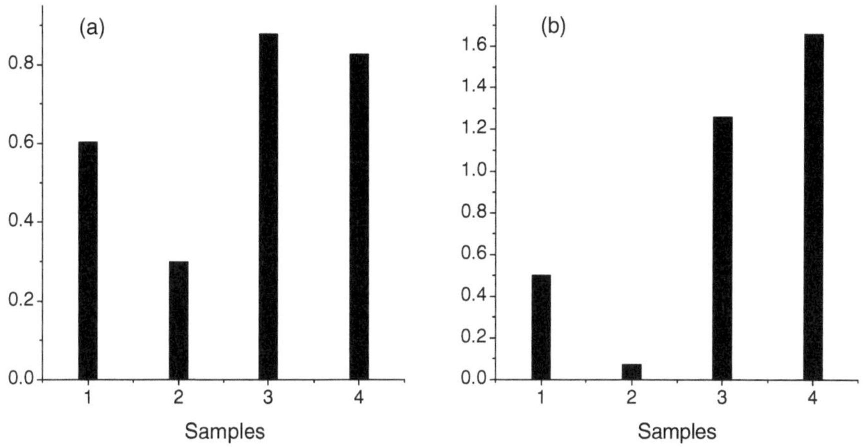

Figure 3. Estimated protecting action of nanocomposite admixture to donor blood (samples 1-4 in Table 1). Discocytes/ghosts ratio was chosen as an index of the protecting effect of the nanocomposite; (a) direct measurement; and (b) specific protecting activity with respect to the amounts of adsorbed sugar.

2.4. TPD MS STUDIES OF BIOCOMPOSITES

Characteristic thermogram patterns were obtained[10] by TPD MS, which allows us to observe details of BSA decomposition and the two monosaccharides in the condensed state. Thermal decomposition patterns for several components of the BSA mass spectrum in condensed and adsorbed states are presented in Figures 4 and 5. After individual adsorption of BSA on fumed silica the pattern of BSA decomposition changed slightly, and could still be used to identify the nanocomposites. The only exception is the temperature dependence of mass line at 34 M/z, which corresponds to the molecular ion H_2S^+ resulting from decomposition of sulfur-containing amino acids of BSA.

Decomposition of the protein starts in both cases with evolution of H_2S^+. The narrow shape of the thermogram for the condensed state in comparison with the adsorbed state, suggests an important role and co-operative nature of thiol-disulphide exchange in BSA thermolysis. This mechanism was discussed elsewhere,[11] as well as the role of hydrogen bonds and water molecules in the formation of protein oligomers. As can be seen from Figure 5, adsorption on silica changes the non-isothermal kinetics of H_2S evolution, and this fact should be taken into account in studies of BSA+sugar containing nanocomposites.

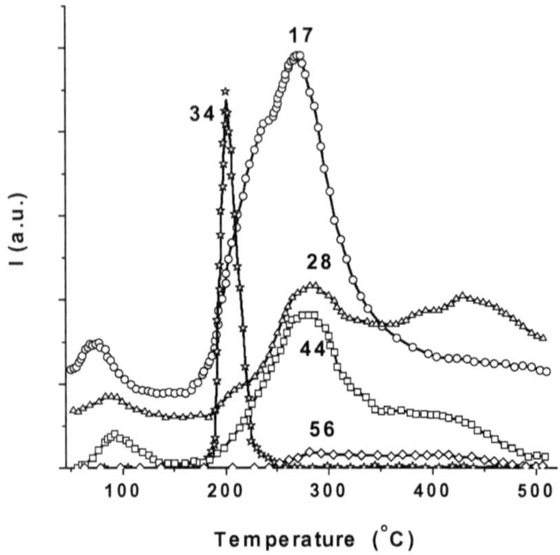

Figure 4. The set of thermograms of BSA in condensed state showing the temperature dependence of the mass spectra for volatile products of protein decomposition at m/z 17 (OH^*), 28 (CO^+), 34 (H_2S), and 44 (CO_2).

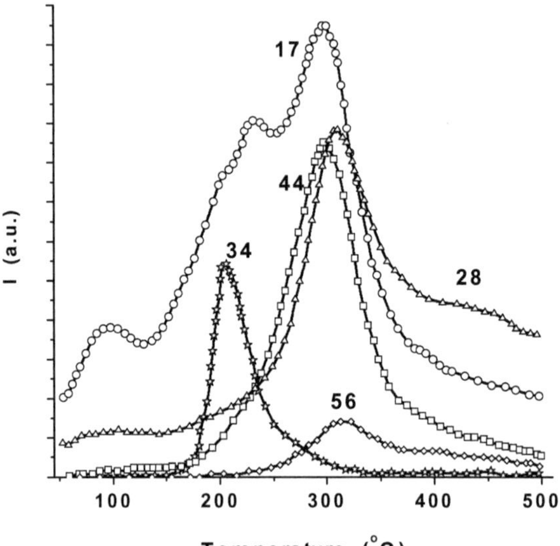

Figure 5. The set of thermograms of BSA in the adsorbed state on the silica surface including the same components as in Figure 4.

Individual thermograms of fructose and glucose adsorbed onto silica could not be obtained because they do not adsorb. After co-operative adsorption with BSA, the characteristic pattern of sugar decomposition in the condensed state was absent in all the cases under investigation, even though the presence of sugar molecules in the adsorption layer was shown by spectrophotometric experiments. Thus the pathways of sugar decomposition on the silica surface modified by adsorbed BSA strongly differ from those without BSA. In the case of fructose adsorption, the characteristic pattern of BSA decomposition repeated the standard pattern similar to that presented in Figure 5. But in the case of glucose the characteristic pattern of BSA decomposition in the adsorbed state was dramatically different. Maxima of albumin decomposition in the adsorbed state were displaced toward high temperatures.

Thus, the TPD MS results show that the adsorption complexes of monosaccharides with BSA change the mechanism of sugar decomposition in comparison with the condensed state. The mechanism of BSA decomposition is completely different in the adsorption complexes with glucose and fructose. Unfortunately, further details of sugar-BSA interaction in the adsorbed state cannot be established reliably with TPD MS because the mass spectra can only be obtained for volatile products. Experimental methods to analyze non-volatile macro-molecules should be applied to achieve an adequate understanding of the composite structure and the mechanism of bio-activity.

3. Discussion

Experimental techniques for the study of cell interactions with fumed silica-based bio-active nanocomposites are important for medical and biotechnological applications. Recent success using matrix assisted laser desorption ionization time of flight (MALDI TOF) techniques allows efficient operation in a mass range up to 100 kDa or more, as seen from Figure 6. The mass spectrum includes the BSA molecular ion, doubly and triply charged molecular ions of BSA, and the charged dimer. In the case of a protein-carbohydrate complex additional carbohydrate-containing components of mass spectra are expected, providing information about the surface complexes of bio-active silica-based nanocomposites.

Development of a wide range of Surface MALDI mass spectrometry methods[12] opens new possibilities in MS investigations of bio-interfacial phenomena, summarized recently as "biological surface science". Thus, matrix-assisted laser desorption ionization mass spectrometry, and its variants, hold promise to help clarify the mechanisms of bio-activity of silica-based nanocomposites.

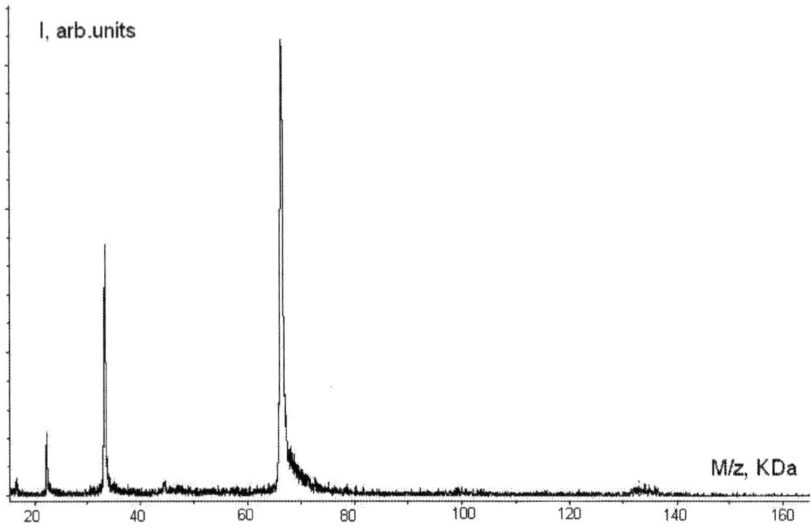

Figure 6. MALDI mass spectrum of individual BSA.

The wide application of newly developed advanced Surface MALDI MS methods, such as laser desorption ionization (LDI), surface-enhanced laser desorption ionization (SELDI), direct ionization on silicon surface (DIOS), etc. have two shortcomings. First, problems with MALDI-TOF MS sample preparation for nano-sized systems have not been completely worked out. Second, the reliability and reproducibility of experimental data concerning similar bio-active systems is lacking. Sometimes molecular, associative and fragment ions cannot be observed in the MALDI mass spectra. We are currently working on overcoming these difficulties with some encouraging results.[13,14]

4. Conclusion

BSA + glucose/silica containing nanocomposites demonstrated much higher bio-activity than BSA + fructose/silica nanocomposites. It is interesting to note that glucose was chosen by evolution as a terminal carbohydrate in oligosaccharide chains used in animal cell membranes receptors, while fructose was not.

Higher biological activity of BSA+glucose/silica containing nanocomposites was observed with respect to both reproductive cells (subsequent adsorption) and erythrocytes (both simultaneous and subsequent adsorption). A broader range of cell species will be used to further investigate this trend.

TPD MS data have shown a dramatic difference between the mechanism of thermal decomposition of bio-active (glucose containing) and less active (fructose containing) nanocomposites in the adsorption layer. Thus, the TPD MS approach is an efficient method to establish an empirical correlation between the structure of the adsorption complexes and the reaction mechanism in the adsorption layer, and the bio-activity of synthesized nanocomposites. This approach cannot provide mechanistic information however, but only provide correlations to guide further research.

Literature analysis and our preliminary experiments with proteins, carbohydrates and other molecules deposited on the surface of nanosilicas suggest that newly developed MALDI MS techniques, in particular Surface MALDI MS, can be adopted as a powerful new tool for surface analysis. Unique capabilities that surpass established techniques in surface analysis of biomaterials using these newly developed methods is an exciting development.

Acknowledgements

The authors wish to thank post-graduate students L. M. Patey, I. V. Gritsenko, and N. S. Nastasienko for their collaboration in the synthesis of nanocomposites and bio-activity measurements.

References

1. L. Hanley, O. Kornienko, E. T. Ada, E. Fuoko, and J. L. Trevor, Surface mass spectrometry of molecular species. *J. Mass. Spectrom.* 34, 70-723 (1999).
2. V. A. Pokrovskiy. Temperature-programmed desorption mass spectrometry of biomolecules, *Rapid Commun. Mass Spectrom.* 9, 588-591 (1995).
3. V. A. Pokrovskiy. Temperature-programmed desorption mass spectrometry, *J. Therm. Anal. Calorim.* 62, 407-415 (2000).
4. V. A. Pokrovskiy and A. A. Chuiko, TPD MS and characteristics of non-isothermal reactions on the surface of ultrafine solids, in: *Chemistry, Physics and Technology of Surfaces*, edited by A. A. Chuiko (Institute of Surface chemistry, Kiev 2001), pp. 79-116.
5. N. P. Galagan, T. V. Kulik, V. V. Vlasenko, I. V. Gritsenko, V. A. Pokrovskiy, and A. A. Chuiko, Contact interaction of ultrafine silicas with cell surface, *Abstr. ICBP 2004, International Conference in Biological Physics* (August 23-27, 2004, Gotheborg, Sweden).
6. N. P. Galagan, V. M. Bogatyrov, I. V. Grytsenko, V. L. Osaulenko, and A. A. Chuiko, Silica-based nano-materials and their effect on cells. Abstr. International Conference *Nanomaterials in Chemistry, Biology and Medicine* (September 15-16, 2005. Kiev, Ukraine), p. 97.
7. V. V. Vlasenko, *Studies on rotatory-translation motion of spermatozoa by laser Doppler spectrometry*, Ph.D. Thesis, Kiev, 1991.
8. V. V. Vlasenko. Studies on photo-sensitivity of unicellular moving micro-algae by method Doppler spectroscopy, *Reports of the National Academy of Sciences of Ukraine* 11, 172-178 (2004).

9. N. P. Galagan, V. V. Vlasenko, N. S. Nastasienko, and A. A. Chuiko. Effect of ultrafine silica modified by polyols on viability of reproductive cells by photon-correlation spectroscopy data, *Proc. Biophys. Karazin National University*, Kharkov, Ukraine (in press).
10. N. P. Galagan, L. M. Patey, N. S. Nastasienko, I. V. Gritsenko, I. L. Orel, O. V. Mischanchuk, V. A. Pokrovskiy, and A. A. Chuiko, Chemical transformations of albumin, glucose and fructose on the surface of ultra fine silica (to be published).
11. V. K. Pogorelyy, V. N. Barvinchenko, and V. V. Turov. Solvation effect on radiolysis of aqueous solutions of human serum albumin in presence of dimethylsulfoxide, *Teoret. Eksperim. Khim.* 1, 107-111 (1990).
12. H. J. Griesser, P. Kingshott, S. L. McArthur, K. M. McLean, G. R. Kinsel, and R. B. Timmons. Surface MALDI mass spectrometry in biomaterials research, *Biomaterials* 25, 86 (2004).
13. V. A. Pokrovskiy, V. M. Kosevich, V. L. Osaulenko, V. V. Chagovets, V. A. Pashinskaya, V. S. Shelkovskiy, V. A. Karatchevtsev, and A. Yu. Naumov, MALDI mass spectrometric study of bis-quaternary ammonium anti-microbial agent decamemetoxinuminum 2,5-dihydroxybenzoic acid, *Mass Spectrom.* 5(12), 183-192 (2005).
14. S. Snigyr, T. Gromovoy, V. Pokrovskiy, and A. Chuiko. MALD TOF mass spectrometry of fullerene C_{60} on silicon and steel targets, Abstr. International Conference *Nanomaterials in Chemistry, Biology and Medicine* (September 15-16, 2005. Kiev, Ukraine), p. 123.

THE ROLE OF NATURE AND STRUCTURE OF SURFACE SITES IN THE BIOLOGICAL RESPONSE TO SILICA PARTICLES

IVANA FENOGLIO, MARA GHIAZZA, RAFFAELLA CESCHINO, FEDERICO GILLIO, GIANMARIO MARTRA, BICE FUBINI*
Dip. di Chimica Inorganica, Chimica Fisica e Chimica dei Materiali and Interdepartmental Centre "G. Scansetti" for Studies on Asbestos and Other Toxic Particulates and Centre of Excellence of Nanostructured Interfaces and Surfaces (NIS) University of Torino, via Pietro Giuria 7, 10125 Torino (Italy)

Abstract. The inhalation of crystalline silica dusts causes lung inflammation and the development of silicosis and lung cancer. Inhalation of amorphous silica dusts seems to elicit no adverse biological response. The large variability in toxicity elicited by quartz dusts having different origins may be assigned to differences in the surface structures of the particles. When a silica particle is in contact with body fluids, adsorption and surface reactions with endogenous molecules (e.g. proteins and pulmonary surfactant) take place. These reactions can modify the surface properties of the silica particles and consequently modulate the pathogenic responses elicited. In this work we study the adsorption of bovine serum albumin (BSA) on different amorphous and crystalline silica dusts, the occurrence of conformational changes of the protein during the adsorption process, and the effect of some surface properties (hydrophilicity, presence at the surface of metal impurities and defects) on the amount of proteins adsorbed.

Keywords: silica, surface, hydrophilicity, protein adsorption

1. Introduction

Silica exists in several crystalline and amorphous forms which greatly differ in their surface characteristics, namely topography (down to a nanometric level), hydrophobicity, silanol distribution and presence of contaminants.

* To whom correspondence should be addressed. B. Fubini, University of Torino, via Pietro Giuria 7, 10125 Torino, Italy E-mail: bice.fubini@unito.it

J.P. Blitz and V.M. Gun'ko (eds.), Surface Chemistry in Biomedical and Environmental Science, 287–298.

The inhalation of crystalline silica dusts (mainly cristobalite and quartz) causes silicosis, lung cancer and several autoimmune diseases; while no adverse effects have so far been associated with the inhalation of artificial amorphous silica forms.[1] Moreover, a large variability in the pathogenic response elicited by quartz dusts having different chemical, mechanical and thermal histories, exists.[1-4] An intense research effort has been recently undertaken to understand the role played by the physicochemical properties and surface reactivity of silica particles in the mechanism of pathogenicity. Several properties have thus far been recognized to influence the biological response towards silica particles: (i) specific surface area (which is related to particle size and morphology) and particle shape; (ii) the hydrophilic character of the surface which is related to the distribution of different silanol types (geminal, H-bonded, isolated) and siloxane bridges (regular and straightened) at the surface, and to the presence of surface impurities; and (iii) the generation of radical species at the surface due to the presence of both redox active metal ions, and silica based surface radicals originating from the rupture of the silica framework during the grinding processes. Each of these properties may influence different steps in the mechanism of action of inhaled particles.[5] Much attention has been given to the role of alveolar macrophages activation, and consequent inflammatory response, caused by particle and/or cell derived oxidants.[6] An additional role, however, may be played by the interaction of the inhaled particles with endogenous molecules present in the extracellular fluids (pulmonary surfactant) e.g. surfactants, antioxidant molecules and proteins.[7] The reactions of such molecules may determine the fate of the particles in the early steps of the pathological process. Here we report a study on the influence of the degree of hydrophilicity/hydropobicity on the adsorption of albumin, a protein which plays important roles in the lung lining layer, at the surface of various finely divided silicas. The employment of microcalorimetry and IR spectroscopy allows a precise evaluation of the hydrophilic or hydrophobic character of the silica powders investigated which include an amorphous silica, and three crystalline polymorphs, tridymite, cristobalite and quartz powders, of different origin. We also report data on albumin adsorption on different amorphous and crystalline dusts, the occurrence of conformational changes of the protein molecules during the adsorption process, and the effect of surface properties on the process.

2. Experimental

2.1. MATERIALS

Silica powder samples were as follows: (a) pyrogenic amorphous silica Aerosil 50 (A50, Degussa); (b) tridymite (Tri), a kind gift of Dr. Stöber (C.I.I.T.,

Triangle Park, NC, USA); (c) cristobalite (Cris, C.E. Mineral Co.), kindly supplied by Dr. D. Hemenway (University of Vermont, USA); (d) α-quartz (Qzm, commercial name Min-U-Sil 5, U.S. Silica), two different lots; (e) one pure quartz (Sigma); (f) two pure quartz dusts: one (Qzp1) obtained by quartz chips purified by melting (99.999% SiO_2, Atomergic), and the second (Qzp2) from a mineral source in Madagascar. Both Qzp dusts were ground in an agate mortar.

2.2. METHODS

2.2.1. Surface modifications

Qzm was degassed in a quartz cell connected with a vacuum line for two hours at 1073 K; 3 g of Qzm was treated with 15 mL of 0.1 M hydrofluoric acid solution for 25 minutes. The dust was then centrifuged, washed four times with distilled water, and dried at 373 K for three hours.

2.2.2. Adsorption calorimetry

The simultaneous measurement of the heat of adsorption and the adsorbed amount of H_2O was performed by means of a Tian-Calvet microcalorimeter, operating at 303 K, connected to a volumetric apparatus. The samples were pretreated in vacuo at the chosen temperature and subsequently transferred into the calorimeter without further exposure to air. Small doses of water were subsequently admitted onto the sample, the pressure being continuously monitored by a transducer gauge (Baratron MKS, 0-100 Torr).

2.2.3. Infrared spectroscopy

For the FTIR measurements (Bruker IFS28; resolution 2 cm^{-1}; MCT detector), the samples, pressed into self-supporting pellets, were placed into a quartz IR cell equipped with KBr windows attached to a vacuum line (residual pressure: 1×10^{-6} Torr). Thermal treatments were carried out *in situ*. To decrease the scattering of the IR beam, severely affecting the transparency of the samples in the in the 4000-3000 cm^{-1} range, a system of condenser mirrors were placed between the IR cell and the detector.

2.2.4. Surface area measurement

The BET surface area was measured by N_2 adsorption at 77 K (BET method) using a "Quantasorb" apparatus (Quantacrome).

2.2.5. *Protein adsorption*

200 mg of each dust was suspended in 20 mL of a solution of bovine serum albumin (BSA) in a 0.01M phosphate buffer (PBS, pH 7.4, 0.138M NaCl, 0.0027M KCl) at different concentrations (10-1000 mg/L). The suspensions were stirred in a thermostatic stirrer at 298 K for 5 minutes. The suspension was filtered through a membrane filter (PVDF, pore diameter 0.45 μm) and the concentration of albumin in the supernatant determined spectrophotometrically (λ = 280 nm). The results are reported as the mean value of two separate determinations ± standard error.

2.2.6. *FTIR studies of proteins adsorbed on silicas*

The conformation of adsorbed protein was studied by means of FTIR spectroscopy. A BSA solution (12.5 mL of 500mg/L) was incubated with 125 mg of silica dusts for 5 min at 298 K. The powder was then separated from the supernatant by centrifugation (4000g for 10 min). The protein not adsorbed was removed from the dusts by washing four times in PBS. The dusts were then re-suspended in 100 μl of PBS 0.01M prepared in D_2O medium and incubated for 1 h at room temperature to exchange the H_2O adsorbed on the sample with D_2O. The exchange was repeated three times.

A slurry of silica particles with adsorbed BSA was placed in a horizontal ATR cell equipped with a ZnSe crystal (5 reflection). The spectrum of a slurry of bare silica was recorded and subtracted as a background.

3. Results

3.1. HYDROPHILIC/HYDROPHOBIC CHARACTER MONITORED BY ADSORPTION CALORIMETRY

Figure 1A shows the differential heat of adsorption versus coverage plots for the adsorption of water on tridymite (Tri), cristobalite (Cris), as well as commercial and pure quartz (Qzm, Qzp1) outgassed at 423 K.

Such treatment causes the complete removal of surface adsorbed molecular water during the storage of the silica powders in air, leaving essentially unaffected the layer of surface hydroxyls (silanols). Following Zettlemoyer *et al.*[8] we define as hydrophilic sites that adsorb H_2O molecules with an enthalpy of adsorption greater than 44 kJ mol^{-1} (latent enthalpy of liquefaction of water) and hydrophobic sites with enthalpy of adsorption less than 44 kJ/mol. Additional insights are obtained from data in Figure 1.

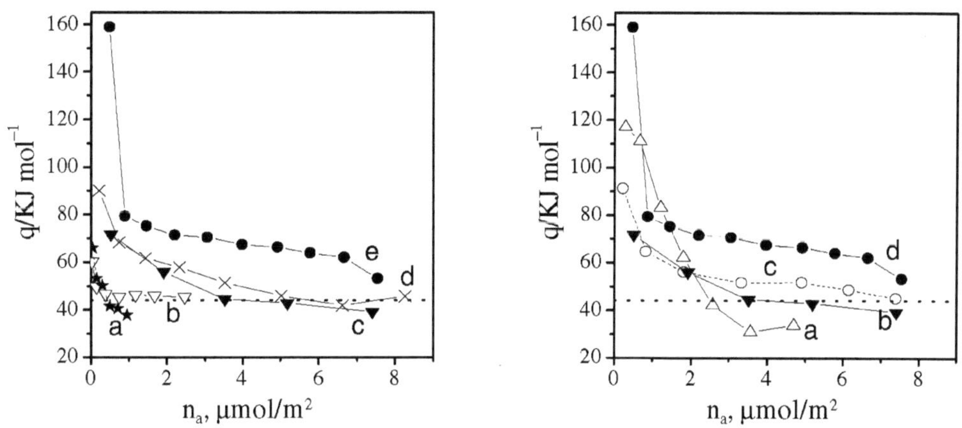

Figure 1. A. Differential heat of adsorption of water at 303 K on: (a) A50, (b) Tri, (c) Qzp1 (d) Cris and (e) Qzm outgassed at 423 K for 2 h. B. Differential heat of adsorption of water at 303 K on: (a) Qzm outgassed at 1073K; (b) Qzp1, (c) QmHF and (d) Qzm. Dotted line indicates the latent enthalpy of liquefaction of H_2O, $-\Delta H_L = 44$ kJ mol^{-1}.

The extrapolated zero-coverage values provide the energy of interaction with the strongest sites; the plateau value provides the energy of interaction in multilayers or on predominantly homogeneous patches on the surface; the extent of coverage in which the heat value decreases before reaching the plateau measures the extent of heterogeneity toward adsorbed molecules. Inspection of the plots indicates that for all samples at the very beginning of the adsorption process, the enthalpy of interaction is greater than 44 kJ mol^{-1}, typical of the interaction of H_2O molecules with hydrophilic sites. However, significant differences occur in the enthalpy values measured for the different samples; in particular, only Qzm has very strong adsorption sites with values greater than 120 kJ/mol. On A50 and Tri the enthalpy of interaction falls below 44 kJ mol^{-1} at very low water coverage, whilst for Cris and Qzp1 the enthalpy falls below this value after much larger amounts of water are adsorbed. This indicates the presence of extended, heterogeneous patches of hydrophilic sites. Such patches overwhelmingly cover the surface of Qzm particles, as for this sample the enthalpy of adsorption is larger than 44 kJ mol^{-1} even at the highest water coverage attained. The five silica dusts in order of decreasing hydrophilicity: Qzm > Cris ≈ Qzp1 > Tri ≈ A50. Two quartz samples hydrophobized in different ways were considered: Qzm outgassed at 1073 K and Qzm treated with dilute hydrofluoric acid. Figure 1B compares the heat of adsorption of water on these modified samples with that of Qzm and pure quartz Qzp1. The thermal treatment causes silanol condensation, and the energy of interaction falls below 44 kJ/mol once 2 μmol/m^2 of water has been adsorbed. At low coverage the very strong adsorbing sites are however still present. Conversely, the sample treated with HF does not exhibit strongly interacting sites at low coverage, but the surface appears to be still fully hydrophilic.

3.2. STRUCTURE OF THE HYDROXYL SURFACE LAYER: IR SPECTRA

Information on the structural features of hydroxyl patches present on the surface of the silica powders was gained by analysing their IR spectra in the 3800-3000 cm^{-1} region. The IR spectra of Cris, Tri, Qzp1 and A50 are shown in Figure 2.

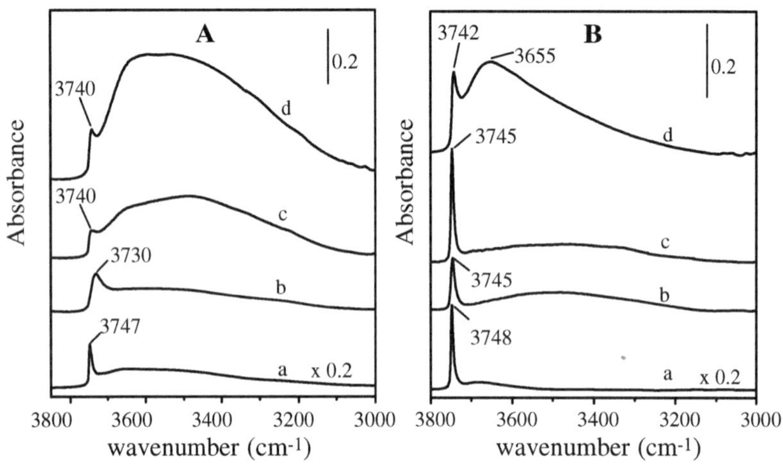

Figure 2. IR spectra, in the 3800-3000 cm^{-1} range, of silicas powders outgassed for 30 min at 423 K (section A) and 773 K (section B): (a) amorphous Aerosil 50; (b) tridymite; (c) cristobalite; and (d) Qzp1. The spectra are reported after baseline correction to eliminate the sloping scattering background.

The spectrum of Qzm is not reported because of the presence of extra-bands, due to structural hydroxyls of kaoline contained in small amounts in this material, heavily interfering with the IR absorption due to surface silanols. The observable signals due to these latter features are similar to those described for Qzp1. For all samples, when carrying their original surface hydroxyl layer (i.e. by desorbing only molecular water by outgassing at 423 K) two main components, a sharper one, in the 3750-3730 cm^{-1} range, and a second very broad and complex, spread over the 3700-3000 cm^{-1} range, were observed (Figure 2A). On the basis of the rich literature on this topic,[9] the narrow peak at 3747 cm^{-1} present in the spectrum of amorphous Aerosil 50 (Figure 2A, a) is due to isolated silanols more than 0.44 nm apart from any nearest neighbour. The broader peak at 3730 cm^{-1} in the spectrum of tridymite (Figure 2A, b), and the poorly resolved component at

3740 cm^{-1} in the spectra of cristobalite and quartz (Figure 2A, c, d, respectively), are assigned to silanols in closer proximity with another. On the same basis, the broad and complex absorption in the 3700-3000 cm^{-1} range is associated with different families of silanols less than 0.30 nm apart interacting through hydrogen bonding. It must be stressed that relevant differences among the spectra of the various silicas are in the relative intensity of the two main components. The absorption due to adjacent silanols interacting through hydrogen bonding dominates the spectra of Cris and Qzp1 by far (Figure 2A, c, d, respectively), indicating that on the surface of these crystalline silicas dense patches of OH groups are more abundant.

Evidence for differences in the features of adjoining OH groups on the various samples was derived from the effect of the treatment at 773 K. After outgassing the band due to H-bonded silanols strongly reduced in intensity in favour of the one due to isolated OH in the case of A50, Tri and Cris (Figure 2B, a, b, c, respectively); indicating that almost all adjoining OH participated in condensation processes, yielding siloxane bridges (which do not contribute to the spectra in the 3800-3000 cm^{-1} region) and new isolated silanols. By contrast, an intense component at 3655 cm^{-1} was still present in the spectrum of quartz (Figure 2B, line d), due to a significant fraction of H-bonded silanols more resistant to the condensation process. Interestingly, this band corresponds to the high frequency part of the broad 3700-3000 cm^{-1} absorption observed for quartz after outgassing at 423 K (Figure 2A, d), which should correspond to silanols experiencing a weak H-bonding, likely because of their larger distance from one another.

3.3. ADSORPTION OF BSA ON AMORPHOUS AND CRYSTALLINE SILICA

The amount of BSA measured in an initial concentration range of 10 to 1000 mg/L adsorbed on Qzm dusts, is compared in Figure 3A with the amount of BSA adsorbed on A50. Qzm and A50 exhibit different affinities for BSA: with increasing protein concentration the coverage on amorphous silica increases up to a value of 500 $\mu g/m^2$, and on quartz the amount of BSA adsorbed reaches 1500 $\mu g/m^2$. In both samples, the amount of BSA adsorbed does not reach the monolayer value (around 2 mg/m^2), assuming that the protein adsorbs sideways-on.

The conformation of protein adsorbed on the two silica samples has been investigated by means of FTIR spectroscopy monitoring two signals: the amide I band ($\nu_{C=O}$ at ca. 1650 cm^{-1}), affected by the feature of the C=O···H–N hydrogen bond in the polypeptidic backbone, and the amide II band (at ca. 1540 cm^{-1}, due to the combination of the C–N stretching and the N–H bending peptide units) remaining after contact with D_2O, monitoring the presence of hydrophobic pockets where the N–H/N–D exchange cannot occur. In Figure 3B the spectra in the region of the amide I and II bands of the native protein in solution (a), BSA adsorbed on A50 (b), and on Qzm (c) are compared.

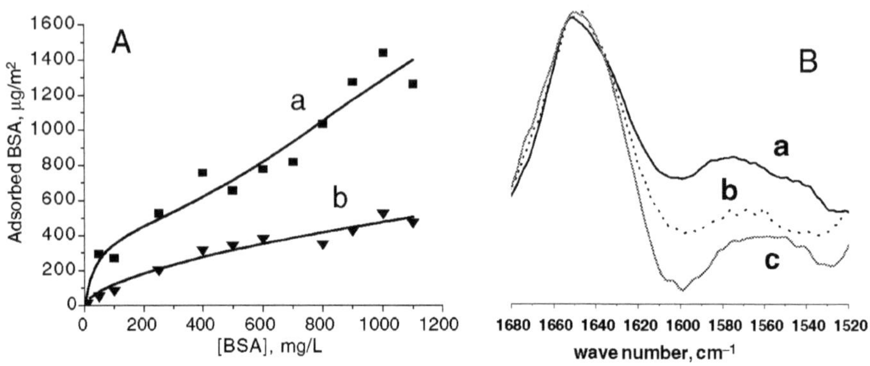

Figure 3. BSA adsorption on quartz and amorphous silica. Amount of protein adsorbed *vs* initial concentration of (a) commercial quartz and (b) amorphous silica. B. FTIR spectra of (a) native BSA, (b) BSA adsorbed on amorphous silica; and (c) BSA adsorbed on quartz.

While limited changes in the features of the amide I band are observed after BSA adsorption on the two silicas; the amide II band, present in the spectrum of the native BSA in solution (curve a) is absent in the spectrum of the protein adsorbed on A50 (curve b), whilst it is partially maintained in that of BSA on Qzm (curve c). This behaviour indicates that the interaction with the surface of the amorphous silicas resulted in an opening of the hydrophobic pockets, allowing the N–H groups therein contained to be converted by contact with D_2O in N–D, producing an IR absorption at lower frequency (ca. 1450 cm^{-1}, not shown). On Qzm the conformational changes are less pronounced.

3.4. EFFECT OF HYDROPHILICITY, SURFACE IMPURITIES AND DEFECTS ON BSA ADSORPITON

Qzm has been treated to decrease the surface hydrophilicity. Three different hydrophobization procedures have been followed: (i) the condensation of silanols into siloxane bridges by outgassing the powder at 1073 K, largely irreversible by re-exposure to water; (ii) the functionalisation of silanols with tryethylsilyl groups; and (iii) treatment with diluted HF. This last treatment dissolves the external disordered layers of particles containing impurities. In Figure 4, the amount of BSA adsorbed on the three treated quartz dusts is compared with untreated samples of pure quartz. All treatments decreased the amount of adsorbed BSA at concentrations higher than 500 mg/L. The amount of BSA adsorbed on HF treated Qzm is lower than that observed for the pure quartz Qzp.

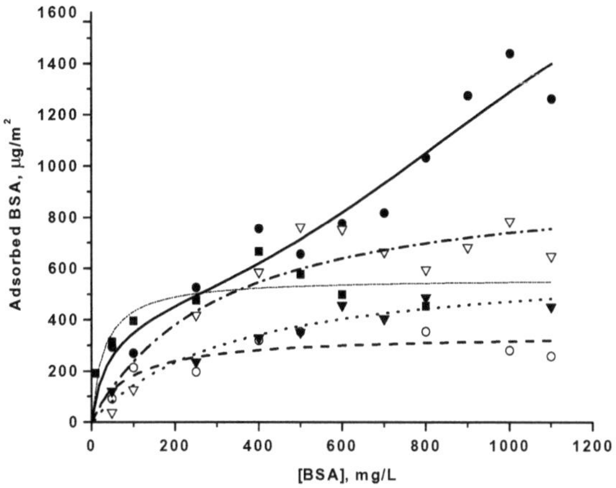

Figure 4. BSA adsorption on modified quartz dusts. amount of protein adsorbed vs initial concentration of Qzm (■); Qzm outgassed at 800° C (▽), silanized (■) and treated with diluted HF (▼) and pure quartz (O).

4. Discussion

4.1. HYDROPHILIC CHARACTER OF THE SURFACE

The combined use of adsorption microcalorimetry and IR spectroscopy allows a precise evaluation of the hydrophilic or hydrophobic character of the silica powders investigated.[10-12] The evolution of the enthalpy of adsorption of water with coverage (Figure 1) indicates the coverage values at which the enthalpy falls below the latent enthalpy of liquefaction (ΔH_L), which can be taken as a reasonable measure of the extent of hydrophilic sites occupied by adsorbed water, before adsorption on the hydrophobic sites begins. Comparison with infrared data confirms that hydrophobicity is associated with isolated or weakly interacting silanols, whereas hydrophilicity corresponds to the presence of hydroxyls close enough to interact with one another through hydrogen bonding.

Apparently, differences in the crystalline or amorphous bulk structure are extended up to the structure of the surfaces of the silica samples considered. At the surface of tridymite, silanols are far enough from one another to interact only weakly with their nearest neighbours, whilst patches of close hydroxyls, interacting through hydrogen bonding, are largely abundant at the surface of cristobalite and quartz. This latter kind of crystalline silica also exhibits adjacent silanols which show a greater stability towards condensation to form siloxane bridges, as a significant fraction of them remain even after outgassing at 773 K.

The interaction energy with water is much higher on commercial quartz than on pure quartz. In particular, at low coverage, the high energy corresponds to a few very strong sites, which appears to be characteristic of commercial quartz. These sites have to be ascribed to the presence of impurities at the dust surface.[4] The most likely impurities are metal ions, such as aluminum and iron, dispersed at the surface by either deposition or grinding. The sample treated with HF, which dissolves the external layer containing defects and metal impurities, is fully hydrophilic but has lost the most energetic sites which adsorb water at low coverage.

4.2. NATURE AND STRUCTURE OF THE SURFACE SITES ON THE AFFINITIES FOR BSA

BSA has been chosen in our experiments because: (i) it has been widely used in several studies as a model protein; (ii) it is similar to human serum albumin, which is the most abundant protein in biological fluids including the pulmonary surfactant; and (iii) work has been reported on the modification of the biological response of cells to silica and silicates following the adsorption of this protein.[13-16]

Several different forces may be involved in protein adsorption at the solid-liquid interface: hydrogen bonding, electrostatic forces, and hydrophobic interactions. Entropic factors such as loss of water, structural deformation of the protein onto hydrophobic patches and dehydration of the protein may drive the adsorption process when there are non favourable electrostatic interactions.

Qzm and A50 have different affinities for BSA. Since the experiment was done at constant pH and ionic strength, the observed differences in the amount of BSA adsorbed, and the different conformation of BSA on the two silica dusts, is ascribed to the intrinsic properties of the two silica forms such as size and morphology (down to a micro- and nano level), surface hydrophilicity, impurities at the surface, and surface charge. Qzm and A50 particles are, in fact, very different entities: they differ in particle dimensions (Qzm mean diameter 1600 nm, A50 40 nm) and bulk structure, which involves different surface features such as silanol population, (quartz exhibits around 5 $SiOH/nm^2$, A50 2-3 $SiOH/nm^2$), hydrophilicity, micromorphology (quartz particles exhibit steps and edges due to the grinding processes, fumed silica is made up by roundish particles) and the presence of metal ion impurities.

The effect of hydrophylicity and the presence of impurities has been evaluated on a series of different quartz dusts which are similar in dimension and micromorphology. The more hydrophilic the surface the greater the amount of BSA adsorbed. Metal impurites also play a role on the adsorption process since pure quartz and quartz treated with HF adsorb less BSA. Since BSA has a IEP of 5.6 and silica has a point of zero charge around 2, at physiological pH they are both negatively charged. Electrostatic forces between the protein and silica surface should then not be the only factor which drives the absorption process, but may influence the different affinities of the various silica forms.

4.3. EFFECT OF THE SURFACE FEATURES ON IN VIVO AND/OR IN VITRO TOXICITY OF THE SILICA DUSTS

The reaction of the inhaled particles with endogenous molecules present in the pulmonary surfactant e.g. surfactant, antioxidant molecules and proteins may determine the fate of the particles in the early steps of the pathological process. Although the silica samples considered in the present work are very pure, with the SiO_2 content great then 98%, they strongly differ in surface properties. The observed different interaction of BSA with the dusts considered is a consequence of such variability. Since the response of macrophages to silica particles may be different depending on the amount of protein adsorbed, the *in vivo* and/or *in vitro* toxicity of the silica dusts may be affected by the different affinities for pulmonary surfactant proteins.

Acknowledgments

The research has been carried out with financial support by the Italian Ministry of Education, University and Research (MIUR), project title: "The interface between silica-based materials and biomolecules and/or cell models".

References

1. IARC Monographs on the Evaluation of the carcinogenic risk of chemicals to humans. *Silica, some silicates, coal dusts para-aramid fibrils* (Lyon, France, vol. 68, 1997).
2. K. Donaldson and P. J. A. Borm, The quartz hazard: a variable entity, *Ann. Occup. Hyg.* 42, 287-294 (1998).
3. B. Fubini, Surface chemistry and quartz hazard, *Ann. Occup. Hyg.,* 42, 521-530 (1998).
4. B. Fubini, I. Fenoglio, R. Ceschino, M. Ghiazza, G. Martra, M. Tomatis, P. Borm, R. Schins, and J. Bruch, Relationships between the state of the surface of four commercial quartz flours and their biological activity in vitro and in vivo. *Int. J. Hyg Environ Health,* 207, 89-104 (2004).
5. B. Fubini and C. Otero-Aréan. Chemical aspects of the toxicity of inhaled mineral dusts, *Chem. Soc. Rev.* 28, 373-381 (1999).

6. B. Fubini and A. Hubbard ROS and RNS Generation by Silica in Inflammation and Fibrosis *Free Rad. Biol. Med.* 34 (12), 1507-1516 (2003).
7. B. Fubini and W. E. Wallace, Modulation of silica pathogenicity by surface processes, in: *Adsorption Silica Surfaces*, edited by E. Papirer (M. Dekker, Mulhouse, France, 1999), pp. 645-664.
8. A. C. Zettlemoyer, F. T. Micale, and K. Klier, in: *Adsorption of Water on Well Characterized Solid Surfaces. Water in Dispersed Systems*, edited by F. Franks (Plenum Press, New York, 1975), vol. 5.
9. A. Burneau and J. P. Gallas, in: *The Surface Properties of Silicas,* edited by A. P. Legrand, (John Wiley & Sons, Chichester, 1998), pp. 147-234.
10. B. Fubini, V. Bolis, M. Bailes, and F.S. Stone, The reactivity of oxides with water vapor, *Solid State Ionics*, 32-33, 258-272 (1989).
11. B. Fubini, V. Bolis, A. Cavenago, and P. Ugliengo Ammonia and water as probes for the surface reactivity of covalent solids: the case of cristobalite and silicon carbide, *J. Chem. Soc. Faraday Trans.* 88, 277-290 (1992).
12. B. Fubini, V. Bolis, A. Cavenago, E. Garrone, and P. Ugliengo, Structural and induced heterogeneity at the surface of some SiO_2 polymorphs evidenced by the adsorption of various probe molecules, *Langmuir* 9, 2712-2721 (1993).
13. A. M. Langer, Crystal faces and cleavage planes in quartz as templates in biological processes, *Quart. Rev. Biophys.* 11, 534-575 (1978).
14. F. Kozin, B. Millstein, G. Mandel, and N. Mandel, Silica induced membranolysis: a study of different structural forms of crystalline and amorphous silica and the effects os protein adsorption, *J. Colloid Interface Sci.* 88, 326-337 (1982).
15. R. K. Merchant, M. W. Peterson, and G. W. Hunninghake, Silica directly increases permeability of alveolar epithelial cells, *J. Appl. Physiol.* 68, 1354-1359 (1990).
16. E. G. Barrett, C. Johnston, G. Oberdorster, and J. N. Finkelstein, Silica binds serum proteins resulting in a shift of the dose-response for silica-indiuced chemokine expression in an alveolar type II cell line, *Toxicol. Appl. Pharm.* 161, 111-122 (1999).

MAGNETICALLY SENSITIVE NANOCOMPOSITES FOR MEDICAL AND BIOLOGICAL APPLICATIONS

P. P. GORBIK,[1] L. P. STOROZHUK,[1] A. A. CHUIKO,[1†] L. Yu. VERGUN,[2] V. F. CHEKHUN[3]
[1]*Institute of Surface Chemistry 17 General Naumov Street, Kyiv 03164,*
[2]*Institute of Hematology and Transfusiology, 12 Berlinskiy Street, Kyiv 04060,*
[3]*Kavetsky Institute of Experimental Pathology, Oncology and Radiobiology, 45 Vasyl'kivska Street, Kyiv 03022, Ukraine*

Abstract. A magnetically sensitive drug carrier based on magnetite with a polyacrylamide layer has been synthesized and its properties have been studied. A model consisting of a Platidiam-containing ferreed drug of cytostatic action has been studied *in vitro*. Interaction of the magnetically sensitive nanocomposites with model enveloped viruses has been investigated for its application as a high-performance decontaminant of donor plasma.

Keywords: magnetically sensitive carrier; magnetite; polyacrylamide layer; ferreed drug; Platidiam; virus decontamination

1. Introduction

Nanotechnology provides a real prospect to obtain unique remedies for medical and biological use.[1-4] Their introduction holds promise in diagnostics and therapy, particularly at the cellular level. Literature data indicates that studies on the elaboration of ferreed drugs of chemotherapeutical, diagnostic, hyperthermic and sorptive action are of current interest, and widely applied in the field of oncomedicine.[5,6] Our results in this area are described below.

The aim of this work is to synthesize and to investigate magnetically sensitive, highly disperse magnetite-based nanocomposites for medical and biological applications by building models of ferreed drug of cytostatic and hyperthermic action and virus decontaminants, as well as modification of the magnetite surface for functionalization.

*To whom correspondence should be addressed. P. P. Gorbik, Institute of Surface Chemistry, 17 General Naumov Street, Kyiv 03164, Ukraine; e-mail: user@surfchem.freenet.kiev.ua

J.P. Blitz and V.M. Gun'ko (eds.), Surface Chemistry in Biomedical and Environmental Science, 299–306.

2. Experimental

Highly disperse magnetite was obtained by coprecipitation as previously described:[7]

$$Fe^{2+} + 2Fe^{3+} + 8NH_4OH \rightarrow Fe_3O_4 + 4H_2O + 8NH_4^+. \quad (1)$$

Drug carriers were prepared using the 30-50 nm fraction corresponding to the monodomain state. The magnetite surface was stabilized by oleic acid in hexane solution. Carbon and hydrogen content in stabilized powder was 2.0 and 1.5 wt.% respectively.

According to high-resolution scanning electron microscopy and atomic force nanoscopy data, magnetite particles of 30-50 nm size were round crystals. Their specific surface area, determined by argon heat of desorption, was S = 90-180 m^2/g. X-ray phase analysis of magnetite was performed using a diffractometer DRON-4-07 with copper anode emission and a Ni-filter in reflected rays via Bregg-Brentano X-ray focusing. Before drug immobilization the magnetite surface was additionally stabilized via modification by polyacrylamide (PAA). Applying acrylic acid (AA) monomer and the coupling agent N,N'-methylene-bis-acrylamide onto the magnetite surface using a rotary evaporator at 30ºC, surface modification was carried out via acrylamide and N,N'-methylene-bis-acrylamide polymerization in a plasma of microwave glow discharge at 10^{-1} Pa and emitter power of 20 and 40 Watt.[8] IR spectra of the magnetite before and after modification by PAA were recorded using a FTIR 1720 × (Perkin Elmer) spectrometer over 400-4000 cm^{-1} range. The content of stabilizing coverage on the surface of initial magnetite was evaluated using a Perkin Elmer C, H, N-analyzer. The coercive force of magnetite-PAA particles was determined by the ballistic method. Platidiam was adsorbed onto magnetite nanoparticles with a polyacrylamide coating to obtain the ferreed drug. The quantity of Platidiam was determined by atomic absorption by analyzing the contact liquids after adsorption. The magnetic carriers were concentrated, and their fixation realized, at a predetermined site in blood vessel models by applying a permanent magnetic field of appropriate induction and gradient. Drug release dynamics was studied by means of the following technique: the sample with adsorbed Platidiam was placed into a model medium (water, 10 mL) then incubated in a thermostat at 37ºC. After 15 min a 5 mL aliquot was taken out for analysis, and water was added to reach the initial volume (10 mL). Pt^{2+} concentration in extracts obtained was determined by atomic absorption. The ferreed drug model of cytostatic action (magnetically sensitive composite with adsorbed Platidiam) was studied *in vitro* using an MCF-7 cell line (human breast cancer) from the cell culture bank of the Kavetsky Institute of Experimental Pathology, Oncology and Radiobiology.

Cells from the MCF-7 line (concentration 10^5 cell/mL and volume 100 μL) were placed into 96-socket plastic plates. Cells were cultivated on a modified Dulbecco – ISCOV (Sigma, Germany) medium with the addition of 10% embryonal veal serum and antibiotic – gentamycin (concentration 40 μg/mL) under standard conditions at 37°C and 5% CO_2 air saturation. After 24 hours in the cell cultivation conditions, the investigated preparations for testing were added (in 4 parallels each, into 100 μL) and the system was incubated under the same conditions. After 24 hours cytotoxicity was determined. The efficiency was evaluated using the MMT-colorimetric test, based on the ability of living cell mitochondrial enzymes to transform 3-[4,5-dimethylthiazol-2-yl]-2,5-diphenyl-tetrazolium bromide (MTT), a yellow-colored salt, into violet crystalline MTT-formazane.[9] For that purpose 20 μL of MTT solution (Sigma) (5 mg/mL of phosphate buffer) was added to the system and was incubated under the same conditions for 3 hours. After centrifugation (1500 rpm, 5 min) the supernatant was removed using a semi-automatic suction pump. Each socket was then filled with 100 μL of dimethylsulfoxide (Serva) to dissolve the formazane crystals. The absorbance was measured with a multisocket spectrophotometer at 540 nm. Application of a magnetic decontamination facility allows simplification of the technology, and to raise the efficiency of donor plasma decontamination. Therefore the ferreed nanocomposites-assisted physicochemical inactivation method was studied. The virus of vesicular stomatitis (vvs) Indiana strain was chosen as the model enveloped virus. Ferric, silicon, titanium, aluminum, magnesium, chrome, manganous and zinc oxides were used as the model adsorbents. The virus was adsorbed on preliminary sterilized oxide materials at room temperature. The virus activity in culture fluid, initially and after exposure to nanosized oxide materials, was determined by titration by the limit dilution method using cell culture Hep – 2c (microplates for cells cultivation Sarstedt G& Co., M96; Nunc). Cells were cultivated in a CO_2-incubator. Infectious titer was calculated using the method of Reed and Muench, and expressed in lg $TCID_{50/cm^3}$.

3. Results and Discussion

The kinetics of AA and N,N'-methylene-bis-acrylamide polymerization with varied contents of monomers (up to 15 carrier wt%) on a magnetite surface at different values of emitter power was studied. The analysis of kinetic curves obtained by monomer residue titration in non-aqueous medium (methanol) indicates that complete polymerization of the biocompatible layer of selected mass is reached in 2 min.

The experimental dependence of specific magnetization on magnetic field intensity revealed that up to 50% mass increment of PAA coating on the magnetite particles surface has a weak influence on its magnetic characteristics.

In modern chemotherapy platinum complex compounds are some of the most efficient anticancer drugs. Platidiam is manufactured in a lyophilizate form in vials, the components ratio is: 10 mg of *cis*-platinum, 100 mg of manitol, and 90 mg of sodium chloride. It is a yellow powder soluble in water, physiological solutions, and dimethylformamide.[10]

The active compound is *cis*-platinum $[Pt(NH_3)_2]Cl_2$, an orangey powder which is weakly water soluble in the presence of chlorine ions. Without Cl^- ions, hydrolysis occurs and toxic aqua-complexes are formed. Therefore it is administered in an animal's peritoneal cavity in the form of an oil suspension or dimethylsulfoxide (DMSO) solution.[10] Platidiam is highly toxic. The lethal dose for rodents is 12-14 mg/kg, for dogs – close to 2.2 mg/kg, for monkeys it is even more toxic. The drug chemotherapeutical index is low as well – about 4-8.[10] Therefore, localization in a lesion site would allow one to obtain a therapeutic effect at significantly lower doses, reducing its toxic effects. Platidiam adsorption (Figure 1) on the surface of magnetite-polyacrylamide particles was carried out from aqueous solution for 12 hours in dynamic mode at room temperature. S-like isotherm character indicates that combined adsorption of solvent (physiological solution) and solute (Platidiam) has occurred. This adsorption behavior may be explained by a stronger interaction between solute molecules and adsorbate. In similar cases, solute molecules tend to displace adsorbate on the surface in the form of chains or clusters.[11]

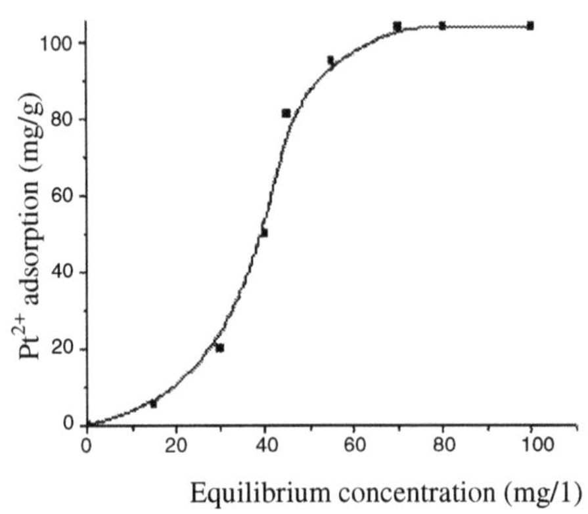

Figure 1. The isotherm of Platidiam adsorption (determined as Pt(II) content) on polyacrylamide-coated magnetite nanoparticles.

Drug release dynamics was studied at constant temperatures of 37, 40, and 45°C. Results are given in Table 1. In the temperature range of 37-45°C within the first 30 min, up to 46% of the administered quantity of Platidiam was released. After 80 min the drug release into the model medium was not significantly changed.

Table 1. Dynamics of the drug release into model medium at different temperatures

No	Time (min)	Quantity of Platidiam released from the surface of the ferreed drug carrier (%)		
		37°C	40°C	45°C
1	15	29.4	30.1	29.9
2	30	45.7	44.9	45.1
3	45	45.9	46.1	45.5
4	60	46.1	45.6	45.9
5	75	46.3	46.0	46.1

The possibility of magnetite use for making the hyperthermic zones in certain parts of an organism the target of nonresonant (thermal) influence of electromagnetic radiation on animal muscular tissues (*in vitro*) was explored. The dependence of muscular tissue temperature upon the time of electromagnetic wave (λ = 3 cm) irradiation was studied. Muscular tissue samples of 5 g mass were used. The irradiation area was about 2 cm^2. It was shown that in the case of disperse magnetite particles introduction into muscular tissues (0.1 wt%), the absorbance of electromagnetic radiation rises, and the warming rate is increased from 3 to 4°C/min at radiation power of 0.5 Watt.

It was shown by previous investigations that use of Platidiam (Ptd) solutions with a concentration of C = 1.5 μg/mL corresponds to an IC_{50} dose for the MCF-7 cell line. Therefore, makeup Ptd solution (water for injections) of mentioned concentration (D_3 dose) and consecutively halved concentrations (D_2 and D_1 doses, respectively) was used in *in vitro* experiments. We also studied the influence of the

- Ptd-L solution (solvent – water for injections) prepared for Platidiam adsorption onto magnetically sensitive nanocomposites (C = 1 mg/l). The 'age' of Ptd-L solution before the experiment was 14 hours;
- magnetically sensitive carrier with biocompatible PAA coating (without Platidiam) FeM;
- samples of V-type (dry substance) – magnetically sensitive carriers with PAA coating and Platidiam adsorbed from Ptd-L solution for 12 hours in dynamic mode at room temperature;

on cell line. Results are given in Table 2. The use of Ptd (Platidiam solution prepared extempore in concentrations D_1, D_2, D_3) inhibits cell growth by 32.5, 44.6, and 57.4% respectively. Inhibition at D_3 concentration corresponds to the IC_{50} index under experimental conditions. The use of Ptd-L solution prepared for adsorption (useful life is 14 hours) in D_1, D_2, and D_3 concentrations inhibits cell growth by 8.7, 19.4, and 24.6%, respectively. Therefore the drug's cytostatic action is decreased with long-term solution storage.

Table 2. The influence of Platidiam adsorbed on ferromagnetic carriers on viability of MCF-7 line cells at different doses

Sample	D_1		D_2		D_3	
	Cell number (%)		Cell number (%)		Cell number (%)	
dead	living	dead		living		dead
Ptd	67.5	32.5	55.4	44.6	42.6	57.4
Ptd-L	91.3	8.7	80.6	19.4	75.4	24.6
FeM	98.2	1.8	94.6	5.4	91.3	8.7
V	90.3	9.7	81.4	18.6	75.8	24.2

Study of the V-type sample (Table 2) has shown that D_1, D_2, and D_3 doses inhibit cell growth by 9.7, 18.6, and 24.2%, respectively. This is close to the effect of Platidiam Ptd-L solution from which the drug adsorption was carried out. Results obtained show that almost all Platidiam is released from the magnetically sensitive carrier into the cultivated medium. Increased cytostatic action of magnetically sensitive nanocomposites (Platidiam carriers) may be achieved by means of solvent, adsorption time, and temperature optimization.

Nanomaterials can be used for purification of plasma, and it is of interest to compare different oxides for this purpose. The results of vvs titration after the contact of virus-containing fluid with various nanomaterials at room temperature are given in Figure 2.

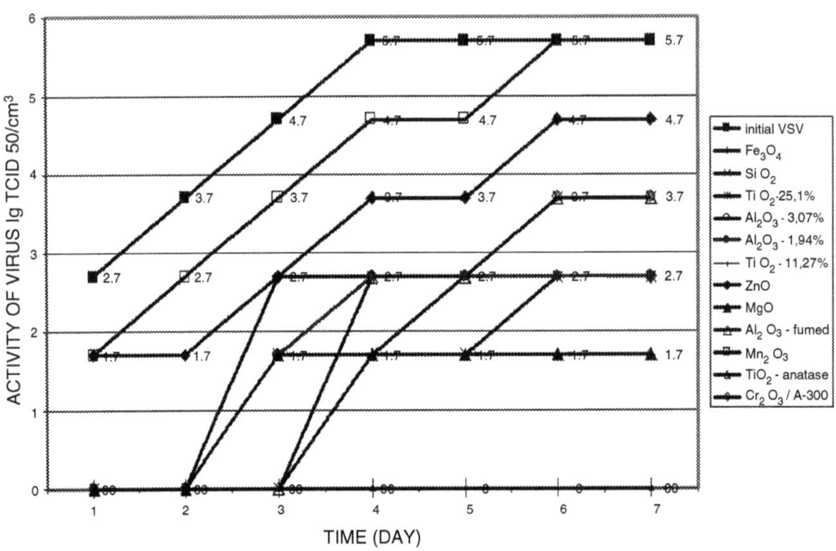

Figure 2. Infectious vvs activity after the contact of virus-containing fluid with nanomaterials at different temperatures.

Adsorption on Mn_2O_3 was not detected. After 7 days the virus amount in culture fluid is no different from that in the initial virus-containing fluid (5.7 lg $TCID_{50/cm^3}$). The virus amount in culture fluid was decreased by one lg after the exposure with ZnO; by two lg – with Al_2O_3, Al_2O_3/SiO_2 (3.07%), SiO_2; by three lg – with Al_2O_3/SiO_2 (1.94%), Cr_2O_3/SiO_2, TiO_2/SiO_2 (25.1%); by four lg – with TiO_2/SiO_2 (11.27%), MgO. Complete vvs adsorption/inactivation was observed after contact with magnetite (Fe_3O_4) and titanium dioxide. A decrease of Al_2O_3 and TiO_2 content on thc surfacc of highly disperse silica leads to an increase of adsorptive capacity of these nanocomposites with respect to vvs.

It should be noted that pH of the initial virus-containing fluid was 10-11 after the sorbent MgO was added. In alkaline medium vvs could be collapsed, in such a pH range adsorption is usually not observed owing to charges of interacting particles. At pH values higher than the isoelectric point the protein will have a negative charge. This negative leads to decreased adsorption due to like charges repulsion.[12-14] As a result we obtained the picture of "survived" virus.

4. Conclusion

Highly disperse magnetite (S = 90-180 m^2/g) was synthesized. Particles of monodomain fraction (d ~ 30-50 nm) were coated with polyacrylamide via polymerization in a microwave plasma, and the drug "Platidiam" was immobilized on the surface of coated particles. The cytotoxic effect of immobilized Platidiam on the human breast cancer cell line MCF-7 was studied. It was shown that the use of Ptd – Platidiam solution prepared extempore in D_2 concentration inhibits cell growth by 44.6%, inhibition at D_3 concentration (by 57.4%) corresponding to an IC_{50} index under experiment conditions. The use of FeM, a magnetically sensitive nanosized carrier with a biocompatible polymeric coating, inhibits cell growth by 1.8, 5.4 and 8.7%, respectively. Thus cytotoxic action is absent (the cytotoxic action criterion is considered to be 25-35%). Examination of nanocomposites carrying Platidiam (V-type sample) has shown that it inhibits cell growth by 9.7, 18.6, and 24.2%, close to the effect of Platidiam Ptd-L solution (useful life is 14 h) from which the drug adsorption was carried out. The results obtained indicate that Platidiam was completely released from the surface of magnetically sensitive carrier into the culture medium. The data obtained indicate that magnetite- and titanium dioxide-based nanocomposites may be useful for application as ferreed virus decontaminants.

Acknowledgement

The work was performed within the framework of Integrated Program of Fundamental Research of Ukrainian National Academy of Sciences "Nano-structured Systems, Nanomaterials and Nanotechnology" (project "Synthesis of Ferreed Medicobiological Nanocomposites and Investigation of Their Properties").

References

1. M. C. Roco, R. S. Williams, and P. Alivisatos, *Nanotechnology research directions. Vision for Nanotechnology R&D in the Next Decade* (Kluwer Academic Publishers, Dordrecht, 2002).
2. R. N. Alyautdin, Y. Kroiter, and D. A. Kharkevitch, Nanoparticles-assisted brain drug delivery, *Experimental and Clinical Pharmacology* 66(2), 65-68 (2003).
3. P. P. Gorbik, I. V. Dubrovin, A. L. Petranovska, O. M. Fedorenko, L. P. Storozhuk, V. F. Chekhun, and O. O. Chuiko, Magnetically sensitive drug carrier based on ultrafine magnetite – polyacrylamide particles, *Reports of NAS of Ukraine* 4, 124-128 (2005).
4. A. L. Petranovska, O. M. Fedorenko, P. P. Gorbik, L. P. Storozhuk, O. O. Chuiko, and V. F. Chekhun, The release dynamics of cytostatic action drug "Platidiam" from the surface of ferreed nanocomposite, *Medical Chemistry* 7(2), 75-78 (2005).
5. *Surface chemistry of silica*, edited by A. A. Chuiko (UkrINTEI, Kiev, 2001).
6. L. Levy, Y. Sahoo, K.-S. Kim, and E. J. Bergey, Nanochemistry: Synthesis and characterization of multifunctional nanoclinics for biological applications, *Chem. Mater* 14, 3715-3721 (2002).
7. V. V. Sviridov, G. A. Popkovitch, and E. I. Vasilevskaya, *Inorganic synthesis* (Minsk University, Minsk, 1996).
8. F. A. Sal'yanov, *Fundamental physics of low-temperature plasma, plasmic devises and technology* (Nauka, Moskow, 1997).
9. T. Mosmann, Rapid colorimetry assay for cellular growth and survival: application to proliferation and cytotoxic assays, *J. Immunol. Methods* 65, 55-63 (1983).
10. E. E. Kris, I. I. Voltchenskova, A. S. Drigor'eva, K. B. Yatsimirskiy, and L. I. Budarin, *Coordination compounds of metals in medicine* (Naukova dumka, Kiev, 1986).
11. *Adsorption form Solutions at the Solid/Liquid Interface*, edited by G. D. Parfitt and C. H. Rochester (Academic Press, London, 1983).
12. *Fundamental virology*, edited by B. N. Fields and D. M. Knipe (Raven Press, New York, 1991).
13. *Medical chemistry and clinical application of silicon dioxide*, edited by A. A. Chuiko (Naukova Dumka, Kiev, 2003).
14. *Principles of Biochemistry,* A. L. Lehninger, D. L. Nelson, and M. M. Cox (Worth, 2004).

HYDROPHOBIZED SILICA NANOCOMPOSITES WITH IMMOBILIZED ANTIOXIDANTS (VITAMINS C AND E)

PAVLO KUZEMA*, OKSANA STAVINSKAYA, OLGA KAZAKOVA, IRINA LAGUTA
Institute of Surface Chemistry of National Academy of Sciences of Ukraine, 17 General Naumov Street, Kyiv, 03164, Ukraine

Abstract. Adsorption of antioxidants (vitamins C and E) from aqueous and ethanol solutions on unmodified and partially hydrophobized nanosilica A-200 was studied using UV spectroscopy and quantum chemical methods with consideration for the solvent effects. Antioxidant power of silica nanocomposites with immobilized vitamins was evaluated by measuring the total polyphenolic index following the Folin-Ciocalteu method. It has been shown that immobilization of vitamins on silica surface leads to their stabilization. Being released from the carrier molecules of vitamins do not lose their antioxidant properties

Keywords: antioxidants, vitamins C and E, partially hydrophobized fumed silica

1. Introduction

Oxidative stress or imbalance between formation and deactivation of free radicals is considered to be one of the main causes of many known human diseases.[1,2] Vitamins E and C are among the most attractive objects for prophylactic drug development.[2,3] Aside from being natural antioxidants of direct action, their combination results in a synergistic effect.[4] However, obtaining stable formulations remains a challenge since these vitamins are readily oxidized in a liquid medium. Immobilization of such biomolecules on solid surfaces may result in enhanced stabilization.

*To whom correspondence should be addressed. P. Kuzema, Institute of Surface Chemistry, 17 General Naumov Street, Kyiv, 03164, Ukraine, e-mail: sci-worker@yandex.ru

J.P. Blitz and V.M. Gun'ko (eds.), Surface Chemistry in Biomedical and Environmental Science, 307–314.

Highly disperse silica is widely used in pharmaceutical formulations as a filler, adsorbent, thickener etc.[5] Their high hydrophilicity and the absence of emulsifying ability restrict their application. In contrast to hydroxylated silica, partially or fully hydrophobized silica may exhibit improved properties as a drug carrier. The main goal of this work is to study hydrophobized silica nanocomposites with immobilized vitamins C and E. Investigations of adsorption-desorption processes which involve silica nanoparticles and the antioxidants are described. Factors affecting the antioxidant stability have also been carefully considered.

2. Experimental

Fumed silica A-200 (Pilot plant at the Institute of Surface Chemistry, Kalush, Ukraine; specific surface area $S_{BET} \approx 200$ m^2/g) was used as the initial adsorbent. Ascorbic acid (vitamin C) and all-rac-α-Tocopheryl acetate (vitamin E acetate) were used as adsorbates. Folin-Ciocalteu's phenol reagent (Merck) was used to measure the total polyphenolic index. Silica samples with different degree of surface silylation were obtained using gas-phase chemical modification of highly disperse silica (A-200) surface by trimethylchlorosilane.[6]

The process of chemical modification was carried out in a reactor with a mixer at 300°C for 1 h.[7] Preliminary thermal treatment of the highly disperse silica surface was made at 200°C for 2 h. After the modification, samples were heated in airflow at 200°C for 2 h to remove the residual modifying agent, as well as the reaction products from the surface.

The degree of substitution of OH groups by trimethylsilyl (TMS) groups (θ) was determined by means of IR spectroscopy using the formula

$$\theta = 1 - \frac{D_x}{D_0}, \qquad (1)$$

where D_0 and D_x is the optical density of the absorption band with maximum at 3750 cm^{-1} (stretching vibrations of the isolated silanols) of the initial and modified samples, respectively.

The IR and UV spectra were recorded using Specord M-80 (IR) and Specord M-40 (UV/vis) (Carl Zeiss, Jena, Germany) spectrophotometers.

The adsorption of vitamins was carried out from ethanol solution at 20°C for 2 h onto unmodified and modified fumed silica. The extent of modification by $-Si(CH_3)_3$ groups is given by the TMS loading $\theta_{TMS} = 0.10$, 0.40 and 0.70. The solution of a vitamin (10 ml) with different initial concentrations C_0 was mixed with an adsorbent (0.05 g). After adsorption for 2 h, the equilibrium concentration of the vitamin (C_{eq}) was determined by UV spectrophotometry. The difference in the maximum of the absorption of the solution before and after adsorption at $\lambda = 284$ nm (Vitamin E) or 250 nm (Vitamin C), was used to determine the extent of adsorption. Desorption experiments were carried out at constant solution volume with the adsorbent/solution ratio of 0.1 g/25 ml.

Antioxidant activity of silica nanocomposites with immobilized vitamin C was tested using the polyphenolic activity index.[8] After adsorption of ascorbic acid on the silica surface and centrifugation, the excess solution was removed to obtain the suspension of a fixed volume (2 ml). Distilled water, sodium carbonate solution, and Folin-Ciocalteu's phenol reagent were subsequently added to suspensions and to the reference Vitamin C solution. The suspensions were then stored for 30 min, and the optical density of supernatant was measured at $\lambda = 750$ nm. The reference solution of ascorbic acid was used to compare antioxidant activity of vitamin-containing nanocomposites with the activity of dissolved vitamin C.

3. Theoretical calculations

Structural and electronic characteristics of model clusters of silica, their complexes with vitamins, and free energy of solvation (ΔG_s) have been calculated using the solvation model SM5.42 with the 6-31G(d) basis set, or the semiempirical method PM3 (GAMESOL program package, Version 3.1).[9] The free energy of solvation of a molecule can be calculated using its geometry determined for the gas phase

$$\Delta G_s(R) = \Delta G_{EP} + \Delta G_{CDS}, \tag{2}$$

where $\Delta G_{EP} = \Delta G_E + \Delta G_P$ is the electrostatic component of ΔG_s; ΔG_E is the energy of deformation of charge density of a molecule resulting from the polarization of solvent; ΔG_P is the energy of interaction of a molecule with the solvent in view of its reorganization; $G_{CDS} = \sum_k A_k \sigma_k$, where A_k is an accessible surface of the k-th atom, σ_k is an atomic superficial tension of the k-th atom (a function of the spatial geometry of a solution and solvent parameters). ΔG_s can be determined with consideration of the geometric relaxation upon solvation

$$\Delta G_s = G(l, \mathbf{R}_e(l)) - G(g, \mathbf{R}_e(g)) = E(l, \mathbf{R}_e(l)) + G_P(\mathbf{R}_e(l)) + G_{CDS}(\mathbf{R}_e(l)) - E(g, \mathbf{R}_e(g)), \tag{3}$$

where indices l and g correspond to liquid and gas phases, respectively, and $\boldsymbol{R}_e$ corresponds to the system's equilibrium geometry. The solvation model SM5.42 has been described in detail elsewhere.[9]

4. Results and Discussion

Methylated silica is less toxic than unmodified silica, and its harmful effect on cell membranes decreases.[10] These data are confirmed by our EPR studies of blood samples after contact with initial and methylsilylated silica samples.[11] The EPR spectra showed a drastic increase in Fe^{3+} content in A-200 – blood system (signal with g-factor of approximately 2.20 and half-width ΔH = 320-350 Gauss). This may be caused by hemoglobin Fe^{2+} oxidation, and indicates that blood components are destroyed. In the case of modified silica, changes in the spectra were not observed.

The IR spectra of silica samples under investigation are shown in Figure 1. A decrease in the intensity of the absorption band at 3750 cm^{-1} and appearance of a band at 2980 cm^{-1} (spectra 2, 3, Figure 1) indicate that substitution of isolated silanol by TMS groups has occured.

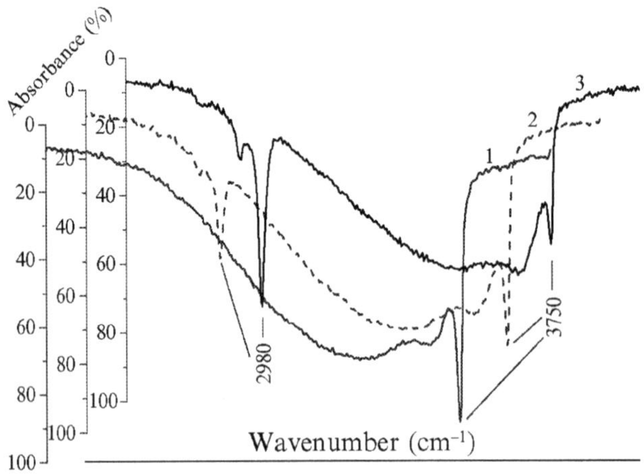

Figure 1. IR spectra of unmodified (1) and modified silica by trimethylchlorosilane (2, 3); modification degree (θ): (2) 0.40 and (3) 0.70.

A comprehensive physicochemical study found that wetting of trimethylsilylated silica samples by water is determined by the amount of grafted groups.[7] In particular, for hydrophilic trimethylsilylated silica (with a degree of surface silylation less than 0.55) in an aqueous medium, the increase of modifycation degree was accompanied by an increase of the structure restoration ability.[7]

Previously we studied the adsorption of vitamins B_1, B_6 and PP on unmodified and silylated silica. These data show that the affinity of adsorbates to the surface of modified silica is probably increased due to dispersive interactions between grafted groups and hydrocarbon fragments of vitamin molecules, as well as hydrophobic interactions. As a consequence, vitamin desorption into physiological solution or water slows down which, in turn, prolongs the action of the nanocomposite.

Unmodified silica was found to have a low adsorptive capacity with respect to vitamin E. In the case of modified silica the quantity of immobilized biomolecules is significantly increased (Figure 2). The adsorption of vitamin E does not prevent interaction of silica with vitamin C (Figure 3). It was found that the adsorption of vitamin C from ethanol solution, on the surface of modified silica with preadsorbed vitamin E, is thermodynamically favourable ($\Delta G_{ads} = -31$ kJ/mol).

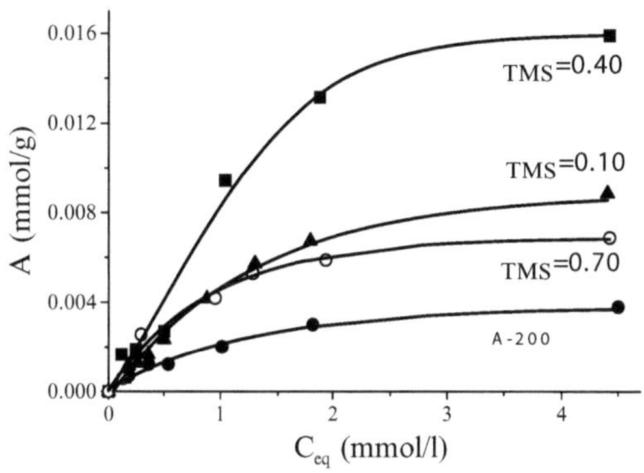

Figure 2. Adsorption isotherms of vitamin E on unmodified and silylated silica.

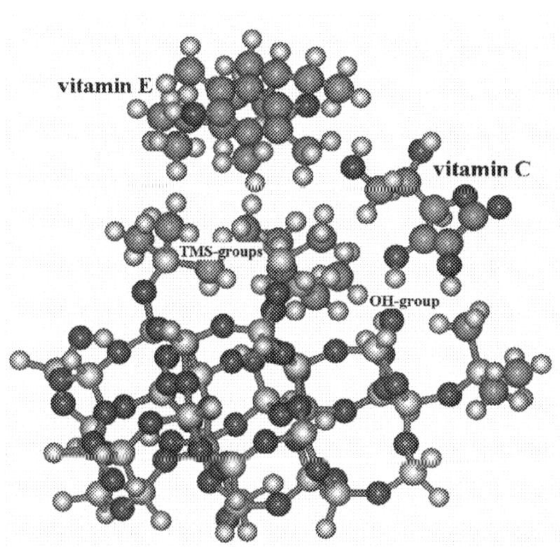

Figure 3. Vitamins C and E molecules on the surface of a modified silica particle modeled by a cluster with TMS and silanol groups.

Interaction of ascorbic acid (vitamin C) with the silica surface was complicated by the fast oxidation of ascorbic acid in aqueous and ethanol solutions because of dissolved oxygen.[12,13] However, both unmodified and modified silica increase the oxidation resistance of vitamin C. In particular, the rate of ascorbic acid oxidation to dehydroascorbic acid was found to be much less in the presence of unmodified or modified silica (Figure 4). Vitamin C is stabilized in the presence of silica, apparently due to interaction of the vitamin with the surface of highly-disperse silica particles, as confirmed by the results of quantum chemical calculations (Figure 5). On interaction of an ascorbic acid molecule with the surface of a silica particle in ethanol solution, the charges of hydrogen atoms (q_{Hi}) of hydroxyls involved in the reaction of ascorbic acid oxidation are decreased. Thus, the decrease in proton-donor ability of OH groups of the adsorbate due to formation of hydrogen bonds with surface silanol groups (Figure 5) probably leads to vitamin C stabilization in the presence of silica in ethanol. Similar processes take place on the surface of modified silica.

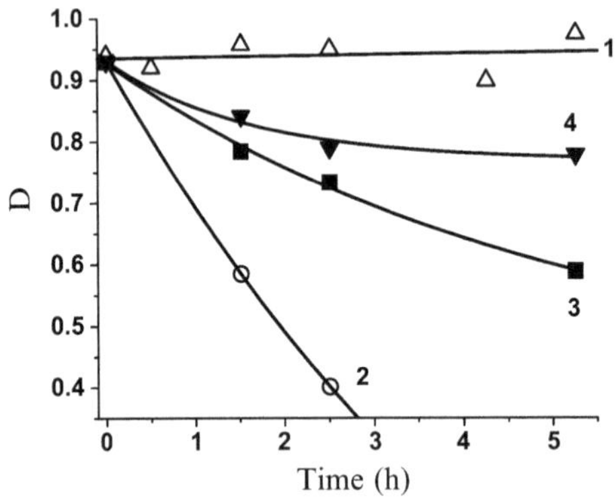

Figure 4. Optical density (at λ = 250 nm) of ethanol solutions of ascorbic acid (C = 10-4 mol/l) (1 – control (makeup) solution, 2 – solution without adsorbent, 3 – solution of ascorbic acid in the presence of initial silica A-200, 4 – solution in the presence of modified silica at $\theta_{TMS} = 0.10$) as a function of time.

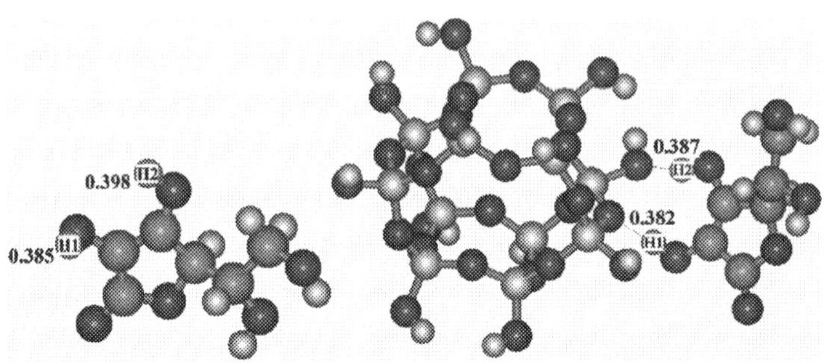

Figure 5. The models of vitamin C (left) and its adsorption complex at the silica surface (right). Charge values (SM5.42/6-31G(d) of hydrogen atoms of OH-groups involved in reactions of oxidation are shown.

To obtain silica nanocomposites with immobilized antioxidants, and to test their antioxidant activity, vitamins C and E were adsorbed onto nanosilica from ethanol solution containing both vitamins. A desorption experiment was then carried out. After the desorption of ascorbic acid from modified silica (θ_{TMS} = 0.40) into aqueous medium, the solvent was changed and the desorption of vitamin E into ethanol was studied. As one can see (Figure 6), both vitamins can be released from the surface for 1-2 hours. It should be noted that no increase in vitamin C concentration was detected in aqueous solution after its contact with A-200 for more than 30 min. Thus, as is the case for vitamins B_1 and B_6, desorption of vitamin C from the surface of modified silica is slower in comparison with unmodified silica. The antioxidant activity of silica nanocomposites with immobilized vitamin C measured by the Folin-Ciocalteu method is shown in Table 1.

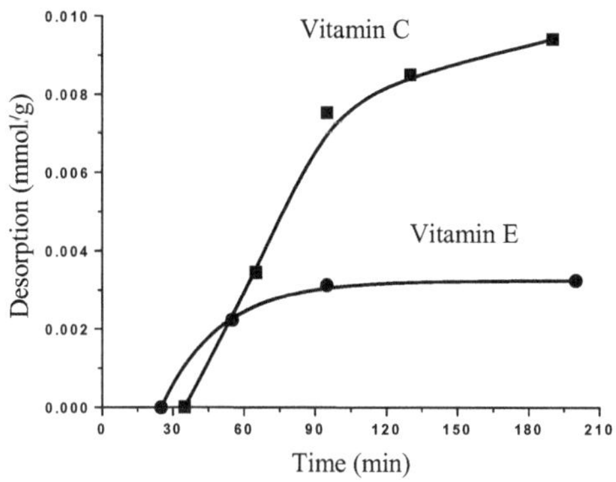

Figure 6. Desorption of vitamin C into aqueous medium and subsequent desorption of vitamin E into ethanol from modified silica (θ = 0.40).

Table 1. Antioxidant activity of silica nanocomposites with immobilized vitamin C

Sample	Reference solution of vitamin C	Initial silica A-200 + vitamin C	Modified silica (θ = 0.40) + vitamin C
Total polyphenolic index	0.8000	1.6077	0.9639

As can be seen from the data, after 30 min of reaction maximum antioxidant power is observed for unmodified silica, with the index being approximately twice as much compared to the control solution. Modified silica also shows an increase in antioxidant power by 30-35% compared to the control solution. The data are in agreement with previous results showing the stabilization of vitamin C on the silica surface, and its slower desorption from the surface of modified samples. Unfortunately, it is not possible to determine the antioxidant activity of vitamin E by the Folin-Ciocalteu method since it is insoluble in aqueous solution.

5. Conclusion

Partial silylation of the highly disperse silica surface enhances the adsorption of vitamin E from ethanol solution, and provides the ability to obtain water-soluble nanocomposites containing vitamin E. Immobilization of vitamin C on the silica surface prevents its oxidation. Its interaction with the adsorbent surface leads to a decrease in proton-donor ability of the OH-groups involved in the oxidation of ascorbic acid. Hydrophobized silica nanocomposites are characterized by a prolonged desorption of immobilized vitamins. It has been shown that vitamin C does not lose its antioxidant properties after desorption.

Acknowledgement

The work was performed within the framework of Integrated Program of Fundamental Research of Ukrainian National Academy of Sciences "Nano-structured Systems, Nanomaterials, Nanotechnology". O. K. is grateful to Prof. V. M. Gun'ko for the use of the GAMESOL (Version 3.1) program package.

References

1. L. Migliore and F. Coppedè, Genetic and environmental factors in cancer and neuro-degenerative diseases, *Mutation Research/Reviews in Mutation Research* 512(2-3), 135-153 (2002).
2. J. M. Tucker and D. M. Townsend, Alpha-tocopherol: roles in prevention and therapy of human disease, *Biomedecine & Pharmacotherapy* 59(7), 380-387 (2005).
3. N. Aksoy, H. Vural, T. Sabuncu, O. Arslan and S. Aksoy, Beneficial effects of vitamins C and E against oxidative stress in diabetic rats, *Nutrition Res.* 25(6), 625-630 (2005).
4. M. E. Carlotti, M. Gallarate, M. R. Gasco, S. Morel, A. Serafino and E. Ugazio, Synergistic action of vitamin C and amino acids on vitamin E in inhibition of the lipoperoxidation of linoleic acid in disperse systems, *Intern. J. Pharmaceut.* 155(2), 251-261 (1997).
5. *Medical chemistry and the clinical application of silicon dioxide*, edited by Chuiko A. A. (Naukova dumka, Kiev, 2003).
6. P. O. Kuzema, *Properties of chemically modified high-disperse silicas with hydrophilic-hydrophobic surface* (PhD Thesis, Kyiv, 2004).
7. P. O. Kuzema and I. V. Laguta, Structural-mechanical properties of hydrophilic-hydrophobic silicas in water and decane, in: *Proceedings of XI International Conference "Technologies of XXI century"* (Alushta, Ukraine, 2004), pp. 49-54.
8. A. M. Alonso, C. Domianguez, D. Guillean, and C. G. Barroso, Determination of Antioxidant Power of Red and White Wines by a New Electrochemical Method and Its Correlation with Polyphenolic Content, *J. Agric. Food Chem.* 50, 3112-3115 (2002).
9. J. D. Xidos, J. Li, T. Zhu, G. D. Hawkins, J. D. Thompson, Y.-Y. Chuang, P. L. Fast, D. A. Liotard, D. Rinaldi, C. J. Cramer, and D. G. Truhlar, *GAMESOL–version 3.1*, University of Minnesota, Minneapolis, 2002, based on the General Atomic and Molecular Electronic Structure System (GAMESS) as described in M. W. Schmidt, K. K. Baldridge, J. A. Boatz, S. T. Elbert, M. S. Gordon, J. J. Jensen, S. Koseki, N. Matsunaga, K. A. Nguyen, S. Su, T. L. Windus, M. Dupuis, and J. A. Montgomery, *J. Comp. Chem.* 14, 1347 (1993).
10. B. Fubini and W. E. Wallace, Modulation of silica pathogenicity by surface processes, *Adsorption on silica surface*, edited by E. Papirer (Marcell Dekker, New York, 2000), Vol. 9, pp. 646-663.
11. I. V. Laguta, O. M. Stavinskaya, P. O. Kuzema, and O. A. Kazakova, Hydrophilic-hydrophobic fumed silicas as carriers of vitamins for biomedical applications, i: *Proceedings of IX Polish-Ukrainian Symposium "Theoretical and Experimental Studies of Interfacial Phenomena and Their Technological Applications"* (Sandomierz, Wolka Milanowska, Poland, September 5-9, 2005) pp. 162-164.
12. W. L. Baker, Ascorbic acid reactionwith disulphide compounds: effect and applications, *Talanta* 52, 425-433 (2000).
13. J. C. B. Fernandes, G. Olivera Neto, and L. T. Kubota, Use of column with modified silica for interfering retention in a FIA spectrophotometric methods for direct determination of vitamin C in medicine, *Analytica chimica Acta* 366, 11-22 (1998).

INTERACTION OF SOME BIOMOLECULES WITH MODIFIED NANOSILICA SURFACES STUDIED BY QUANTUM CHEMISTRY

O. TSENDRA,* A. DATSYUK, V. LOBANOV, A. GREBENYUK, A. CHUIKO†
Institute of Surface Chemistry, 17 General Naumov Street, Kiev 03164, Ukraine

Abstract. The equilibrium geometries and heat of adsorption values for some complexes of amino acids and the helical section of human serum albumin on a silica surface, in both gaseous phase and water solutions, have been calculated using quantum chemical methods. In this study a suitable method is proposed for calculating the interaction energy. A probable mechanism has been suggested for the interaction between a hydroxylated silica nanoparticle, with negative surface charge, and lipid molecules. As both interacting species have the (negative) charge, and experiment gives evidence of their cohesion, in order to provide converging clusters we believe that hydronium ions are stationed between these species. If the total charge of this supramolecular structure is assumed to be zero, an optimization of geometric parameters results in the formation of a water-bonded complex.

Keywords: silica nanoparticle, sulfur-containing amino acid, human serum albumin, phospholipids, adsorption mechanism, quantum chemistry

1. Introduction

Highly disperse silica was first applied in medical practice as an effective enterosorbent. It was then found to have a therapeutic influence, and that it can be used as a carrier for medicinal purposes.[1,2]

A silica nanoparticle falling into a mouth cavity is immediately affected by a solution with a significant amount of biological activity.[3] Adherence of a

*Author to whom correspondence should be addressed: Oksana Tsendra, Institute of Surface Chemistry, 17 General Naumov Street, Kyiv 03164, Ukraine; e-mail: oksynka@mail.ru

J.P. Blitz and V.M. Gun'ko (eds.), Surface Chemistry in Biomedical and Environmental Science, 315–324.

water-organic mix (whose components are amino acids, dipeptides, proteins, and carbohydrates) takes place on the silica surface. When a SiO_2 nanoparticle surface (bearing grafted or adsorbed drugs) contacts a biological medium, many uncontrolled processes take place.[2] When in the gastrointestinal channel, biologically active matter earlier adsorbed on the silica surface (including pharmaceuticals) can be displaced from the surface by similar compounds with greater affinity to silica.[2]

As a rule, most medical preparations begin to act only after reaching the stomach. Taking into account this circumstance, it is impossible to give a justified answer to the urgent question, "Which form of preparations, chemically grafted or adsorbed on the silica surface, shows activity?"

There exists an obstruction mechanism within animal organisms which prevents mechanical damage of cell membranes from dust microparticles inhaled with air. It is possible to suggest that a non-specific protection mechanism was developed during the evolution of animal organisms against damage caused by small and very small mineral particles. As both the surface of silica nanoparticles and that of cells (in particular, those of erythrocytes) are negatively charged, and experiments testify to their agglutination, an electrostatic interaction between them should be facilitated by chemical aspects capable to promote their attraction at relatively large distances.

The above-stated ideas stimulate activities in detection and analysis of subtle but important mechanisms of interaction between silica nanoparticle surfaces and the environment of a cell membrane.

2. Computational Methods

Experimental methods often provide the integral characteristics of a system, but do not detect the influence of some factors. Quantum chemical methods provide an opportunity to perform calculations such that the role of some factors becomes apparent through the entire multistage mechanism of interaction between a silica nanoparticle and biologically active molecules.[4]

This paper presents results obtained using quantum chemistry tools of different complexity levels, including semiempirical (PM3)[5] and ab initio computational schemes within a cluster model.[6] The surface was modeled as a fragment of an edge (111) of β-cristobalyte with structural motifs characteristic of the large part of an amorphous silica surface.[7] The edge valences of the cluster were saturated with hydrogen atoms.[8]

3. Results

3.1. ADSORPTION OF SULFUR-CONTAINING AMINOACIDS

Sulfur-containing amino acids have a significant role in the functioning of organisms. Their high concentrations must be removed from the human organism, a task which is possible with silica enterosorbents.[9-11]

In Figure 1 the equilibrium spatial structures of adsorption complexes of sulfur-containing amino acids: cysteine, methionine, homocysteine and cystine on a silica surface obtained from gas phase are shown.

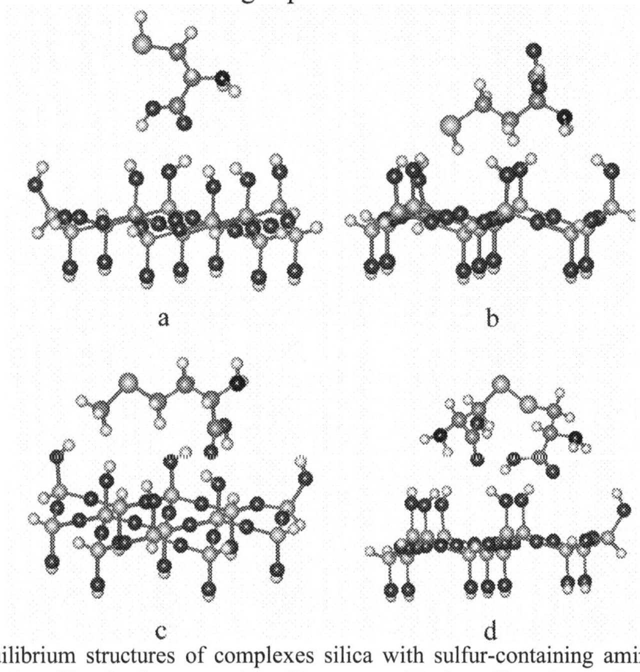

Figure 1. Equilibrium structures of complexes silica with sulfur-containing aminoacids in gas phase: a – cysteine-silica; b - homocysteine-silica; c - methionine-silica; d – cystine-silica.

Analysis of the equilibrium configurations of the models of adsorption complexes reveal that their existence is due to hydrogen bonding between carboxyl or amino groups of sulfur-containg amino acids, and surface silanol groups. There are no thiol groups binding with the surface. The stability of adsorption complexes derived by amino groups of all sulfur-containing acids is somewhat lower than that of similar complexes involving carboxyl groups.

When simulating the interaction of amino acids with a silica surface in a biological medium, the zwitterionic form as well as hydration should also be taken into consideration.[12] Shielding of functional groups (amino, carboxyl, and thiol) of sulfur-containing amino acid molecules by water molecules results in averaging their potential capabilities to adsorb on silica surface. The bonding of zwitterions with the surface occurs through water molecules forming hydrogen bonds with functional groups of amino acids and silanol groups of silica.

Tables 1 and 2 contain the formation heat values for adsorption complexes of sulfur-containing amino acids on a silica surface from gas phase and for their zwitterions with a hydrated environment.

Table 1. Enthalpy of formation (kJ/mol) of equilibrium complexes of sulfur-containing amino acids with silica surface in gas state (PM3).

Amino acid	–COOH	$-NH_2$	–S–	Average value
cysteine	–47.8	–52.8	–55.7	–52.1
homocysteine	–35.6	–19.3	–36.1	–30.3
methionine	–33.1	–15.9	–20.1	–23.0
cystine	–77.3	–50.3	–57.4	–61.7

Table 2. Gibbs free energy of the formation of equilibrium complexes of sulfur-containing amino acids with silica surface in liquid state (SM5.42R / 6-31+G).

Zwitterion of sulfur-containing amino acid	ΔG (кJ/mol)
Cysteine	–13.5
Homocysteine	–21.1
Methionine	4.6
Cystine	22.8

When adsorbed from the gas phase sulfur-containing acids form the most stable neutral complexes, whereas a similar process in solution is accompanied by rearrangement into a zwitterionic structure. It should be noted that adsorbed cystine molecules preserve their disulfide bridges, indicating that their native structure remains under interaction with the silica surface.

3.2. ADSORPTION OF FRAGMENT OF HUMAN SERUM ALBUMIN MOLECULE

Silica is known to have unique adsorption properties with respect to proteins, which makes it an ideal enterosorbent for protein-based toxins.[13,14] A study was thus undertaken to investigate the origin of the affinity of albumin with silica.

The albumin/silica system is too large to model completely. Thus a small fragment of is selected to examine the main features of its adsorption. The helical section of human serum albumin (HSA) has been chosen as a model consisting of a 16 fragment amino acid chain placed at the periphery of a protein globule.

After calculating the HAS fragment spatial and electronic structure, the first hydrate shell was built. Water molecules were situated around the α-helical section near its active centers (–NH_2 and –COOH groups). A calculation was then carried out with the following components: a segment of a peptide molecule, a hydrate shell, and a cluster simulated for a fragment of a silica surface. Table 3 shows the heats of formation of the components of the adsorption complex HSA on the silica surface. The heat of adsorption is found to be 108 kJ/mol, six times greater than a typical hydrogen bond (Figure 2).

Table 3. Heats of formation of the components of adsorption complex HSA on silica surface (PM3).

Object	$-Q_{form}$ (кJ/mol)
cluster I (α-helix of HAS)	3195
solvent shell (10 water molecules)	2234
cluster II (cluster I + solvent shell)	5592
β-cluster (surface (111) β-cristobalite)	14450
complex (cluster II + β-cluster)	20150

A detailed analysis of the structure of this adsorption complex shows that the water molecules, which are arranged between the surface and the protein molecule, act as channels for electron density transferred either from the protein molecule to the surface or vice versa. There is also no loss of structure in the model with only one helical fragment, indirect evidence of the absence of a destructive effect of a silica surface on protein molecules.

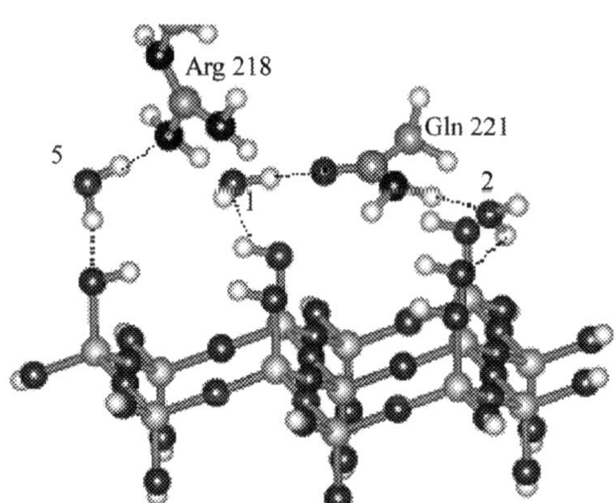

Figure 2. Equilibrium structure of α-helix of human serum albumin on silica surface.

3.3. INTERACTION OF A SILICA NANOPARTICLE WITH A LIPID FRAGMENT OF A CELL MEMBRANE

The use of nanosilica particles as drug carriers raises questions concerning the nature of forces controlling the adsorption interaction of cells with SiO_2 nanoparticles.[15] The average diameter of human erythrocytes is 7500 nm[16] while typical dimensions of silica nanoparticles with specific area of 200 to 300 m^2/g commonly used for water dispersions are from 20 to 10 nm, with dissipation of 4 to 17 nm (within a 90 per cent interval).[2,8]

Cell membranes preventing free access of dissolved molecules and nanoparticles to the endocellular medium consist of lipids, proteins, and carbohydrates.[17] The ratios between the principal components of cell membranes are different in plant and animal cells, but the lipid content can reach 80 per cent for erythrocytes.[15]

The lipid layer consists predominantly of molecules of phosphatidic acids, phosphatidylcholine, phosphatidylserine, phosphatidylethanolamine, and cholesterol.[18] As two-base phosphatidic acids are strong acids (pK_1 = 3.9 and pK_2 = 8.3), there is always a negative charge at the cell membrane surface, caused by dissociation of phosphatidic acids and the C-edges of membranous proteins. This charge is compensated largely by positively charged groups $[-N(CH_3)_3]^+$ of phosphatidylcholine (Figure 3).

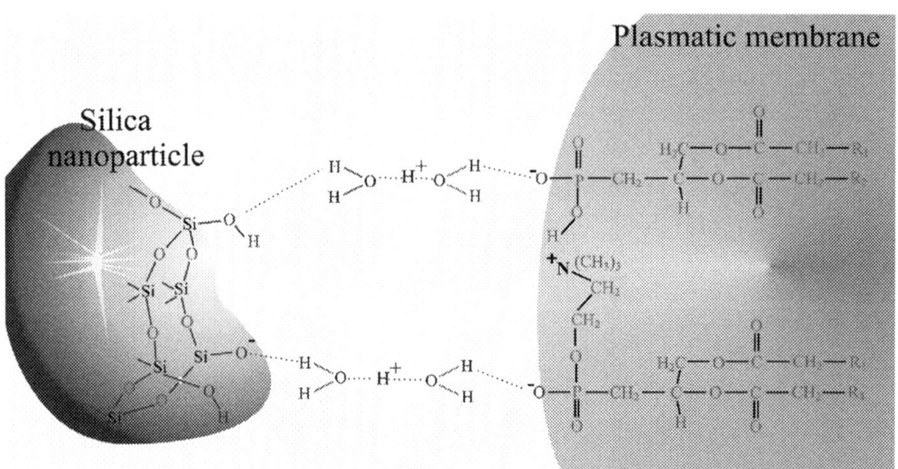

Figure 3. Interaction of silica surface with a cell membrane lipid fragment at distances of 5-7Å.

Despite the mosaic structure of the external surface of a cell membrane (an alternation of positively and negatively charged sites), the surface of an erythrocyte is charged negatively as a whole.[16] Silica nanoparticles are known to bear a negative charge due to dissociation of surface silanol groups.[8] However, experimental data testify a rather strong interaction between them; so that agglutination occurs on addition of highly disperse silica into an aqueous cellular suspension. Simple calculations within the framework of an electrostatic model fail to describe an attractive interaction between the negatively charged site of a SiO_2 nanoparticle and the positively charged group $-[-N(CH_3)_3]^+$ of a phosphatedylcholine molecule.

In a rough approximation, the SiO_2 nanoparticles/erythrocytes suspension may be related to a lyophobic system, but erythrocytes have a higher affinity to water than silica. This circumstance complicates applications of DFLO theory[17] to the system considered. In addition, use of this theory to explain the interactions within weakly-charged and highly disperse colloids, on the basis of long-range forces of an electrostatic nature, fails at distances where the overlap of the atomic orbitals of periphery atoms of the interacting objects is essential.

A simple model has been proposed to ascertain some chemical aspects of the contact of silica nanoparticle with membrane. To represent a cell membrane a cluster consisting of a complex of phosphatydic acid and phosphatidylcholine molecules was chosen.[15,16] To simplify calculations, the number of carbon atoms in fatty radicals was shortened to 4. The complex formed by phosphatydic acid and phosphatydilcholin molecules was also considered as negatively charged, due to dissociation of the phosphatydic acid molecule. The silica fragment was simulated with a cluster of six silicon-oxygen tetrahedral including 3 silanol groups. One silanol group proton in the beta-cluster was removed, so that the total charge was equal to –1.

The blood plasma and intercellular space contain relatively high concentrations of Na^+,Ca^{2+} Cl^-, K^+, Mg^{2+}, hydrogen ions[18] and cations from hydration such as $[H_2OH]^+$ and $[H_2OHOH_2]^+$. Calculation of the interaction energy in the model system was carried out for a joint cluster with two cations $[H_2O\cdots H\cdots OH_2]^+$ as binding agents.

The total charge of the system containing beta-cluster, two cations $[H_2OHOH_2]^+$, the deprotonated complex of phosphatydic acid and phosphatydilcholine is zero. The initial spatial structure placed the distances between the periphery atoms of the cluster and of $[H_2OHOH_2]^+$ particles, as well as those between cations and the deprotonated complex of phospholipids, was somewhat greater than at equilibrium to reconstruct a situation excluding an immediate contact between SiO_2 nanoparticles and the cell membrane.

The heat effects related to reactions due to interactions of silica nanoparticles with membrane phospholipids are shown in Table 4.

These data show that the complex formation between phosphatydic acid and phosphatydilcholine results in decreasing the proton affinity as compared with phosphatydic acid alone. This can be an indirect explanation of the lipid composition of cell membranes. Secondly, proton affinity values for silica and complex anions considerably exceed those for hydroxonium and cations $[H_2OHOH_2]^+$.

Table 4. The heat effects calculated for the reactions occurring due to an interaction of phospholipids with a silica nanoparticle (PM3 method).

№	Reaction	ΔH (kJ/mol)
1.	$SiO_2 \rightarrow SiO_2^- + H^+$	1624.2
2.	$PHA \rightarrow [PHA]^- + H^+$	1515.5
3.	$PHA \cdot PHCH \rightarrow [PHA \cdot PHCH]^- + H^+$	1252.3
4.	$SiO_2^- + H_3O^+ \rightarrow SiO_2 + H_2O$	–992.0
5.	$H_2O + H^+ \rightarrow H_3O^+$	–590.4
6.	$H_2O + H_2O \rightarrow (H_2O)_2$	–14.2
7.	$H_3O^+ + H_2O \rightarrow [H_2OHOH_2]^+$	–80.8
8.	$(H_2O)_2 + H^+ \rightarrow [H_2OHOH_2]^+$	–656.9
9.	$SiO_2^- + 2[H_2OHOH_2]^+ + [PHA \cdot PHCH]^-$	
	$\rightarrow SiO_2 \cdots 4(H_2O) \cdots PHA \cdot PHCH$	–151.5

PHA – phosphatidic acid
PHCH– phosphatidylcholin

It is surmised that the following chemical events on the interaction SiO_2 nanoparticles and erythrocytes are plausible. At large distances between them, both silica nanoparticles and an erythrocyte surface bear negative charges as a result of proton dissociation into solution. These protons are hydrated, and the derived cations are placed between negatively charged clusters, promoting their interaction. When the distance between objects reaches some critical value, the anions of SiO_2 and the lipid complex saturate the localization of negative charges with protons from H_3O^+ and $[H_2OHOH_2]^+$. The derived neutral components of the disperse system interact, water molecules being involved which are intermediates of a grid of hydrogen bonds (Figure 4).

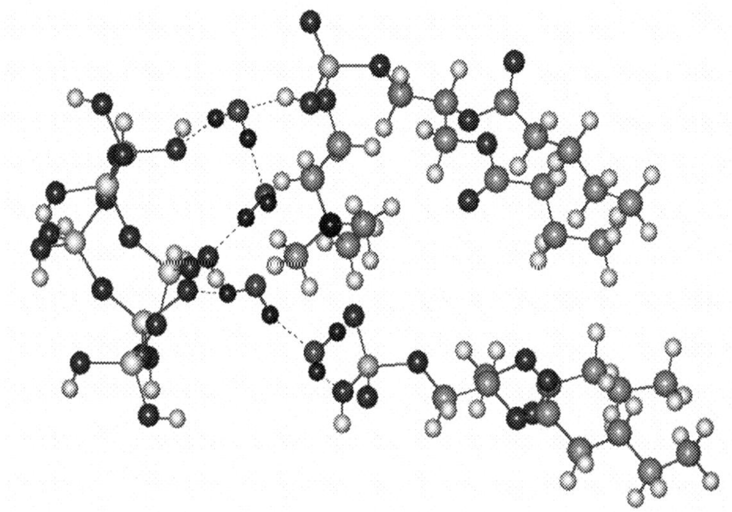

Figure 4. Equilibrium spatial structure of the complex where water molecule provide binding between a fragment of SiO_2 surface and lipid area of cell membrane.

4. Conclusion

The adsorption of sulfur-containing aminoacids from the gaseous phase occurs in the molecular form, while sorption from solution is more favorable in zwitterion forms saving the hydration shell.

The results obtained for the helical section of the human serum albumin model testify to the stability of its conformation on the adsorption on silica surface.

The interaction of silica nanoparticles with a lipid fragment of a cell membrane at distances of 5 to 7 Å is mediated with charged electrolyte particles, whereas a net with hydrogen bonds is formed close to the surface where the orbital overlap of edge atoms is essential.

Acknowledgments

The authors would like to thank Professor V. M. Gun'ko for valuable discussions. We also thank Dr. O. O. Kazakova who made some calculations on adsorption complexes of sulfur-containing amino acids with silica surface within the SM5.42R solvation model.

References

1. R. K. Iler, *The Chemistry of Silica,* (Jon Wiley, New York, 1979).
2. *Medical chemistry and clinical applications of silica dioxide*, edited by A. A. Chuiko, Naukova Dumka (Kiev, 2003).
3. C. A. Haynes and W. Norde, Structures and stabilities of adsorbed proteins, *J. Colloid Interface Sci,* 169(2), 313-328 (1995).
4. V. V. Lobanov, Yu. I. Gorlov, A. A. Chuiko, V. M. Pinchuk, Yu. S. Sinekop, and Yu. I. Yakimenko, *Role of electrostatic interactions in adsorption on solid oxide surfaces,* (Vek+ Ltd, Kiev, 1999).
5. J. J. P. Stewart, MOPAC: A semiempirical molecular orbital program, *J. Computer-Aided Mol. Design.* 4, 1-105 (1990).
6. G. M. Zhidomirov, and V. B. Kazansky, Quantum-chemical cluster models of acid-base sites of oxide catalysts, *Adv. Cat.* 34, 131-201 (1986).
7. A. P. Legrand, *The Surface Properties of Silicas*, (John Wiley & Sons, Chichester, 1998).
8. *Surface Chemistry of Silica*, edited by A. A. Chuiko (Institute of Surface Chemistry, Kiev, 2001).
9. A. R. Ivanov, I. V. Nazimov, and L. Baratova, Determination of biologically-active low-molecular-mass thiols in human blood. I. Fast qualitative and quantitative, gradient and isocratic reversed-phase high-performance liquid-chromatography with photometric and fluorescence detection, *J. Chromatogr. A.* 895, 157-166 (2000).
10. A. R. Ivanov, I. V. Nazimov, and L. Baratova, Determination of biologically-active low-molecular-mass thiols in human blood. II. High-performance capillary electrophoresis with photometric detection, *J. Chromatogr. A* 895, 167-171 (2000).
11. T. Tamura, S. M. Bergman, and S. L. Morgan, Hyperhomocysteinemia as a cause of vascular occlusion in end-stage-renal disease, *Int. J. Artificial Organs.* 21, 72-74 (1998).
12. A. Terney, *Modern organic chemistry* (Mir, Moscow, 1981).
13. V. M. Gun'ko, A. V. Klyueva, Yu. N. Levchuk, and R. Leboda, Proton correlation spectroscopy investigations of proteins, *Adv. Colloid Interface Sci.* 105, 201-328 (2003).
14. A. Kondo and H. Fukuda, Effects of adsorption conditions on kinetics of protein adsorption and conformational changes at ultrafine silica particles, *J. Colloid Interface Sci.* 198, 34-41 (1998).
15. A. A. Boldyrev, *Introduction to Membrane Biochemistry*, (Vysshaya Shkola, Moscow, 1980).
16. R. Gennis, *Biomembranes* (Splinger-Verlag, New York, 1989).
17. B. V. Deryagin, N. V. Churaev, and V. M. Muller, *Surface Forces* (Nauka, Moscow, 1985).
18. J. Koolman and K-H. Röhm, *Taschenatlas der Biochemie* (Georg Thieme Verlag Stuttgart, New-York, 1998).

SYNTHESIS AND CHARACTERIZATION OF ORDERED MESOPOROUS MATERIALS FOR REMOVAL OF HEAVY METAL IONS

MIETEK JARONIEC* AND OKSANA OLKHOVYK
Department of Chemistry, Kent State University, Kent, Ohio 44242, USA

Abstract. Synthesis and characterization of novel multifunctional adsorbents for mercury ions from aqueous solutions, based on ordered mesoporous materials (OMMs), is reviewed. Ordered mesoporous silicas (OMSs) with chemically bonded surface ligands and periodic mesoporous organosilicas (PMOs) with incorporated bridging groups are compared with respect to their adsorption properties and affinity towards heavy metal ions. Various types of OMSs possessing a hexagonal or cubic arrangement of ordered mesopores, grafted with multifunctional 1-benzoyl-3-propylthiourea and 2,5-dimercapto-1,3,4-thiadiazole ligands, are compared to PMOs containing specific bridging groups incorporated directly into the silica framework. PMOs having a single type of organic functionality (e.g., tris[3-(trimethoxysilyl)propyl]isocyanurate) in the framework, or multiple framework and surface functionality (e.g., tris[3-(trimethoxysilyl) propyl]isocyanurate and 3-mercaptopropyl, respectively) are discussed. Nitrogen adsorption was a main method used to monitor changes in the porous structures of OMMs upon surface and framework modification.

Keywords: heavy metal ions; MCM-41; MCM-48; mercury adsorption; mesoporous organosilicas; ordered mesoporous silicas; SBA-15

1. Introduction

Ecosystems that are severely disrupted by invasion of human activity attract a lot of attention among environmentalists. Pollution of aqueous systems by heavy metal ions is considered a major environmental problem that requires immediate solution. Heavy metal ions such as mercury, cadmium, and lead are

*To whom correspondence should be addressed. Mietek Jaroniec, Chemistry Department, Kent State University, Kent, Ohio 44242, USA; e-mail: jaroniec@kent.edu

J.P. Blitz and V.M. Gun'ko (eds.), Surface Chemistry in Biomedical and Environmental Science, 325–336.

extremely harmful to living cells. Hazards of these environmental pollutants are known and well documented.[1] High affinity of heavy metals to thiol groups is a primary cause of cell metabolism malfunction, and constitutes the high risk of intoxication to living organisms. Thus reduction or elimination of these persistent pollutants in ecosystems, and their strict monitoring, are important issues established by EPA to protect these systems from degradation.

Porous materials with chemically modified surfaces have been extensively studied as adsorbents for heavy metal ions from water (see the review by Jal et al.[2] and references therein). There is a continuously growing demand for adsorbents which are non-swelling, thermally and hydrothermally stable, exhibit large adsorption capacity, fast kinetics and high affinity towards heavy metal ions. Discovery of self-assembled ordered mesoporous silicas (OMSs)[3] opened enormous opportunities for the design and synthesis of highly selective and efficient adsorbents for heavy metal ions.

A simple method to create a self-assembled monolayer of thiol groups on the mesopore walls of OMSs afforded materials with large surface area, insoluble and stable framework, tunable pore size and relatively high adsorption affinity towards heavy metal ions.[4,5] Covalently bonded thiol monolayers are easy to graft because mesoporous hosts possess high amount of silanols. Up to 75 percent of total surface silanols can be substituted by thiol groups. Mesosized and well-defined porosity assures fast diffusion and mass transfer of ions. The aforementioned advantages of these OMSs along with highly specific functionality were recognized in the synthesis of mercury adsorbents.[6] Up to 0.6g of Hg^{2+} per gram of thiol functionalized OMS was reported.[7] The MCM-41 and MCM-48 silicas exhibiting a hexagonal or cubic arrangement of mesopores, as well as SBA-15 with hexagonally arranged large mesopores interconnected by small micropores, were extensively examined to design highly effective adsorbents for heavy metal ions.[8,9] Also, chemically modified mesoporous solids, used in sensing devices, were reported to be useful for electrochemical monitoring of mercury ions.[10-13]

The performance of adsorbents for heavy metal ions depends strongly on the type of attached functional groups. Since silanol concentration and surface area limit ligand coverage in the resulting adsorbents, the idea of introducing multifunctional ligands was shown to be beneficial for increasing their adsorption capacity. The soft acid-soft base theory was elegantly employed in chemical modification of OMS to mimic naturally occurring elements in living organisms, which are targeted by heavy metal ions during the intoxication process. This approach was utilized in chemical modification of OMSs to design adsorbents with high affinity multifunctional ligands towards heavy metal ions. Such ligands not only selectively preconcentrate mercury ions from aqueous media, but also their interactions are week enough to facilitate regeneration of adsorbents. 1-Allyl-3-propylthiourea,[14] 1-benzoyl-3-propylthiourea,[15] and 2,5-dimercapto-

1,3,4-thiadiazole ligands[16] grafted onto mesopores of MCM-41, MCM-48 and SBA-15 showed unprecedented mercury loading capacities (up to 6 mmol Hg^{2+} per gram of the material), high distribution coefficients (~10^9) and high degree of regeneration (up to 70%). Weaker interactions with heavy metal ions allowed for mild regeneration of these adsorbents using acidic thiourea solutions instead of concentrated hydrochloric acid regeneration employed previously for adsorbents with thiol functionality.[17,18]

A desire to design high capacity adsorbents resulted in the synthesis of hybrid materials with multifunctional groups introduced via direct co-condensation of proper organosilanes instead of less effective post-synthesis silanization.[19,20] One-pot synthesis through self-assembly mechanism allows the creation of active sites inside mesopores, that after template removal become accessible for adsorption.[21] An alternative method for chemical grafting of OMSs, known as template displacement synthesis, permits a simultaneous template removal and attachment of functional groups.[22]

Enormous opportunities in the design of novel materials with desired surface and structural properties appeared recently due to the discovery of periodic mesoporous organosilicas (PMOs).[23-25] This type of material possesses advantageous characteristics because of the possibility to incorporate various bridging groups directly into the framework. Cyclam units,[26,27] crown ethers,[28] and highly selective thioether groups[29,30] were successfully used as bridging elements to design high capacity adsorbents for heavy metal ions. Large pore size, high surface area and relatively high concentration of incorporated organics, homogeneously distributed within the PMO framework, favor high performance of these materials as adsorbents for mercury ions.

Recently, PMO with a large heterocyclic molecular ring of isocyanurate was studied.[31] A substantial incorporation (*ca.* 28%) of isocyanurate ring into the PMO afforded materials with an ordered mesopore network of hexagonal P6mm crystallographic symmetry. Large pore size (up to 8 nm) was achievable due to the use of a polymeric template as a structure directing agent. A successful attempt of the synthesis of PMO with binary functionality resulted in obtaining materials with large adsorption capacity and high affinity towards mercury ions.[32] Significant progress has been made in the synthesis and characterization of ordered mesoporous hybrid adsorbents for heavy metal ions, and their environmental applications in adsorption and chemical sensing.[33-40] Especially, smart nanostructures that have the ability to detect specific heavy metal ions are of great importance for environmental analysis. The aim of this article is two-fold: first of all, an extensive comparison of synthesis methods will be given with special emphasis put on the design of multifunctional adsorbents for heavy metal ions adsorption. Secondly, nitrogen adsorption characterization of ordered mesoporous materials will be explored as an effective tool for monitoring the synthesis of adsorbents for heavy metal ions.

2. Synthesis and Characterization of OMMs with Mercury-Specific Ligands

2.1. POST-SYNTHESIS MODIFICATION OF OMSs

A major task in the synthesis of ordered mesoporous materials for adsorption of heavy metal ions is to introduce a large amount of surface groups that exhibit high affinity towards these ions. The well-known surfactant-templated MCM-41 and MCM-48 materials possess relatively small pore size (~3-6 nm). This factor limits significantly the choice of ligands for surface modification, after which sufficient pore space should remain for adsorption. The state of self-assembled monolayers of functional ligands within mesopores, and changes in the porous structure were studied by various methods.[7,41,42] Except for the simple thiol surface ligand, the attachment of more complex ligands requires the initial introduction of reactive amino- or chloro-groups on the silica surface, which would subsequently react with other molecules to form surface ligands of desired functionality.[15] Conventional modification involves immersion of the material in silane with addition of solvent and/or pyridine, and subsequent reaction at elevated temperature under reflux conditions.[43] After introduction of reactive groups another modification can be carried out as illustrated in Figure 1.

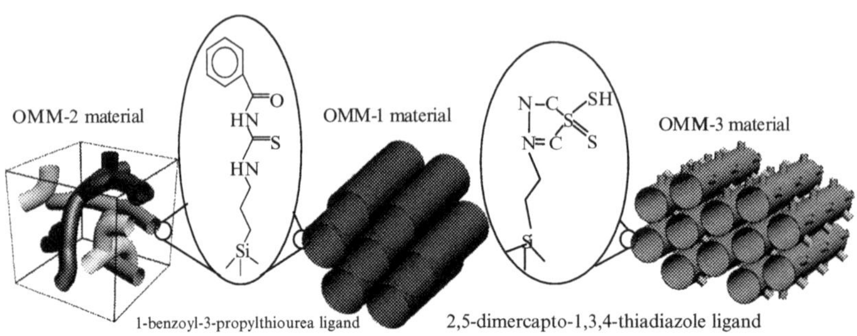

Figure 1. Schematic representation of OMSs with modified by multifunctional ligands.

The calcined OMSs used in post-synthesis modification possessed a hexagonal (MCM-41 and SBA-15) or cubic (MCM-48) arrangement of mesopores. Benzoyl isothiocyanate was reacted with aminopropyl-modified surfaces of MCM-41 and MCM-48 to give the OMM-1 and OMM-2 samples, respectively.[15,44]

The polymer-templated OMSs such as SBA-15[45] possess pores about 8-12 nm, which are more spacious and can be decorated with more complex ligands. The SBA-15 sample studied exhibited ordered mesopores about 10 nm, and irregular micropores in the mesopore walls as shown in Figure 1. Initially chloropropyl-SBA-15 material was obtained and subsequently grafted with 2,5-dimercapto-1,3,4-thiadiazole ligand[16] (OMM-3 material in Figure 1). These studies showed that the surface concentration of silanol groups and pore architecture are important factors in achieving high ligand coverage.

Table 1. Adsorption parameters for the OMMs studied with mercury-specific functionalities.

Sample	BET* specific surface area (m^2/g)	Total pore volume* (cm^3/g)	Pore width* (nm)	Mercury adsorption capacity (g/g)	C_{lig} (mmol/g)**	Reference to data
OMM-1	380	0.45	3.0	1.00	1.5	15
OMM-2	505	0.47	2.8	1.34	1.6	44
OMM-3	556	0.50	4.9	1.70	2.7	16
PMO-1	622	0.81	7.9	0.41	0.8	31
PMO-2	515	0.54	7.3	1.77	1.4	31
PMO-3	496	0.62	6.3	1.13	0.7 (0.9)	32
PMO-4	514	0.54	7.3	0.85	1.0 (0.6)	32

* Nitrogen adsorption isotherms for the OMMs studied were recorded at 77K using a Micromeritics ASAP 2010 adsorption analyzer. All samples prior to adsorption analysis were degassed at 120°C for 2h under vacuum. The BET specific surface area was calculated from the adsorption data in the range of the relative pressure from 0.04 to 0.2 according to the BET method.[46] The total pore volume was estimated from the amount adsorbed taken at the relative pressure about 0.99.[47] The pore width was estimated at the pore size distribution maximum obtained by the KJS method.[48]

**Quantitative estimation of the ligand coverage was done by CHNS analysis. The content of carbon, nitrogen and sulfur in all samples was determined using a L.ECO elemental analyzer (Model CHNS-932) from St. Joseph, MI. In the case of PMO-3 and PMO-4 the values in brackets denote concentrations of mercaptopropyl groups, while the remaining values refer to the concentration of bridging groups.

Changes in the porous structure upon surface modification can be monitored and evaluated by gas adsorption. A reduction in the pore size upon silanization indicates a successful attachment of ligands onto the mesopore surface. Differences in chemical modification methods and the type of mesoporous support affect the ligand coverage and consequently adsorption properties of the resulting material. Nitrogen adsorption was used to investigate the influence of the aforementioned factors on the surface properties of OMMs with multi-functional 1-benzoyl-3-propylthiourea and 2,5-dimercapto-1,3,4-thiadiazole ligands. The template displacement synthesis was found to be an effective procedure for obtaining adsorbents of desired characteristics, e.g., OMM-3 contained 2.7 mmoles of 2,5-dimercapto-1,3,4-thiadiazole per gram of material, accessible pore size of 5 nm (after modification) and good mercury adsorption capacity (1.7 g of mercury per gram of adsorbent). However, in the case of 1-benzoyl-3-propylthiourea-modified MCM-41 and MCM-48 materials our study suggests that the bi-continuous 3D pore structure of MCM-48 provides better access for adsorbate species, and therefore, larger adsorption capacity was obtained in comparison to the MCM-41-type material with a 2D pore network (Table 1). The distribution coefficient (indicator of the ligand affinity towards heavy metal ion) exceeded the value of 10^9.

Nitrogen adsorption analysis revealed a great difference in adsorption behavior of OMSs upon introduction of various mercury-specific ligands.

Figure 2 shows a comparison of nitrogen adsorption isotherms at 77 K for the organic-inorganic hybrid materials developed for adsorption of mercury ions. The shape of the nitrogen adsorption isotherms indicates that the mesostructure was preserved after chemical modification of the surface. Evaluation of the shift of the capillary condensation step and reduction of its height provide information about formation of the chemically bonded layer inside mesopores.

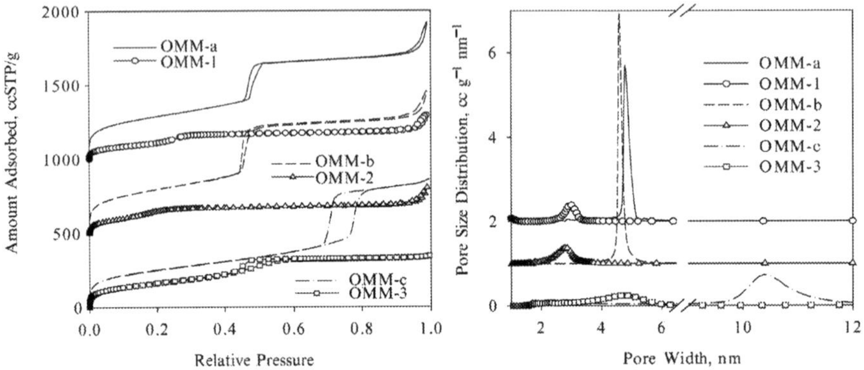

Figure 2. Nitrogen adsorption isotherms at 77K for the OMM samples with multifunctional mercury-specific ligands: 1-benzoyl-3-propylthiourea (OMM-1 and OMM-2), 2,5-dimercapto-1,3,4-thiadiazole (OMM-3) and the corresponding differential pore size distributions. OMM-a, OMM-b and OMM-c refer to the calcined MCM-41, MCM-48 and SBA-15 samples, respectively. For clarity the isotherms are offset vertically by 1000 ccSTP/g for OMM-a and OMM-1 samples and by 500 ccSTP/g for OMM-b and OMM-2 samples. The PSD plots are offset vertically by 2 $ccg^{-1}nm^{-1}$ for OMM-a and OMM-1 and by 1ccg^{-1} nm^{-1} for OMM-b and OMM-2 samples.

After attachment of surface ligands the pore size decreased as expected. In addition to small-angle X-ray scattering and transmission electron microscopy, which are used to obtain information about structural ordering, nitrogen adsorption provides important information about solids under study such as BET surface area, pore size, pore volume, microporosity and surface heterogeneity. Pore size analysis for OMMs is especially crucial and it was performed by the KJS method,[48] which was elaborated especially for ordered mesostructures. Table 1 presents adsorption parameters for the materials under study and provides details about calculations.

2.2. ONE-POT SYNTHESIS OF MERCURY-SPECIFIC PMO

The co-condensation of two or more appropriate organosilanes allows one to tailor surface and framework properties of OMMs. A controlled hydrolysis of bis(trialkoxysilyl) organics of $(R'O)_3SiRSi(OR')_3$ structure, where R' denotes a hydrolysable group and R denotes a bridging organic group permits the incorporation of an organic functionality into a silica framework without diminishing the pore space. Simple aliphatic or aromatic bridging groups have been introduced into PMOs.[23-25] There are almost unlimited possibilities for incorporation of various bridging groups into a silica framework, bi- or multi-functional PMOs have already been reported.[49-53]

In PMOs with incorporated metal-chelating bridging groups the pore accessibility and ligand functionality determine their adsorption capacity. Often the addition of tetraethoxyorthosilicate (TEOS) is required in the co-condensation synthesis to introduce a more complex functionality. This approach assures the periodicity of the mesoporous network, which could undergo significant deterioration if high amount of organic bridging units is introduced into PMO. For instance, Zhang et al.[29] reported introduction of about 15 percent of thioether into PMO.

Recently we reported PMO materials with tris[3-(trimethoxysilyl)propyl] isocyanurate bridging groups[31] possessing highly ordered mesopores of the size about 10 nm, BET surface area of *ca.* $600 m^2/g$, and high thermal stability (see PMO-1 in Figure 3). It was possible to tailor the structure of these materials, from ordered hexagonal to disordered worm-like, by varying the percentage of tris[3-(trimethoxysilyl)propyl]isocyanurate (ICS) in relation to TEOS used in the co-condensation synthesis in the presence of triblock-copolymer (Pluronic 123). PMO-1 and PMO-2 materials differ in the percentage of the ICS moiety incorporated, which is higher for PMO-2. Up to 28% of heterocyclic organics was possible to introduce and retain honeycomb structure. Addition of another functionality into PMO materials with isocyanurate bridging groups, e.g., mercaptropropyl ligands, afforded bi-functional PMO adsorbents with ICS and mercaptopropyl moieties (PMO-3 and PMO-4 samples in Figure 3, where the ratio of ICS to mercaptopropyl was higher in the PMO-4 material). The latter PMOs possess a high mercury adsorption capacity of 4-6 mmoles of Hg^{2+} per gram of material (Table 1). PMOs bearing functional groups within the pore framework possess a great advantage over surface-modified OMSs as their pore size is determined by the template used, and not significantly affected by introduction of organic functionality. A successful incorporation of bridging groups into PMOs can be monitored by elemental analysis, IR spectroscopy, NMR, thermogravimetric analysis as well as adsorption. The co-condensation synthesis assures a homogeneous distribution of bridging groups within the framework that is usually considered a major advantage of PMOs versus surface modified OMSs.

Figure 4 illustrates the behavior of PMOs upon incorporation of various amounts of isocyanurate groups into the material's framework. For higher amounts of the ICS used, the capillary condensation step shifted in the direction

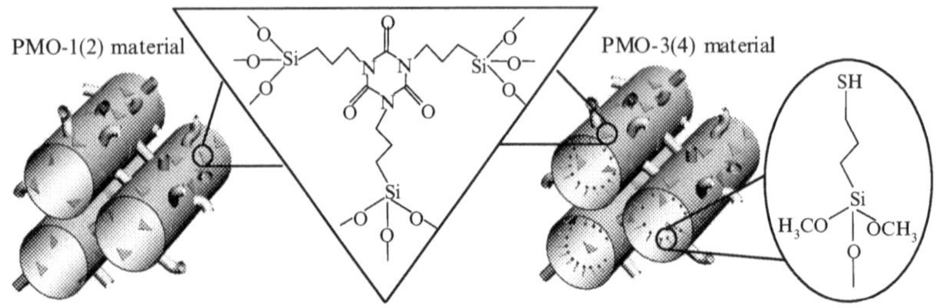

Figure 3. Schematic representation of PMOs with multifunctional organic bridging groups and bifunctional PMOs with multifunctional bridging groups and mercaptopropyl surface ligands.

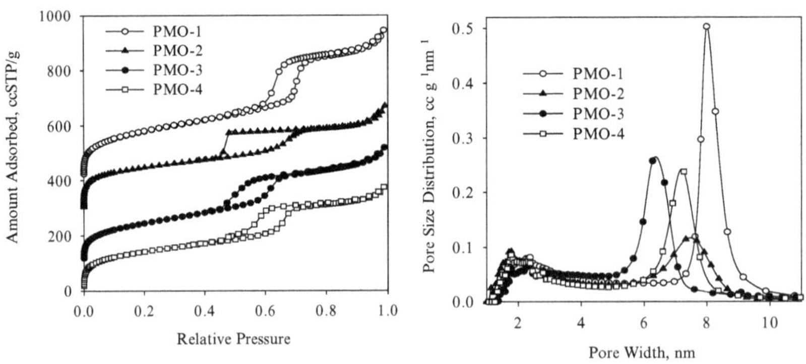

Figure 4. Nitrogen adsorption isotherms recorded at 77K for the PMO samples with multifunctional groups and the corresponding pore size distributions. PMO-1 and PMO-2 contains tris[3-(trimethoxysilyl)propyl]isocyanurate bridging groups only, while PMO-3 and PMO-4 contains tris[3-(trimethoxysilyl)propyl]isocyanurate bridging groups and mercaptopropyl surface ligands. For clarity the isotherms are offset vertically by 400 ccSTP/g for PMO-1, 300 ccSTP/g for PMO-2 and by 100 ccSTP/g for PMO-3.

of lower relative pressures, and the height of the capillary condensation step decreased significantly. This indicates that the structure deteriorates with increasing ICS concentration. The BET surface area and the total pore volume of the materials studied were reduced significantly, while the pore size remained in the range of 8-10 nm. For moderate amount of incorporated ICS (PMO-1) the hexagonal arrangement of mesopores was retained. For high amount of ICS the structure was worm-like (PMO-2). However, higher mercury adsorption capacity was observed for PMO-2 because it contained a high concentration of ICS groups; the adsorption capacity being about 1.8 g of Hg^{2+} per gram of material (Table 1).

For bi-functional PMOs with isocyanurate bridging groups and thiol surface functionality, an addition of the thiol group caused a change in the symmetry group from hexagonal to cubic (Figure 5). Three reflections indexed as (110), (220), and (321) were found on the SAXS patterns (Figure 5) for bi-functional

PMO materials with ICS bridging groups and mercaptopropyl surface ligands indicating of a $I4_132$ group.[32]

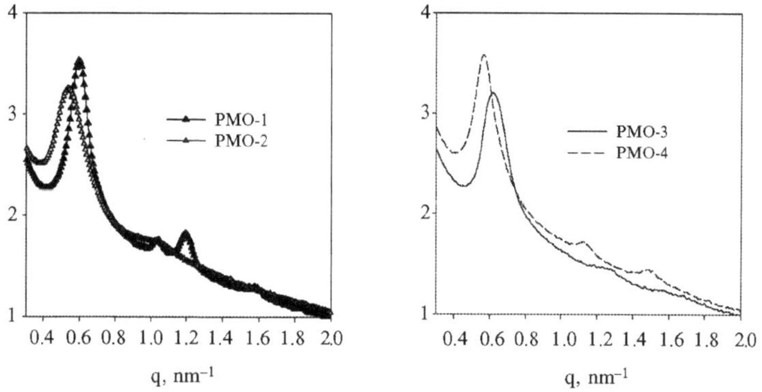

Figure 5. Small angle X-ray scattering (SAXS) patterns for PMOs with tris[3-(trimethoxysilyl)propyl]isocyanurate bridging groups only (PMO-1 and PMO-2) and for PMOs that contain tris[3-(trimethoxysilyl)propyl]isocyanurate bridging groups and mercaptopropyl surface ligands (PMO-3 and PMO-4). The SAXS measurements were conducted using the NanoSTAR system (Bruker AXS) with pinhole collimation and a two-dimensional detector (HiSTAR) mounted on a micro focus X-ray tube with copper anode and equipped with crossed Göbel mirrors. The sample-to-detector distance was 650 mm. The exposure time for a single frame was, 5000 – 10000s. All measurements were performed at room temperature.

PMOs with two functional groups capable of attracting mercury ions showed good adsorption capacity of 1.2 g of Hg^{2+} per gram of material. Nitrogen adsorption isotherms (Figure 4) for bi-functional PMOs feature well-defined adsorption–desorption hysteresis loops characteristic for SBA-15-type materials, and quite steep capillary condensation steps at a relative pressure of about 0.6. It is evident that the capillary condensation step is not as steep as for the PMO-1 material with a similar degree of incorporation of ICS; it is much lower and occurs at lower values of relative pressure because of expected structural changes after introduction of the second organic into PMO.

3. Conclusion

Nitrogen adsorption is a useful tool for monitoring synthesis of ordered mesoporous materials with organic functionalities. Attachment of multi-functional organic ligands to the mesopore walls is reflected by a shift of the capillary condensation step on the adsorption isotherm in the direction of low pressures, which indicates the reduction of the pore size upon surface modification. However, no significant shift of this step was observed on the isotherms for PMOs with bridging groups.

It is shown that OMSs and PMOs with multifunctional organic ligands possess a high capacity towards mercury ions, which is due to the accessible pore network and high concentration of multifunctional ligands. Especially

PMOs with proper multifunctional bridging and surface groups seem to be very promising adsorbents for heavy metal ions from water.

Acknowledgment

The authors acknowledge support by NSF Grants CTS-0086512 and CHE-0093707, and thank Professor S. Pikus from M. Curie-Skłodowska University (Poland) for providing SAXS data.

References

1. L. Jaerup, Hazards of heavy metal contamination, *British Med. Bull.* 68, 167-182 (2003).
2. P. K. Jal, S. Patel, and B. K. Mishra, Chemical modification of silica surface by immobilization of functional groups for extractive concentration of metal ions, *Talanta* 62, 1005-1028 (2004).
3. J. S. Beck, J. C. Vartuli, W. J. Roth, M. E. Leonowicz, C. T. Kresge, K. D. Schmitt, C. T.-W. Chu, D. H. Olson, E. W. Sheppard, S. B. McCullen, J. B. Higgins, and J. L. Schlenker, A new family of mesoporous molecular sieves prepared with liquid crystal templates, *J. Am. Chem. Soc.* 114, 10834-10843 (1992).
4. X. Feng, G. E. Fryxell, L. Q. Wang, A. Y. Kim, J. Liu, and K. Kemner, Functionalized monolayers on ordered mesoporous supports, *Science* 276, 923-930 (1997).
5. L. Mercier and T. J. Pinnavaia, Access in mesoporous materials: Advantages of a uniform pore structure in the design of a heavy metal ion adsorbent for environmental remediation, *Adv. Mater.* 9, 500-503 (1997).
6. J. Brown, L. Mercier, and T. J. Pinnavaia, Selective adsorption of Hg^{2+} by thiol-functionalized nanoporous silica, *Chem. Commun.* 69-70 (1999).
7. J. Liu, X. Feng, G. E. Fryxell, L. Q. Wang, A. Y. Kim, and M. Gong, Hybrid mesoporous materials with functionalized monolayers, *Adv. Mater.* 10, 1-5 (1998).
8. K. Moller and T. Bein, Inclusion chemistry in periodic mesoporous hosts, *Chem. Mater.* 10, 2950-2963 (1998).
9. A. M. Liu, K. Hidajat, S. Kawi, and D. Y. Zhao, A new class of hybrid mesoporous materials with functionalized organic monolayers for selective adsorption of heavy metal ions, *Chem. Commun.* 1145-1146 (2000).
10. A. Walcarius and J. Bessiere, Electrochemistry with mesoporous silica, selective mercury(II) binding, *Chem. Mater.* 11, 3009-3011 (1999).
11. A. M. Bond, W. Miao, T. Smith, and J. Jamis, Voltammetric reduction of mercury(II), silver(I), lead(II) and copper(II) ions adsorbed onto a new form of mesoporous silica, *Analytica Chimica Acta* 396, 203-213 (1999).
12. W. Yantasee, Y. Lin, X. Li, G. E. Fryxell, T. S. Zemanian, and V. V. Viswanathan, Nanoengineered electrochemical sensor based on mesoporous silica thin-film functionalized with thiol-terminated monolayer, *Analyst* 128, 899-904 (2003).
13. W. Yantasee, Y. Lin, G. E. Fryxell, and B. J. Busche, Simultaneous detection of cadmium, copper, and lead using a carbon paste electrode modified with carbamoylphosphonic acid self-assembled monolayer on mesoporous silica (SAMMS), *Analytica Chimica Acta* 502, 207-212 (2004).
14. V. Antochshuk and M. Jaroniec, 1-Allyl-3-propylthiourea modified mesoporous silica for mercury removal, *Chem. Commun.* 258-259 (2002).
15. V. Antochshuk, O. Olkhovyk, M. Jaroniec, I.-S. Park, and R. Ryoo, Benzoylthiourea-modified mesoporous silica for mercury(II) removal, *Langmuir* 19, 3031-3034 (2003).

16. O. Olkhovyk and M. Jaroniec, Ordered mesoporous silicas with 2,5-dimercapto-1,3, 4-thiadiazole ligand, high capacity adsorbents for mercury ions, *Adsorption* (2005) in press.
17. G. E. Fryxell and J. Liu, Designing surface chemistry in mesoporous silica, in: *Adsorption on Silica Surfaces*; edited by E. Papirer (Marcel Dekker, New York, 2000), pp. 665-687.
18. R. I. Nooney, M. Kalyanaraman, G. Kennedy, and E. J. Maginn, Heavy metal remediation using functionalized mesoporous silicas with controlled macrostructure, *Langmuir* 17, 528-533 (2001).
19. J. Brown, R. Richer, and L. Mercier, One-step synthesis of high capacity mesoporous Hg^{2+} adsorbents by non-ionic surfactant assembly, *Microporous Mesoporous Mater.* 37, 41-48 (2000).
20. B. Lee, Y. Kim, H. Lee, and J. Yi, Synthesis of functionalized porous silicas via templating method as heavy metal ion adsorbents, the introduction of surface hydrophilicity onto the surface of adsorbents, *Microporous Mesoporous Mater.* 50, 77-90 (2001).
21. A. Bibby and L. Mercier, Mercury(II) ion adsorption behavior in thiol-functionalized mesoporous silica microspheres, *Chem. Mater.* 14, 1591-1597 (2002).
22. V. Antochshuk and M. Jaroniec, Simultaneous modification of mesopores and extraction of template molecules from MCM-41 with trialkylchlorosilanes, *Chem. Commun.* 2373-2374 (1999).
23. S. Inagaki, S. Guan, Y. Fukushima, T. Ohsuna, and O. Terasaki, Novel mesoporous materials with a uniform distribution of organic groups and inorganic oxide in their framework, *J. Am. Chem. Soc.* 121, 9611-9614 (1999).
24. B. J. Melde, B. T. Holland, C. F. Blanford, and A. Stein, Mesoporous sieves with unified hybrid inorganic/organic frameworks, *Chem. Mater.* 11, 3302-3308 (1999).
25. C. Yoshina-Ishii, T. Asefa, N. Coombs, M. J. MacLachlan, and G. A. Ozin, Periodic mesoporous organosilicas, POMs, fusion of organic and inorganic chemistry 'inside' the channel walls of hexagonal mesoporous silica, *Chem. Commun.* 2539-2540 (1999).
26. G. Dubois, R. J. P. Corriu, C. Reye, S. Brandes, F. Denat, and R. Guilard, First organic-inorganic hybrid materials with controlled porosity incorporating cyclam units, *Chem. Commun.* 2283-2284 (1999).
27. G. Dubois, C. Reye, R. J. P. Corriu, S. Brandes, F. Denat, and R. Guilard, Coordination chemistry in the solid, study of the incorporation of CuII into cyclam-containing hybrid materials, *Angew. Chem., In. Ed.* 40, 1087-1090 (2001).
28. C. Chuit, R. J. P. Corriu, G. Dubois, and C. Reye, Hybrid organic-inorganic materials. Preparation and properties of dibenzo-18-crown-6 ether-bridged polysilsesquioxanes, *Chem. Commun.* 723-724 (1999).
29. L. Zhang, W. Zhang, J. Shi, Z. Hua, Y. Li, and J. Yan, A new thioether functionalized organic-inorganic mesoporous composite as a highly selective and capacious Hg^{2+} adsorbent, *Chem. Commun.* 210-211 (2003).
30. J. Liu, J. Yang, Q. Yang, G. Wang, and Y. Li, Hydrothermally stable thioether-bridged mesoporous materials with void defects in the pore walls, *Adv. Funct. Mater.* 15(8), 1297-1302 (2005).
31. O. Olkhovyk and M. Jaroniec, Periodic mesoporous organosilica with large heterocyclic bridging groups, *J. Am. Chem. Soc.* 127, 60-61 (2005).
32. O. Olkhovyk, S. Pikus, and M. Jaroniec, Bifunctional periodic mesoporous organosilica with large heterocyclic bridging groups and mercaptopropyl ligands, *J. Mater. Chem.* 15, 517-1519 (2005).
33. Y.-K. Lu and X.-P. Yan, An imprinted organic-inorganic hybrid sorbent for selective separation of cadmium from aqueous solution, *Anal. Chem.* 76, 453-457 (2004).
34. T. Kang, Y. Park, and J. Yi, Highly selective adsorption of Pt^{2+} and Pd^{2+} using thiol-functionalized mesoporous silica, *Ind. Eng. Chem. Res.* 43, 1478-1484 (2004).
35. T. Kang, Y. Park, K. Choi, J. S. Lee, and J. Yi, Ordered mesoporous silica (SBA-15) derivatized with imidazole-containing functionalities as a selective adsorbent of precious metal ions, *J. Mater. Chem.* 14, 1043-1049 (2004).

36. A. Sayari, S. Hamoudi, and Y. Yang, Applications of pore-expanded mesoporous silica. 1. Removal of heavy metal cations and organic pollutants from wastewater, *Chem. Mater.* 17, 212-216 (2005).
37. L. C. Cides da Silva, G. Abate, N. Andrea, M. C. A. Fantini, J. C. Masini, L. P. Mercuri, O. Olkhovyk, M. Jaroniec, and J. R. Matos, Microwave synthesis of FDU-1 silica with incorporated humic acid and its application for adsorption of Cd^{2+} from aqueous solutions, *Stud. Surf. Sci. Catal.* 156, 941-950 (2005).
38. K. H. Nam and L. L. Tavlarides, Synthesis of a high-density phosphonic acid functional mesoporous adsorbent, application to Chromium(III) removal, *Chem. Mater.* 17, 1597-1604 (2005).
39. T. Balaji, M. Sasidharan, and H. Matsunaga, Optical sensor for the visual detection of mercury using mesoporous silica anchoring porphyrin moiety, *Analyst* 130, 1162-1167 (2005).
40. R. Metivier, I. Leray, B. Lebeau, and B. Valeur, A mesoporous silica functionalized by a covalently bound calixarene-based fluoroionophore for selective optical sensing of mercury(ii) in water, *J. Mater. Chem.* 15, 2965-2973 (2005).
41. A. Walcarius, M. Etienne, and B. Lebeau, Rate of access to the binding sites in organically modified silicates. 2. Ordered mesoporous silicas grafted with amine or thiol groups, *Chem. Mater.* 15, 2161-2173 (2003).
42. A. Stein, B. J. Melde, and R. C. Schroden, Hybrid inorganic-organic meso-porous silicates-nanoscopic reactors coming of age, *Advanced Materials* 12, 1403-1419 (2000).
43. C. P. Jaroniec, M. Kruk, and M. Jaroniec, Tailoring surface and structural properties of MCM-41 silicas by bonding organosilanes, *J. Phys. Chem. B* 102(28), 5503 –5510 (1998).
44. O. Olkhovyk, V. Antochshuk, and M. Jaroniec, Benzoylthiourea-modified MCM-48 mesoporous silica for mercury(II) adsorption from aqueous solutions, *Colloids Surf. A* 236, 69-72 (2004).
45. D. Zhao, Q. Huo, J. Feng, B. F. Chmelka, and G. D. Stucky, Nonionic triblock and star copolymer and oligomeric syntheses of highly ordered, hydrothermally stable, mesoporous silica structures, *J. Am. Chem. Soc.* 120, 6024-6036 (1998).
46. S. Brunauer, P. H. Emmett, and E. Teller, Adsorption of gases in multimolecular layers, *J. Am. Chem. Soc.* 60, 309-319 (1938).
47. S. J. Gregg and K. S. W. Sing, *Adsorption, Surface Area and Porosity* (Academic Press, London, 1982).
48. M. Kruk, M. Jaroniec, and A. Sayari, Application of large pore MCM-41 molecular sieves to improve pore size analysis using nitrogen adsorption measurements, *Langmuir* 13, 6267-6273 (1997).
49. S. R. Hall, C. E. Fowler, B. Lebeau, and S. Mann, Template-directed synthesis of bi-functionalized organo-MCM-41 and phenyl-MCM-48 silica mesophases, *Chem. Commun.* 201-202 (1999).
50. T. Asefa, M. Kruk, M. J. MacLachlan, N. Coombs, H. Grondey, M. Jaroniec, and G. Ozin, Novel bifunctional periodic mesoporous organosilicas, BPMOs, synthesis, characterization, properties and in-situ selective hydroboration – alcoholysis reactions of functional groups, *J. Am. Chem. Soc.* 123, 8520-8530 (2001).
51. M. C. Burleigh, M. A. Markowitz, M. S. Spector, and B. P. Gaber, Direct synthesis of periodic mesoporous organosilicas, functional incorporation by co-condensation with organosilanes, *J. Phys. Chem. B.* 105, 9935-9942 (2001).
52. Q. Yang, M. P. Kapoor, and S. Inagaki, Sulfuric acid-functionalized mesoporous benzene – silica with a molecular scale periodicity in the walls, *J. Am. Chem. Soc.* 124, 9694-9695 (2002).
53. W.-H. Zhang, X.-B. Lu, J.-H. Xiu, Z.-L. Hua, L.-X. Zhang, M. Robertson, J.-L. Shi, D.-S. Yan, and J. D. Holmes, Synthesis and characterization of bifunctionalized ordered mesoporous materials, *Adv. Functional Mater.* 14, 544-552 (2004).

FUNCTIONALIZED SURFACES: SILICA STRUCTURE AND METAL ION ADSORPTION BEHAVIOR

IAN P. BLITZ, JONATHAN P. BLITZ*, VLADIMIR M. GUN'KO[1], DANIEL J. SHEERAN
Eastern Illinois University, Charleston, IL 61920 USA
[1]Institute of Surface Chemistry 03164 Kiev, Ukraine

Abstract. Two silica gels with different sized mesopores and one nanoparticulate fumed silica, were modified with amino- and mercapto-functional silanes. One aminofunctional silane used was monofunctional preventing polymerization, the other trifunctional which permits surface polymerization. Structural and textural characteristics of these silicas before and after surface modification were probed using N_2 adsorption. The trifunctional aminosilane resulted in a material with less porosity than the monofunctional silane, which affected adsorption capacity of tested Cu(II) and Pb(II) from aqueous solution. Surface functionalization with the mercaptofunctional silane resulted in the smallest loss in porosity. A narrow pore size results in a more favorable ΔG_{ads}, but a larger pore size insures greater access of surface groups to adsorbed ions resulting in enhanced adsorption capacity. A comparison of the adsorption behavior of amino and mercapto groups for Cu(II) and Pb(II) shows that Cu(II) adsorbs to a greater extent on amino-functionalized than mercapto-functionalized surfaces. Pb(II) adsorption is not as greatly affected by the type of surface functionalization under the conditions studied.

Keywords: silica, organosilane, metal adsorption, isotherms, surface modification

1. Introduction

1.1. GENERAL OVERVIEW

An abundance of clean, uncontaminated water is a fundamental requirement for human life and environmental protection. As a result of various technological

*To whom correspondence should be addressed. J. P. Blitz, Eastern Illinois University, Charleston, IL 61920 USA E-mail jpblitz@eiu.edu

J.P. Blitz and V.M. Gun'ko (eds.), Surface Chemistry in Biomedical and Environmental Science, 337–348.

activities in the past century including mining, nuclear energy and the burning of fossil fuels, metal ions such as Sr^{2+}, Cs^{+}, Hg^{2+}, Cu^{2+}, Pb^{2+} and Cd^{2+} have been released into the environment at unprecedented levels. Considerable effort has gone into searching for the best ways to rid aqueous solutions of these unwanted solutes.[1,2]

Materials that contain functional groups to complex dissolved metal ions in aqueous solutions can be very effective. The use of an inorganic matrix, onto which a variety of functional groups can be chemically immobilized, has signficant advantages over conventional organic/polymeric supports. These advantages include a high surface area to enhance adsorption capacity, and greater physical and chemical robustness to withstand a variety of harsh environments. Whereas clays are a natural form of support to use,[3,4] various synthetic forms of silica provide a greater ability to vary the structure of these materials to optimize adsorption properties.

Highly structured mesoporous silicas using a surfactant template provide a well characterized, ultra-high surface area material.[5] In this volume Jaroniec provides an excellent summary of various novel materials based on this chemistry for Hg^{2+} adsorption. In this report we discuss results from a recent study utilizing two types of silica gels and a nanoparticulate fumed silica, all surface modified with complexing functional groups, for the adsorption of Cu^{2+} and Pb^{2+}.

1.2. BRIEF SCIENTIFIC BACKGROUND

Leyden and coworkers[6] were the first to use organosilane-modified silica surfaces to chelate metal ions, as an analytical preconcentration tool for subsequent analysis by EDXRF. While much has been learned since this time, fundamental questions remain concerning important factors which affect the metal adsorption behavior on these types of materials. The role of three main factors will be investigated in this work:

1. how the potential for the organosilane to form a cross-linked polymeric type of modified surface affects adsorption;
2. how the structural/textural characteristics of the modified surface affects metal adsorption behavior;
3. how the chemical functionality (mercapto versus amino) of the organosilane-modified surface affects adsorption of Cu^{2+} and Pb^{2+} from aqueous solution.

The first two factors above are linked since cross-linking on the surface and inside the pores of a silica gel can signficantly affect porosity. This cross-linking is varied by using organosilanes with either one or three hydrolyzable groups for surface modification. Organosilanes with one hydrolyzable group are unable to crosslink;

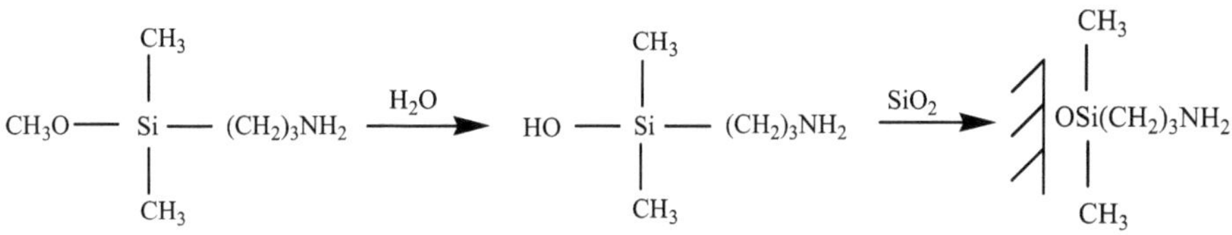

whereas organosilanes with three hydrolyzable groups are able to form polymeric phases on the surface as depicted below.

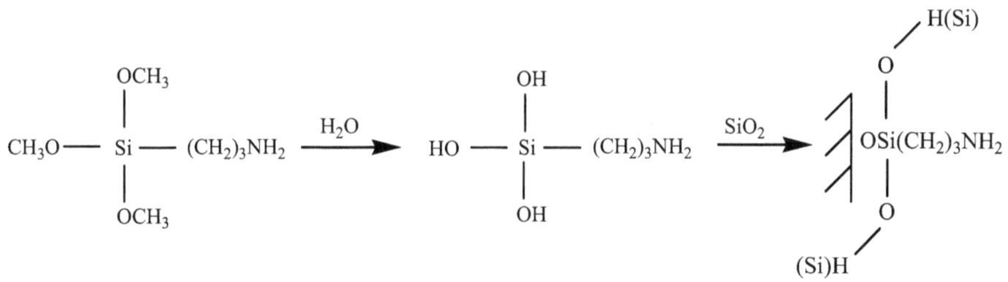

Porosity and pore size distributions, both before and after surface modification, are analyzed from N_2 adsorption data. Both capacity as well as adsorption free energies are obtained from metal ion adsorption isotherms from aqueous solution as described in the experimental section.

2. Experimental

2.1. MATERIALS

The nanoparticulate fumed silica Cab-O-Sil HS5 was provided from the Cabot Corporation, silica gels HP39 and 200DF were provided by Crosfield, and used as received. The organosilanes 3-aminopropyltriethoxysilane, 3-aminopropyldimethylmethoxysilane, and 3-mercaptopropyltrimethoxysilane were purchased from Gelest and used as received. All aqueous solutions were made from Millipore® water. Standard Cu^{2+} solutions were prepared using Cu wire dissolved in HNO_3 (pH subsequently adjusted to 5.0 by dropwise addition of NaOH), and Pb^{2+} solutions using primary standard $PbNO_3$ with a natural solution pH very close to 5.

2.2. METHODS

2.2.1. Materials synthesis

Four grams of silica was reacted with 20 mmoles of organosilane in 200 mL of refluxing toluene for a minimum of 2 hours. The slurry was filtered and washed with toluene, then dried at 150°C in vacuum for 2 hours. Reactions with the mercaptosilane were carried out in the presence of 100 μl of n-butylamine catalyst. The trialkoxysilane-modified silicas were resuspended in millipore water for 30 minutes, then once again dried in vacuum at 150°C to facilitate cross-linking. The extent of aminosilane reaction was determined by combustion analysis using a CE440 Microanalyzer (±0.06%), and mercaptosilane reaction using S analysis by ICP with a Perkin-Elmer 2000DV. Surface modification was verified by FTIR spectroscopy as characteristic NH_2 and SH functional group frequencies are readily detected.

2.2.2. Nitrogen adsorption data collection and analysis

Approximately 0.3 g of sample was outgassed at 110°C prior to analysis. Nitrogen adsorption-desorption isotherms were recorded at 77.35 K using a Micromeritics ASAP 2010 adsorption analyzer at $p/p_0 > 10^{-6} - 10^{-5}$, where p and p_0 denote the equilibrium pressure and saturation pressure of nitrogen at 77.35 K, respectively.

Pore size distributions (PSDs) (differential PSD $f_V(R_p) \sim dV_p/dR$) of the studied adsorbents were calculated using modified overall adsorption equation in the form proposed by Nguyen and Do (ND method)[7,8] for slitshaped pores of carbons and modified for other materials with cylindrical pores and pores as gaps between spherical particles.[9-16] The nitrogen desorption data were utilized to compute the $f_V(R_p)$ distribution functions using a modified regularization procedure[17] under non-negativity condition ($f_V(R_p) \geq 0$ at any R_p) with a fixed regularization parameter $\alpha = 0.01$.

The $f_V(R_p)$ functions linked to pore volume can be transformed to the $f_S(R_p)$ distribution functions with respect to surface area using the corresponding pore models

$$f_S(R_p) = \frac{w}{R_p}(f_V(R_p) - \frac{V_p}{R_p}) \tag{1}$$

where $w = 1, 2, 3$, and 1.36 for slitlike, cylindrical, spherical pores and gaps between spherical particles packed in the cubic lattice. The $f_V(R_p)$ and $f_S(R_p)$ functions were used to calculate contributions of nanopores (V_{nano} and S_{nano} at the pore radius $R_p < 1$ nm), mesopores (V_{mes} and S_{mes} at 1 nm $\leq R_p \leq$ 25 nm), and macropores (V_{mac} and S_{mac} at $R_p > 25$ nm) to the total pore volume and the specific surface area (Table 2). The values of S_{nano}, S_{mes}, and S_{mac} were corrected such that $S_{nano} + S_{mes} + S_{mac} = S_{BET}$.

For a pictorial presentation of the pore size distributions, the $f_V(R_p)$ functions were re-calculated to incremental PSDs (IPSDs)

$$\Phi_V(R_{p,i}) = 0.5(f_V(R_{p,i}) + f_V(R_{p,i-1}))(R_{p,i} - R_{p,i-1}). \tag{2}$$

2.2.3. *Metal ion adsorption data collection and analysis*

Various concentrations (10.00 mL) of Cu^{2+} or Pb^{2+} solutions were exposed to 0.0250 g of silica in sealed conical flasks for 16 hours in a wrist action shaker. The slurries were centrifuged, solution aliquots then obtained and analyzed by atomic absorption (AA, Perkin-Elmer 2380). The extent of metal ion adsorption on silica was obtained by subtracting the amount of metal ion present in solution after adsorption from the amount of metal ion in the original solution.

To calculate the free energy distributions ($f(\Delta G)$) of ion adsorption, the Langmuir equation was used as the kernel of the Fredholm integral equation of the first kind

$$\Theta = \int \frac{z\exp\left(-\frac{\Delta G - \gamma}{R_g T}\right)}{1 + z\exp\left(-\frac{\Delta G - \gamma}{R_g T}\right)} f(\Delta G)d(-\Delta G) \tag{3}$$

where $\Theta = a/a_m$ is the relative adsorption, a_m is the monolayer capacity; $z = C_{ads}/C_{eq}$, and R_g is the gas constant. Eq. (3) was solved using the regularization procedure under non-negativity condition ($f(\Delta G) \geq 0$ at any ΔG) with a fixed regularization parameter $\alpha = 0.01$.

3. Results and Discussion

Results of the extent of surface modification reaction of the three silicas with the three organosilanes studied are shown in Table 1.

Table 1. Elemental analysis results of organosilane-modified silicas. APDMS = 3-aminopropyldimethylmethoxysilane; APTS = 3-aminopropyltriethoxysilane; MPTMS = 3-mercaptopropyltrimethoxysilane.

Organosilane	mmol N(S)/g Cab–O–Sil HS5	mmol N(S)/g Crosfield HP39	mmol N(S)/g Crosfield 200DF
APDMS	1.684	1.963	1.920
APTS	1.224	1.781	1.281
MPTMS	(0.823)	(1.101)	(1.319)

The extent of surface modification for both Cab-O-Sil and HP39 follows the trend APDMS > APTS > MPTMS. More MPTMS reacts with 200DF silica gel than APTS, which may be explained by data below concerning the porosity of these silicas before and after surface modification. These three silicas were chosen for study because of their varying structural/textural characteristics, characterized by N_2 adsorption. As illustrated in Table 2 and Figures 1 and 2, the surface area and porosity of these silicas are quite different.

Table 2. Structural characteristics of silicas before surface modification.

Silica	S_{BET} (m^2/g)	S_{nano} (m^2/g)	S_{meso} (m^2/g)	S_{mac} (m^2/g)
HS5	326	152	129	45
HP39	467	125	272	70
200DF	540	208	312	20

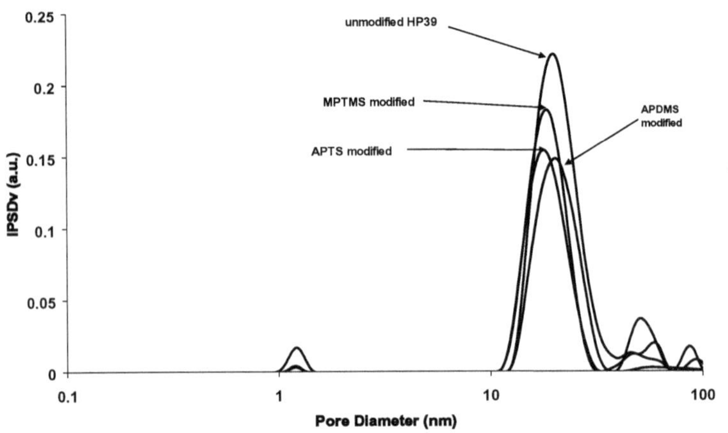

Figure 1. Incremental pore size distributions of HP39 before and after surface modification with APDMS, APTS, and MPTMS.

It is worth mentioning that the surface area of Cab-O-Sil fumed silica is derived from its very small particle size, and its porosity is determined by the arrangement of these primary particles into aggregates and larger agglomerates. Thus the surface area of this nanoparticulate fumed silica is subject to large changes dependent upon its physical environment and history. These effects are beyond the scope of this paper, but have been addressed in other contributions in this volume and elsewhere.[13,14,16] Effects of surface modification on porosity are thus addressed only for the silica gels.

It is not surprising that in all cases surface modification results in a reduction in pore volume. The average pore diameter of 20 nm for HP39 is unchanged as a result of modification by APDMS. However, reaction with trifunctional APTS or MPTMS results in a reduced average pore size of approximately 18 nm, indicating some surface polymerization may result in a ~10% reduction in pore diameter.

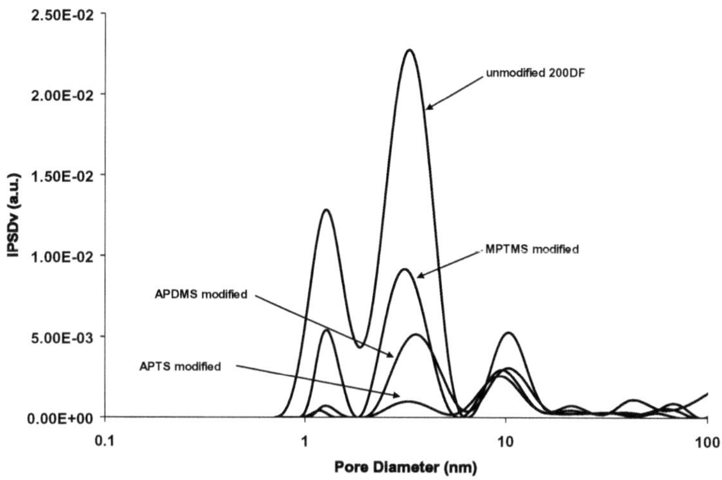

Figure 2. Incremental pore size distributions of 200DF before and after surface modification with APDMS, APTS, and MPTMS.

More significant changes in 200DF porosity are seen as a result of surface modification, since the average pore diameter of this silica gel is much smaller to begin with. Surface modification with 3-mercaptopropyltrimethoxysilane (MPTMS) results in the smallest reduction in porosity, whereas APTS surface modification results in almost complete loss of porosity below 10 nm. This APTS result indicates complete blockage of the smaller pores by APTS, which may explain why less APTS than MPTMS reacts with this silica gel. Note that the smallest diameter 1 nm pores are blocked to an equal extent with APTS or APDMS, but largely remain after MPTMS reaction. The smaller pores, 5 nm or less in diameter, can be blocked by surface polymerization. A second contribution to pore blockage can be from hydrogen bonding interactions between amino groups and residual silanols with APTS and APDMS, while these hydrogen bonding interactions do not occur with MPTMS thiol groups and residual silanols. This may explain why the smallest diameter pores remain to a much larger extent with MPTMS surface modification.

Adsorption isotherms for Cu^{2+} on APTS-modified silicas are shown in Figure 4, and for APDMS-modified silicas in Figure 5. These data show that HP39 > Cab-O-Sil > 200DF with respect to total Cu^{2+} adsorption capacity. Note that more APTS is bonded to 200DF than Cab-O-Sil (Table 1), indicating that a substantial portion of APTS is unavailable for adsorption on 200DF probably because of inaccessibility inside blocked pores. HP39, with the larger pore size, shows the greatest accessibility of Cu^{2+} adsorption with bonded APTS.

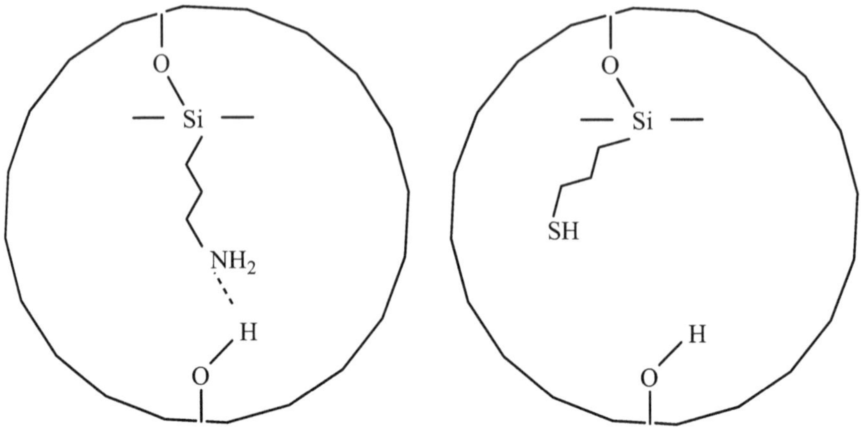

Figure 3. Depiction of hydrogen bonding of an aminosilane with an unreacted silanol inside a small pore (left). Hydrogen bonding is not possible between a mercaptosilane and a residual silanol (right).

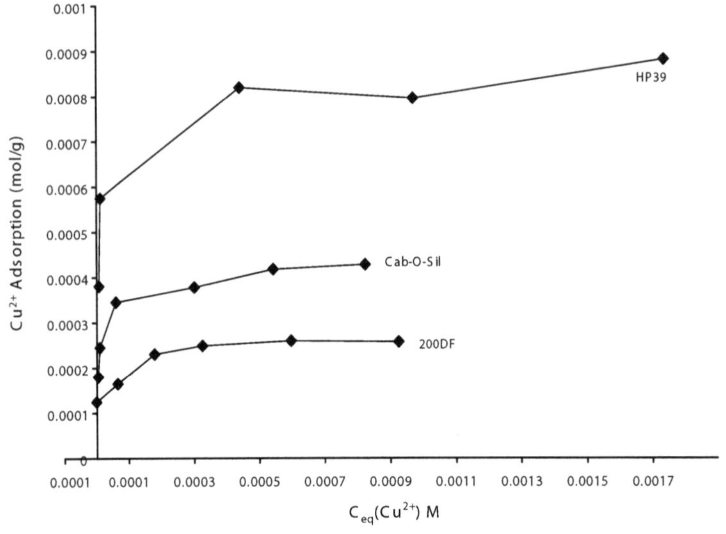

Figure 4. Cu^{2+} adsorption isotherms of APTS-modified silicas.

For APDMS-modified silicas 200DF has at least as large a Cu(II) adsorption capacity as Cab-O-Sil. This is consistent with results in Figures 1 and 2, showing that for APDMS-modified silicas significant porosity remains in the 5 nm pore diameter range, permitting greater accessibility to Cu(II) than for APTS-modified 200DF.

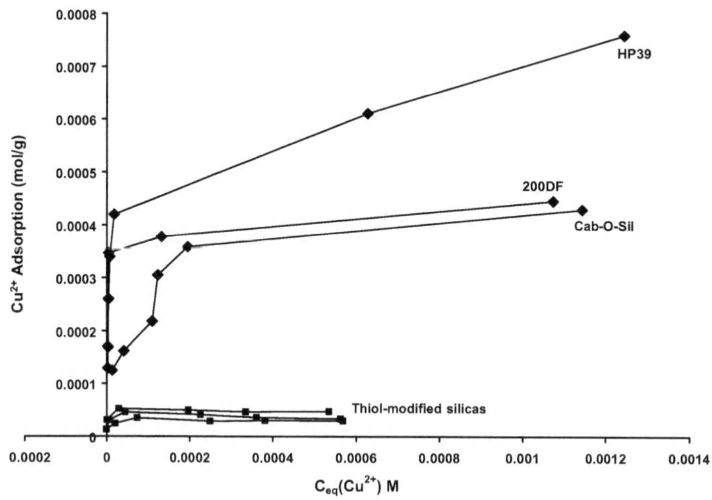

Figure 5. Cu^{2+} adsorption isotherms of APDMS and thiol-modified silicas.

Adsorption free energy distribution functions can be calculated from the ion adsorption isotherms. Figure 6 shows a plot of this data for Cu(II) adsorption on APDMS-modified silicas.

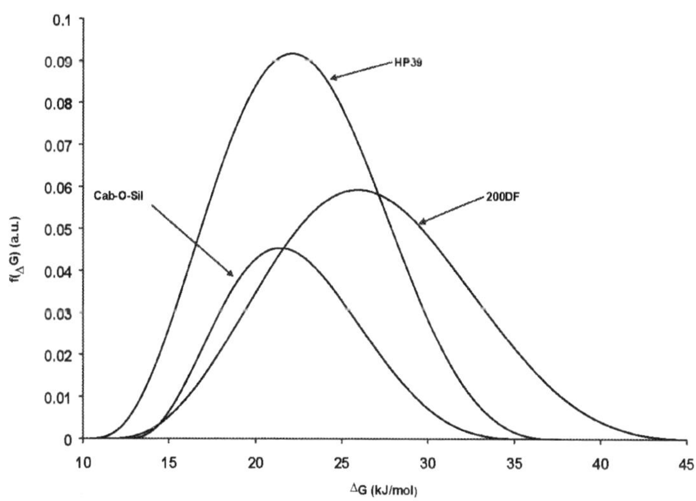

Figure 6. Distribution function of changes in Gibbs free energy of Cu(II) adsorption on APDMS silicas.

The shift to greater negative Gibbs free energy of Cu(II) adsorption on APDMS modified 200DF silica is readily apparent. This trend does not hold for APTS-modified silicas (not shown). The presence of small mesopores in APDMS-modified 200DF silica gel may favor the free energy of adsorption (Figure 2). These types of pores are not present after APTS surface modification (Figure 1), resulting in no enhancement of $-\Delta G_{ads}$. It would thus appear that a compromise must be struck between a favorable free energy of adsorption promoted by small mesopores on 200DF, and enhanced accessibility to surface adsorption sites inside larger mesopores on HP39.

Data in Figure 5 illustrates the large differences in Cu(II) adsorption capacity for thiol- and amino-modified silicas, the former being much less than the latter. It is interesting to note that calculated adsorption free energies for Cu(II) on amino- and thiol-modified silicas are similar, even though their adsorption capacities are quite a bit different.

The large differences seen in Cu(II) adsorption capacity for thiol- and amino-modified silicas does not occur for Pb(II) adsorption (Figure 7). While it is tempting to argue these differences in adsorption behavior based on hard-soft acid-base theory the hardness of Cu^{2+} and Pb^{2+} are very similar, both generally categorized as intermediate cases.[18] It is more likely that Pb(II) adsorption behavior under the conditions of this study is dictated largely by solution behavior and speciation of Pb^{2+} in aqueous solution at pH 5. Indeed it was found that these solutions were unstable and prone to precipitation slowly over time under these conditions. Further work will be needed as a function of pH, surface functionality, and metal ion to determine the role of various factors on adsorption behavior of these systems.

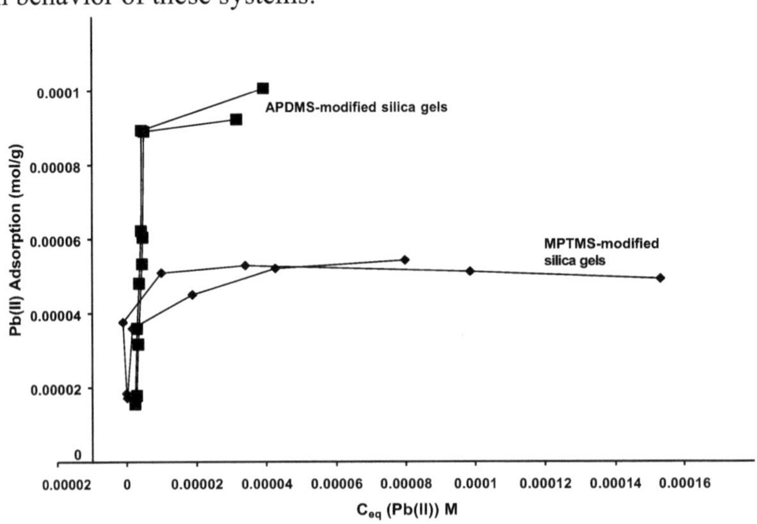

Figure 7. Pb(II) adsorption isotherms on 3-aminopropyldimethylmethoxysilane (APDMS) modified silica gels, and 3-mercaptopropyltrimethoxysilane (MPTMS) modified silica gels.

4. Conclusion

Surface functionalization of high surface area silicas with Lewis bases for the coordination of metal cations, is an effective means to rid aqueous solutions of these unwanted species. Maximum adsorption capacity is achieved using a mesoporous silica with relatively large (~20 nm) pore diameter reacted with a cross-linkable aminopropylsilane. The maximum driving force, measured by ΔG_{ads}, is obtained using a non-crosslinkable aminosilane-modified mesoporous silica with a much smaller average pore diameter (~5 nm after surface modification). This suggests an approach using a silica gel with a relatively large pore diameter which, when modified with functional groups for metal ion adsorption, results in a material with a smaller pore diameter providing both maximum adsorption capacity and large negative ΔG_{ads}.

5. Acknowledgments

We thank Dr. Steve Augustine for collection of the N_2 adsorption data. JPB and VMG thank NATO for funding of the Advanced Research Workshop where this work was presented. This work was supported in part by a grant from the Petroleum Research Fund administered by the American Chemical Society.

References

1. C. Kantipuly, S. Katragadda, A. Chow, and H. D. Gesser, Chelating polymers and related supports for separation and preconcentration of trace metals, *Talanta* 37(5), 491-517 (1990).
2. P. K. Jal, S. Patel, and B. K. Mishra, Chemical modification of silica surface by immobilization of functional groups for extractive concentration of metal ions, *Talanta* 62, 1005-1028 (2004).
3. R. Celis, M. Carmen Hermosin, and J. Cornejo, Heavy metal adsorption by functionalized clays, *Environ. Sci. Technol.* 34, 4593-4599 (2000).
4. L. Mercier and T. J. Pinnavaia, A functionalized porous clay heterostructure for heavy metal ion (Hg^{2+}) trapping, *Micropor. Mesopor. Mater.* 20(1-3), 101-106 (1998).
5. K. J. Edler, Current understanding of formation mechanisms in surfactant-templated materials, *Aust. J. Chem.* 58, 627-643 (2004).
6. D. E. Leyden and H. G. Luttrell, Preconcentration of trace metals using chelating groups immobilized via silylation, *Anal. Chem.* 47(9), 1612-1617 (1975).
7. C. Nguyen and D. D. Do, A new method for the characterization of porous materials, *Langmuir* 15(10) 3608-3615 (1999).

8. C. Nguyen and D. D. Do, Effects of probing vapors and temperature on the characterization of micro- and mesopore size distribution of carbonaceous materials, *Langmuir* 16(18), 7218-7222 (2000).
9. V. M. Gun'ko, J. Skubiszewska-Zięba, R. Leboda, K. N. Khomenko, O. A. Kazakova, M. O. Povazhnyak, and I. F. Mironyuk, Influence of morphology and composition of fumed oxides on changes in their structural and adsorptive characteristics on hydrothermal treatment at different temperatures, *J. Colloid Interface Sci.* 269, 403-424 (2004).
10. V. M. Gun'ko, E. F. Voronin, I. F. Mironyuk, R. Leboda, J. Skubiszewska-Zięba, E. M. Pakhlov, N. V. Guzenko, and A. A. Chuiko, The effect of heat, adsorption and mechanochemical treatments on stuck structure and adsorption properties of fumed silicas, *Colloids Surf. A* 218(1-3), 125-135 (2003).
11. V. M. Gun'ko, V. I. Zarko, E. F. Voronin, V. V. Turov, I. F. Mironyuk, I. I. Gerashchenko, E. V. Goncharuk, E. M. Pakhlov, N. V. Guzenko, R. Leboda, J. Skubiszewska-Zięba, W. Janusz, S. Chibowski, Yu. N. Levchuk, and A. V. Klyueva, Impact of some organics on structural and adsorptive characteristics of fumed silica in different media, *Langmuir* 18(3), 581-596 (2002).
12. E. F. Voronin, V. M. Gun'ko, N. V. Guzenko, E. M. Pakhlov, L. V. Nosach, M. L. Malysheva, J. Skubiszewska-Zięba, R. Leboda, M. V. Borysenko, and A. A. Chuiko, Interaction of poly(ethylene oxide) with fumed silica, *J. Colloid Interface Sci.* 279, 326-340 (2004).
13. V. M. Gun'ko, D. J. Sheeran, S. M. Augustine, and J. P. Blitz, Structural and energetic characteristics of silicas modified by organosilicon compounds, *J. Colloid Interface Sci.* 249, 123-133 (2002).
14. V. M. Gun'ko, V. I. Zarko, D. J. Sheeran, J. P. Blitz, R. Leboda, W. Janusz, and S. Chibowski, Characteristics of modified Cab-O-Sil in aqueous media, *J. Colloid Interface Sci.* 252, 109-118 (2002).
15. M. V. Borysenko, V. M. Gun'ko, A. G. Dyachenko, I. Y. Sulim, R. Leboda, and J. Skubiszewska-Zięba, *Appl. Surf. Sci.* 242, 1-12 (2005).
16. V. M. Gun'ko, I. F. Mironyuk, V. I. Zarko, E. F. Voronin, V. V. Turov, E. M. Pakhlov, E. V. Goncharuk, Yu. M. Nicheporuk, T. V. Kulik, B. B. Palyanytsya, S. V. Pakhovchishin, N. N. Vlasova, P. P. Gorbik, O. A. Mishchuk, A. A. Chuiko, J. Skubiszewska-Zięba, W. Janusz, A. V. Turov, and R. Leboda, Morphology and surface properties of fumed silicas, *J. Colloid Interface Sci.* 289, 427-445 (2005).
17. S. W. Provencher, A constrained regularization method for inverting data represented by linear algebraic or integral equations, *Comp. Phys. Comm.* 27, 213-227 (1982).
18. G. L. Miessler and D. A. Tarr, *Inorganic Chemistry, 2nd Edition* (Prentice-Hall, New Jersey, 1998).

HETEROGENEOUSLY CATALYZED ALKANE ISOMERIZATION – TOWARDS 100% SELECTIVITY

HELMUT KNÖZINGER
Department Chemie und Biochemie, Universität München
Butenandtstrasse 5-13, Haus E, 81377 München, Germany,
e-mail: helmut.knoezinger@cup.uni-muenchen.de

Abstract. Tungstated zirconia catalysts are stable and highly selective catalytic materials for the isomerization of alkanes when promoted by platinum and a transition metal oxide and when dihydrogen is present in the feed. Physical properties and the catalytic performance of these solids for the isomerization of n-pentane are discussed.

Keywords: alkanes-isomerization; iron-promoter; n-pentane-isomerization; platinum-promoter; selectivity; tungsten oxide; zirconia-tungstated-acidity; zirconia-tungstated-redox properties

1. Introduction

The isomerization of naturally abundant straight-chain light alkanes is a technologically very important process for the production of branched alkanes, which are good gasoline components. Isobutane is also an important raw material for alkylation of alkenes.

The most widely applied isomerization catalysts are platinum-promoted solid acids, namely chlorinated alumina[1,2] and mordenite.[3-5] The former catalysts require the continuous addition of chlorine-containing compounds, are moisture-sensitive and are poisoned by sulfur impurities.[6] They are also highly corrosive and hence, environmentally hazardous. Zeolite catalysts are significantly less active and have to be operated at higher temperatures (500-550 K). As a consequence of thermodynamic limitations only relatively low yields of the target branched alkanes can be achieved under these conditions.

Sulfated zirconia (SZ) catalysts have first been proposed by Hohn and Bailey[7] as alternative environmentally friendly candidate catalysts, which are active for the isomerization of light straight-chain alkanes at low temperatures

J.P. Blitz and V.M. Gun'ko (eds.), Surface Chemistry in Biomedical and Environmental Science, 349–358.

(300-420 K).[8-10] The promotion of SZ catalysts by the addition of noble metals and transition metal oxides improved their stability, activity and selectivity even further. However, SZ catalysts suffer from several problems that impede their practical application.[11]

Consequently, investigators, beginning with Hino and Arata,[12] sought alternatives to SZ, turning to tungstated zirconia (WZ). Although WZ catalysts are markedly less active than SZ, they have superior stability under both reducing and oxidizing conditions and appear to be more suitable for industrial applications. The catalytic activity of WZ, like that of SZ, is greatly improved by promotion with platinum and with transition metal oxides. Pt/WZ catalysts are significantly more selective than Pt/SZ for formation of branched alkane isomers, as shown by results characterizing the conversion of *n*-heptane[13] and *n*-octane.[14] Pt-promoted WZ catalysts appear to have found commercial application already.[15]

We review here a standard synthesis of unpromoted and promoted WZ catalysts, and their physical characteristics as related to their catalytic performance in the isomerization of n-pentane to isopentane.

2. Preparation of Tungstated Zirconia Catalysts

We prepared tungstated zirconia catalysts by impregnation of $ZrO_x(OH)_{4-2x}$ in aqueous suspensions containing ammonium metatungstate followed by drying and calcination at 923 K.[16] The resulting materials, containing approximately one monolayer of the tungsten species (17 or 19 wt% WO_3 on ZrO_2), were the most active catalysts in a series obtained by variation of the preparation parameters. These unpromoted catalysts are denoted 17WZ or 19WZ, respectively.

We also prepared WZ catalysts promoted by both platinum and iron.[17] Similar catalysts were reported in a patent by Chang et al.[18] $Fe(NO_3)_3$ or $FeSO_4$ was introduced in the impregnation step mentioned above, giving catalysts referred to as FeWZ(N) and FeWZ(S), each containing 17.7 wt% WO_3 and an Fe content corresponding to 1 wt% Fe_2O_3. Subsequently, the calcined FeWZ materials were impregnated with a platinum component to give samples denoted PtFeWZ(N) and PtFeWZ(S) containing 1 wt% Pt.[19]

3. Stuctural Characterization

3.1. STRUCTURE OF THE ZIRCONIA SUPPORT

XRD patterns of the calcined catalysts show the formation of tetragonal ZrO_2 in the presence of WO_x surface compounds (see below).[16] This is supported by the appearance of bands in the Raman spectrum at 146, 270, 316, 460, and 645 cm^{-1}.[16] The WO_x surface species stabilize the metastable tetragonal ZrO_2 phase and inhibit the sintering of ZrO_2 crystallites, as indicated by the much larger surface area of approximately 100 m^2/g of WO_x-promoted ZrO_2 as compared to that of similarly treated ZrO_2 itself.[16]

3.2. STRUCTURE OF SURFACE TUNGSTEN SPECIES

Raman spectra of calcined 19WZ in the dehydrated state show bands at 830, 910 and 1020 cm^{-1}.[20] The bands at 830 and 1020 cm^{-1} are assigned to W-O-W and to W = O stretching vibrations, respectively, characteristic for polyoxotungstates,[20] and suggest the formation of interconnected tungsten species with W^{6+} in octahedral oxygen environment. The broad band at 910 cm^{-1} was tentatively assigned to W-O-Zr stretching vibrations characterizing the bond anchoring WO_x species to the support.[20] There is no spectroscopic evidence for the formation of WO_3 as long as the tungsten loading remains below the monolayer capacity of ca. 19 wt% WO_3 on ZrO_2. This structural description of the surface WO_x species is confirmed by the position of the absorption edge position in UV-visible absorption spectra.[16,21]

The W^{6+} ions in the surface polyoxotungstates are supposed to be octahedrally coordinated, and are therefore coordinatively saturated. This is indicated by the fact that W^{6+}-CO complexes were not detected by IR spectroscopy when CO was added to the sample.[16,21]

We propose that three-dimensional polyoxotungstate clusters are anchored to the surface of tetragonal zirconia. We infer that the structure and the size of these clusters, and their specific interactions with the zirconia support largely determine the acidic, redox, and catalytic properties of the tungstated zirconia WZ, as also suggested by others.[22] The same polyoxotungstate clusters are also present on the promoted catalysts containing 1 wt% Pt or 1 wt% Pt and 1 wt% Fe_2O_3.

4. Surface Properties of Tungstated Zirconia Catalysts

4.1. SURFACE ACIDITY

The acid strength of the catalyst was probed by CO adsorption and IR spectroscopy.[23] Low-temperature CO adsorption on oxidized WZ indicates the presence of both Lewis and Brønsted acid sites.[20,21] Lewis acidity is attributed to coordinatively unsaturated Zr^{4+} sites, present both on pure ZrO_2 and on WZ. The acid strengths of these centers are enhanced by the presence of polytungstate species.[20]

OH groups of the catalyst are characterized by bands or shoulders at 3670 and about 3620-3630 cm^{-1}.[16] Low-temperature CO adsorption leads to hydrogen bonding between OH groups and the weak base CO, which results in a red shift of the O-H stretching frequencies. The frequency shift, $\Delta\upsilon_{OH}$, is a measure of the hydrogen bond energy and it has been suggested that it correlates with the Brønsted acid strength.[23]

Table 1. Frequency shifts, $\Delta\upsilon_{OH}$, of O-H stretching frequencies induced by low-temperature CO adsorption on zirconia, tungstated zirconia, and zeolite HZSM-5

Sample	$\Delta\upsilon(OH)$ (cm^{-1})
ZrO_2	60
19WZ	160
HZSM-5	320

A comparison of the $\Delta\upsilon OH$ values of several materials is shown in Table 1. The frequency shift characterizing 19WZ is 160 cm^{-1}, which is significantly greater than the shift of 60 cm^{-1} recorded for pure ZrO_2.[20] The presence of polytungstate species on the surface thus enhances the Brønsted acidity, and the results suggest that larger WO_x domains create stronger acid sites.[13,15,16] The $\Delta\upsilon_{OH}$ value of 160 cm^{-1} representing 19WZ is significantly less than the values reported for the zeolites HZSM-5[24,25] and H-mordenite.[24]

We conclude that there is no evidence for WZ catalysts having superacidic properties or sites with the acidic character that would be necessary for initiation of catalysis by alkane protonation. In as much as WZ catalysts are some four orders of magnitude more active than zeolites for alkane isomerization,[26] it is clear that there is no one-to-one correlation between acid strength of WZ and its catalytic activity. We therefore infer that although the acidity of WZ catalysts is important in alkane conversion catalysis, the reaction is most likely initiated by a reaction other than protonation of the alkane by the catalyst or a species formed from it.

4.2. REDOX PROPERTIES

We investigated the reduction in H_2 of 19WZ and 1Pt19WZ by EPR spectroscopy.[27,28] The results show that W^{5+} can be detected on 19WZ after reduction at 470 K and higher temperatures,[27] whereas the maximum possible degree of reduction of the platinum-promoted catalyst is achieved even at room temperature.[28] Reduced tungsten sites were also detected on WZ when NO was used as an IR probe molecule; the results show that the reductive treatment of the unpromoted WZ catalyst at temperatures above 523 K leads to the formation of W^{5+}- and W^{4+}-containing centers.[29] The formation of W^{5+} centers is accompanied by the creation of new OH groups, as evidenced by IR spectra.[27,28]

Low-temperature CO adsorption measurements indicate that the strength of Brønsted acid sites on reduced catalysts are lower than those characteristic of the oxidized catalyst.[27] This finding is not unexpected, as it is well known that the acid strength of inorganic oxyacids depends on the oxidation state of the central atom.[30]

EPR experiments have shown that the redox ability of WZ catalysts is sufficient to initiate a homolytic cleavage of C-H bonds in alkanes. Exposure of a WZ catalyst to *n*-pentane at 523 K led to the formation of W^{5+} species and organic radicals on the surface.[27] The formation of organic radicals also occurred when WZ catalysts interacted with other hydrocarbons, including benzene.[31] We therefore infer that one-electron transfer, although it is not regarded as a step in the catalytic cycle, can initiate catalysis by a proccss that leads to the formation of the carbenium ion chain carriers,[27] as also occurs in acidic solutions.[32] We emphasize that a strong redox reactivity is necessary but not sufficient for the catalytic activity of WZ.

5. State of Platinum and Iron in Promoted Catalysts

The promoted WZ catalysts containing 1wt% Pt and/or 1wt% Fe_2O_3 were characterized by CO chemisorption and by transmission electron microscopy (TEM) in combination with EDX.[33] The state of iron and its redox behavior was also analyzed by x-ray absorption spectroscopy (XANES and EXAFS at the Fe K-edge), by in situ EPR spectroscopy and by Mössbauer spectroscopy.

CO chemisorption on reduced PtWZ catalysts gave dispersions of the order of 5% suggesting either the formation of large Pt particles or encapsulation of smaller particles. TEM clearly demonstrated that large Pt particles with average diameters of up to 50 nm were present depending on the calcination temperature.

In contrast, TEM of Fe-containing catalysts (FeWZ and PtFeWZ) shows that the iron in the oxidized catalysts is homogeneously distributed on the catalytic materials. The XANES results did not support the formation of Fe_2O_3 clusters, but rather suggested that isolated Fe^{3+} ions in octahedral coordination are present in the catalysts.[34] EXAFS analysis showed that next-nearest neighbors of Fe are Zr-atoms, while Fe next-nearest neighbors remained undetectable. This result together with the XANES data is highly suggestive for the formation of a surface solid solution of Fe^{3+} in the zirconia matrix.[34]

A g-value of 4.25 was observed by EPR which is consistent with the distorted octahedral coordination of Fe^{3+} ions.[34] However, due to the high sensitivity of EPR a small signal at $g = g_e$ was also observed in the oxidized samples which is most likely attributed to a small concentration of α-Fe_2O_3 clusters. Mössbauer spectroscopy at 4.2 K also confirms the dominant formation of a (surface) solid solution $Fe_xZr_{1-x}O_{2-(x/2)}$ in the oxidized state of the catalysts, with admixture of approximately 10 % of a magnetically split Fe^{3+} species which would correspond to α-Fe_2O_3 as detected by EPR.[34] EPR and Mössbauer spectroscopy also demonstrated that both iron phases are easily reducible in H_2, the formation of Fe^{2+} already beginning at ca. 300 K and being completed at about 470 K. Quantitative reoxidation in O_2 occurs also near 300 K.[34] During H_2 reduction a portion of the W^{6+} ions are reduced to W^{5+} (and some small percentage of W^{4+}). This reduction process is strongly favoured for Pt-containing catalysts, and is completely reversible under mild reoxidation conditions.

6. Isomerization of n-Pentane

The catalytic isomerization of *n*-pentane was carried out in a once-through fixed bed flow reactor using helium as a carrier gas, at a standard temperature of 523 K in the absence and presence of dihydrogen in the feed. Characteristic rate and selectivity data are compared in Table 2 for WZ and PtWZ.[21,35]

The unpromoted WZ catalyst develops a low activity characterized by conversions $\leq 1\%$ under the conditions of our experiment. The rate of the total *n*-pentane conversion depends strongly on the time-on-stream (TOS). The selectivity for isopentane formation remains roughly constant at 40-60%, with the products isobutene, propane, and *n*-butane and traces of unsaturated hydrocarbons also being observed. A detailed analysis of the dependence of the rate on TOS, of the distribution of side products and of UV-Vis,[35] Raman-spectra[36] and EPR-spectra[27] of the functioning catalyst, suggests that the reaction is initiated by a dehydrogenation step forming pentenes. These pentenes can be isomerized via carbenium ion intermediates and subsequently hydrogenated to yield the target product isopentane. Cracking reactions are responsible for the low selectivity, and carbon deposits lead to catalyst deactivation. The presence of dihydrogen in the feed slightly enhances the activity but reduces the selectivity to about 30%.

Table 2. Comparison of the catalytic performance of Pt-promoted tungstated zirconia and comparison with Pt-free material.

Catalyst	H_2 in feed?	Maximum rate ($mol\ s^{-1}\ m^{-2}$)	selectivity (%)
WZ	yes	$8.0x10^{-10}$	~ 30
	no	$3.0x10^{-10}$	40-60
PtWZ	yes	$(4.8x10^{-8})^a$	95
	no	$2.4x10^{-9}$	15-35

[a] Equlibrium conversion

The platinum-promoted PtWZ catalyst is significantly more active than WZ, and almost reaches equilibrium composition for *n*- and isopentane under the conditions of our experiments in the presence of dihydrogen in the feed.[35] Dihydrogen also has a dramatically positive effect on the selectivity, which is essentially constant with TOS at about 95%. This observation demonstrates the potential of PtWZ catalysts, and it shows that the presence of dihydrogen is essential for the performance of the catalyst. The small amount of side products contain methane and butanes, as well as ethane and propane in equimolar ratios.[35] This suggests that the side products result from direct hydrogenolysis of C_5 species, likely on platinum surfaces.[37] Catalyst activity and selectivity remained essentially stable with TOS under the conditions of our experiments. We infer that the platinum provides atomic hydrogen, which permits a fast desorption of unsaturated species and reactive intermediates, and suppresses oligomerization and polymerization reactions as well as coke deposition. Thus, a clean monomolecular isomerization via carbenium ion intermediates is made possible with high selectivities.

When iron is added as a second promoter, the performance of PtFeWZ catalysts is dramatically improved in the presence of dihydrogen in the feed.[19,21] Under identical reaction conditions, PtFeWZ(S) is characterized by an *n*-pentane isomerization rate of 9×10^{-8} mol s^{-1} m^{-2}. Whereas the PtWZ catalyst is characterized by a nearly stable selectivity of about 95% (see Table 2), the PtFeWZ(S) catalyst develops a selectivity (increasing with TOS) of up to 98%, and PtFeWZ(N) shows a stable selectivity greater than 99%. The suppression of the hydrogenolysis products, which are formed on the platinum in PtWZ by the addition of iron as a second promoter, might be a consequence of the suppression of the formation of metallic platinum. Furthermore, the redox properties of the Fe^{3+}/Fe^{2+} pair in the surface solid solution (see above) might be involved, and could influence the reducibility of the tungsten phase and its dehydrogenation/hydrogenation activity.

7. Conclusion

The results suggest that the doubly promoted WZ catalysts are bifunctional materials which possess dehydrogenation/hydrogenation and acidic properties. There is no evidence for superacidic sites in these materials. The reduced surface polyoxotungstate species provide dehydrogenation/hydrogenation activity and are the origin of medium-strong protonic acidity. The platinum dissociates dihydrogen and catalyzes hydrocracking reactions, while iron appears to suppress this side reaction thus improving the selectivity. Furthermore, iron may be involved in the redox properties of the tungsten phase.

The catalysts based on tungstated zirconia are stable and highly selective. Their activity, however, must be further improved so that reaction temperatures can be reduced and product yields improved. It is expected that this goal can be achieved by variations of the preparation conditions and modification of the catalyst composition. Recent experiments gave promising results with new WZ catalysts working below 470 K.

Acknowledgements

The author expresses his thanks to his colleagues and coworkers for a very efficient collaboration: Professors M. Che, B. C. Gates, R. K. Grasselli, F. E. Wagner and Drs. X. Carrier, P. Concepción-Heydorn, F. C. Jentoft, S. Kuba, P. Lukinskas, M. Scheithauer, L. Stiefano, B. Tesche and P. Vijayanand. This research was supported by the Deutsche Forschungsgesellschaft, the Fonds der Chemischen Industrie and the Alexander-von-Humboldt-Gesellschaft and Max-Planck-Gesellschaft (Max-Planck-Forschungspreis to H. K.). M. Che, B. C. Gates and R. K. Grasselli are indebted to the Alexander-von-Humboldt-Gesellschaft for Senior Scientist Awards, and P. Lukinskas for a research fellowship. P. Vijayanand thanks the German Academic Exchange Service for a grant.

References

1. D. H. Beldon, Penex and Platforming term up, *Petrol. Refiner* 35, 149–152 (1956).
2. B. W. Burbridge and J. K. R. Rolfe, *Hydrocarbon Process. Petrol. Refiner* 45, 168 (1966).
3. H. W. Kouwenhoven and W. C. van Zijill Langhout, Shell´s hydroisomerization process, *Chem.- Eng. Progr.* 67, 65–70 (1971).
4. H. W. Kouwenhoven and W. C. van Zijill Langhout, *Oil Gas J. March 8*, 44 (1971).

5. J. G. Santiesteban, D. C. Calabro, C. D. Chang, J. C. Vartuli, T. J. Fiebig, and R. D. Bastian, The Role of Platinum in Hexane Isomerization over $Pt/FeO_y/WO_x/ZrO_2$, *J. Catal.* 202, 25–33 (2001).
6. S. T. Sie, in: *Handbook of Heterogeneous Catalysis*, Vol. 4, edited by G. Ertl, H. Knözinger, and J. Weitkamp (Wiley-VCH, Weinheim, 1997), pp. 1998–2017.
7. V. C. F. Holm and G. C. Bailey (Phillips Petroleum Co.), US Patent 3,032,599 (1962).
8. V. Adeeva, J. W. de Haan, J. Jaenchen, G. D. Lei, V. Schünemann, L. J. M. van Veen, W. M. H. Sachtler, and R. A. van Sauten, Acid Sites in Sulfated and Metal–Promoted Zirconium Dioxide Catalysts, *J. Catal.* 151, 364–372 (1995).
9. C. Y. Hsu, C. R. Heimbuch, C. T. Armes, and B. C. Gates, A highly active solid superacid catalyst for n – butane isomerization: a sulfated oxide containing iron, manganese and zirconium, *J. Chem. Soc. Chem. Commun.* 1645–1646 (1992).
10. M. Signoretto, F. Pinna, G. Strukul, P. Chies, G. Cerrato, S. Di Ciero, and C. Morterra, Platinum – Promoted and Unpromoted Sulfated Zirconia Catalysts Prepared by a One – Step Aerogel Procedure, *J. Catal.* 167, 522–532 (1997).
11. G. Larsen, E. Lotero, R. D. Parra, L. M. Petkovic, H. S. Silva, and S. Raghavan, Characterization of palladium supported on sulfated zirconia catalysts by DRIFTS, XAS and n – butane isomerzation reaction in the presence of hydrogen, *Appl. Catal A* 130, 213–226 (1995).
12. M. Hino and K. Arata, Synthesis of solid superacid of tungsten oxide supported on zirconia and its catalytic action for reactions of butane and pentane, *J. Chem. Soc. Chem. Commun.* 1259–1260 (1988).
13. E. Iglesia, D. G. Barton, S. L. Soled, S. Miseo, J. E. Baumgartner, W. E. Gates, G. A. Fuentes, and G. D. Meitzner, Selective Isomerization of Alkanes on Supported Tungsten Oxide Acids, *Stud. Surf. Sci. Catal.* 101, 533–542 (1996).
14. J. M. Grau, J. C. Yori, and J. M. Parera, Hydroisomerization – Cracking of n – octane on Pt/WO_4^{2-} - ZrO_2 and Pt/SO_4^{2-} - ZrO_2: Effect of Pt load on catalyst performance, *Appl. Catal. A* 213, 247 (2001).
15. C. D. Chang, J. G. Santiesteban, and D. L. Stern, US Patent 6,080,904 (1996).
16. M. Scheithauer, R. K. Grasselli, and H. Knözinger, Genesis and Structure of WO_x/ZrO_2, *Langmuir* 14, 3019–3029 (1998).
17. M. Scheithauer, R. E. Jentoft, B. C. Gates, and H. Knözinger, n – Pentane Isomerization Catalyzed by Fe – and Mn – containing Tungstated Zirconia Characterized by Raman Spectroscopy, *J. Catal.* 191, 271–274 (2000).
18. C. D. Chang, C. T. Kresge, J. G. Santiesteban, and J. C. Vartuli, US Patent 6,124,232 (1998).
19. S. Kuba, B. C. Gates, P. Vijayanand, R. K. Grasselli, and H. Knözinger, An active and selective alkane isomerization catalyst: iron – and platinum – promoted tungstated zirconia, *J. Chem. Soc. Chem. Commun.* 321–322 (2001).
20. M. Scheithauer, T. K. Cheung, R. E. Jentoft, R. K. Grasselli, B. C. Gates, and H. Knözinger, Characterization of WO_x/ZrO_2 by Vibrational Spectroscopy and n – Pentane Isomerization Catalysis, *J. Catal.* 180, 1 (1998).
21. S. Kuba, P. Lukinskas, R. K. Grasselli, B. C. Gates, and H. Knözinger, Structure and Properties of Tungstated Zirconia Catalysts for Alkane Conversion, *J. Catal.* 216, 353–361 (2003).
22. D. G. Barton, S. L. Soled, G. D. Meitzner, G. A. Fuentes, and E. Iglesia, Structural and Catalytic Properties of Solid Acids Based on Zirconia Modified by Tungsten Oxide, *J. Catal.* 181, 57–72 (1999).
23. H. Knözinger in: *Handbook of Heterogeneous Catalysis*, Vol. 2, edited by G. Ertl, H. Knözinger, and J. Weitkamp (Wiley-VCH, Weinheim, 1997), pp. 707–732.

24. L. M. Kustov, V. B. Kazansky, S. Béran, L. Kubelkova, and P. Jiru, Adsorption of carbon monoxide on ZSM – 5 zeolites: infrared spectroscopic study and quantum – chemical calculations, *J. Phys. Chem.* 91, 5247–5251 (1987).
25. I. Mirsojew, S. Ernst, J. Weitkamp, and H. Knözinger, Characterization of acid properties of [Al] – and [Ga] – HZSM – 5 zeolites by low temperature Fourier transform infrared spectroscopy of adsorbed carbon monoxide, *Catal. Lett.* 24, 235–248 (1994).
26. J. C. Vartuli, J. G. Santiesteban, P. Traverso, N. Cardona-Martinez, C. Chang, and S. A. Stevenson, Characterization of the Acid Properties of Tungsten/Zirconia Catalysts Using Adsorption Microcalorimetry and n – Pentane Isomerization Activity, *J. Catal.* 187, 131–138 (1999).
27. S. Kuba, P. Concepción Heydorn, R. K. Grasselli, B. C. Gates, M. Che, and H. Knözinger, Redox properties of tungstated zirconia catalysts: Relevance to the activation of n – alkanes, *Phys. Chem. Chem. Phys.* 3, 146–154 (2001).
28. S. Kuba, M. Che, R. K. Grasselli, and H. Knözinger, Evidence for the Formation of W^{5+} Centers and OH Groups upon Hydrogen Reduction on Platinum – Promoted Tungstated Zirconia Catalysts, *J. Phys. Chem. B* 107, 3459–3463 (2003).
29. K. Hadjiivanov, P. Lukinskas, and H. Knözinger, Detection of reduced W^{n+} sites on WO_3 – ZrO_2 and Pt/WO_3 – ZrO_2 catalysts by infrared spectroscopy of adsorbed NO, *Catal. Lett.* 82, 73–77 (2002).
30. L. Pauling, in: *General Chemistry*, 3rd ed., (Cornell University Press, 1960), p. 409.
31. F. R. Chen, G. Coudurier, J. F. Joly, and J. C. Vedrine, Superacid and Catalytic Properties of Sulfated Zirconia, *J. Catal.* 143, 616–626 (1998).
32. D. Farcasiu and P. Lukinskas, The two modes of reaction of hexane catalyzed by trifluoromethanesulfonic acid, *J. Chem. Soc. Perkin Trans.* 2, 2715–2718 (1999).
33. P. Lukinskas, S. Kuba, B. Spliethoff, R. K. Grasselli, B. Tesche, and H. Knözinger, Role of promoters on tungstated zirconia catalysts, *Top. Catal.* 23, 163–173 (2003).
34. X. Carrier, P. Lukinskas, S. Kuba, L. Stievano, F. E. Wagner, M. Che, and H. Knözinger, The State of the Iron Promoter in Tungstated Zirconia Catalysts, *Chem. Phys. Chem.* 5, 1191–1199 (2004).
35. S. Kuba, P. Lukinskas, R. Ahmad, F. C. Jentoft, R. K. Grasselli, B. C. Gates, and H. Knözinger, Reaction pathways in n – pentane conversion catalyzed by tungstated zirconia: effects of platinum in the catalyst and hydrogen in the feed, *J. Catal.* 219, 376–388 (2003).
36. S. Kuba and H. Knözinger, Time – resolved in situ Raman spectroscopy of working catalysts: sulfated and tungstated zirconia, *J. Raman Spectrosc.* 33, 325–332 (2002).
37. J. C. Yori, C. R. Vera, and J. M. Parera, n – butane isomerization on tungsten oxide supported on zirconia, *Appl. Catal. A* 163, 165–175 (1997).

LOW TEMPERATURE OXIDATION OF CARBON MONOXIDE OVER ALKALI-METAL PROMOTED PALLADIUM-TIN OXIDE CATALYSTS

BURCU MIRKELAMOGLU, GURKAN KARAKAS*
Department of Chemical Engineering, Middle East Technical University, 06531 Ankara, Turkey

Abstract. The CO oxidation activity of sol-gel prepared tin dioxide based palladium and sodium-palladium promoted catalysts were investigated by XRD, XPS, DRIFTS and temperature programmed desorption techniques. XPS data suggests that alkali-metal promotion results in the formation of super-oxide species and promotes the surface segregation of Pd atoms, increasing the number of active sites for CO adsorption/oxidation. The formation of a ternary oxide of sodium and palladium is postulated. Through sub-ambient TPD experiments, the noble metal promoted catalysts have been identified to exhibit CO oxidation activity below ambient temperature. Steady-state CO conversion and rate of reaction data revealed that improvements in the CO oxidation activity of palladized tin dioxide could be achieved under both oxygen-rich and oxygen-lean conditions by sodium-promotion of the catalyst.

Keywords: alkali-metal promotion; tin dioxide catalysts; CO oxidation; temperature-programmed characterization

1. Introduction

Monitoring pollutants in a variety of composition ranges in motor vehicle and chemical process exhaust gases is a major area of research in pollution abatement technology. Low-temperature CO oxidation catalysts are needed for zero emission vehicles, CO gas sensors, selective oxidation of CO in H_2 rich streams in fuel cell applications,[1,2] and in closed-cycle CO_2 lasers used for remote sensing in space applications.[3-5] Effective oxidation of CO during

* To whom correspondence should be addressed. Gurkan Karakas, Department of Chemical Engineering, Middle East Technical University, 06531 Ankara, Turkey; e-mail: gkarakas@metu.edu.tr

J.P. Blitz and V.M. Gun'ko (eds.), Surface Chemistry in Biomedical and Environmental Science, 359–368.

engine cold-start period is critically important for the performance of catalytic converters. Drastic improvements in terms of the light-off temperatures, especially for CO oxidation, are achieved by the addition of palladium.[6]

Tin dioxide is a multi-functional material which is the most widely applied semiconducting metal oxide in gas sensing with solid-state sensor devices.[7] Addition of noble-metals to tin dioxide has been observed to enhance the sensitivity of the sensor towards different reducing gases, and decrease the optimum operating temperature of these sensors.[8] Although research is focused mainly on the sensor applications of palladized/platinized tin dioxide, these catalysts are effective red-ox catalysts for CO,[9] CH_4,[10] and NO_X.[11] Tin dioxide oxidizes chemisorbed CO through a mechanism of lattice oxygen abstraction with the participation of oxygen in the surface layers at temperatures above 300°C.[12,13] Different reaction routes have been proposed for CO oxidation over tin dioxide surfaces as the oxidation of chemisorbed CO to CO_2 proceeds either by reaction with ionosorbed oxygen species to yield carboxylate species,[14] or reaction with hydroxyl species to yield either formate[15] and/or carbonate species.[5,14] Palladized/platinized tin dioxide have attracted particular attention having more activity than the noble metals and the semiconductor individually, catalyzing the reaction between CO and O_2 at –27°C.[3] This synergy has been documented by various studies and the effect has been attributed to: (i) spillover of activated species from the noble metal centers to the oxide,[13,16-18] (ii) reverse spillover of oxygen species from tin dioxide to the metal,[17,19] (iii) formation of Pt/Sn alloy,[4,5,18] and (iv) localized temperature increase in the vicinity of the metal centers leading to enhanced catalytic activity.[20]

Alkali-metals are frequently used in heterogeneous catalysis to modify adsorption of diatomic molecules over transition metals through the alteration of relative surface coverages and dissociation probabilities of these molecules.[21] Alkali-metals are electropositive promoters for red-ox reactions; they are electron donors due to the presence of a weakly bonded s electron, and thus they enhance the chemisorption of electron acceptor adsorbates and weaken chemisorption of electron donor adsorbates.[22] The effect of alkali-metal promotion over transition metal surfaces was observed as the facilitation of dissociation of diatomic molecules, originating from alkali mediated electron enrichment of the metal phase and increased basic strength of the surface.[23] The increased electron density on the transition metal results in enhanced back-donation of electrons from Pd-3d orbitals to the antibonding π-molecular orbitals of adsorbed CO, and this effect has been observed as a downward shift in the IR spectra of CO adsorbed on Na-promoted Pd catalysts.[24] Alkali-metal-promotion has previously been applied to a number of supported transition metal systems, and it was observed to facilitate the weakening of C-O and N-O bonds, upon the chemisorption of these diatomic molecules over alkali-metal promoted surfaces.[25,26]

In our earlier studies CO oxidation activity of bare and palladium promoted tin dioxide catalysts were investigated, and alkali-metal promotion was applied to PdO/SnO_2 system. Through temperature-programmed adsorption/desorption studies, tin dioxide was identified to be an active CO oxidation catalyst above 200°C, and the addition of palladium was observed to contribute another active site for CO oxidation which is active at much lower temperatures. Furthermore, promotion of tin dioxide with sodium and palladium was observed to have a positive effect on the activity of tin dioxide, and this influence has been attributed to the spillover of activated species from metal centers to the oxide surface. Sodium promotion of PdO/SnO_2 was observed to enhance the CO oxidation activity by reducing the activation energy, when the catalyst was activated in reducing atmosphere.[27] Additionally, temperature-programmed reaction spectroscopy was utilized to investigate the effect of alkali-metal promotion on the CO oxidation rate, and the light-off temperatures over PdO/SnO_2. It was observed that improvements in the CO oxidation activity of PdO/SnO_2 can be achieved by the addition of 0.1 wt. % Na. Moreover, surface titration experiments utilizing a pulse technique showed that oxygen diffusion from the bulk of the catalyst effectively promoted the alkali-metal promoted catalyst, whereas it was not sufficient to promote the activity of alkali-free catalyst. These effects have been attributed to the formation of ternary palladate compounds ($Na_XPd_3O_4$) to stabilize PdO on the surface, and suppress its diffusion through the tin dioxide matrix, increasing both the oxygen storage capacity of the catalyst and the number of reaction sites for CO oxidation.[9]

2. Experimental

1 wt.% PdO/SnO_2 and 0.1 wt% Na – 1 wt.% PdO/SnO_2 catalysts were synthesized through a sol-gel route, using the appropriate amounts $SnCl_4.5H_2O$, $Pd(CH_3CO_2)_2$ and NaCl as precursors.[27] A tin alkoxide solution was obtained *in situ* by dissolving $SnCl_4.5H_2O$ in isopropyl alcohol. The hydrolysis step was initiated by slow addition of water to the alkoxide solution, and the sol was continuously stirred and aged overnight at room temperature. Aerogels were obtained by drying the gel at 100°C for 6 hours and calcining at 600°C for 6 hours under a continuous flow of dry air. The SEM examination revealed that the average particle size of the samples was 80 nm, and SnO_2, PdO/SnO_2, and Na-PdO/SnO_2 samples had BET surface areas of 32, 52 and 54 m^2/g, respectively.[27]

The X-ray powder diffraction pattern was collected (Philips, PW 1840) with a Cu target and Ni filter ($\lambda_{CuK\alpha}$ = 0.15418 nm). The IR spectra were collected with a FTIR spectrometer (Bruker, Equinox 55) equipped with a diffuse reflectance sampling stage (Praying Mantis, Harrick) utilizing the ambient chamber. The samples were analyzed by an UNI-SPECS ESCA system with a monochromator (Specs GMbh.) utilizing Mg Kα ($\lambda = 1253.6$ eV) X-ray source to examine the valence states of palladium in the powder samples. The collected spectra were referenced to the binding energy of Au $4f_{7/2}$ (84.0 eV).

Prior to dynamic testing all catalyst samples were subjected to *in situ* pretreatment procedures. To assure oxygen saturation at the surfaces samples were treated with 1.3% O_2/He at 150°C for 20 minutes, and samples subjected to this type of pretreatment procedure are referred as oxidized samples. Samples were also tested after *in situ* reduction pretreatment with 5% H_2/He at 150°C for 30 minutes, and these samples are referred to as the pre-reduced samples. Each TPD experiment consisted of a 20 minute isothermal CO and O_2 co-adsorption stage at –68°C. This temperature is 4°C higher than the dew point of CO_2 under the experimental conditions. During the co-adsorption stage, 160 ppm CO and 160 ppm O_2 in He was fed to the reactor, and this stage was followed by a temperature-programmed desorption stage to reach 550°C at a rate of 12°C/min during which 200 ppm of O_2/He was present in the gas phase. The steady-state CO oxidation activities of the catalysts were tested at 30°C, 50°C, 100°C, 150°C, 200°C and 250°C. The reported reaction rates and CO conversion values represent the activity of the catalysts after 45 minutes on stream. The weight of the catalyst was 150 mg for each run and the flow rate was 50 cm^3 (STP)/min, which gave a space velocity of 10,870 hr^{-1}. The catalytic activity was tested by maintaining two different CO-to-O_2 feed ratios, namely 0.55 and 0.45, representing the oxygen rich and lean conditions.

3. Results and Discussion

3.1. CHARACTERIZATION

To observe the effect of the addition of the alkali metal on the crystal structure, surface morphology and chemical structure, the catalysts were analyzed by XRD, DRIFTS and XPS. The XRD pattern of SnO_2 suggests that the major crystallographic phase formed was tin dioxide in a cassiterite structure (Figure 1a). Elaboration of the XRD pattern of Na-PdO/SnO_2 revealed that the addition of the noble metal did not induce any structural change in the crystal structure of SnO_2 (Figure 1b). The only effect of addition of palladium was a shoulder observed at 60.8°, which could be assigned to the {200} plane of palladium oxide. The presence of any other phase that could be associated with the presence of sodium has not been observed.

To determine the effect of alkali-metal promotion on the surface morphology, the catalysts were investigated by IR spectroscopy.[9] DRIFT spectra revealed two intense reflectance bands at 760–680 cm^{-1} and 650–570 cm^{-1} for the sodium promoted sample. Moreover, an absorption band at 535 cm^{-1}, which has previously been assigned to PdO_3,[28] was observable in the DRIFT spectra of oxidized Na-PdO/SnO_2.

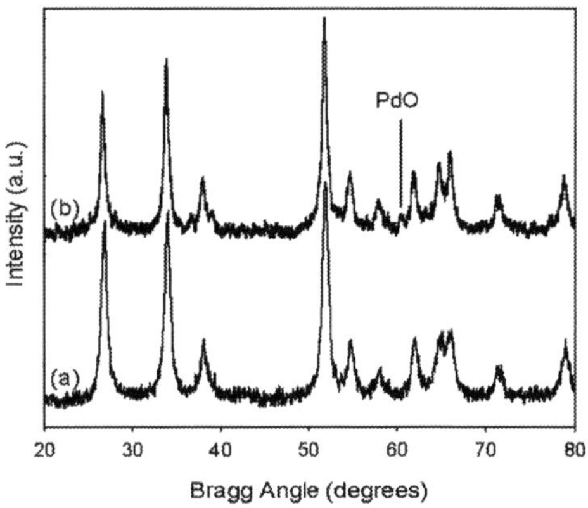

Figure 1. XRD pattern of (a) oxidized SnO_2 and (b) oxidized Na-PdO/SnO_2.

The binding states of palladium in PdO/SnO_2 and Na-PdO/SnO_2 catalysts were investigated with XPS. Analysis of the Pd 3d envelope of a PdO/SnO_2 sample suggested the presence of three different states of Pd for which, the $3d_{5/2}$ was located at 336.6 eV (Pd^{2+}), 337.7 eV (Pd^{4+}) and 339.1 eV (Pd^{6+}).[9] Pd^{4+} and Pd^{6+} species were also observed in the XPS spectra of Na-PdO/SnO_2. In addition to those, another state located at 340.6 eV was present in Na-PdO/SnO_2 sample, assigned to super oxide species formed in the presence of sodium. The quantification of XPS data to assess the relative concentrations of Pd and Sn at the surface layers probed by XPS showed that Pd concentration at the surface was doubled upon sodium promotion, which suggested that Na-promotion resulted in the surface segregation of Pd atoms. It has been postulated that the surface segregation of Pd atoms in the presence of Na resulted in the enhanced activity of Na-PdO/SnO_2 catalysts, increasing the number of active sites for CO oxidation.[9] Although the diffusivity of sodium compounds in tin dioxide is unknown, the formation of a ternary oxide of Pd and Na might be postulated to lead to surface segregation of Pd atoms through stabilizing PdO species at the surface, and suppressing the diffusion of these species through the tin dioxide matrix. Such ternary compounds of alkali-metals and platinum group metals have already been reported in the literature,[29,30] and formation of a ternary oxide of sodium and rhenium ($NaReO_4$) has been observed to alter the catalytic activity of alkali-metal promoted Re_2O_4/TiO_2 and Re_2O_4/Al_2O_3.[31]

3.2. DYNAMIC TESTING

The steady-state CO oxidation activities of SnO_2, PdO/SnO_2, and Na-PdO/SnO_2 catalysts were tested under both oxygen rich (CO-to-O_2 ratio = 1:0.55) and oxygen lean conditions (CO-to-O_2 ratio = 1:0.45). During the activity tests of PdO/SnO_2 and Na-PdO/SnO_2 samples, an initial CO oxidation activity was observed at 30°C under both feed conditions. However this appeared to be a transient activity, and the CO oxidation reaction was observed to proceed at an insignificant rate below 150°C over both catalysts (Figures 2a and 2b). Under the oxygen rich feed condition, the rate of CO oxidation reaction was calculated as 0.531 μmol/min.m^2 cat. over PdO/SnO_2 at 150°C, and approximately a two-fold increase in the reaction rate to 0.900 μmol/min.m^2 cat. was observed upon sodium promotion of the catalyst. At the same temperature the rate of reaction was insignificant over bare SnO_2, calculated as 0.072 μmol/min.m^2 cat. Although the steady-state conversion of CO converges to the same value at 200°C and 250°C for both of alkali-metal promoted and alkali-free catalysts, the alkali-metal promoted catalyst performed better than the alkali-free counterpart at temperatures below 200°C.

Upon running the same analysis under oxygen lean conditions, the alkali-metal promoted catalyst performed much better than the alkali-free catalyst in catalyzing the reaction between CO and O_2. A steady-state conversion of 90%, which was the maximum theoretical conversion of CO that could be reached, was attained above 200°C over Na-PdO/SnO_2. The maximum theoretical conversion could not be reached over PdO/SnO_2. Under the experimental conditions, the rate of reaction at 150°C was 1.571 μmol/min.m^2 cat. over Na-PdO/SnO_2, and 0.915 μmol/min.m^2 cat. over PdO/SnO_2. Unlike the oxygen rich feed condition, where the steady-state activity of both of the Pd promoted catalysts converged to the same value, under oxygen lean condition Na-PdO/SnO_2 was observed to exhibit superior catalytic activity compared with the alkali-free counterpart at all temperatures studied. The superior activity of the alkali-metal promoted catalyst can be attributed to the surface segregation of Pd atoms in the presence of the alkali, leading to a greater number of active sites for CO oxidation and increasing the rate of reaction.

Another interesting observation from the activity tests of PdO/SnO_2 and Na-PdO/SnO_2 powders was the self sustained oscillations in the species concentration profiles (Figure 3a and 3b) at temperatures above 150°C in the vicinity of complete conversion. Such sustained kinetic oscillations during noble metal catalyzed oxidation reactions are well established features of the kinetic behavior of CO oxidation over supported palladium/platinum catalysts.[32,33] The effect of alkali-metal promotion on these kinetic oscillations was observed as a significant decrease in the amplitudes of oscillations when the catalyst was promoted with sodium. In both cases, no oscillations were observed in the CO concentration (m/z) = 28 under oxygen rich conditions. Further investigations on the kinetic oscillations of CO oxidation over the PdO/SnO_2 system are necessary to arrive at conclusions on the exact nature and factors influencing these regimes.

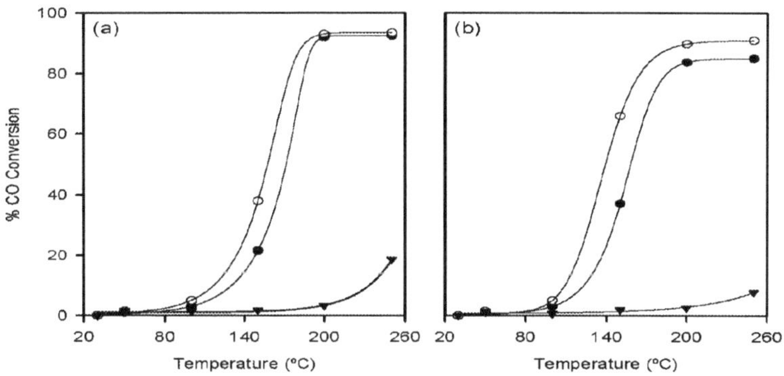

Figure 2. Conversion rate of CO over SnO_2 (▼) PdO/SnO_2 (●) and Na-PdO/SnO_2 (○) under different CO-to-O_2 ratios: (a) 1:0.55 and (b) 1:0.45 (SV=10,780 hr^{-1}).

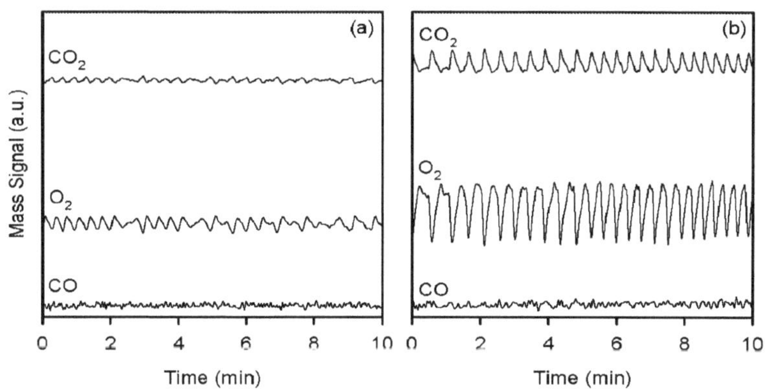

Figure 3. Oscillatory profiles of CO_2 and O_2 species during CO oxidation at 200 °C over (a) oxidized Na-PdO/SnO_2 and (b) oxidized PdO/SnO_2.

In a previous study,[27] palladium promoted SnO_2 catalysts have been observed to exhibit an initial high CO oxidation activity transient at room temperature, and this activity has been regarded as an indication of sub-ambient CO oxidation activity of the catalysts. For identification of the active sites at sub-ambient temperatures, sub-ambient temperature programmed desorption experiments were conducted. The CO_2 evolution profiles showed that a high temperature CO_2 desorption band was present in the 280-500°C region (T_{max} = 400°C) on bare SnO_2 catalysts. No significant differences were observed in CO_2 evolution profiles of both oxidized and reduced SnO_2 catalysts, which means that the activity of SnO_2 was unaffected by the oxidation and reduction pretreatments (Figure 4a and 4b). This high-temperature activity, which has been attributed to the activity of SnO_2, was also observable on PdO/SnO_2 and Na-PdO/SnO_2 catalysts. In addition to the high temperature activity of the bare support, two other desorption bands were observed in CO_2 evolution profiles of oxidized and pre-reduced palladium promoted samples (Figures 4c and 4d). These bands should be isolated from the activity of the bare support and attributed to the presence of noble metal in the structure. The first of these bands, which has also been observed in a previous study,[27] corresponds to the 100°C-200°C region. This has been assigned to the activity of PdO centers which vanish after pre-reduction treatment. The other band corresponds just to the onset of the TPD at –68°C. Alkali-metal promotion was observed to shift the CO_2 evolution band to lower temperatures and increase its intensity (Figures 4e and 4f). The decrease in temperature of the CO_2 evolution band can be attributed to a decrease in the activation energy of CO to CO_2 conversion in the presence of the alkali-metal, which has also been postulated by previous TPD studies.[27] Moreover, the increase in the intensity of the band could be attributed to the increased number of Pd atoms on the surface, which has been observed through XPS measurements.

The sub-ambient CO_2 evolution band was observed to increase in intensity after subjecting Na-PdO/SnO_2 to reduction pretreatment (Figure 4f). The effect could be attributed to the competitive adsorption of CO and O_2 over palladium surfaces. When the surface is initially free of ionosorbed oxygen species, as in the case of pre-reduced Na-PdO/SnO_2, oxygen species would not block the chemisorption of CO. Alternatively, when ionosorbed oxygen species are abundant over the surface of oxidized Na-PdO/SnO_2, these species would hinder the chemisorption of CO, decreasing the CO concentration on the surface.

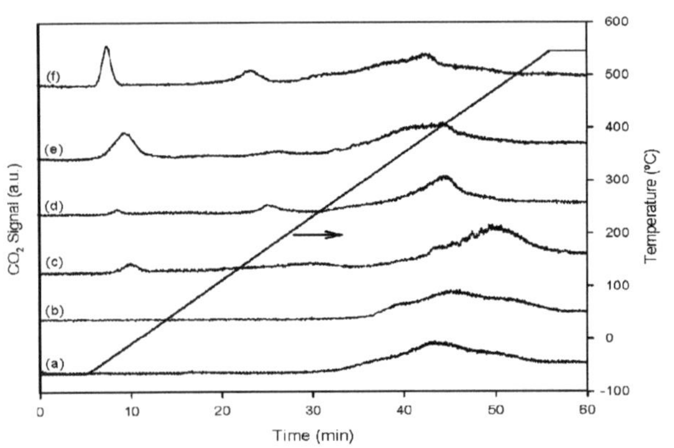

Figure 4. CO_2 evolution profiles during post CO-O_2 co-adsorption He TPD (a) oxidized SnO_2, (b) pre-reduced SnO_2, (c) oxidized PdO/SnO_2, (d) pre-reduced PdO/SnO_2, (e) oxidized Na-PdO/SnO_2, and (f) pre-reduced Na-PdO/SnO_2.

4. Conclusion

SnO_2-based catalysts were prepared by a sol-gel route and characterized with surface-spectroscopic and dynamic methods. XPS data suggested the segregation of palladium atoms at the surface layer. The formation of sodium oxopalladate ($Na_XPd_3O_4$) was postulated to result in the increased surface concentration of palladium atoms, possibly by stabilizing Pd at the surface and suppressing the diffusion of PdO through the tin dioxide matrix during the calcination stage. Formation of super-oxide species was observed over Na-PdO/SnO_2, and these species were suggested to increase the oxygen storage capacity of the alkali-metal promoted catalysts.

The catalysts were identified as active for CO oxidation below room temperature, studied by utilizing the temperature-programmed adsorption/desorption technique to observe the presence of a phase active below ambient temperature. Steady-state reaction and CO conversion rates suggest that the reaction proceeded at an insignificant rate below 150°C. Alkali-metal promoted PdO/SnO_2 was observed to be superior to PdO/SnO_2 as a CO oxidation catalyst. Under both oxygen-rich and oxygen-lean conditions, significant improvements in the rate of reaction and CO conversion were achieved upon Na-promotion. This fact has been attributed to the presence of an increased number of reaction sites in the alkali-metal promoted samples. A decrease in the activation energy of CO to CO_2 conversion was also postulated.

Acknowledgements

This work was supported by Middle East Technical University under grant BAP 2003-03-04-03.

References

1. J. N. Armor, The multiple roles for catalysis in the production of H_2, *Appl. Catal. A* 176, 159-176 (1999).
2. H. Tanaka, S. Ito, S. Kameoka, K. Tomishige, and K. Kunimori, Catalytic performance of K-promoted Rh/USY catalysts in preferential oxidation of CO in rich hydrogen, *Appl. Catal. A* 250, 255-263 (2003).
3. D. S. Stark and M. R. Harris, Catalysed recombination of CO and O_2 in sealed $CO_{2\ \mathrm{TEA}}$ laser gases at temperatures down to –27°C, *J. Phys. E: Sci. Instrum.* 16, 492-496 (1983).
4. E. Drawdy, G. B. Hoflund, S. D. Gardner, E. Yngvadottir, and D. R. Schryer, Effect of pretreatment on a platinized tin oxide catalyst used for low-temperature CO oxidation, *Surf. Interface Anal.* 16, 369-374 (1990).
5. D. R. Schryer, B. T. Upchurch, J. D. van Norman, K. G. Brown, and J. Schryer, Effects of pretreatment conditions on a Pt/SnO_2 catalyst for the oxidation of CO in CO_2 lasers, *J. Catal.* 122, 193-197 (1990).
6. M. Skoglundh, and E. Fridell, Strategies for enhancing low-temperature activity, *Top. Catal.* 28(1-4), 79-87 (2004).
7. D. Kohl, Surface processes in the detection of reducing gases with SnO_2-based devices, *Sens. Actuators* 18, 71-113 (1989).
8. P. Siciliano, Preparation, characterization and applications of thin films for gas sensors prepared by cheap chemical method, *Sens. Actuators B* 70, 153-164 (2000).
9. B. Mirkelamoglu and G. Karakas, The role of alkali metal promotion on CO oxidation over PdO/SnO_2 catalysts, submitted to *Appl. Catal. A*. (2005) in press.
10. P. Gélin and M. Primet, Complete oxidation of methane at low temperature over noble metal based catalysts, a review, *Appl. Catal. B* 39, 1-37 (2002).

11. D. Amalric-Popescu and F. Bozon-Verduraz, SnO_2-supported palladium catalysts, activity in deNOx at low temperature, *Catal. Lett.* 64, 125-128 (2000).
12. M. J. Fuller and M. E. Warwick, The catalytic oxidation of carbon monoxide on tin(IV) oxide, *J. Catal.* 29, 441-450 (1973).
13. S. Emiroglu, N. Bârsan, U. Weimar, and V. Hoffmann, In situ diffuse reflectance infrared spectroscopy study of CO adsorption on SnO_2, *Thin Solid Films* 391, 176-185 (2001).
14. M. M. Gadgil, R. Sasikala, and S. K. Kulshreshtha, CO oxidation over Pd/SnO_2 catalyst, *J. Mol. Catal.* 87, 297-310 (1994).
15. S. H. Hahn, N. Bârsan, U. Weimar, S. G. Ejakov, J. H. Visser, and R. E. Soltis, CO sensing with SnO_2 thick film sensors, role of oxygen and water vapour, *Thin Solid Films* 436, 17-24 (2003).
16. M. Sheintuch, J. Schmidt, Y. Lecthman, and G. Yahav, Modelling catalyst-support interactions in carbon monoxide oxidation catalysed by Pd/SnO_2, *Appl. Catal.* 49, 55-65 (1989).
17. S. K. Kulshreshtha and R. Sasikala, Pd/SnO_2 and Pt/SnO_2 catalysts, spill over effects, *Indian J. Chem.* 33A, 115-119 (1994).
18. T. Takeguchi, O. Takeoh, S. Aoyama, J. Ueda, R. Kikuchi, and K. Eguchi, Strong chemical interaction between Pd and SnO_2 and the influence on catalytic combustion of methane, *Appl. Catal. A* 252, 205-214 (2003).
19. K. Grass and H. G. Lintz, The kinetics of carbon monoxide oxidation on tin(IV) oxide supported platinum catalysts, *J. Catal.* 172, 446-452 (1997).
20. N. D. Gangal, N. M. Gupta, and R. M. Iyer, Microcalorimetric study of the interaction of CO, O_2 and CO + O_2 with Pt/SnO_2 and SnO_2 catalysts, *J. Catal.* 126, 13-25 (1990).
21. G. A. Somorjai, The catalytic hydrogenation of carbon monoxide. The formation of C_1 hydrocarbons, *Cat. Rev. - Sci. Eng.* 23, 189-202 (1981).
22. C. G. Vayenas, S. Bebelis, C. Pliangos, S. Brosda, and D. Tsiplakides, *Electrochemical Activation of Catalysis* (Kluver Academic/Plenum, New York, 2001), p. 35.
23. H. Wu, L. Liu, and S. Yang, Effects of additives on supported noble metal catalysts for oxidation of hydrocarbons and carbon monoxide, *Appl. Catal. A* 211, 159-165 (2001).
24. L. F. Liotta, G. A. Martin, and G. Deganello, The influence of alkali metal ions in the chemisorption of CO and CO_2 on supported palladium catalysts, a Fourier transform infrared spectroscopic study, *J. Catal.* 164, 322-333 (1996).
25. N. Macleod, J. Isaac, and R. M. Lambert, Sodium promotion of Pd/γ-Al_2O_3 catalysts operated under simulated "three-way" conditions, *J. Catal.* 198, 128-135 (2001).
26. S. J. Pratt and D. A. King, Coverage dependent promoter action, K adsorption and reactions with CO and CO_2 on Pd{110}, *Surf. Sci.* 540, 185-206 (2003).
27. B. Mirkelamoglu and G. Karakas, CO oxidation over palladium- and sodium-promoted tin dioxide, catalyst characterization and temperature-programmed studies, *Appl. Catal. A* 281, 275-284 (2005).
28. J. M. Tura, P. Regull, L. Victori, and M. D. de Castellar, XPS and IR (ATR) analysis of Pd oxide films obtained by electrochemical methods, *Surf. Interface Anal.* 11, 447-449 (1988).
29. J. J. Scheer, A. E. van Arkel, and R. D. Heyding, Oxide complexes formed in the systems platinum metals, alkali carbonates, oxygen, *Canad. J. Chem.* 33, 683-686 (1955).
30. B. L. Dubey, J. A. Gard, F. P. Glasser, and A. R. West, Synthesis, structure and stability of phases in the system Li_2O-Pd-O_2, *J. Solid State Chem.* 6, 329-334 (1973).
31. B. Mitra, X. Gao, I. E. Wachs, A. M. Hirt, and G. Deo, Characterization of supported rhenium oxide catalysts, effect of loading, support and additives, *Phys. Chem. Chem. Phys.* 3, 1144-1152 (2001).
32. G. Ertl, P. R. Norton, and J. Rüstig, Kinetic oscillations in the platinum-catalyzed oxidation of CO, *Physical Review Letters* 49(2), 177-180 (1982).
33. V. V. Gorodetskii, A. V. Matveev, E. A. Podgornov, and F. Zaera, Study of the low-temperature reaction between CO and O2 over Pd and Pt surfaces, *Topic. Catal.* 32(1-2), 17-28 (2005).

NANOPARTICLES OF ALUMINIUM SALTS HYDROLYSIS PRODUCTS IN WATER TREATMENT AND DISINFECTION

S. BARANY,[1,*] J. GREGORY,[2] A. SHCHERBA,[3] I. SOLOMENTSEVA[3]
[1] *University of Miskolc, Institute of Chemistry, H-3515 Miskolc-Egyetemváros, Hungary*
[2] *University College London, Gower Street, London, WC1E 6BT, UK*
[3] *Institute for Electrodynamics, National Academy of Sciences of Ukraine, Kiev, Ukraine*

Abstract. The size, surface area, electrokinetic potential, hydration, as well as rate and degree of aggregation of aluminium salts hydrolysis product particles (HPP) as a function of salt dosage, pH, alkalinity, water ionic strength, and OH/Al ratio in the salt molecule has been determined. A relationship between surface characteristics and the degree and mechanisms of HPP aggregation is established. The efficiency of aluminium salts HPP in removal of dispersed particles and bacteria from water is demonstrated. A new method of electric spark discharge for producing aluminium-hydroxide HPP and its use in water treatment is described.

Keywords: aluminium salts, hydrolysis products, size, electrokinetic potential, hydration, aggregation rate, salt dosage, pH, alkalinity, ionic strength, water treatment, disinfection, electric sparks

1. Introduction

Aluminium salts are widely used as coagulants in water treatment. They are effective in removing a broad range of impurities from water, including colloidal particles and dissolved organic substances. Their mode of action is generally explained in terms of two distinct mechanisms: charge neutralisation of typically negatively charged colloids present in water by cationic hydrolysis products, and incorporation of impurities in an amorphous hydroxide precipitate

*To whom correspondence should be addressed. S. BARANY, University of Miskolc, Institute of Chemistry, H-3515 Miskolc-Egyetemváros, Hungary, e-mail: akmbsab@gold.uni-miskolc.hu

J.P. Blitz and V.M. Gun'ko (eds.), Surface Chemistry in Biomedical and Environmental Science, 369–382.

('sweep flocculation').The relative importance of these mechanisms depends on factors such as pH, coagulant dosage, and the composition and content of electrolytes in water. The key processes while using aluminium salts in water treatment are[1-3] (1) hydrolysis of coagulant with formation of monomeric units:

$$Al^{3+} \rightarrow Al(OH)^{2+} \rightarrow Al(OH)_2^{+} \rightarrow Al(OH)_3 \rightarrow Al(OH)_4^{-}$$

various oligomeric and polymeric hydrolysis products may be formed, the best known of which is $Al_{13}O_4(OH)_{24}^{7+}$, as well as an amorphous hydroxide precipitate; (2) adsorption of Al^{3+} and polyions on negative particles (charge neutralisation); (3) adsorption of organics on freshly precipitated $Al(OH)_3$; (4) heterocoagulation of positive $Al(OH)_3$ and negative particles of contaminants; and (5) mechanical incorporation of contaminants into growing $Al(OH)_3$ precipitates ("sweep" coagulation).

The pH for precipitation of $Al(OH)_3$ is 4.8-5.0, and the minimum solubility of precipitate is observed in the pH 6-7 region.[1-3] Above pH 8.5 dissolution of the precipitate takes place. The size of primary hydrolysis product particles (HPP) and polymeric species strongly depends on the concentration of the coagulant, its OH/Al ratio (in the case of pre-hydrolysed basic salts), and water parameters such as pH, alkalinity, electrolyte content, temperature, etc. Depending on the parameters, the size of HPP varies from several nanometers (3-5 nm) to dozens of nanometers, and even to micrometers (d = 3-50 nm→2-3 μm→10-30 μm).[1,3] Highly polymerised hydroxo-complexes can form chain-like clusters of nanometer size, for example the radius of an Al13 cluster reaches ~40 nm.[1] Hydrolysis product particles arising immediately after hydrolysis quickly aggregate. The rate and mechanisms of this process, i.e. whether particles aggregate in the primary or secondary minimum, the size and strength of aggregates formed, are a function of the system parameters that determine the colloid-chemical properties of aggregating particles. Consequently, the colloid-chemical characteristics of HPP have a profound effect on the mode and efficiency of water treatment. There is limited literature data devoted to elucidation of the role of HPP parameters in the water treatment processes.

The aim of our study is to establish a relationship between the mechanism of formation and colloid-chemical properties of hydrolysis product particles (HPP) from traditional coagulant- aluminium sulphate (AS), novel prehydrolysed forms - basic aluminium sulphates (BAS), and basic aluminium chlorides (BAC) with different OH/Al ratio. Also, the efficiency of products of hydrolysis of these coagulants for removal of inorganic impurities and biological contaminants from water has been investigated. Such an approach enables control of the coagulation process by changing system pH, type, concentration and dose of coagulant, solution ionic strength, and other parameters that affect the main colloid-chemical characteristics of the disperse phase. To characterise the HPP we have performed a complex study including determination of the size, surface area, electrokinetic potential and degree of hydration of AS, BAS, BAC hydrolysis product particles as a function of the aforementioned variables. Some previous results have been published.[3-7]

2. Materials and Methods

We have studied the properties of hydrolysis products formed from aluminium sulphate (AS), basic aluminium sulphates (BAS) and basic aluminium chlorides (BAC) in water as a function of the concentration of the salt, its dosage, pH, alkalinity and ionic strength of the solution. Samples characteristics are given in Table 1.

Table 1. Characteristics of AS, BAS and BAC studied.

AS/BAS Sample	OH/Al mol. ratio	Al-oxide content (%)	No of BAC Sample	OH/Al mol. ratio	Al-oxide content (%)
1	0	16.73	1	1.0	7.30
2	0.25	18.00	2	2.0	14.60
3	0.5	19.33	3	2.5	22.75
4	0.75	20.93			
5	1.0	23.29			

The methods of determining the HPP parameters including degree of aggregation, aggregate properties, and water purification procedures will be briefly described in the following sections.

3. Size and Surface Area of Hydrolysis Product Particles

To get information about the size and surface area of HPP, as well as aggregation kinetics, we have used two methods: (1) adsorption of methylene blue[5] dye to estimate the accessibility of the surface of the hydrolysis product particles during formation of primary aggregates; and (2) the laser diagnostic method[8] that gives information about the size-to-density distribution of aggregates in the course of their successive enlargement.

Table 2 shows the effect of initial pH_o and alkalinity on the average radius of primary particles (a_0) of the BAC HPP, with different OH/Al molar ratios, as well as on the decrease of specific surface area in the course of aggregation (ΔS_{rel},%) The particles radius was calculated from the specific surface area assuming nonporous spherical particles. It is seen that an increase of pH and alkalinity, as well as a decrease in the OH/Al ratio, cause the formation of primary particles of larger radius. The parameter ΔS_{rel} was calculated as the relation between the difference of the specific surface area (determined by adsorption of methylene blue) of the HPP at the moment of formation for $\tau = 0$, and that for the aggregate/formation time in excess of 10 min. A decrease in S takes place for approximately 10 min, i.e. primary aggregates of the hydrolysis products are likely to be formed within this period of time from the particles having units of nanometers (Table 2).

Table 2. Effect of solution pH_0 and alkalinity on the size of primary particles and decrease of specific surface area of BAC HPP with different OH/Al ratio.

pH_o	Alkalinity (mmol/L)	OH/Al = 2.5		OH/Al = 2.0		OH/Al = 1.0	
		a_o (nm)	ΔS_{rel} (%)	a_o (nm)	ΔS_{rel} (%)	a_o (nm)	ΔS_{rel} (%)
7.5	2.33	3.1	18.0	3.8	24.2	5.0	29.6
8.3	2.33	3.7	22.5	4.3	27.0	5.6	32.0
9.1	2.33	5.1	28.0	5.7	31.0	6.5	35.5
10.4	2.33	7.7	32.3	8.0	42.6	8.7	44.4
9.1	1.17	3.1	27.3	3.9	29.7	4.9	34.5
9.1	2.33	5.1	28.0	5.7	31.0	6.5	35.5
9.1	3.50	5.8	31.4	6.3	34.3	7.5	37.6
9.1	5.35	6.9	34.2	7.4	37.4	8.1	40.3

It is also seen that an increase of pH_0 and alkalinity leads to an increase in the ΔS_{rel} value. These data agree with the established (see below and Refs. 5and 6) relationships of the electrokinetic properties and hydration of HPP. It was shown that an increase of system pH causes a decrease in the ζ-potential of BAC hydrolysis products, and an increase in their hydration. We consider that an increase of ΔS_{rel} in the studied pH range and alkalinity is related to a predominant effect of the electrokinetic potential of particles on the properties of primary aggregates formed from them. At the same time, an increase in ΔS_{rel} when decreasing the OH/Al ratio of the BAC (Table 2) should be related to an increase in the density of primary aggregates formed during their hydrolysis in the same direction. The latter is possible if we take into account that the ζ-potential and the hydration of HPP decreases when passing from high-to low-basic coagulants. This contributes to the formation of less diffuse aggregates. Similar dependencies, i.e. an increase in the size of primary particles formed during the hydrolysis process with increasing alkalinity and water pH, and decreasing the OH/Al ratio of the BAS have been found. Typically, dimensions of these particles originating from basic aluminium sulfate are bigger compared to those from basic aluminium chlorides at the same conditions. In the range of pH and alkalinity studied, the minimum and maximum sizes of primary particles from BAS hydrolysis reach 3.0 nm and 11.8 nm, and from BAC 3.1 nm and 8.7 nm. This corresponds to the minimum sizes of aluminium hydroxide particles.[1,3,5]

4. Electrosurface Properties and Hydration of HPP

The influence of major variables indicated above on the electrokinetic potential and degree of hydration of AS, BAS and BAC hydrolysis product particles has been studied in detail. The electrophoretic mobility of particles was measured by microelectrophoresis. As a measure of hydration of HPP the time of spin-spin relaxation of protons of adsorbed/adherent water (T_1) was used. The T_1 values were measured by NMR with a pulse relaxometer by the null method.[5] The results obtained can be summarised as follows.

(1) With increasing BAS dosages over the range studied, the ζ-potential of the colloidal HPP is lowered (Figure 1), while for BAC the opposite occurs: increasing BAC dosage from 3 to 40 mg/L results in a doubling of the ζ-potential. This dependence is similar for all coagulant samples, but for BAS with higher OH/Al ratio the electrokinetic potential values are higher. Our investigations show that hydrolysis products from BAC are less hydrated than those from BAS, and their hydration decreases as the dosage is raised. This should bring about a shift of the slip plane towards the particle surface and a corresponding rise in the ζ- potential. The hydrolysis products of BAS are more hydrated, and the decrease in ζ- potential is associated with the slip plane's shift towards the solution.

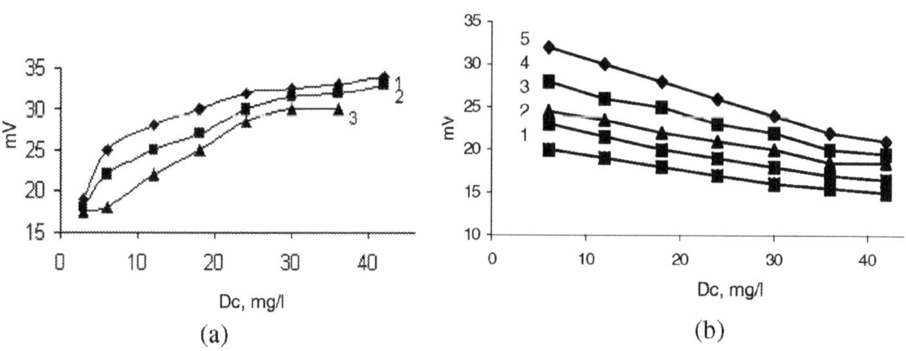

Figure 1. Dependence of the electrokinetic potential of BAC (a) and BAS (b) HP on the coagulant dose. OH/Al ratio in molecule for BAC: 2.5 (1), 2.0 (2), 1 (3) and for BAS: 0 (1), 0.25 (2), 0.5 (3), 0.75 (4) and 1.0 (5). pH 8.3.

(2) The system pH also has a profound influence on the electrokinetic potential of BAS and BAC hydrolysis product particles (Figure 1). Between pH 6.5 and 8.7 , the ζ-potential for aluminium sulphate hydrolysis product fell from 40 to 22 mV, and from 23 mV at pH 7.5 to 14 mV at pH 8.7 when using the coagulant with highest OH/Al ratio. There is a difference between BAC and BAS: the sign of particle charge is reversed at higher pH values with basic aluminium sulfate compared with basic aluminium chloride. Also the minimum pH ranges at which microscopically visible spherical primary aggregates appear

are different between BAS and BAC; around 6.5 with BAS and pH 7.5 with BAC.

(3) The relationship between the ζ- potential of the hydrolysis product particles and the ionic strength (I) of the solution reveals non-trivial behaviour. A monotonic increase in the electrokinetic potential of BAS hydrolysis products is seen with increasing electrolyte concentration in the region of -logI between 3.5 and 1.5. For BAC HPP a passage of the zeta-potential vs. ionic strength has been detected (see more in Refs. 5 and 6). The variations in ζ-potential of BAS HPP, and the differences compared to the BAC behaviour, can be explained under the assumption that the relevant parameters are influenced by two groups of factors. The first are pH effects in both the coagulant solutions themselves and the water under treatment. Secondly, the presence of sulfate-ions in the system can lead to the sorption of some sulphate-ions in the Stern layer (promoting a reduction in ζ-potential and enhancing hydration), and/or to a salting-out effect, whereby particle hydration is reduced and the slip plane is shifted towards the surface.

Hydration of HPP. All factors mentioned above affect the T_1 value, showing the change in the state of water to be the result of a change in the composition of the water and the surface properties of the disperse phase. Aluminium hydrolysis products are mostly particles of $Al(OH)_3$ with aluminium hydroxyl complexes adsorbed on them, so the change in T_1 can be associated with an altered nature and number of hydrophilic centres. It has been shown[5,6] that in all cases the spin-spin relaxation time of water protons decreases with increasing OH/Al ratio in the coagulant molecule. This is an evidence of increased hydration of particles surface in this direction.

A higher dose of coagulants reduces the hydration of BAC HPP, T_1 increases and increases the ζ-potential. This can be attributed to the shift of the slip plane toward the surface as a result of adsorption of lesser hydrated hydrolysis products and reduced chemisorption of aluminium polyhydroxyl complexes. A rise in the pH from 7 to 9.3 is accompanied by a gradual reduction in the T_1 time indicating increased hydration is in this direction because of increasing adsorption of hydroxyl complexes. This is accompanied by a decrease of the ζ-potential. It is also shown that there is a significant reduction in the hydration of particles surface with rising ionic strength from 10^{-3} to 10^{-1}.

5. Aggregation of HPP

Analysis of oscillograms (plots of the fluctuation amplitude of light flux vs. time[5,8]) testifies to the rapid occurrence of the HPP aggregation: fluctuation entities can be observed practically immediately (within a few seconds) after the starting of oscillogram records (30 s after mixing the system components).

From the size distribution of hydrolysis product particles obtained after fixed time intervals (Figure 2), it follows that 1 min after adding the reagents to the simulated water, an appearance of a large number of primary aggregates with radius of about 1-2 μm is registered.

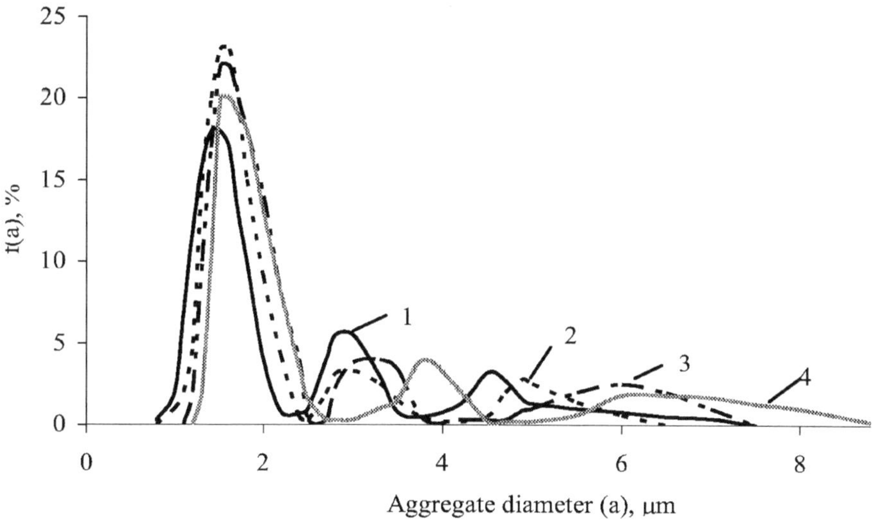

Figure 2. Distribution curves of aggregates of AS HPP by their size at different time of flocs formation. Time, min: 1 (1); 3 (2); 10 (3); and 20 (4).

The amount of these aggregates vs. time reveals a maximum. In most cases each curve has several maxima which differ in height and duration. This indicates a stepwise aggregate formation in the system, i.e. the presence of several predominant aggregate fractions at each time interval. The oscillograms also show that the packing density of particles in aggregates is higher for BAS hydrolysis products compared to BAC. During the stepwise aggregation process, the density of subsequent aggregates decreases. This testifies to the non-uniform density distribution of aggregates in the volume. There are two possible models to explain this fact: (1) the existence of a more dense core in the aggregate that is formed at the very beginning of the process, and of a more diffuse peripheral part that is formed from aggregates of higher orders at the later stages of the process; (2) a mosaic fluctuation of density in the aggregates volume due to a different type of binding of primary (first order) aggregates into secondary ones, of the secondary (second order) aggregates into tertiary ones, and so on. In the second case aggregates contain more dense and less dense fractions. We can not make a definite choice in favour of one of these models but some of our additional laser diagnostic experiments suggest the second model.

From the size distribution curves, derivative dependencies of the predominant fraction radius of primary aggregates of the BAS hydrolysis products on alkalinity, pH, coagulant dose and ionic strength, have been constructed.[3] The results obtained can be summarised as follows: (i) in all cases (for the exception of high coagulant doses) the size of primary aggregates for BAS are less than that for BAC (compare with Ref. 5); (ii) with increasing the OH/Al ratio of the BAS coagulant, the size of primary (and successive) aggregates increases; (iii) an increase in the alkalinity and system pH results in larger HPP aggregates; (iv) with increasing ionic strength of the solution the radius of predominant fraction of primary aggregates goes through a maximum; and (v) an increase in the coagulant dose results in an increased aggregate size for BAS hydrolysis products, but a decrease for BAC hydrolysis products.[5] The latter is in line with a decrease of the electrokinetic potential of BAS HPP with increasing reagent dosage in water (see above).

6. Removal of Dispersed Particles

According to classical ideas of colloid stability, destabilisation can be brought about by either: (i) an increase in ionic strength, giving some reduction in the zeta- potential and a decreased thickness of the diffuse part of the electrical double layer, or (ii) specific adsorption of counterions to neutralise the particle charge. Aluminium salts give cationic hydrolysis products that are strongly adsorbed on typically negatively charged contaminant particles, resulting in effective destabilisation and coagulation mainly due to the second mechanism. Results on coagulation of kaolin suspensions with aluminium sulfate at pH 7 show (Figure 3) that the optimum coagulation dosage corresponds with the condition where the zeta potential of the particles is close to zero, indicating that charge neutralisation is responsible for the destabilisation of the clay particles.

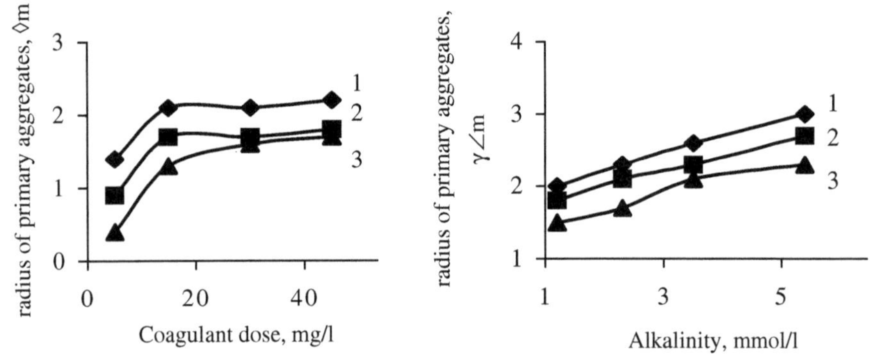

Figure 3. Dependence of radius of primary aggregates of BAS HPP on the coagulant dose and system alkalinity. OH/Al ratio: 2.5 (1); 2.0 (2); and 1.0 (3).

If charge neutralisation is the predominant destabilisation mechanism, then there should be a stoichiometric relationship between the particle concentration and the optimum coagulant dosage. At low particle concentrations, low coagulant dosages should be required. Under these conditions coagulation rates can be very low, which causes problems in water treatment. Another practical difficulty is that the optimum coagulant dosage range can be quite narrow, which means that rather precise dosing control is needed.

Both of these difficulties can be overcome by using higher coagulant dosages, where extensive hydroxide precipitation occurs, giving *sweep flocculation.* In this case optimal removal of particles from water is achieved under conditions of rapid and extensive hydroxide precipitation. In the case of aluminium coagulants, optimum pH values are around 7, close to the minimum solubility of $Al(OH)_3$ and close enough to the i.e.p. to give fairly rapid aggregation of the colloidal precipitate particles. At these conditions impurity particles are enmeshed in a growing hydroxide precipitate and effectively removed from suspension. Sweep flocculation generally gives considerably improved particle removal than when particles are destabilised just by charge neutralisation. At least part of the reason is the greatly improved rate of aggregation, because of the increased solids concentration. Hydroxide precipitates tend to have a rather open structure, so that even a small mass can give a large effective volume concentration, and hence a high probability of capturing other particles. It is also possible that binding of particles by precipitated hydroxide may give stronger aggregates. Increasing the coagulant dosage in the sweep region gives progressively larger volumes of sediment[9] but, beyond the operational optimum dosage, there is little further improvement in particle removal.

The different mechanisms outlined above have led to the definition of four *zones* of coagulant dosage, with the following consequences for negatively charged particles: **Zone 1:** Very low coagulant dosage; particles still negative and hence stable; **Zone 2:** Dosage sufficient to give charge neutralisation and hence coagulation; **Zone 3:** Higher dosage giving charge neutralisation and restabilisation; and **Zone 4:** Still higher dosage giving hydroxide precipitate and sweep flocculation.

The results of jar tests and electrophoretic mobility (EM) measurements for kaolin suspensions with AS at pH 7 (Figure 4) show that below about 8 μM Al there is essentially no reduction in turbidity, since the particles are negatively charged and colloidally stable (Zone 1). There is a narrow range of lowered turbidity in the region of 15 μM Al, which is close to the dosage where the EM is reduced to zero (Zone 2). By 20 μM Al, the particles are positively charged and completely restabilised, since the residual turbidity is no lower than that for the original clay suspension (Zone 3). Beyond about 60 μM Al the turbidity falls again as a result of sweep flocculation (Zone 4). It is very significant that a substantial change in residual turbidity occurs in a region of alum dosage where the EM of the particles is still positive, and shows no appreciable reduction.

Although there is a gradual reduction in EM as the alum dosage is increased, this is not obviously related to the degree of turbidity removal. It is also worth noting that the residual turbidity in Zone 4 is significantly lower than in Zone 2, indicating a much greater degree of clarification by sweep flocculation.

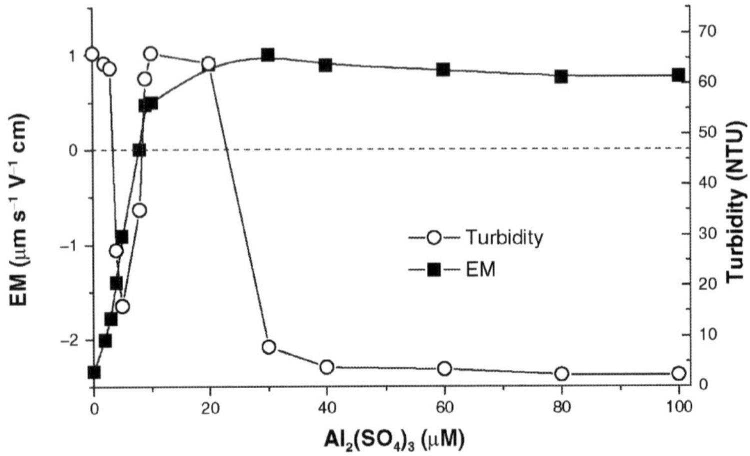

Figure 4. Electrophoretic mobility (EM) and residual turbidity for kaolin suspensions (50 mg/L) vs. dosage of aluminium sulfate at pH 7.[3]

7. Disinfection by Aluminium Salts

We have investigated the removal of *E. coli* cells (strain 1257) from model water using coagulation by aluminium sulfate and basic aluminium sulfate with OH/Al molar ratio in the salt molecule of 1.0. A detailed description of the experimental procedure is given in Ref. 7. The contact time of the coagulant with bacteria was 1 hour. The solution was then filtered through a paper filter, throwing away the first 100 ml of filtrate. The remained solution was tested on the amount of survived bacteria. The results were calculated as the ratio of bacteria remaining in solution (N_1) to their initial quantity (N_0).The ionic strength of the model water changed during our experiments from 1×10^{-3} to 2×10^{-2} g-ion/L; the alkalinity of the initial water varied between 0.5 and 5 mg-equiv/L. The doses of aluminum sulfate and basic aluminum sulfate coagulants varied between 5 and 40 mg/L, solution pH was 7.0 and 8.0. Kinetic experiments showed that the majority of microorganisms are removed from the water during the first 10 min (98% or 1.8 order of 5) of the coagulant contact with cells. Almost the same time (10-15 min) is needed to reach the final (adsorbing) surface area of AS and BAS hydrolysis products particles: 90 and 120 m^2/g, respectively, several times less than the specific surface area for particles formed during the first seconds of the process. The main regularities of *E. coli* cells removal from water by aluminium salts HPP can be summarised as follows (example: Figure 5): (1) the highest degree of *E. coli* removal was observed at low aluminium sulphate concentrations, an increase in the coagulant dose from 5 to 40 mg/L for 10^4 CFU/mL content (CFU is the colony forming unit, the number of living cells) is accompanied by a 1-1.5 order decrease of the removal efficiency of micro-organisms from the water. (2) The higher the bacteria concentration in the water, the higher is the coagulant dose necessary to reach a certain degree of clarification.

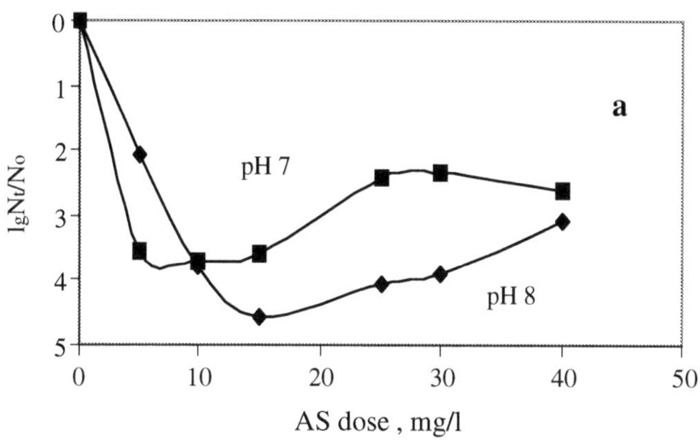

Figure 5. Dependence of *E. coli* extraction by AS vs. coagulant dose at pH 7.0 and 8.0.

Thus, at water contamination 10^4 CFU/ml, the highest *E. coli* removal of 99.996% (4.35 order) is observed at 5 mg/L of $Al_2(SO_4)_3$, while at water contamination 10^5 CFU/ml, 15 mg/L of $Al_2(SO_4)_3$ were required to reach the same level of *E. coli* removal. When polluting model water with 10^3, 10^4 and 10^5 CFU/ml of *E. coli*, the degree of removal of microorganisms at these conditions reaches on average 2° (99%), 3.5° and 4.6°, respectively. (3) The degree of cell removal from waters in the presence of AS is higher than by BAS, the highest removal is observed at BAS and AS dosages 5 and 15 mg/L, respectively. (4) A decrease in the pH (alkalinity 1 mg-equiv/L) of water containing 10^5 CFU/ml from 8.0 to 7.0 worsens the efficiency of the bacteria removal using AS by one degree, from 99.997 to 99.97, respectively. Analogous dependencies have been observed for BAS. At an optimal BAS dose (5 mg/L) the degree of *E. coli* removal from water with pH 8.0 is 2.8 fold, and in water with pH 7.0 – 3.51 (or 99.9% removal). (5) The dependence of the cell extraction vs. the system alkalinity both for BAS and AS reveal a minimum and a maximum. Optimal alkalinity for bacteria removal using AS is 0.5-1 mg-equiv/L: under this condition 99.8% of the added cells (4×10^5 CFU/ml) can be removed, and only a few single cells remain in solution. An increase of solution alkalinity leads to a significant decrease in purification efficiency. The dependence of water purification using BAS on system alkalinity is similar to that for AS, but shifted to higher removal efficiency values. At a BAS dose of 5 mg/L, the optimal alkalinity to remove test microorganisms is 1.0 mg/L. The degree of cell removal in this case reaches 4.8 order or 99.998%. The unusually high efficiency of bacteria removal (five orders) from water by small doses of AS and BAS coagulants should be stressed. There are no data on similar effects while using hydrolysing salts for purification of water from inorganic colloids. The mechanism of purification from *E. coli* cells is likely *sweep coagulation.*

8. Method of Electric Sparks Discharge for Producing Aluminium-Hydroxide HPP and its Use in Water Treatment

Electric sparks arising between metal granules of Al or Fe in electric fields can be used for producing micro- and nano-sized hydroxide particles. These hydroxides possess a unique structure, and properties that can be efficiently used for water treatment and disinfection.[10,11]

The feature of this novel method is the formation of multi-channel, quickly migrating volume electrospark discharges between the particles of coagulant-forming metal submerged into water, producing highly-active coagulant $Al(OH)_3$ with particle sizes in the 3-70 micrometer range. The particle size and their concentration depend on the quantity of the electric power supplied to the technological unit, slope, amplitude and duration of the discharge current pulse, the heat dissipation rate from sparking contacts zone, and on the physical properties of the current conducting material, the thermal capacity and temperature of the treated water, etc.

Table 3 shows the parameters of $Al(OH)_3$ particles received by the electric spark discharge method.

Table 3. Parameters of aluminium hydroxide received by electric spark erosion.

Pulse power P_P (kW)	Sediment weight (g/L)	Specific surface area (m^2/g)	Measured radius of particles (nm)
16.3	0.21	115	7.5
29.26	0.55	48	15
42.65	0.87	38	18
57.3	1.14	24	24
81.3	6.67	5	40

The conditions are capacitor voltage 100-500 V, capacitance C = 12.5 μF, pulse duration 20-80 μsec, frequency f = 25 Hz, distance between electrodes $l = 52$ mm, width of the granules layer $b = 24$ mm, height $h = 25$ mm, mean diameter of a granule in the layer of 4 mm, alkalinity of the water in cell 2 mmol/L, and ionic strength: $I = 10^{-3}$ mol/L. The high efficiency of the electric spark method of water treatment and disinfection is due to a highly active coagulant with a highly developed surface, formed in the water treatment process in situ. Short-range action effects due to the employment of branched spark discharge include multiple physical effects enhancing the action of hydrolysis products such as: (a) treatment by UV irradiation, (b) treatment by a high temperature (several thousand centigrade) microplasma in many discrete points of the reactor, (c) physiological action of the current pulses of several thousand amperes; (d) treatment by the volume micro electric hydraulic shock; (e) ultrasonic irradiation treatment; and (f) action of different oxidisers (hydroxyl or per-oxyl radicals, ozone) emerging during the discharge process. The efficiency of bacteria removal (*E. coli*, *Bacillus subtilis*, *B. antracoides*, *B. mesentericus*, *B. mycoides*, *Staphylococcusaureus*, *B. antracoides*) from water with a coagulant formed by electric spark discharge method has been studied (Figure 6). It has been shown that the majority of bacteria is removed, independently of the coagulant concentration and species affiliation, during the first 8-10 min of the process. The removal of spore forms is more efficient than that of vegetative forms.

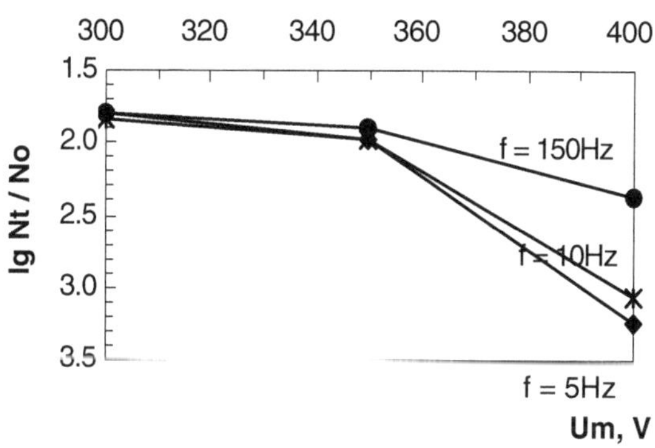

Figure 6. Dependence of the anti-microbe effect (ratio of the *E. coli* 1257 living cell number at given time N_t to the total number of cell N_0) of the electric spark treatment as a function of the external field strength U_m and electric pulse frequency f.

Acknowledgements

This study has been supported in part by Hungarian OTKA agency, grant 2040043 and by Hungarian-Portuguese TeT project No 11/03.

References

1. L. Kulsky, *Theoretical Foundation and Technology of Water Treatment* (Naukova Dumka, Kiev, 1983).
2. A. Amirtharajah, and C. R. O'Melia, in: *Water Quality and Treatment*, edited by F. W. Pontius, (AWWA, McGraw-Hill, 1990) pp. 1-269.
3. J. Duan and J. Gregory, Coagulation by hydrolyzing metal salts, *Adv. Colloid Interface Sci.* 100-102, 475-502(2003).
4. J. Gregory and J. Duan, The effect of dissolved silica on the action of hydrolysing metal coagulants, *Wat. Sci. Tech.* No 6, 113-120 (1998).
5. I. M. Solomentseva, N. G. Gerasimenko and S. Barany, Surface Properties and Aggregation of Basic Aluminium Chloride Hydrolysis Products, *Colloids Surf. A* 151, 113-126 (1999).
6. I. Solomentseva, S. Barany and J. Gregory, Electrokinetic potential and hydration of basic aluminium sulphate hydrolysis product particles, *Colloids Surf.* A 230(1-3), 117-131 (2003).
7. S. Barany, V. Goncharuk, J. Gregory, I. Solomentseva, O. Savluk, and T. Levadna, in: *Role of Interfaces in Environmental Protection*, edited by S. Barany (NATO ARW series, Kluwer, Dordrecht, 2003), pp. 457-470.
8. I. M. Solomentseva and V. V. Teselkin, Study of the kinetics of particles aggregation by non-linear laser diagnostic method, *Russian Colloid. J.* 57, 407-415 (1995).
9. J. Gregory and V. Dupont, Properties of flocs produced by water treatment coagulants, *Water Sci. Technol.* 44, 231-243 (2001).
10. A. A. Shcherba, A. D. Podoltsev, and I. N. Kucherjavaya, Spark erosion of conducting granules in a liquid: analysis of electromagnetic, thermal and hydrodynamic processes, *Technical Electrodynamics*, No. 6, 4-17 (2004).
11. A. A. Shcherba, O. S. Savluk, N. G. Potapchenko, S. Zakharchenko, V. Kosinova, Method of Complex Water Treatment with Volume Electrospark Discharhge, *CECECO-2003 International Conference on Carpathian Euroregion Ecology, Proceedings*, edited by S. Barany (University of Miskolc, Hungary, 2003), pp. 259-267.

CO-ADSORPTION OF THE LOW MOLECULAR CARBOXYLIC ACIDS AND CADMIUM IONS AT THE METAL OXIDE/ ELECTROLYTE INTERFACE

WŁADYSŁAW JANUSZ*, MARLENA MATYSEK
Department of Radiochemistry and Colloid Chemistry
Maria Curie Skłodowska University
pl. M. C. Skłodowskiej 3, 20-031 Lublin, Poland

Abstract. Oxalate and citrate anions can form with Cd^{2+} cations two types of complexes, but the main complexed form in the measured pH range is CdC_2O_4 and CdHCit, respectively. The presence of oxalate or citrate anions (0.001 and 0.002 M) changes the cadmium adsorption on titania and goethite samples, causing an increase of Cd^{2+} adsorption at low pH and a decrease of Cd^{2+} adsorption at high pH. Results are explained with adsorption – desorption processes of oxalate (or citrate) anions on the metal oxide surfaces. Carboxylic ions adsorbing on titanium dioxide or goethite at low pH values change the surface affinity to metal cations. They have a higher affinity to Cd^{2+} than surface hydroxyl groups, and we observe a higher adsorption of cadmium. An increased pH leads to desorption of organic compounds, and their concentration in electrolyte solution increases. The organic compounds form complexes with cadmium cations and weaken Cd^{2+} adsorption at high pH. On the other hand, the presence of cadmium ions leads to increased citrate adsorption on silica, and oxalate adsorption on goethite and titania surfaces at high pH. This is caused by oxalate or citrate adsorption on the adsorbed cadmium cations.

Keywords: co-adsorption, cadmium ions, carboxylic acids, titania, goethite, silica

1. Introduction

The metal oxide/electrolyte solution interface is one of the most well-known systems existing in our environment. The coadsorption of inorganic and organic compounds has been studied from both theoretical and practical points of view.

* To whom correspondence should be addressed: Władysław Janusz, Dept. of Radiochemistry and Colloid Chemistry, Maria Curie Skłodowska University, pl. M. C. Skłodowskiej 3, 20-031 Lublin, Poland; e-mail: wjanusz@hermes.umcs.lublin.pl

J.P. Blitz and V.M. Gun'ko (eds.), Surface Chemistry in Biomedical and Environmental Science, 383–392.

Carboxylates (i.e. ligands containing –COOH functional groups, like oxalic, phtalic or citric acids) are commonly found in nature as well as industrial processes.[1] These organic substances are present in soils and soil solutions as a product of microbiological metabolism, the decomposition of plant or animal remains, and rhizosphere activity.[2,3] The concentration of low molecular weight (LMW) organic acids in most soil systems are usually in the low micromolar range (1 to 50 μM), but the highest concentration of these acids is found in soils rich in organic matter and natural fertilizer.[4] The levels of LMW organic acids are often greatest in the organic horizon and decline in deeper layers.[5] Organic acids are complexing agents for many essential metals like Zn, Al, and Fe, but also for toxic heavy metals like Cd and Hg. The complexation reactions have an effect on the mobility of metals in soils.[4] Compounds with carboxylic groups show a maximum adsorption at low pH, whereas metal ions adsorb at high pH.[6] Organic acids, adsorbing onto metal oxides at low pH, change the adsorption affinity of surface sites and the structure of the electrical double layer at the metal oxide/electrolyte solution interface. Organic acids change the surface charge from positive or neutral to negative, and create adsorption sites for metal ions. Under basic conditions organic acids can compete with the negatively charged surface sites for metal ions in solution, and enhance metal mobilities.[7]

2. Materials and Methods

Titania, goethite and silica were used as adsorbents. Titanium dioxide (anatase) and silica powders were obtained commercially. Goethite was prepared from an $Fe(NO_3)_3$ solution by the precipitation of ferrihydrite.[8] The suspension was held in a closed flask at 70°C for 60 hours. During this period the red brown suspension of ferrihydrite was converted to a yellow brown goethite. These oxides were washed with double distilled water to remove impurities, until the supernatant conductivity was below $2\mu S \times cm^{-1}$.

The specific surface area, determined by adsorption and desorption of nitrogen for titania, goethite and silica samples were 13.5, 42.0, and 261.7 m^2/g, respectively.

The adsorption of Cd(II) and oxalate or citrate ions was calculated from an uptake of their concentration from solution. The ion concentration in solution was determined by the radiotracer method. For labelling the solution, ^{14}C or ^{115}Cd isotopes were used. Radioactivity of the solutions before and after adsorption was measured using a liquid scintillation counter (LS5000 TD Beckmann). Co-adsoprtion of Cd(II) and carboxylate ions was determined simultaneously. The adsorption of measured ions was conducted in a pH range from 3.0 to 10.0.

Electrical double layer properties at the solid/electrolyte solution interface were analyzed by potentiometric titration and electrophoresis measurements. Potentiometric titration and electrokinetic measurements were performed for three different concentrations: 1×10^{-3}, 1×10^{-2}, and 1×10^{-1} M of $NaClO_4$ solutions. The initial concentrations of Cd(II) and oxalate or citrate ions were 1×10^{-6}, 1×10^{-5}, 1×10^{-4}, and 1×10^{-3} M, respectively. Double distilled water was used to prepare all solutions. All reagents used for experiments were analytical gradc.

3. Results and Discusion

The electrical double layer at the metal oxide/electrolyte solution interface can be described by characteristic parameters such as surface charge and electrokinetic potential. Metal oxide surface charge is created by the adsorption of electrolyte ions and potential determining ions (H^+ and OH^-).[9] This phenomenon is described by ionization and complexation reactions of surface hydroxyl groups, and each of these reactions can be characterized by suitable constants such as pK_{a1}, pK_{a2}, pK_{An} and pK_{Ct}. The values of the point of zero charge (pH_{pzc}), the isoelectric point (pH_{iep}), and all surface reaction constants for the measured oxides are collected in Table 1.

Table 1. The values of pH_{pzc}, pH_{iep} and surface reaction constants for TiO_2, FeOOH and SiO_2.

System	pH_{pzc}	pH_{iep}	Surface reaction constants
$TiO_2/NaClO_4$	5.5	5.2	$p^*K_{a1} = 2.16$; $p^*K_{a2} = 8.26$; $p^*K_{Cl} = 1.18$; $p^*K_{Na} = 7.04$
$FeOOH/NaClO_4$	8.3	8.0-9.2	$p^*K_{a1} = 3.3$; $p^*K_{a2} = 9.37$; $p^*K_{Cl} = 4.74$; $p^*K_{Na} = 8.06$
$SiO_2/NaClO_4$	3.4	3.0	$p^*K_{a1} = 5.5$; $p^*K_{a2} = 10.9$; $p^*K_{Cl} = 7$; $p^*K_{Na} = 9.45$

Figure 1 shows adsorption of Cd^{2+} for an initial cadmium concentration of 1×10^{-5} M as a function of pH on the metal oxide/0.001 M $NaClO_4$ solution interface. The adsorption of Cd(II) is pH dependent because the process of cadmium adsorption proceeds through exchange of hydrogen from surface hydroxyl groups. The plot of Cd(II) adsorption vs pH is characteristic for the adsorption of divalent cations on metal oxides and is called the 'edge of adsorption'.[10]

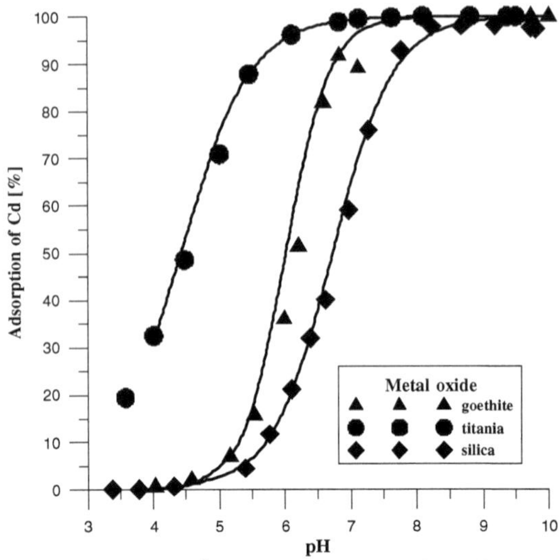

Figure 1. The adsorption of 1×10^{-5} M Cd(II) as a function of pH at the metal oxide/0.001 M $NaClO_4$.

Based on the adsorption data, parameters characterizing the adsorption edge were calculated with results shown in Figure 2. As can be seen, the adsorption edge position characterized by the parameter $pH_{50\%}$ shifts with an increase in cadmium concentration for all oxides. The slope of the adsorption edge is almost constant for an initial concentration of Cd^{2+} ranging from 1×10^{-6} to 1×10^{-4} M. For the highest Cd concentration (0.001 M) the adsorption edge becomes more flat.

The adsorption affinity of Cd ions on the adsorbents following the order titania > goethite > silica. The adsorption of Cd(II) ions was also investigated for three selected cadmium concentrations: 1×10^{-6}, 1×10^{-4} and 1×10^{-3} M, and the same sequence was observed.

The presence of Cd(II) ions leads to an increase in the number of negatively charged sites at the metal oxide surfaces. This is the most significant influence for the highest Cd(II) concentration 0.001M (Figure 3).

Previous workers concluded that cadmium ions adsorb onto the goethite surface by the formation of bidentate surface complexes.[11,12] However, the cadmium adsorption data and our calculations (see Figure 3) shows that cadmium adsorbs on goethite mainly on one surface hydroxyl group, forming monodentate complexes, according the following reaction:

$$\equiv FeOH + Cd^{2+} \leftrightarrow \equiv FeO^{-}Cd^{2+} + H^{+} \qquad (1)$$

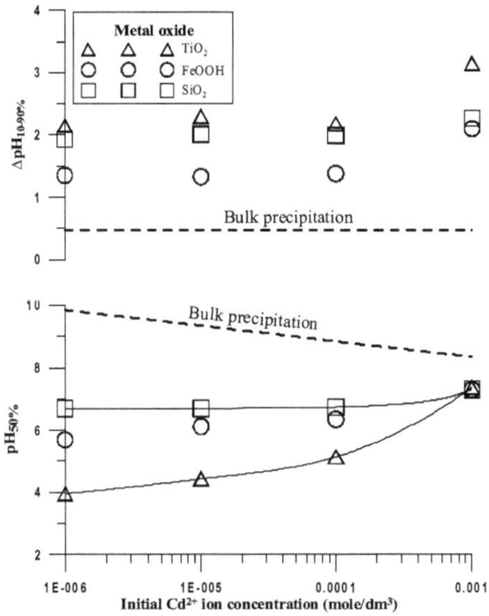

Figure 2. The adsorption edge parameters as a function of initial concentration of cadmium ions for metal oxide/electrolyte solution, containing cadmium, interface.

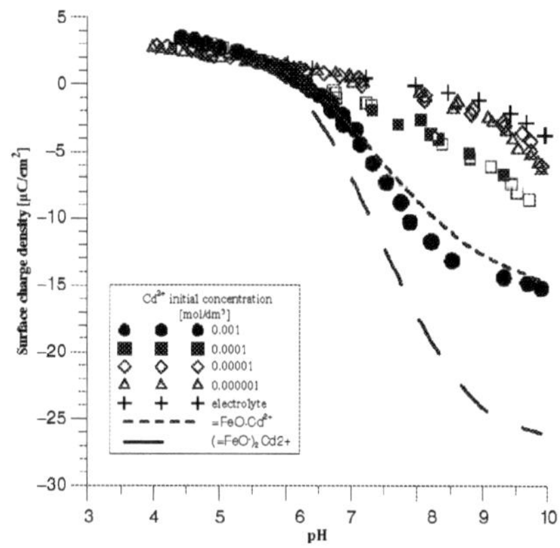

Figure 3. The surface charge density at the goethite/electrolyte interface a function of pH in a presence and absence of cadmium ions.

The adsorption of hydrolysable cations like cadmium may have a strong impact on the zeta potential. Figure 4 shows the zeta potential as a function of pH, and Cd(II) concentration at the silica/0.001 M $NaClO_4$ solution interface.

The ζ potential decreases with an increase of Cd(II) concentration. For 0.001 M Cd(II) concentration a charge reversal point (CR2) at pH = 8.6 is observed. It is connected with an overcharging of the compact part of the electrical double layer on the metal oxide surface. A similar dependence was also observed for the anatase/0.001 M $NaClO_4$ with Cd(II) system.

Anion adsorption at the metal oxide/electrolyte interface proceeds through anion exchange with hydroxyl anions from surface groups, or reaction of anions and a proton with an –OH group. Since ligand adsorption leads to a release of hydroxyl ions or consumption of H^+ ions, anion adsorption is favored by lower pH values and is causing an increase in the concentration of positively charged groups at the surface.

The adsorption of oxalate ions on metal oxides is presented in Figure 5. The adsorption of oxalate ions is characteristic for anion adsorption on metal oxides, and decreases with increasing pH. The following sequence of oxalate adsorption affinity is observed: goethite > titania > silica. At the silica/electrolyte interface, the adsorption density of oxalate (or citrate) anions is small. This is a result of hydrogen bonding interactions between hydrogen atoms of carboxylic groups and surface hydroxyl groups.

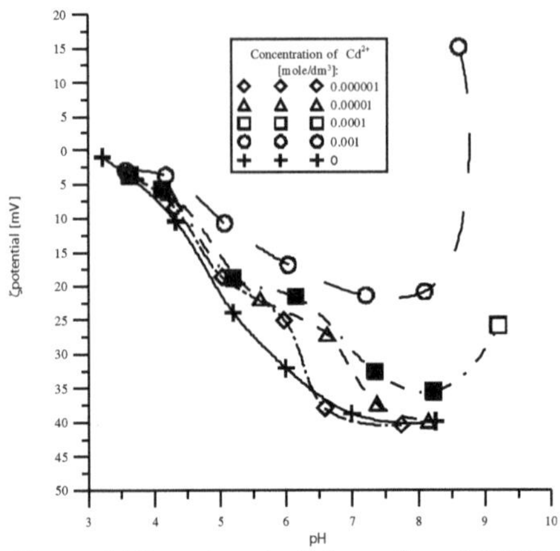

Figure 4. The zeta potential vs. pH and Cd(II) concentration at SiO_2/0.001M $NaClO_4$ electrolyte interface.

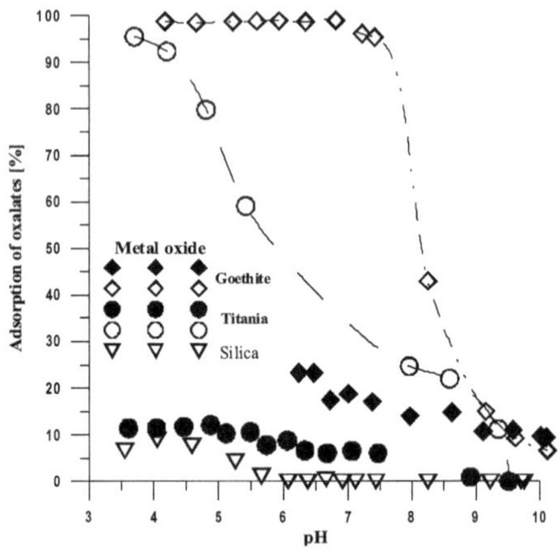

Figure 5. The adsorption of oxalate ions as a function of pH on metal oxide surfaces: (1×10^{-5} M ○–TiO_2; ◇–FeOOH; ▽–SiO_2; 1×10^{-3} M ●–TiO_2; ◆–FeOOH).

Such behavior is called an 'adsorption envelope'.[13] The adsorption of oxalate anions on silica is minimal, because silicon dioxide has mainly negatively charged sites in the measured pH range (from 3.0 to 10.0). Oxalate ions do not adsorb nonspecifically on such an interface.

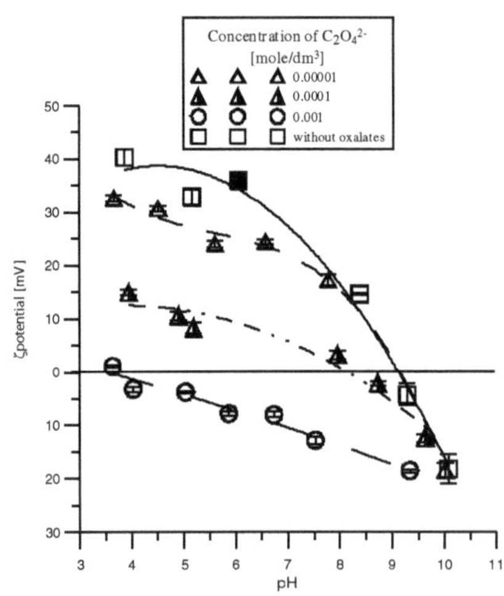

Figure 6. The ζ potential as a function of pH and oxalate ions concentration in the system: FeOOH/0.001 M $NaClO_4$.

The adsorption of oxalate anions onto metal oxide surfaces changes the ζ potential in the system. This is shown by data in Figure 6. As one can see, the ζ potential for the goethite /electrolyte solution system in the presence of oxalates has negative values in the entire measured pH range for the highest oxalate ions concentration (0.001 M), and decreases with an increase of oxalate ion concentration.

Oxalate anions can form with Cd^{2+} cations two types of complexes: CdC_2O_4 and $Cd(C_2O_4)^{2-}$. The main complexed form in the measured pH range is CdC_2O_4.

The presence of oxalate anions (0.001 and 0.002 M) changes the cadmium adsorption on titania and goethite samples. Oxalate anions cause an increase of Cd^{2+} adsorption at low pH range, and a decrease of Cd^{2+} adsorption at high pH range (Figures 7 and 8). Cadmium (II) ion adsorption may be explained with adsorption – desorption processes of oxalate anions on the metal oxide surfaces. Oxalate ions adsorbing on titania and goethite surfaces at low pH values changes the surface affinity to metal cations. These anions have a higher affinity to Cd^{2+} than surface hydroxyl groups, and cause an increase of Cd adsorption. Increase of pH leads to desorption of oxalates, and their concentration in the electrolyte solution increases.

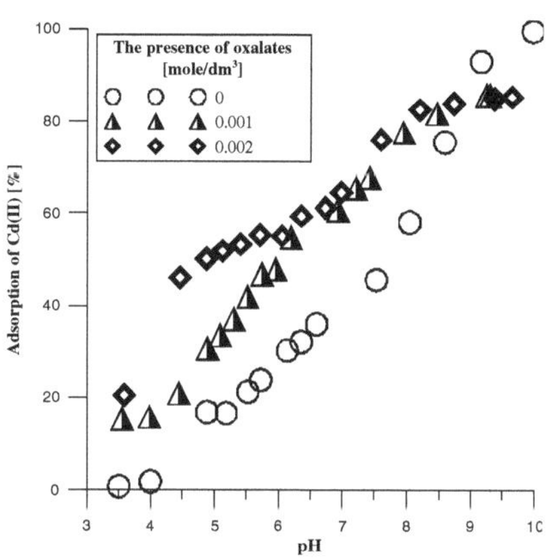

Figure 7. A percentage of cadmium adsorption (0.001 M) at the anatase / electrolyte interface in a presence of oxalates ions (0.001 and 0.002 M).

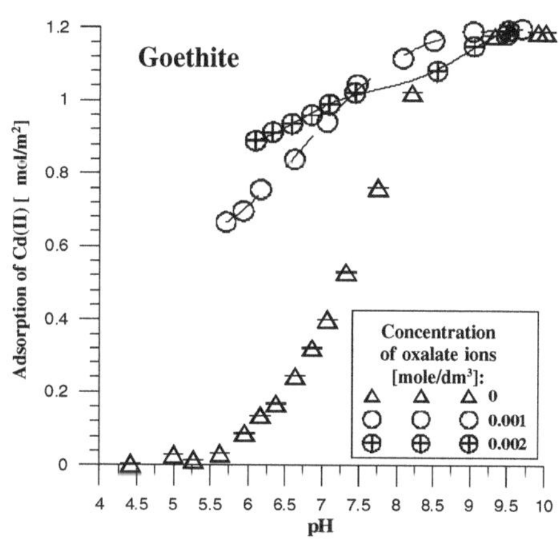

Figure 8. The adsorption of 0.001 M Cd(II) ions at goethite/electrolyte interface in a presence of oxalate ions (0.001 and 0.002 M).

Thus they form complexes with cadmium cations and weaken Cd^{2+} adsorption at high pH range.

On the other hand the presence of cadmium ions leads to an increase in oxalate adsorption on metal oxide surfaces at high pH. This is caused by oxalate adsorption on the adsorbed cadmium cations.

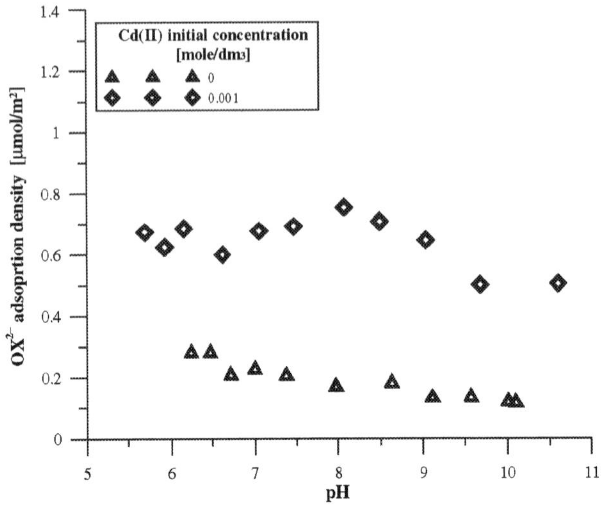

Figure 9. Adsorption of oxalates (0.001 M) in a presence (0.001 M) and absence of cadmium cations at the FeOOH/0.001 M $NaClO_4$ solution interface.

A similar phenomenon was observed for citrate adsorption on a silica surface in the presence of 0.001M cadmium cations (Figure 11). Increasing pH causes an increase in citrate adsorption on adsorbed cadmium ions.

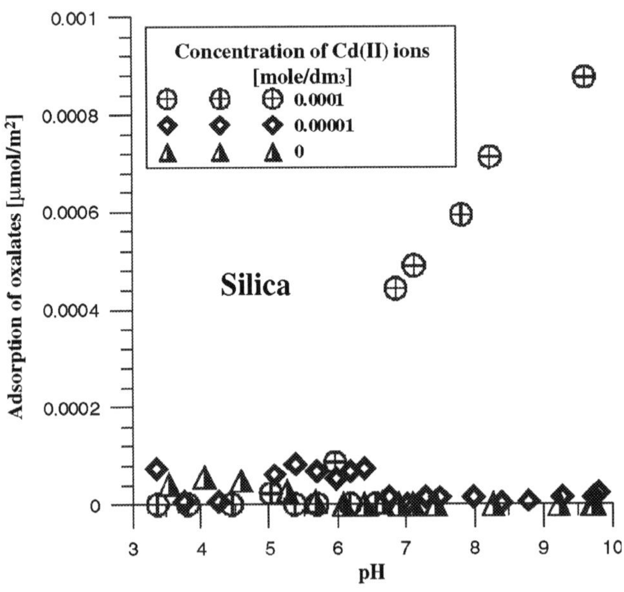

Figure 10. Adsorption of oxalate ions (0.001 M) in a presence (0.00001 and 0.0001 M) and absence of cadmium ions on silica surface.

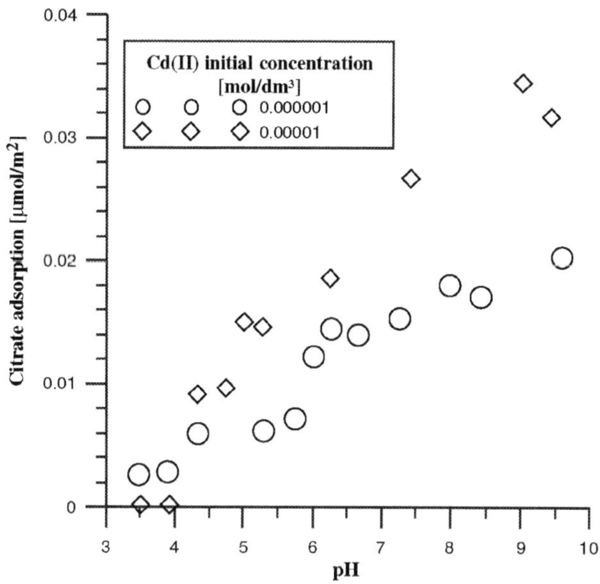

Figure 11. Adsorption of citrate ions (0.001 M) in a presence (0.00001 and 0.0001 M) of cadmium ions on SiO_2.

4. Conclusion

The adsorption curve of cadmium ions as a function of pH on metal oxide surfaces is characteristic for such systems, and is called the adsorption edge. The adsorption of cadmium increases with increasing pH, and is almost complete (~100%) at high pH. The adsorption edge shifts with an increase in the initial concentration of Cd^{2+} ions towards higher pH values. The adsorption of cadmium cations causes an increase of the ζ potential, with an increase of total concentration of these ions in the system.

The course of adsorption of oxalate (or citrate) ions at the metal oxide interface is characteristic for multivalent anion adsorption on such interfaces, and is called an "envelope adsorption". The adsorption of oxalate ions is complete at low pH range and then decreases to ~0% as pH increases.

The presence of oxalate (or citrate) ions causes changes in the Cd^{2+} adsorption edge, increasing the adsorption at low pH and in the case of titania decreasing the adsorption at high pH. The observed relationships are connected with adsorption – desorption processes of oxalate or citrate ions on the metal oxide surface. Oxalate and citrate ions change surface affinity to metal cations, and have a higher affinity to Cd^{2+} ions than surface hydroxyl groups. On the other hand, an increase of oxalate and citrate adsorption in the presence of cadmium ions is observed. This is caused by oxalate (or citrate) adsorption on the adsorbed cadmium cations.

References

1. J. Rosenqvist, K. Axe, S. Sjöberg, and P. Person, Adsorption of dicarboxylates on nano-sized gibbsite particles: effects of ligand structure on bonding mechanisms, *Colloids Surf. A* 220, 91-104 (2003).
2. H. Q. Hu, J. Z. He, X. Y. Li, and F. Liu, Effect of several organic acids on phosphate adsorption by variable charge soils of central China, *Environment International* 26, 353-358 (2001).
3. P. A. W. van Hees, S. I. Vinogradoff, A. C. Edwards, D. L. Godbold, and D. L. Jones, Low molecular weight organic acid adsorption in forest soils: effects on soil solution concentrations and biodegradation rates, *Soil Biol. Biochem.* 35, 1015-1026 (2003).
4. B. W. Strobel, Influence of vegetation on low-molecular-weight carboxylic acids in soil solution - a review, *Geoderma* 99, 169-198 (2001).
5. P. A. W. van Hees, U. S. Lundström, and R. Giesler, Low molecular weight organic acids and their Al-complexes in soil solution—composition, distribution and seasonal variation in three podzolized soils, *Geoderma* 94, 173-200 (2000).
6. J. D. Filius, D. G. Lumsdon, J. C. L. Meeussen, T. Hiemstra, and W. H. van Riemsdijk, Adsorption of fulvic acid on goethite, *Geochim. Cosmochim. Acta* 64, 51-60 (2000).
7. I.-F. Boily and J. B. Fein, Experimental study of cadmium-citrate co-adsorption onto α-Al_2O_3, *Geochim. Cosmochim. Acta* 60, 2929-2938 (1996).
8. U. Schwertmann and R. M. Cornell, *Iron oxides in the laboratory. Preparation and Characterization* (VCH, 1991).
9. J. A. Davies, R. O. James, and J. O. Leckie, Surface Ionization and Complexation at the Oxide/Water Interface. I. Computation of Electrical Double Layer Properties in Simple Electrolytes, *J. Colloid Interface Sci.* 63, 480-499 (1978).
10. A. P. Robertson and J. O. Leckie, Cation Binding Predictions of Surface Complexation Models: Effects of pH, Ionic Strength, Cation Loading, Surface Complex, and Model Fit, *J. Colloid Interface Sci.* 188, 444 (1997).
11. S. R. Randall, D. M. Sherman, K. V. Ragnarsdottir, and C. R. Collins, The mechanism of cadmium surface complexation on iron oxyhydroxide minerals, *Geochim. Cosmochim. Acta* 63, 2971-2987 (1999).
12. L. Spadini, A. Manceau, P. W. Schindler, and L. Charlet, Structure and stability of Cd^{2+} surface complexes on ferric oxides, *J. Colloid Interface Sci.* 168, 73-86 (1994).
13. S. Goldberg, Inconsistency in the triple layer model description of ionic strength dependent boron adsorption, *J. Colloid Interface Sci.* 285, 509-517 (2005).

INTERACTION OF PHENOLIC POLLUTANTS WITH PNIPA HYDROGELS

KRISZTINA LÁSZLÓ,* KATALIN KOSIK, ERZSÉBET WILK,
Department of Physical Chemistry, Budapest University of Technology and Economics, Budapest 1521, Hungary

ERIK GEISSLER
Laboratoire de Spectrométrie Physique CNRS UMR5588, Université J. Fourier de Grenoble, BP 87, 38402 St Martin d'Hères, France

Abstract. Poly(N-isopropylacrylamide) (PNIPA) gels swollen in pure water exhibit a volume phase transition (VPT) at 34°C, above which the solvent is expelled. In contact with aqueous phenol solutions, these gels may display a VPT at 20°C at aromatic acid concentrations, depending on the number of OH-substitutions. This observation makes PNIPA a promising actuator for phenols in water, and also offers opportunities for water treatment applications. Dynamic light scattering (DLS) data show that the product of the collective diffusion coefficient and the Rayleigh ratio of the thermodynamic fluctuations is independent of the aromatic composition below the critical concentration. It follows, as is also confirmed by small angle neutron scattering (SANS) experiments, that the phenols are not directly in contact with the network chain.

Keywords: volume phase transition (VPT); PNIPA; phenol; resorcinol; phloroglucinol; swelling; uptake isotherm; dynamic light scattering (DLS); small angle neutron scattering (SANS)

* To whom correspondence should be addressed. Krisztina László, Department of Physical Chemistry, Budapest University of Technology and Economics, Budapest 1521, Hungary, e-mail: klaszlo@mail.bme.hu

J.P. Blitz and V.M. Gun'ko (eds.), Surface Chemistry in Biomedical and Environmental Science, 393–402.

1. Introduction

Phenol and other aromatic pollutants are among the most frequent anthropogenic contaminants of water. These pollutants appear as a consequence of degradation of organic compounds used as intermediates in the synthesis of dyes, pesticides, insecticides, explosives, etc. Although most procedures designed to remove these molecules from water use activated carbons, responsive soft materials have recently emerged as a topic of intense research activity as their potential to act as sensors of environmental signals is gradually being recognized.

The operating principle of such devices relies on materials whose volume changes abruptly, i.e., that display a volume phase transition (VPT), when an appropriate stimulus is applied. Thermosensitive poly(N-isopropylacrylamide) (PNIPA) hydrogel, where the cross-links are chemical, namely peptide bonds, is widely used for this purpose since its VPT in pure water is close to 34°C.[1-3]

Below this temperature the three-dimensional swollen network of PNIPA is transparent and, as it can hold about forty times its own weight of water, is highly permeable to the swelling liquid. The degree of swelling depends on the 3D behaviour of the gel, which can be characterized by the density of the cross-links.

The collapse of the fully hydrated network is induced by dehydration of the hydrophobic polymer chain as a result of transfer of the hydrated water molecules to the bulk. Above the VPT temperature the network shrinks and becomes white, with concomitant expulsion of the solvent. Generally, similar VPTs can be triggered by further physical (pressure, electromagnetic radiation, external electric or magnetic field) and chemical means (added small molecules). Molecules dissolved in the mobile phase may move together with the solvent, offering the possibility of stimulus controlled swelling (membranes of tunable permeability) or chemical delivery (e.g., controlled drug release). Nevertheless, dissolved small molecules also modify the VPT, e.g., by shifting its temperature, as they reduce the chemical potential of water in the bulk phase.

Gels swollen in binary solvents have been the subject of many investigations for their ability to modify the polymer-solvent interaction.[4-7] Several authors[8-10] have reported that gels made from PNIPA, swollen in aqueous solutions of aromatic compounds, collapse at aromatic concentrations above a certain threshold. This phenomenon has been attributed to the approach of the solubility limit of the host molecule in water.[9]

The system discussed in this paper also has biological relevance. Owing to the C(=O)–NH– bonds, the polymer chain of the PNIPA can be considered as a model protein, while phenol is a frequent substituent of amino acids or other biologically active molecules.[11]

Our research into a phenol-sensitive device for sensor and/or adsorption applications yielded results that are difficult to interpret in terms of the current literature. We therefore undertook a systematic investigation of the interaction of phenols with PNIPA hydrogels by applying a multi-method approach. The observations reported here involve measurements of swelling degree, aromatic uptake, dynamic light scattering (DLS), as well as small angle neutron scattering (SANS).

2. Experimental

Poly(N-isopropylacrylamide) (PNIPA) gels of [NIPA]/[BA] molar ratio of 150 were prepared from N-isopropylacrylamide (NIPA) and N,N'-methylenebisacrylamide (BA) by radical polymerization.[13] Gel films of 2 mm thickness were polymerized and dialysed in water to remove unreacted chemicals. The films were cut into discs (diameter 7 mm), then dried and stored above concentrated sulfuric acid.

To determine the equilibrium uptake of the polymers, the dry discs were placed in contact with the appropriate aqueous solutions for 48 hrs, a period determined from preliminary kinetic measurements. Aqueous solutions of high purity phenols (Merck) were used. Swelling was determined by weighing, while the uptake of the small aromatic molecules was calculated from the initial and equilibrium concentrations, c_0 and c_e, of the phenols in the free liquid.[13]

Light scattering measurements were made with an ALV DLS/SLS 5022F goniometer equipped with optical fiber coupling and an avalanche diode, with a 22 mW HeNe laser working at 632.8 nm and an ALV 5000E multi-tau correlator.

Small angle neutron scattering measurements were made on the KWS-1 instrument at the IFF Forschungszentrum, Jülich, Germany. The neutron velocity selector was set for wavelength λ = 8Å ($\Delta\lambda/\lambda = 0.2$). For contrast matching with the PNIPA gels a water mixture containing 18% D_2O and 82% H_2O by volume was used as the swelling medium. Deuterated phenol (D6, Polymer Source Inc.) dissolved in the mixture was employed to detect possible decoration or clustering effects of this molecule in the vicinity of the polymer chains. The wavevector range explored in the measurements was 2.3×10^{-3} Å$^{-1} \leq q \leq 0.2$ Å^{-1}.

3. Results and Discussion

Selected properties of the aromatic pollutants used in these studies are listed in Table 1.

Table 1. Characteristic data of the phenols.

	Phenol	Resorcinol	Phloroglucinol
M	94.11	110.11	126.11
Solubility in water, g/l	82	1470	$20.8_{25°C}$
pK_a	$9.89_{20°C}$	$9.81_{25°C}$	8.8; 9.0; $14.1_{20°C}$

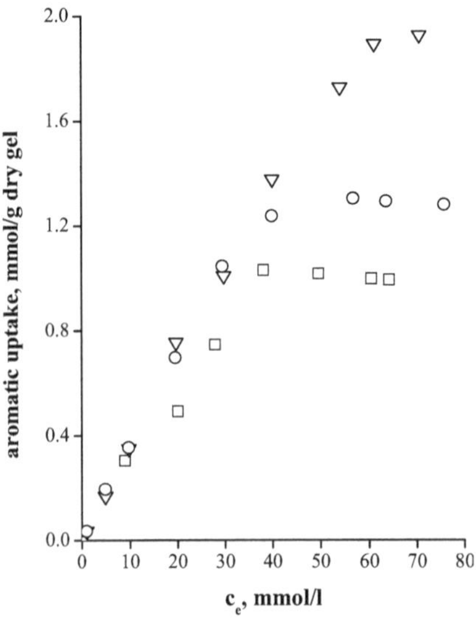

Figure 1. Equilibrium uptake of phenol at 20 °C; ▽: phenol, O: resorcinol, and □: phloroglucinol.

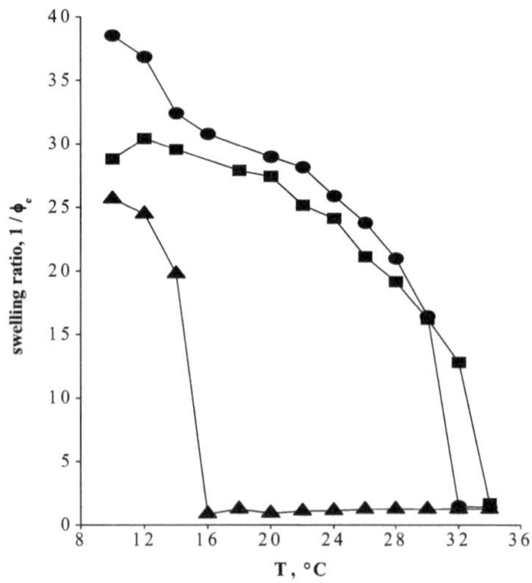

Figure 2. Equilibrium swelling ratio of PNIPA gel in water (■), 5 (●) and 50 (▲) mM aqueous phloroglucinol solutions.

In the low concentration range the isotherms describing the aromatic uptake (Figure 1) can be fitted to the Henry model (Table 2). At higher concentration, dependent on the degree of OH-substitution, the curves deviate and saturation occurs. The position (equilibrium concentration) of the saturation as well as the limit of the uptake depends on the OH-groups. In the saturation range we also estimated the aromatic and OH/monomer unit ratios, respectively. In all cases the calculated values are far from 100%. It is worth mentioning, however, that the phenol uptake of the PNIPA at saturation is comparable to that of commercial activated carbons developed for phenol removal (2-3 mmol/g).

The equilibrium swelling ratio $1/\phi$ of PNIPA gels[†] of a specific [monomer]/[crosslinker] ratio depends on the temperature and the concentration

[†] ϕ is the volume fraction of the polymer

of the aromatic contaminant (Figure 2). At low concentration, i.e., in the fully swollen state this behaviour is similar to pure water. At a critical pollution level however, the swelling ability is already slightly reduced at low temperatures and the VPT temperature is significantly lowered. In isothermal conditions on the other hand, a phase transition much more abrupt than in pure water occurs.

Table 2. Parameters obtained from the equilibrium uptake isotherms in Figure 1.

	Slope of the initial section	R^2 of the linear fit	Uptake at saturation		Phenyl-group	OH-group
			mmol/g dry gel	mg/g dry gel	per 100 monomer units	
Phenol	0.036	0.9968	1.90	178.6	21.4	21.4
Resorcinol	0.033	0.9789	1.31	144.1	14.8	29.6
Phloroglucinol	0.025	0.9767	1.02	128.5	11.5	34.5

The swelling below this concentration, as well as the critical concentration and the sharpness of the transition, are influenced by the number of OH-groups decorating the aromatic ring (Figure 3).

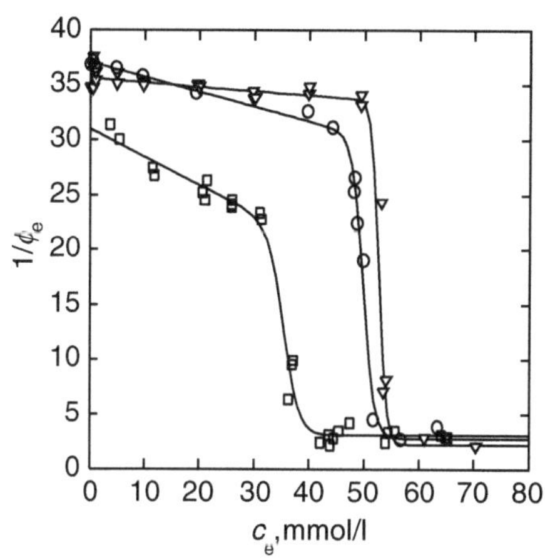

Figure 3. Influence of the concentration of various phenols on the equilibrium swelling behaviour of PNIPA gel at 20°C. Symbols are the same as in Figure 1.

From these results we conclude that the Lundelius rule, i.e., that the uptake of a solute is inversely proportional to its solubility in the solvent, does not apply to the uptake of the phenols by the PNIPA chains.

Murase et al.,[9] by fitting their adsorption curves to a Langmuir equation, deduced that the aromatic molecule investigated (benzyl alcohol) was adsorbed on the PNIPA chains – as shown in the left cartoon of Figure 4 – and the collapse of the gel occurred in the vicinity of the solubility limit of the aromatic molecule in water. The suggested mechanism of phenol sorption is site-binding through the amide-group.[11] If adsorption did occur, however, it should increase the cross-section of the polymer chains and hence the friction coefficient. This would be reflected in a reduction of the collective diffusion coefficient.

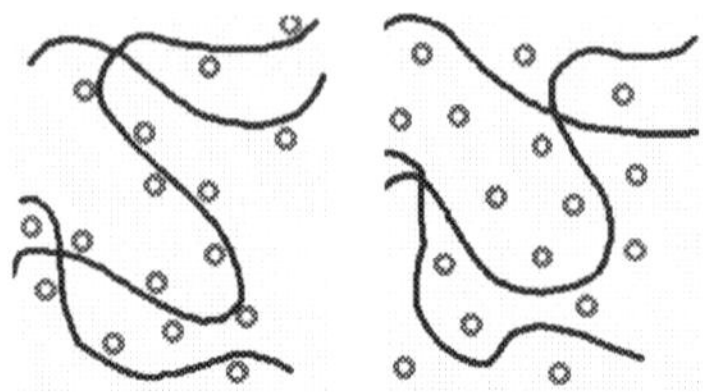

Figure 4. Possible distribution of the aromatic molecules along the PNIPA chains.

Results from DLS measurements [Figure 5 (left)], show the collective diffusion coefficient $D_c = \Gamma/q^2$ for the PNIPA gels swollen in aqueous solutions at different equilibrium concentrations c_e of phenols, where Γ is the characteristic relaxation rate of the DLS correlation function. With this technique systems can be measured only below the phase transition. Samples equilibrated with phloroglucinol cannot therefore be studied above ca 30 mM at room temperature. All the systems display a similar response, with a noticeable, albeit limited, downturn as the limiting concentration is approached.

In semi-dilute solutions, the diffusion coefficient describing the thermal motion of the polymer chains that take part in an osmotic fluctuation in the surrounding solvent may be defined as follows

$$D_c = \frac{\partial\Pi / \partial\phi}{f}, \quad (1)$$

where f is the friction coefficient of the polymer chain moving in the surrounding fluid, and ϕ the volume fraction of polymer at a given state of swelling. In a gel, owing to the permanent elasticity, the term $\partial\Pi/\partial\phi$ is replaced by M_{os}/ϕ, where the osmotic modulus is[12]

$$M_{os} = \phi\partial\omega / \partial\phi + 4G/3, \quad (2)$$

ω is the swelling pressure. The elastic modulus G of the gels was determined from stress–strain measurements[13]. (At small strains G follows the expected behavior of standard network theory, i.e., $G \propto \phi^{1/3}$.) Similarly, the Rayleigh ratio of light scattered by the thermal fluctuations is given by

$$R_{dyn}(q=0) = K\frac{k_B T\phi^2}{M_{os}}, \quad (3)$$

k_B is Boltzmann's constant, T the absolute temperature, and K is the optical contrast factor.[13] In the present measurements no angular variation of R_{dyn} was observed. The values of R_{dyn} are displayed in Figure 5 (right). In this case the opposite behavior to that of D_c is observed, i.e., the intensity of the scattered light increases moderately as c_e increases. The intensity data, however, exhibit no sign of a divergence at the transition.

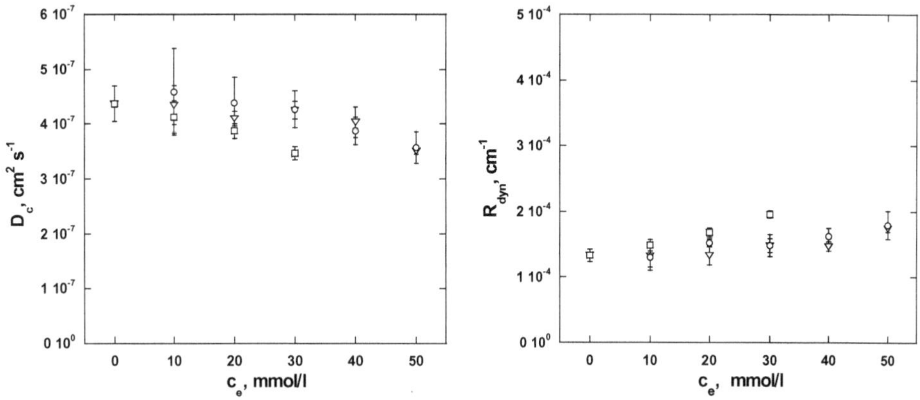

Figure 5. Collective diffusion coefficient $D_c = \Gamma/q^2$ (left) and Rayleigh ratio R_{dyn} (right) from dynamic light scattering intensity in PNIPA gels at 20°C at equilibrium with aqueous solutions containing different concentrations c_e of phenols. Symbols are the same as in Figure 1.

The weak increase in the experimental value of R_{dyn}, displayed in Figure 5 (right), corresponds to a decrease in the osmotic modulus as the transition is approached. This in turn, by virtue of equations 1 and 2, causes a reduction in D_c. The absence of a divergence at the transition, however, shows that it is of first-order type, in which the osmotic susceptibility undergoes a discontinuity. This finding is consistent with the step-like volume change of Figure 3, another characteristic of first order transitions. It is also consistent with the observations of Suzuki et al.,[8] who found the addition of an aromatic compound (hydroquinone) lowers the transition temperature and enlarges the volume discontinuity.

From expressions 1 and 3 it may be remarked that the product $D_c R_{dyn}/\phi$ is proportional to $k_B T/f$ and is therefore independent of the osmotic susceptibility. This quantity is displayed in Figure 6 where, within experimental error, $D_c R_{dyn}/\phi$ is indeed independent of aromatic content in the case of phenol and resorcinol. Only phloroglucinol exhibits a marked deviation at the highest concentration measurable by this method. This finding implies that the friction coefficient f of the polymer chains moving in the solvent is independent of the nature of the solvent and the aromatic content below the critical concentration. It may therefore be concluded that the cross-section of the chains is unaltered, and that the observed decrease in the diffusion coefficient of the gel at higher aromatic content is of purely thermodynamic origin.

A further finding from the observations presented here is that no significant difference appears between the effects of the different phenols in the swollen region. At the transition, however, the repulsion towards the PNIPA network increases with the OH-substitution.

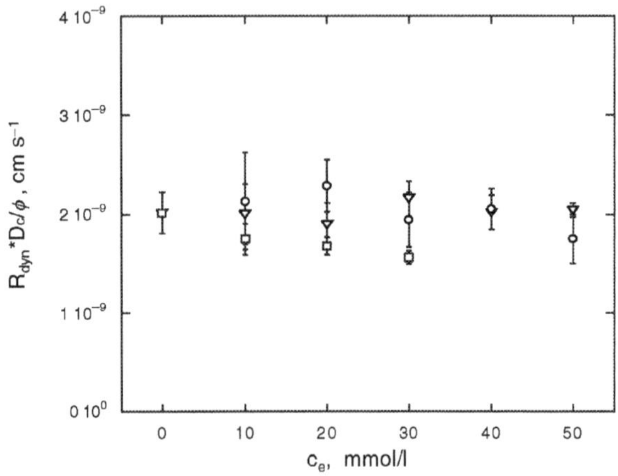

Figure 6. Product D_cR_{dyn}/ϕ as a function of aromatic content c_e in the surrounding liquid; symbols are the same as in Figure 1.

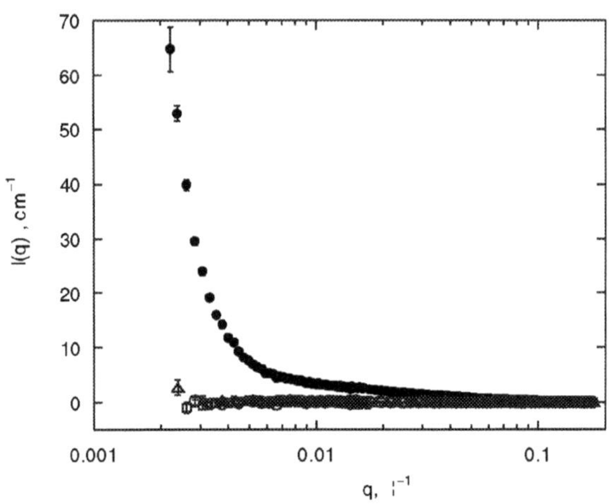

Figure 7. SANS signal from PNIPA in fully deuterated water (●) at 20 °C and in contrast matched conditions (18% D_2O:82% H_2O) with different amounts of deuterated phenol (O 0 mM, □ 20 mM, and Δ 40 mM).

SANS experiments were undertaken to confirm the DLS observations. Results obtained with phenol solutions are shown in Figures 7 and 8. In the conditions of contrast matching employed here, where the scattering length density of the water was adjusted to be equal to that of the polymer, clustering of the phenol can in principle be detected if its contrast is sufficiently different from that of the solvent (or polymer). Figure 7 shows the signal of the PNIPA gels in pure D_2O, and at the contrast match point (18% D_2O/82% H_2O v/v mixture), with different amounts of deuterated phenol. Figure 8 shows the scattering signal when fully deuterated phenol is added to the contrast matched solvent. No measurable signal at low q corresponding to phenol clustering is visible. It may be concluded within experimental error that the phenol molecules do not cluster significantly around the polymer chains but are distributed uniformly in the solvent (Figure 4, right).

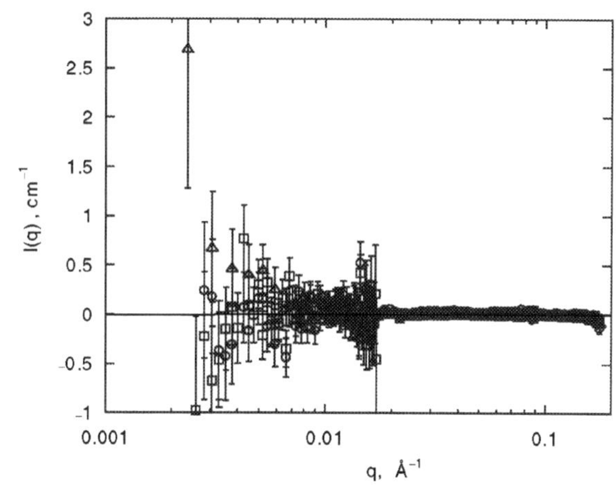

Figure 8. Close-up view of contrast-matched signals. The error bars are much larger for the large sample-detector distances d, because the signal intensity decreases as $1/d^4$. The signal from the phenol is zero within the level of the noise. The phenol is therefore mainly distributed uniformly in the liquid. Symbols are the same as in Figure 7.

4. Conclusion

The volume transition of PNIPA gels swollen at 20 °C in aqueous solutions of phenols is studied by observations of swelling equilibrium, elastic modulus and dynamic light scattering. The collective diffusion coefficient D_c decreases at high aromatic content, while the dynamic component of the scattering intensity, R_{dyn}, increases. No divergence of R_{dyn} close to the transition is observed however, as expected in a first order transition. The light scattering response of the gels swollen in both solvents is very similar. Within experimental error the product $D_c R_{dyn}/\phi$ is independent of aromatic concentration. It follows that the friction coefficient of the polymer chains is not modified in the presence of the binary solvent and that the reduction in the value of D_c is solely due to the reduction in the osmotic modulus of the system. SANS observations also indicate that the majority of aromatic molecules do not establish a binding relationship with the chains.

Acknowledgements

The authors are indebted to the IFF Forschungszentrum, Jülich for access to the KWS-1 instrument and to V. Pipich for his invaluable help. Grateful thanks are also extended to E. Fülöp and G. Bosznai for their technical assistance. This research was supported by a joint EU - Hungarian Government fund (GVOP - 3.2.2 - 2004 - 07 - 0006/3.0).

References

1. T. Tanaka, Collapse of gels and critical endpoint, *Phys. Rev. Lett.* 40(12), 820-823 (1978).
2. M. Shibayama, Y. Shirotani, H. Hirose, and S. Nomura, Simple scaling rules on swollen and shrunken polymer gels, *Macromolecules* 30(23), 7307-7312 (1997).
3. H. Kawasaki, S. Sasaki, and H. Maeda, Effects of the gel size on the volume phase transition of poly(N-isopropylacrylamide) gels: A calorimetric study, *Langmuir* 14(4), 773-776 (1998).
4. E. Geissler, and A. M. Hecht, Dynamic light scattering from gels in a poor solvent, *J. Physique (Paris)* 39(9), 955-960 (1978).
5. T. Amiya, Y. Hirokawa, Y. Hirose, Y. Li, and T. Tanaka, Reentrant phase-transition of N-isopropylacrylamide gels in mixed-solvents, *J. Chem. Phys.* 86(4), 2375-2379 (1987).
6. C. G. Varelas, and C. A. Steiner, Bulk and microscopic properties of copolymer networks in mixed aqueous solvents, *J. Polym. Sci. B: Polym. Phys. Ed.* 30(11), 1233-1283 (1992).
7. F. Horkay, P. J. Basser, A. M. Hecht, and E. Geissler, Osmotic and SANS observations on sodium polyacrylate hydrogels in physiological salt solutions, *Macromolecules* 33(22), 8329-8333 (2000).
8. Y. Suzuki, N. Suzuki, Y. Takasu, and I. Nishio, A study on the structure of water in an aqueous solution by solvent effect on a volume phase transition of N-isopropylacrylamide gel and low-frequency Raman spectroscopy, *J. Chem. Phys.* 107(15), 5890-5897 (1997).
9. Y. Murase, T. Onda, K. Tsujii, and T. Tanaka, Interactions of a dye and a solvent with poly(N-isopropylacrylamide) gel in relation to its phase-transition behavior, *Bull. Chem. Soc. Jpn.* 73(11), 2543-2547 (2000).
10. S. Koga, S. Sasaki, and H. Maeda, Effect of hydrophobic substances on the volume-phase transition of N-isopropylacrylamide gels, *J. Phys. Chem. B* 105(19), 4105-4110 (2001).
11. T. Kawashima, S. Koga, M. Annaka, and S. Sasaki, Roles of Hydrophobic Interaction in a Volume Phase Transition of Acrylamide Gel Induced by the Hydrogen-Bond-Driving Alkylphenol Binding, *J. Phys. Chem. B* 109(2), 1055-1062 (2005).
12. T. Tanaka, L. O. Hocker, and G. B. Benedek, Spectrum of light scattered from a viscoelastic gel, *J.Chem. Phys.* 59(9), 5151-5159 (1973).
13. K. László, K. Kosik, C. Rochas, and E. Geissler, Phase transition in poly (N-isopropylacrylamide) hydrogels induced by phenols, *Macromolecules* 36(20), 7771-7776 (2003).

NANOSCALE OXIDES AS DESTRUCTIVE SORBENTS FOR HALOGENATED HYDROCARBONS

ALEXANDER M. VOLODIN,[1,*] ALEXANDER F. BEDILO,[1] DAVID S. HEROUX,[2] VLADIMIR I. ZAIKOVSKII,[1] ILYA V. MISHAKOV,[1] VLADIMIR V. CHESNOKOV,[1] KENNETH J. KLABUNDE[3]
[1]*Boreskov Institute of Catalysis, Novosibirsk, 630090, Russia;*
[2] *Department of Natural Sciences, University of Maine at Farmington, Farmington, ME, USA;*
[3]*Department of Chemistry, Kansas State University, Manhattan, KS, USA, 66506*

Abstract. Destructive adsorption of halocarbons on nanocrystalline oxides has been studied. The effect of nanoparticle size and phase composition on the reaction kinetics is discussed. The reactivity of nanocrystalline oxides has been found to increase after deposition of a permeable carbon coating. The possibility of synthesis of new nanocrystalline halogenated materials using nanoscale oxides as precursors has been demonstrated.

Keywords: nanocrystalline oxides; aerogels; destructive sorbents; halocarbons; MgO; TiO_2; titanium oxydifluoride; carbon coating

1. Introduction

Although nanoscale materials promise to revolutionize many of our industries including electronics, health care, energy and more, the near term uses are in environmental remediation and green chemistry applications. One reason for this is that nanomaterials possess unique properties as adsorbents and catalysts, because: (1) they possess high surface areas with large surface to bulk ratios so that the nanomaterial is used efficiently; (2) nanocrystals have

*To whom correspondence should be addressed. A. M. Volodin, Boreskov Institute of Catalysis, Novosibirsk, 630090, Russia; e-mail: volodin@catalysis.nsk.su

J.P. Blitz and V.M. Gun'ko (eds.), Surface Chemistry in Biomedical and Environmental Science, 403–412.

unusual shapes and possess high surface concentrations of reactive edge, corner, and defect sites that impart intrinsically higher surface reactivities; (3) a wide range of Lewis acid-base properties and oxidation/reduction potentials can be engineered into the nanomaterials since the periodic table of the elements (and their oxides, sulfides, etc.) becomes a literal playground for their design, and in a sense becomes three-dimensional since nanocrystal size is important as well as chemical potential.

Nanocrystals of common metal oxides such as MgO, CaO, ZnO, TiO_2, Al_2O_3, and Fe_2O_3, have been shown to be highly efficient and active adsorbents for many toxic chemicals including air pollutants, chemical warfare agents, and acid gases.[1-7] In most cases, destructive adsorption takes place on the surface of the nanocrystals, so that the adsorbate is chemically dismantled and thereby made nontoxic. In particular, aerogel-prepared (AP) nanocrystalline MgO has been shown to have small average particle size (~4 nm), high surface area (>500 m^2/g) and high reactivity.[1,5] At elevated temperatures, chlorocarbon reactions can be driven to stoichiometric proportions, especially if small amounts of transition metal catalysts are added. Thus, CCl_4 reacts stoichiometrically with nano-CaO if a mono layer of Fe_2O_3 is placed on the CaO.[8]

Nanoparticles of MgO, CaO, SrO, Al_2O_3, and intimately mixed oxides MgO-Al_2O_3, and SrO-Al_2O_3 (all prepared by modified aerogel procedures-MAP) have proven effective in ambient temperature detoxification of chemical warfare agents (organophosphorus nerve agents and mustard gas), and their simulants (paraoxon, 2-chloroethylethylsulfide, organophosphorus fluorides and others).[7,8,10] Adsorption capacities are high compared with other more common sorbents such as activated carbon or Ambersorb. Chemical warfare agents are chemically dismantled (destructively adsorbed) into non-toxic fragments.[7,8,10] For example, at room temperature paraoxon [$(EtO)_2P(O)OC_6H_4NO_2$] suffers bond cleavage of all three P-OR bonds as time goes on (as followed by solid state ^{31}P NMR).

Results with intimately mixed oxides show that further enhancement of reactivity for these detoxification reactions can be achieved. Mixed metal oxide systems of AP-MgO-Al_2O_3 and AP-CaO-Al_2O_3 are better at destructively adsorbing paraoxon than AP-MgO, AP-CaO and AP-Al_2O_3 by themselves.[10] Sulfated mixed metal oxides also show further improved adsorption suggesting that increasing the acidity of the sample enhances adsorption.[10]

Nanoscale metal oxides also exhibit biocidal properties due to their abrasive nature, alkaline surfaces, oxidizing power (when elemental halogens are preadsorbed), and the fact that their average particle charge (positive) attracts bacteria (which generally carry overall negative charge).[11] In fact, nano-MgO is biocidal by itself for vegetative cells such as e-coli or vegetative bacillus cereus. However, for spores, which possess a tough protective outer shell, preadsorbed halogens are necessary. Raman and UV-vis spectra suggest that halogen-halogen bonds are weakened upon adsorption, which apparently is the reason for this enhanced reactivity toward organics and bacteria.[12]

The use of these nanocrystalline metal oxides is limited under conditions where liquid water or water vapor is present due to their tendency to adsorb water, and thereby be partially deactivated toward adsorption of the target pollutants. Although the target pollutants are usually subjected to conversion in the presence of water as well, relatively large amounts of water can mitigate against the adsorption of the target adsorbate decreasing the efficiency of the destructive adsorbent. Recently we have shown that coating destructive sorbent with an intelligent carbon layer significantly improves the stability of such nanocrystalline materials in the presence of water, and the possible time of their storage in air in comparison with non-coated nanoscale oxides without a considerable loss of activity.[13] For practical application of nanoscale oxides as destructive sorbents, it is extremely important that the carbon coating makes such materials hydrophobic.

In the current publication we shall discuss the specific features of the reactions of nanoscale MgO and TiO_2 with chlorofluorocarbons and the effect of carbon on the performance of the destructive sorbents.

2. Experimental

Three different MgO samples were used in the experiments. AP-MgO with a surface area of 385 m^2/g was prepared by a sol-gel technique involving high-temperature supercritical drying described in detail earlier.[4,13] CP-MgO (281 m^2/g) was obtained by decomposition of $Mg(OH)_2$ prepared by hydration of commercial MgO. In both cases the final preparation step was overnight evacuation at 500°C followed by storage under ambient conditions. Their performance was compared to that of a commercial low surface area sample CM-MgO (10 m^2/g).

A flow reactor equipped with a MacBain spring balance and an on-line gas chromatograph was used in the flow experiments. This made it possible to monitor both the composition of the products and the sample weight during the reaction. The sensitivity in the determination of the sample weight was 10^{-4} g. CF_2Cl_2 and $CFCl_3$ of "pure" grade and Ar of "ultra pure" grade were used in the experiments. In the latter case, the argon content was no less than 99.99%, with oxygen concentration less than 0.001 vol.% and water vapor content less than 0.1 g/m^3. The halocarbon was passed through the reactor with a volume rate of

3 L/h. The MgO sample weight was 0.05-0.1 g. The samples were activated in argon flow at 500°C for 30 min.

The same flow reactor equipped with a MacBain spring balance was used for “in situ” carbon formation. AP-MgO was placed in a quartz basket and placed on a spring. This allowed for measuring of the sample weight change and accurate determination of carbon percentage in the sample. Then, the sample was heated under a flow of argon, at 500°C for 1 hour. After that butadiene, used as a carbon source, was added to the flow. The flow rates were: butadiene - 7.5 L/h; argon - 75 L/h. Initial carbon formation occurred at a rate of approximately 2 wt. % per hour. After the desired amount of carbon was obtained, the butadiene flow was stopped and the product was cooled to room temperature under an argon flow.

Two TiO_2 samples were used in the experiments: Degussa P-25 and Nanoactive TiO_2 (Nanoscale Materials). Before the experiments, the samples were calcined in an argon flow at 500°C for 1 h. The temperature was then decreased to the desired reaction temperature, and the halocarbon flow was turned on. The standard reaction time was 30 min. The sample was finally cooled in the argon flow to room temperature. The products were analyzed using a Shimadzu GCMS-QP5000 equipped with a 30 m Restec capillary column with XTI-5 phase. XRD studies were performed using a Brucker D8 instrument with CuK_α monochromatic irradiation.

3. Results and Discussion

3.1. CF_2Cl_2 DESTRUCTIVE ADSORPTION OVER AP-MGO

CF_2Cl_2 is a colorless and odorless gas, which is chemically very inert and has a high thermal stability. It appeared to react with nanoscale MgO at temperatures as low as 325°C to give a significant gain of the sample weight as evidenced by kinetic curves presented in Figure 1. The presence of an obvious induction period, after which an intense gain of the sample weight starts, is the most distinctive and interesting feature of the kinetic curves. The duration of the induction period depends on the reaction temperature. At 350°C it exceeds two hours (curve 3), while at 450°C the sample weight gain starts almost immediately (curve 1). The induction period becomes more than 8.5 hours long when the reaction temperature is further decreased to 325°C (curve 4).

The induction period phenomenon is reproduced with ca. 5-10% precision when all conditions are kept the same. The existence of an induction period on the kinetic curve must be connected with gradual accumulation of certain active sites or defects on the MgO surface, which eventually lead to complete transformation of the nanoparticles. As soon as enough active sites are

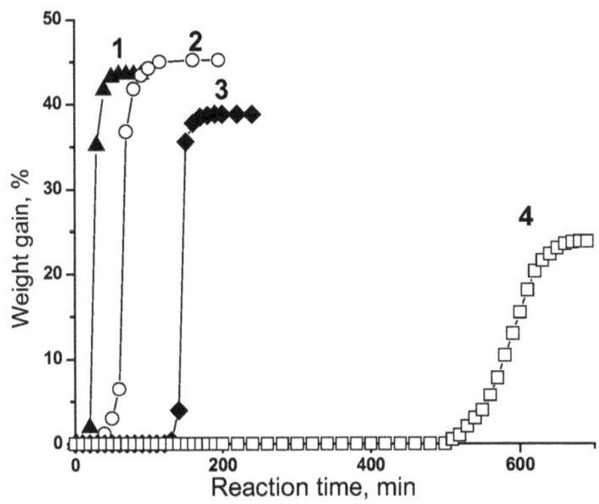

Figure 1. Kinetic weight gain curves of AP-MgO treated with CF_2Cl_2 at 450 (1), 400 (2), 350 (3) and 325°C (4).

accumulated, intense bulk substitution of oxygen by fluorine starts. At lower temperatures the rate of the halocarbon reaction with the oxide surface is significantly lower. This makes the induction period much longer. Note that the weight gain stops at values close to 40-45% indicating the end of the reaction. The maximum theoretical weight gain corresponding to complete transformation of the oxide to fluoride is 55%. This indicates that about 80% transformation is achieved in our experiments.

In order to evaluate the role of small crystals of AP-MgO in the reaction with CF_2Cl_2, similar experiments were performed over other MgO samples prepared by different methods, having lower surface areas and different crystal sizes and shapes. An experiment with CP-MgO showed that the induction period at 400°C was twice as long over the sample with the lower surface area (40 min vs. 20 min) and the maximum MgO conversion to MgF_2 was lower (65% vs. 80%). The kinetics of the process proved to be very different over CM-MgO, which had a surface area 40 times less than AP-MgO. CM-MgO did not react much with CF_2Cl_2, giving a weight gain of only 1.5%. Such a low conversion degree indicates that large particle surfaces are not halogenated even at temperatures as high as 500°C.

The main products resulting from the AP-MgO reaction with CF_2Cl_2 are MgF_2, CO_2 and CCl_4[14] formed according to overall reaction (1). The former is identified as the only solid phase according to XRD and HRTEM data. CCl_4 appears to result from CF_2Cl_2 exchange reaction with the partially chlorinated MgO/MgF_2 surface. An MgO excess is required to convert all CF_2Cl_2 to CO_2 and solid products according to reaction (2). Indeed, we managed to achieve complete CF_2Cl_2 mineralization at 450°C and a MgO/CF_2Cl_2 ratio of 2.5:1. The

XRD spectrum of the solid sample after reaction shows the presence of both MgF_2 and $MgCl_2 \cdot 6\,H_2O$, in addition to unreacted MgO.

$$2MgO + 2CF_2Cl_2 \rightarrow 2MgF_2 + CO_2 + CCl_4 \quad (1)$$

$$2\,MgO + CF_2Cl_2 \rightarrow MgF_2 + MgCl_2 + CO_2 \quad (2)$$

3.2. EFFECT OF CARBON COATING ON AP-MGO REACTIONS WITH HALOCARBONS

Recently we have shown that carbon-coated AP-MgO has improved stability in the presence of water, and the possible storage time in air, in comparison with not coated nanoscale MgO.[13] Our EPR investigation of the basic active sites present on the surface of AP-MgO, partially covered with carbon, showed that samples containing 10% carbon or less retain significant amounts of open active sites that are available for destructive adsorption reactions.[15] So, it was interesting to study how introduction of the carbon coating affects the AP-MgO reactivity with respect to destructive adsorption of halocarbons. Unexpectedly we observed that the carbon coating improves the reactivity of AP-MgO in destructive adsorption of CF_2Cl_2. Its reaction with AP-MgO is characterized by a prolonged induction period that is nearly 1 h at 400°C (Figure 2).

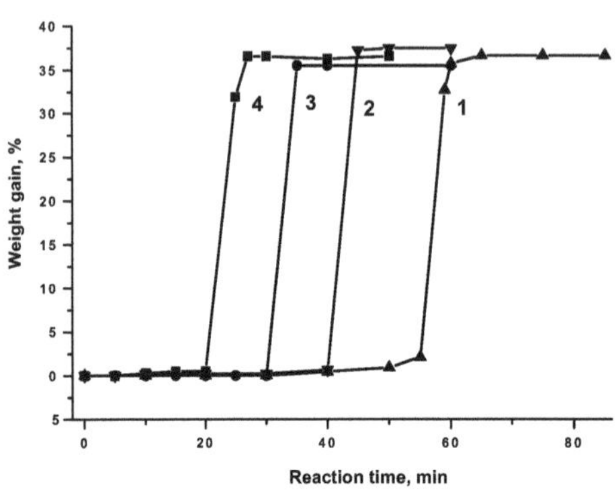

Figure 2. Kinetic weight gain curves of AP-MgO (1), 1% C/AP-MgO (2), 3% C/AP-MgO (3) and 10% C/AP-MgO (4) treated with CF_2Cl_2 at 400°C.

The addition of only 1 wt.% carbon significantly shortens the induction period, while 10% C/AP-MgO has an induction period three times less than AP-MgO. Similar results were obtained in the reaction with another halocarbon $CFCl_3$ (Figure 3).

No well-defined induction period was observed in this case. We attribute this fact to a different Cl/F ratio in the reactant that leads to less sample overheating during reaction. Sample overheating by as much as 50°C was observed during the fast stage of the AP-MgO reaction with CF_2Cl_2.

Interestingly, at all temperatures between 350 and 400°C the carbon coating increased the reactivity of AP-MgO, making the reaction both faster and more complete. The maximum weight gain of 32% could be obtained in this temperature range only in the presence of carbon. Still this weight gain corresponds only to ~60% MgO conversion, considerably lower than in the reaction with CF_2Cl_2. Lower conversion in this case might be another consequence of lower sample temperature during the fast reaction stage.

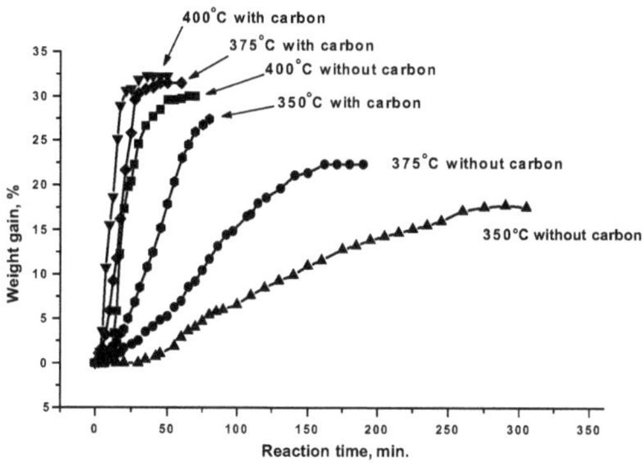

Figure 3. Kinetic weight gain curves of AP-MgO and 3% C/AP-MgO treated with $CFCl_3$ at various temperatures

We are not yet sure what causes the positive effect of the carbon coating on the performance of AP-MgO in the reactions with halocarbons. The carbon coating may adsorb some of the reaction products or impurities in the feed, that allows for a faster reaction of the halocarbons with the oxide core. Further studies including mass-spectrometric analysis of the gas phase composition are required.

3.3. PHASE SELECTIVITY IN SOLID STATE TRANSFORMATION OF NANO-TiO_2 TO $TiOF_2$

It has been shown above that the kinetics of the MgO reaction with halocarbons, forming of solid MgF_2, depend on the size and morphology of MgO particles. Phase composition of the oxide nanoparticles appears to be another parameter having a significant effect on such reactions. The phase composition and size of the nanoparticles can be readily regulated for titanium dioxide. We have studied the reaction of nanocrystalline TiO_2 samples with CHF_2Cl, resulting in the formation of a new chemical compound titanium oxyfluoride:

$$TiO_2 + CHF_2Cl \rightarrow TiOF_2 + CO + HCl \quad (3)$$

The reaction was studied in halocarbon flow at different temperatures. The kinetics appear to be significantly dependent on the size of TiO_2 nanoparticles and their crystalline phase. Figure 4 presents the XRD spectra of the solid reaction products of Degussa P-25 TiO_2 with CHF_2Cl at different temperatures. The initial TiO_2 sample (spectrum 1) consists of a mixture of anatase (~75%) and rutile (~25%) with a particle size of about 25 nm. Reaction with CHF_2Cl leads to the formation of a new solid product identified by XRD as $TiOF_2$, and a decrease in the intensity of anatase lines (spectra 2 and 3). Note that even after complete transformation of anatase the lines of rutile remain unchanged (spectrum 3). No reaction of rutile with CHF_2Cl was observed even when the reaction temperature was increased to 400°C.

The observed phase selectivity phenomenon vividly shows that the crystalline phase of titania nanoparticles has a major effect on their reactivity. A similar phenomenon had been recently described for dissolution of a similar TiO_2 sample in HF.[16] This reaction can be used for the preparation of a pure nanoscale rutile phase for use as a photocatalyst.

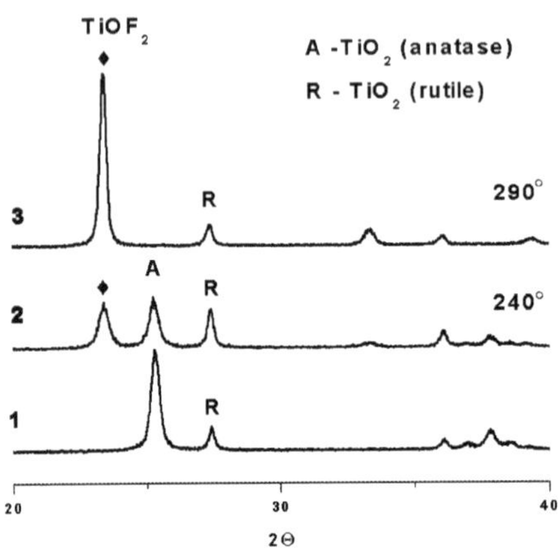

Figure 4. XRD spectra of Degussa P-25 TiO_2 before reaction (1) and after reaction with CHF_2Cl at 240°C (2) and 290°C (3).

Using nanoactive TiO_2 prepared by sol-gel technology, it is possible to prepare samples with different particle size and phase composition by simply varying the heat treatment temperature. For example, after calcination at 500°C for 1 h, the average size of nanoparticles according to the XRD data is about 15 nm, and anatase is the only identified crystalline phase (Figure 5, spectrum 1). Its reaction with CHF_2Cl at temperatures as low as 240°C results in almost complete TiO_2 transformation to $TiOF_2$ (Figure 5, spectrum 2). This temperature is 50°C lower than the temperature required for complete conversion of 25 nm anatase nanoparticles (Figure 4). Thus, similar to reactions of halocarbons with MgO, a decrease of the TiO_2 particle size allows for a significant decrease of the temperature necessary for complete conversion of solid phase nanoparticles.

According to the XRD data, the particle size of reaction product $TiOF_2$ is similar to that of the nanoparticles of the initial TiO_2. The surface area of the material also does not change much. These results prove that such reactions can be used for synthesis of new halogenated nanocrystalline materials using nanocrystalline oxides as precursors.

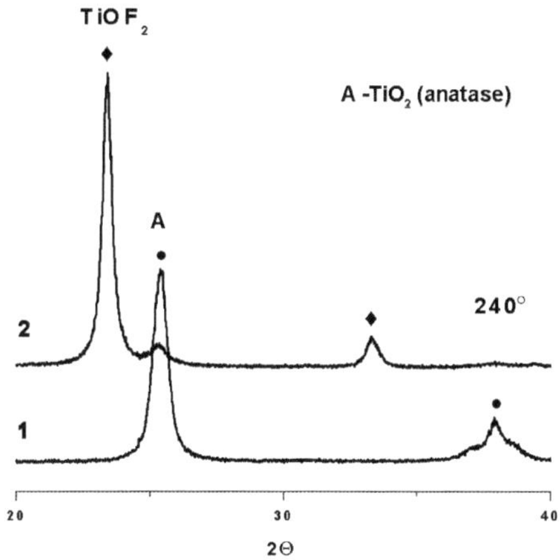

Figure 5. XRD spectra of Nanoactive TiO_2 before reaction (1) and after reaction with CHF_2Cl at 240°C (2).

4. Conclusion

Nanocrystalline oxides have exceptionally high reactivity with respect to halocarbons. First of all, the temperature of their reaction with halocarbons is much lower than observed for bulk analogs. It is thus possible to carry out destructive adsorption reactions at fairly low temperatures. This is especially important for the development of effective destructive sorbents for neutralization of toxic compounds at mild temperatures. We have also demonstrated that nanocrystalline oxides can be used as precursors for the synthesis of novel halogenated nanocrystalline materials.

Of special interest are composite materials consisting of carbon-coated nanocrystalline oxides. Such permeable carbon coating not only provides high stability of the destructive sorbents under atmospheric conditions, but in some cases shown in this publication considerably increases their reactivity. Elucidation of the mechanism of this interesting phenomenon will require further research.

Acknowledgment

This study was supported by the Russian Foundation for Basic Research (Grant 05-03-33000).

References

1. A. Khaleel, P. N. Kapoor, and K. J. Klabunde, Nanocrystalline metal oxides as new adsorbents for air purification, *Nanostruct. Mater.* 11, 459-468 (1999).
2. E. M. Lucas, and K. J. Klabunde, Nanocrystals as destructive adsorbants for mimcs of chemical warfare agents, *Nanostruct. Mater.* 12, 179-182 (1999).
3. O. B. Koper, I. Lagadic, A. Volodin, and K. J. Klabunde, Alkaline-earth oxide nanoparticles obtained by aerogel methods. Characterization and rational for unexpectedly high surface chemical reactivities, *Chem. Mater.* 9, 2468-2480 (1997).
4. K. J. Klabunde, J. Stark, O. Koper, C. Mohs, D. G. Park, S. Decker, Y. Jiang, I. Lagadic, and D. J. Zhang, Nanocrystals as stoichiometric reagents with unique surface chemistry, *J. Phys. Chem.* 100, 12142-12153 (1996).
5. Y. X. Li, and K. J. Klabunde, Nanoscale Metal-Oxide Particles As Chemical Reagents - Destructive Adsorption of A Chemical-Agent Simulant, Dimethyl Methylphosphonate, on Heat-Treated Magnesium-Oxide, *Langmuir* 7, 1388-1393 (1991).
6. G. W. Wagner, P. W. Bartram, O. Koper, and K. J. Klabunde, Reactions of VX, GD, and HD with nanosize MgO, *J. Phys. Chem.* B 103, 3225-3228 (1999).
7. G. W. Wagner, O. B. Koper, E. Lucas, S. Decker, and K. J. Klabunde, Reactions of VX, GD, and HD with nanosize CaO: Autocatalytic dehydrohalogenation of HD, *J. Phys. Chem. B* 104, 5118-5123 (2000).
8. S. Decker, and K. J. Klabunde, Enhancing effect of Fe2O3 on the ability of nanocrystalline calcium oxide to adsorb SO_2, *J. Am. Chem. Soc.* 118, 12465-12466 (1996).
9. R. M. Narske, K. J. Klabunde, and S. Fultz, Solvent effects on the heterogeneous adsorption and reactions of (2-chloroethyl) ethyl sulfide on nanocrystalline magnesium oxide, *Langmuir* 18, 4819-4825 (2002).
10. K. J. Klabunde, G. Medine, A. Bedilo, P. Stoimenov, and D. Heroux, Nanocrystalline metal oxides: A new family of mesoporous inorganic materials useful for destructive adsorption of environmental toxins, *ACS Symp. Ser.* 890, 272-276 (2005).
11. P. K. Stoimenov, R. L. Klinger, G. L. Marchin, and K. J. Klabunde, Metal oxide nanoparticles as bactericidal agents, *Langmuir* 18, 6679-6686 (2002).
12. P. K. Stoimenov, V. Zaikovski, and K. J. Klabunde, Novel halogen and interhalogen adducts of nanoscale magnesium oxide, *J. Am. Chem. Soc.* 125, 12907-12913 (2003).
13. A. F. Bedilo, M. J. Sigel, O. B. Koper, M. S. Melgunov, and K. J. Klabunde, Synthesis of carbon-coated MgO nanoparticles, *J. Mater. Chem.* 12, 3599-3604 (2002).
14. I. V. Mishakov, V. I. Zaikovskii, D. S. Heroux, A. F. Bedilo, V. V. Chesnokov, A. M. Volodin, I. N. Martyanov, S. V. Filimonova, V. N. Parmon, and K. J. Klabunde, CF_2Cl_2 decomposition over nanocrystalline MgO: Evidence for long induction periods, *J. Phys. Chem. B* 109, 6982-6989 (2005).
15. D. S. Heroux, A. M. Volodin, V. I. Zaikovski, V. V. Chesnokov, A. F. Bedilo, and K. J. Klabunde, ESR and HRTEM study of carbon-coated nanocrystalline MgO, *J. Phys. Chem. B* 108, 3140-3144 (2004).
16. T. Ohno, K. Sarukawa. and M. Matsumura, Photocatalytic activities of pure rutile particles isolated from TiO_2 powder by dissolving the anatase component in HF solution, *J. Phys. Chem. B* 105, 2417-2420 (2001).

DYNAMIC ADSORPTION OF TERT-BUTYLBENZENE, CYCLOHEXANE AND WATER VAPOURS ON FIXED ACTIVATED CARBON/MOLECULAR SIEVE BEDS

D. PALIJCZUK,[a,*] R. SZMIGIELSKI,[a] V. M. GUN'KO, [b] R. LEBODA[c]
[a] *Military Institute of Chemistry and Radiometry, 00-910 Warsaw, Poland,*
[b] *Institute of Surface Chemistry, 17 General Naumov Street, 03164 Kiev, Ukraine*
[c] *Faculty of Chemistry, Maria Curie-Skłodowska University, 3 Maria Curie-Skłodowska Square, 20-031 Lublin, Poland*

Abstract. Activated carbon Norit R 08 Extra, and molecular sieve type 4A, were investigated using dynamic (tert-butylbenzene (TBB), cyclohexane (CHX) and water vapour) adsorption methods. The TBB, CHX and water breakthrough plots for fixed activated carbon – molecular sieve beds were analyzed. It was found that the type of bed composition with mechanically mixed activated carbon with molecular sieve, or separated activated carbon and molecular sieve layers, affects the dynamic adsorption characteristics.

Keywords: activated carbons; molecular sieve; breakthrough dynamics; tert-butylbenzene adsorption; cyclohexane; water; breakthrough time

1. Introduction

Absorption of harmful organic compounds by activated carbons from gas and liquid media is of interest and importance for human and environmental protection purposes.[1-21] The influence of the texture of carbon granules (size and volume of pores, specific surface area, granule size *d*, and carbon bed depth *L*), gas stream humidity and velocity, and amounts of pre-adsorbed water are investigated on adsorption of organics in different media.[1-21]

*To whom correspondence should be addressed. D. Palijczuk, Military Institute of Chemistry and Radiometry, 00-910 Warsaw, Poland e-mail: dorota.palijczuk@wichir.waw.pl

J.P. Blitz and V.M. Gun'ko (eds.), Surface Chemistry in Biomedical and Environmental Science, 413–420.

In a previous paper,[22] we investigated the breakthrough dynamics of tert-butylbenzene (TBB) using different activated carbons in dry and wetted beds, and dry and wet airstream. It was shown that the availability of a developed network of mesopores (and macropores as transport pores), a low nonuniformity of pores and surfaces, and a low affinity with respect to water (low content of polar surface sites), are necessary conditions for a high effectiveness of carbon adsorbents in the adsorption of organics characterized by an extension of the critical breakthrough time.

Recently,[23] we investigated the influence of structural characteristics of nanoporous and nano/mesoporous activated carbons after different pretreatments (e.g. wetting, freezing-defreezing) on TBB breakthrough parameters. Several generalized adsorption and breakthrough equations, allowing calculations of the corresponding distribution functions, were utilized.

It was found[22,23] that noticeable changes in the adsorptive capacity of carbons, such as reduction of the breakthrough time for organics, occur when water is present in the air stream. To protect the carbon bed from water vapour, molecular sieve was used. The aim of this work is to study the adsorption of TBB, CHX and water vapour on activated carbon/molecular sieve beds. TBB is an interesting substance to test the breakthrough dynamic parameters, and can be used as a model adsorbate simulating toxic compounds such as sarin, since the breakthrough plots for sarin and TBB under the same conditions are similar.[9] The physical properties (molecular weight, molecular diameter, vapour pressure, boiling point etc.) of TBB are close to those of sarin. TBB is an aromatic compound with nonpolar 3D-aliphatic group. CHX is a standard test organic substance (EN 14387:2004) used in industrial respiratory protection devices (e.g. full face gas mask). CHX is used as a simulant of middle boiling point toxic substances such as chloropicrin.

2. Experimental

2.1. MATERIALS

Two different adsorbents, activated carbon Norit R 0.8 Extra (Norit N.V., The Netherlands) and molecular sieve (type 4A, Merck), were used to study tert-butylbenzene, cyclohexane, and water vapour breakthrough dynamics. Structural parameters of the carbon adsorbent were calculated from benzene vapour adsorption-desorption isotherms measured gravimetrically at 293 K using a McBain-Bakr quartz microbalance, and nitrogen adsorption-desorption isotherms recorded at 77.4 K using a Micromeritics ASAP 2405N analyzer described in detail elsewhere.[22,24] Activated carbon Norit has a cylindrical regular form of d = 0.8 mm. Molecular sieve type 4A was in the form of spherical particles with a mean diameter of 1 mm.

Tert-butylbenzene (BDH Chemicals Ltd Poole, England) and cyclohexane (POCH, Poland) were used as probe organic compounds in the breakthrough experiments.

2.2. TBB BREAKTHROUGH

Compressed air was purified using columns with activated carbon, molecular sieves and silica gel. The TBB vapour was generated using an infusion pump with a syringe. Dry air, spiked with required amounts of TBB vapour, was supplied through a glass tube (internal section 5.187 cm^2) filled with the tested carbon and molecular sieve. An initial concentration of TBB was $c_0 = 1 \pm 0.1$ mg dm^{-3}. A detailed description of the experimental set-up and breakthrough measurement procedure was described elsewhere.[22] The experiments were carried out in two bed configurations: 1) a layer of molecular sieve type 4A (1.5 cm in thickness) and a second layer with activated carbon (1.5 cm) (labelled "S"); and 2) mechanically mixed activated carbon with molecular sieve in equal volumetric amounts (labelled "M").

The carbon/molecular sieve bed was temperature-controlled at 293 K. The volumetric flow rate was $Q = 5.19$ dm^3 min^{-1} (1.0 dm^3 min^{-1} cm^{-2}), resulting in the linear flow rate $v_L = 1000$ cm min^{-1}. The gas flow was controlled by several flow meters. The outlet concentrations were analyzed in cycles of 3 min with a CP 9001 CHROMPACK gas chromatograph with a flame ionization detector. The breakthrough time was determined at the outlet by a TBB concentration (behind the carbon bed) $c_x = 10^{-5}$ mg dm^{-3} ($c_x/c_o = 10^{-5}$). To study the water influence on TBB breakthrough, water vapour was added to reach 50% relative humidity (RH) of the air flow. The measurements were performed with dry carbon/molecular sieve beds using dry or wet air.

To compute parameters such as kinetic saturation capacity W_e, and overall mass-transfer coefficient β_e*, shown in Tables 1 and 2, the equation for breakthrough was used in the following form

$$\lg\left(\frac{c_x}{c_o}\right) = \frac{0.434 c_o \beta_e^*}{\rho_b W_e} t_b - \frac{0.434 \beta_e^*}{v_L} L$$

where c_x and c_0 are outlet and inlet concentrations respectively, and t_b is the breakthrough time at given outlet concentration.

2.3. CYCLOHEXANE BREAKTHROUGH

Dry air, spiked with required amounts of cyclohexane vapour, was supplied through a glass tube (internal section 20 cm^2) filled with the tested carbon and molecular sieve. An initial concentration of cyclohexane was $c_0 = 3.6 \pm 0.1$ mg dm^{-3} (0.1% V/V). The experiments were carried out with two bed configurations: 1) a separated (layer of molecular sieve type 4A 2 cm in height, and a layer of activated carbon 2 cm); and 2) mechanically mixed activated carbon with molecular sieve in equal volumetric amounts.

The carbon/molecular sieve bed was temperature-controlled at 293 K. The volumetric flow rate was $Q = 7$ dm^3 min^{-1} (0.35 dm^3 min^{-1} cm^{-2}), resulting in a linear flow rate $v_L = 350$ cm min^{-1}. The outlet concentrations were analyzed in cycles of 3 min by using a CP 3800 VARIAN gas chromatograph with a flame ionization detector. The breakthrough time was determined at the outlet CHX concentration (behind the bed) $c_x = 3.6 \times 10^{-5}$ mg dm^{-3} ($c_x/c_o = 10^{-5}$). To study water influence on the CHX breakthrough, water vapour was added to reach 50% relative humidity (RH) of the air flow. The RH behind the carbon/molecular was controlled by using a thermo-hygrometer. Measurements were performed with dry carbon/molecular sieve beds using dry or wet air.

3. Results and Discussion

The β_e^* and W_e values were calculated from the breakthrough curves of TBB and CHX (Tables 1 and 2) using the procedure described elsewhere.[22,23] The shape of log-scaled TBB breakthrough plots vs time for "M" and "S" beds are very similar (Figures 1 and 2). The TBB (or TBB with water vapour) breakthrough curve for the "S" bed shifts toward longer time (Figure 1) and the t_b value is two times larger than that for the "M" bed (Table 1). The parameters β_e* and W_e, characterizing bed capacity are higher for "S" activated carbon/molecular sieve beds than that for the "M" bed. Therefore, better performance of complex activated carbon/molecular sieve beds with respect to TBB breakthrough, is achieved when separate layers with different adsorbents is used. The presence of water vapour in the airstream is much more detrimental for the t_b value in the "M" bed than for the "S" bed (Table 1), as the gap between the corresponding curves increases (Figures 1 and 2). Perhaps this effect can be linked to water distribution along the bed. For instance, larger amounts of water adsorbed onto the first layer, but TBB is more strongly adsorbed by the second layer of the "S" bed. The water vapour breakthrough plots vs time (Figure 3) for the mixed "M" bed is insignificantly displaced toward shorter time in comparison with the "S" bed. Consequently, the bed construction ("M" or "S") is unimportant in the case of dynamic adsorption

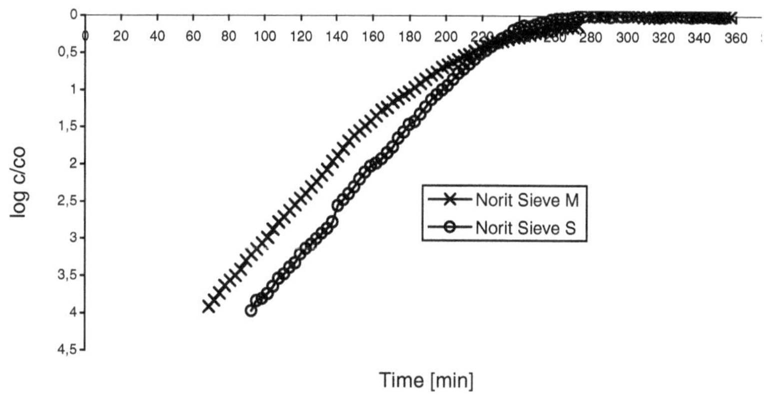

Figure 1. Log-scaled TBB breakthrough plots vs. time for mixed "M" and separated "S" beds of activated carbon and molecular sieve without water vapour (RH = 0 %).

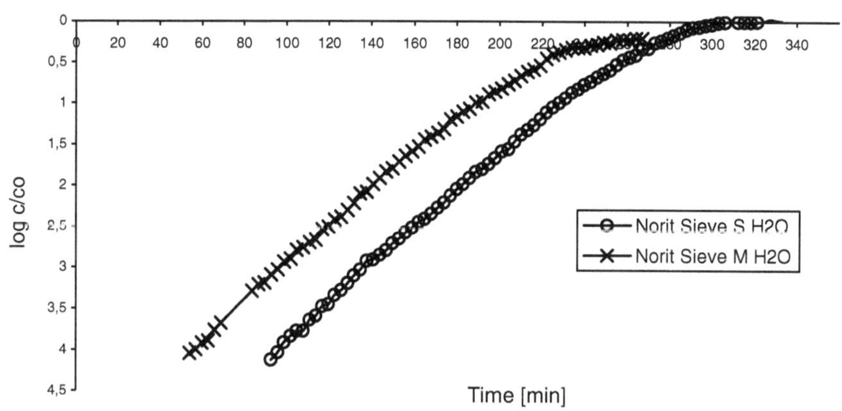

Figure 2. Log-scaled TBB breakthrough plots vs. time for mixed "M" and separated "S" beds of activated carbon and molecular sieve with water vapour (RH = 50%).

of water under these conditions with simultaneous breakthrough of TBB. However, the distribution of water along the bed differs for the "S" and "M" beds, which affects the breakthrough of TBB. The CHX (or CHX with water vapour) breakthrough curves for "M" beds are displaced toward longer time (Table 2, Figure 4). The β_e^* and W_e values are higher for "M" activated carbon/molecular sieve beds than for "S". Therefore, better performance of activated carbon/molecular sieve beds with respect to CHX, is obtained with a mixture of these materials.

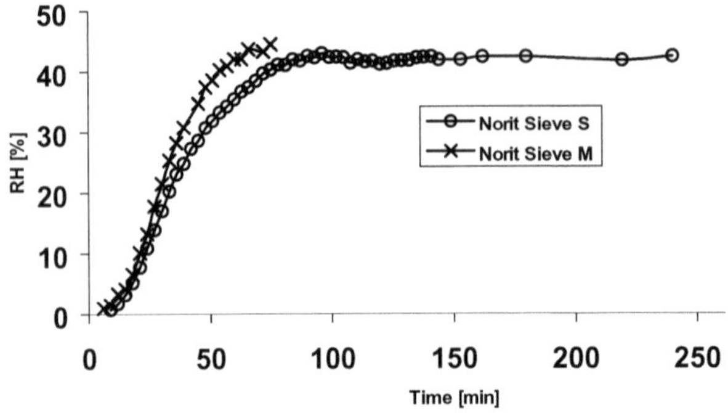

Figure 3. Water vapour breakthrough plots vs time for mixed “M” and separated “S” beds of activated carbon and molecular sieve measured simultaneously with TBB breakthrough.

Table 1. Dynamic adsorption of TBB on carbon/molecular sieve beds.

Sample	t_b (min)	β_e^* (min^{-1})	W_e ($mg\ g^{-1}$)
Norit/Sieve, S RH = 0 %	59	10300	381
Norit/Sieve, M RH = 0 %	30	9360	345
Norit/Sieve, S, H_2O RH = 50 %	52	9580	442
Norit/Sieve, M, H_2O RH = 50 %	13	8170	374

Note: β_e* is the overall mass transfer coefficient; m_{MS} = 6.224 g or m_{AC} = 3.128 g are the weights of sieve or carbon; ρ_{bMS} = 0.809 g/cm^3 and ρ_{bAC} 0.402 g/cm^3 are bulk density of molecular sieve or carbon bed, respectively.

Table 2. Dynamic adsorption of cyclohexane.

Sample	t_b (min)	β_e^* (min^{-1})	W_e ($mg\ g^{-1}$)
Norit/Sieve, S RH = 0 %	37	3190	158
Norit/Sieve, M RH = 0 %	55	3470	205
Norit/Sieve, S, H_2O RH = 50 %	6	2130	162
Norit/Sieve, M, H_2O RH = 50 %	45	3250	186

Note: m_{MS} = 32.36 g or m_{AC} = 16.08 g.

The linear flow rate v_L for TBB breakthrough experiments was about three times larger than for CHX. The differences in the TBB and CHX breakthrough behaviour for “M” and “S” activated carbons/molecular sieve beds could be connected with the magnitude of this linear rate, and the nature of these organics. CHX is not aromatic with a 3D-structure, in contrast to the aromatic TBB with only a 3D-tail (Figure 5). These molecules can interact differently not only with the carbon adsorbent, but also with the sieve. This interaction depends on the presence and distribution of water in the beds. Additionally, a decrease in the v_L value can cause penetration of smaller water molecules deeper into long and narrow pores which can affect the capacity of pores with respect to CHX.

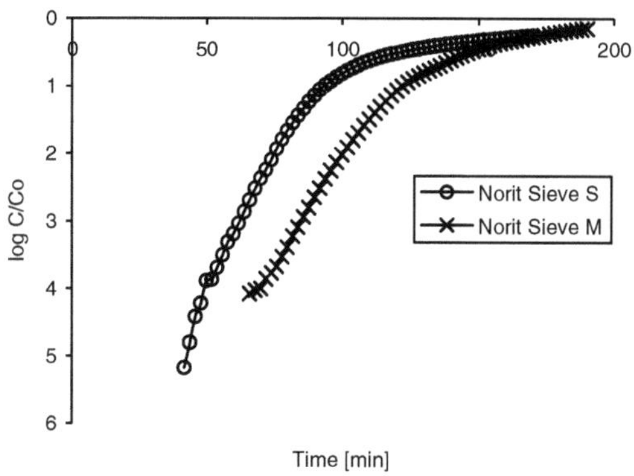

Figure 4. Water vapour breakthrough plots vs time for mixed "M" and separated "S" beds of activated carbon and molecular sieve measured simultaneously with TBB breakthrough.

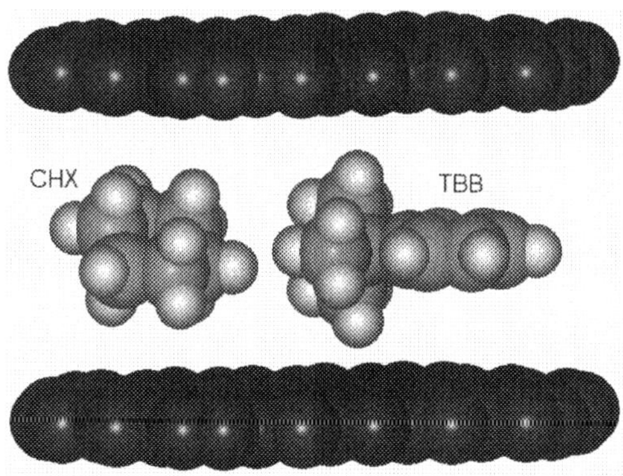

Figure 5. TBB and CHX molecules in a model carbon nanopore (1 nm in width).

4. Conclusion

The presence of water in the airstream leads to negative changes in the TBB and CHX breakthrough, namely diminution of the breakthrough time and the effective adsorption rate coefficient.

Comparing the log-scaled TBB breakthrough plots vs time for mixed "M" and separated "S" beds of activated carbon and molecular sieve without or with water vapour, it can be affirmed that separated activated carbon/molecular sieve bed ("S") is more effective than mixed ("M"). In the case of cyclohexane breakthrough a negative effect caused by mixing of activated carbon with molecular sieve is observed. This effect is probably caused by the different linear flow rates for TBB and CHX on the breakthrough experiments.

References

1. S. J. Gregg and K. S. W. Sing, *Adsorption, Surface Area and Porosity* (2nd ed., Academic Press, London, 1982).
2. A. W. Adamson and A. P. Gast, *Physical Chemistry of Surface* (6th ed., Wiley, New York, 1997).
3. V. B. Fenelonov, *Porous Carbon* (Nauka, Novosibirsk, 1995).
4. J. K. Brennan, T. J. Bandosz, K. T. Thomson, and K. E. Gubbins, Water in porous carbons, *Colloids Surf. A* 187–188, 539-568 (2001).
5. *Proceedings of the Sixth Conference on Fundamentals of Adsorption*, presqu'ile de Giens, (France, 24–26 May 1998, Elsevier, Paris, 1998).
6. P. Lodewyckx and E. F. Vansant, Influence of humidity on adsorption capacity from the Wheeler-Jonas model for prediction of breakthrough times of water immiscible organic vapors on activated carbon beds, *Am. Ind. Hyg. Assoc. J.* 60, 612-617 (1999).
7. P. Lodewyckx and E. F. Vansant, The influence of humidity on the overall mass transfer coefficient of the Wheeler-Jonas equation, *Am. Ind. Hyg. Assoc. J.* 61, 461-468 (2000).
8. P. Lodewyckx and E. F. Vansant, Estimating the overall mass transfer coefficient k_v of the Wheeler-Jonas equation: a new simple model, *Am. Ind. Hyg. Assoc. J.* 501-505 (2000).
9. K. M. Nikolaev, *Facilities of Individual and Collective Protection* (Military Academy of Chemical Protection, Moscow, 1976).
10. L. A. Jonas, Reaction steps in gas sorption by impregnated carbon, *Carbon* 16, 115-119 (1978).
11. L. A. Jonas and J. A. Rehrmann, The kinetics of adsorption of organo-phosphorus vapors from air mixtures by activated carbon, *Carbon* 10, 657-663 (1972).
12. L. A. Jonas and J. A. Rehrmann, Predictive equations in gas adsorption kinetics, *Carbon* 11, 59-64 (1973).
13. L. A. Jonas and J. A. Rehrmann, The rate of gas adsorption by activated carbon, *Carbon* 12, 95-101 (1974).
14. A. Wheeler and A. J. Robell, Performance of fixed-bed catalytic reactors with poison in the feed, *J. Catal.* 13, 299-305 (1969).
15. L. A. Jonas and W. J. Svirbely, The kinetics of adsorption of carbon tetrachloride and chloroform from air mixtures by activated carbon, *J. Catal.* 24, 446-459 (1972).
16. J. A. Rehrmann, L. A. Jonas, Dependence of gas adsorption rates on carbon granule size and linear flow velocity, *Carbon* 16, 47-51 (1978).
17. G. O. Wood, J. F. Stampfer, Adsorption rate coefficients for gases and vapors on activated carbons, *Carbon* 31, 195-200 (1993).
18. G. O. Wood, Activated carbon adsorption capacities for vapors, *Carbon* 30, 593-599 (1992).
19. O. Busmundrud, Vapour breakthrough in activated carbon beds, *Carbon* 31, 279-286 (1993).
20. Y. H. Yoon, J. H. Nelson, Application of gas adsorption kinetics I. A theoretical model for respirator cartridge service life, *Am. Ind. Hyg. Assoc. J.* 45, 509-516 (1984).
21. N. V. Beck, S. E. Meech, P. R. Norman, and L. A. Pears, Characterisation of surface oxides on carbon and their influence on dynamic adsorption, *Carbon* 40, 531-540 (2002).
22. D. Palijczuk, V. M. Gun'ko, R. Leboda, J. Skubiszewska-Zięba, and S. Ziętek, Porous structure of activated carbons and tert-butylbenzene breakthrough dynamics, *J. Colloid Interface Sci.* 250, 5-17 (2002).
23. V. M. Gun'ko, D. Palijczuk, R. Leboda, J. Skubiszewska-Zięba, R. Szmigielski, and S. Ziętek, Influence of pore structure and pretreatments of activated carbons and water effects on breakthrough dynamics of tert-butylbenzene, *J. Colloid Interface Sci.* 294(1), 53-68 (2006).
24. V. M. Gun'ko, V. V. Turov, J. Skubiszewska-Zięba,R. Leboda, M. D. Tsapko, and D. Palijczuk, Structural characteristics of a carbon adsorbent and influence of organic solvents on interfacial water, *Appl. Surf. Sci.* 214(1-4), 178-189 (2003).

INFLUENCE OF PRE-ADSORBED NaCl ON BREAKTHROUGH DYNAMICS OF TBB AND DMMP ON ACTIVATED CARBON AT DIFFERENT AIRSTREAM HUMIDITIES

R. SZMIGIELSKI,[a,]* D. PALIJCZUK,[a] R. LEBODA,[b]
V. M. GUN'KO,[c] J. SKUBISZEWSKA-ZIEBA[b]
[a] *Military Institute of Chemistry and Radiometry, 00-910 Warsaw, Poland,*
[b] *Faculty of Chemistry, Maria Curie-Skłodowska University, 3 Maria Curie-Skłodowska Square, 20-031 Lublin, Poland*
[c] *Institute of Surface Chemistry, 17 General Naumov Street, 03164 Kiev, Ukraine*

Abstract. The influence of an inert impregnant (NaCl) on the adsorption properties of activated carbon Norit R 0.8 Extra was studied on breakthrough dynamics of tert-butylbenzene (TBB) and dimethylmethylphosphonate (DMMP). Pre-adsorbed NaCl (5-20 wt.%) strongly affects both structural (e.g. volume of nanopores and mesopores) and adsorption (adsorption potential, breakthrough time, kinetic saturation capacity, etc.) characteristics.

Keywords: activated carbon; pre-adsorbed NaCl; breakthrough dynamics; tert-butylbenzene adsorption; dimethylmethylphosphonate; breakthrough time

1. Introduction

Activated charcoal is one of the most popular adsorbents utilized in many applications[1] such as respiratory protection against highly toxic substances.[2] For these purposes activated carbons are used in the form of a fixed bed with a high-efficiency glass fiber aerosol filter.[3] For military purposes carbon adsorbents are impregnated with different inorganic salts, used not only for physical adsorption but also as a catalyst support. Whetlerite consists of an

*To whom correspondence should be addressed. R. Szmigielski, Military Institute of Chemistry and Radiometry, 00-910 Warsaw, Poland e-mail: radoslaw.szmigielski@wichir.waw.pl

J.P. Blitz and V.M. Gun'ko (eds.), Surface Chemistry in Biomedical and Environmental Science, 421–428.

activated carbon and a mixture of different salts of Cu-Cr-Ag. For testing of this type of material, tert-butylbenzene (TBB) and dimethylmethylphosphonate (DMMP) are used as simulants of nerve agents such as sarin and soman.[2] In previous papers[4,5] the effects of porosity and carbon treatments on the dynamic behaviour of TBB adsorption were studied in dry and wet air.

The aim of the present work was to investigate the influence of pre-adsorbed NaCl (as an impregnant) on the adsorption and breakthrough dynamics under real-world conditions. The breakthrough dynamics was studied at relative humidities of 0, 50 and 80%. The outlet concentration of water vapour was measured simultaneously with organics. Investigations of the effects of pre-adsorbed NaCl as an impregnant can help to explain the behaviour of certain catalysts adsorbed on a carbon surface. The interaction of carbon and a catalyst, especially after thermal treatment during the preparation process, can change their structure and state. The thermal stability and chemical inertness of both carbon and NaCl provides grafting of NaCl on the carbon surface without changes in the chemical structure of the matrix and deposits.

2. Experimental

2.1. MATERIALS

Activated carbon Norit R 0.8 Extra was impregnated with NaCl by incipient wetness. This activated carbon has a cylindrical regular form with $d = 0.8$ mm. Commercial carbon was previously de-ashed with concentrated HCl and methanol. Impregnated carbon samples had 0, 5, 15, and 20 wt.% of NaCl (C_{NaCl}) (Table 1). The structural and adsorption parameters of the samples were calculated with benzene vapour adsorption-desorption isotherms measured gravimetrically at 293 K using a McBain-Bakr quartz microbalance. Nitrogen adsorption-desorption isotherms were also recorded at 77.4 K using a Micrometrics ASAP 2405N analyzer. Water adsorption-desorption isotherms were measured gravimetrically at 293 K in the same way as the benzene isotherms. Calculations of the incremental pore size distributions based on nitrogen desorption data, and the adsorption potential distributions with respect to water, benzene and nitrogen are previously described.[4-11]

Tert-butylbenzene (BDH Chemicals Ltd Poole, England) and dimethyl-methylphosphonate (Fluka Chemie GmbH) were used as probe organic compounds in the breakthrough experiments.

Table 1. Parameters of breakthrough of TBB and DMMP at different humidity on carbon beds with pre-adsorbed NaCl.1000

Organics	C_{NaCl} (wt.%)	RH (%)	t_b (min)	β_E^* (min^{-1})	W_e ($mg\ g^{-1}$)	m_o (g)	ρ_b ($g\ cm^{-3}$)
TBB	0	0	64	10550	372	3.262	0.42
TBB	0	50	58	10100	398	3.262	0.42
TBB	0	80	27	8734	360	3.262	0.42
TBB	5	0	70	9050	364	3.371	0.43
TBB	5	50	55	10500	364	3.371	0.43
TBB	5	80	6	7920	237	3.371	0.43
TBB	15	0	0	6990	392	3.506	0.45
TBB	15	50	3	7830	250	3.506	0.45
TBB	15	80	0	7420	177	3.506	0.45
DMMP	0	0	161	13310	633	3.262	0.42
DMMP	5	0	72	9660	538	3.371	0.43
DMMP	15	0	26	9940	532	3.506	0.45

Note. t_b is the breakthrough time at the outlet concentration $c_x = 10^{-5}$ mg/dm^3; W_e is the kinetic adsorption capacity; β_e* is effective overall adsorption rate coefficient; m_o is the weight of bed, and ρ_b is the bulk density of carbon bed.

2.2. BREAKTHROUGH OF TBB AND DMMP

Compressed air was purified using columns with activated carbon, molecular sieves and silica gel. The TBB vapour was generated using an infusion pump with a syringe. Dry air, challenged with required amounts of TBB or DMMP vapour, was supplied through a glass tube (internal section 5.187 cm^2) filled with a test carbon. An initial concentration of TBB or DMMP was $c_0 = 1 \pm 0.1$ mg dm^{-3}. A carbon bed with a constant depth of L = 1.5 cm was temperature-controlled at 293 K. The volumetric flow rate was Q = 5.19 dm^{3}min^{-1} (1.0 dm^{3}min^{-1}cm^{-2}), resulting in a linear flow rate v_L = 1000 cm min^{-1}. The outlet concentrations were analyzed using a CP 9001 CHROMPACK gas chromatograph with a flame ionization detector. The breakthrough time was determined from the outlet concentration (behind the carbon bed) of the tested vapour at $c_x = 10^{-5}$ mg dm^{-3} (i.e. $c_x/c_0 = 10^{-5}$).[4] To study the water influence on TBB breakthrough, water vapour was added to reach 50% or 80% relative humidity (RH) of the airstream. The RH behind the bed was controlled with a thermo-hygrometer. DMMP and TBB breakthrough dynamics were analyzed using a modified Wheeler-Jonas model[3] describing the breakthrough time versus the carbon bed depth and gas flow parameters.

3. Results and Discussion

Grafting of NaCl leads to an increase in the adsorption potential for water (Figure 1). On the other hand, the adsorption potential for nonpolar nitrogen and benzene diminishes with increasing NaCl content (Figure 2).

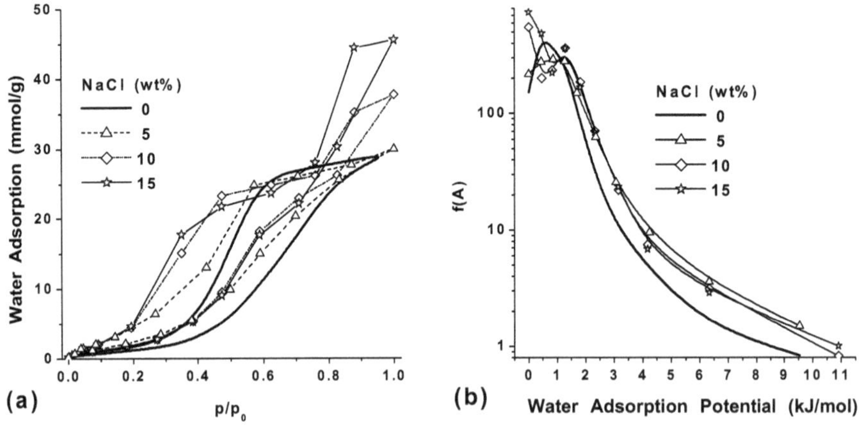

Figure 1. (a) Adsorption-desorption water isotherms at 293 K, and (b) adsorption potentials for water adsorbed onto carbon Norit at different C_{NaCl} values.

The reduced adsorption potential for nitrogen is significant at C_{NaCl} = 15 or 20 wt.%, and is characteristic for nanopores ($A > 3$ kJ/mol), narrow mesopores ($1 < A < 3$ kJ/mol) and broad pores ($A < 1$ kJ/mol) (Figure 2a). A similar reduction is observed at C_{NaCl} = 5 and 10 wt.%, but only for narrow pores. It is concluded that at low amounts of NaCl are located in narrow pores. This assumption is confirmed by the IPSD showing a reduced contribution of nanopores (Figure 3a) with a simultaneous increase in mesoporosity (Figure 3b). It should be noted that deposition of NaCl occurs in different pores, because the IPSD depends nonlinearly on C_{NaCl} over the total range of the pore sizes (Figure 3). Significant reduction of the porosity from NaCl can be caused by two factors: (i) pore filling by NaCl, and (ii) increased bulk density of NaCl/carbon (NaCl can place a role of ballast even it does not fill pores but forms segregated and relatively large particles). This is confirmed by an AFM image of the sample with maximal C_{NaCl} value (Figure 4) showing formation of the NaCl phase at the outer surface of a carbon granule. These structural features of NaCl/carbon samples might affect the breakthrough dynamics of TBB and DMMP.

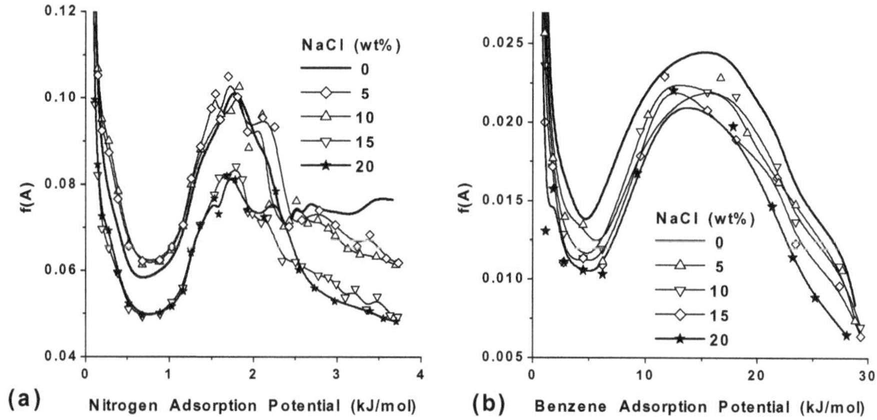

Figure 2. Adsorption potential for (a) nitrogen and (b) benzene adsorbed onto Norit with different amounts of NaCl.

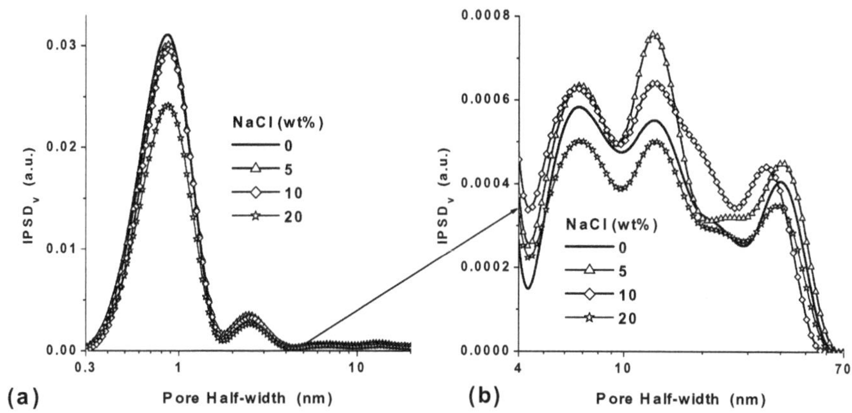

Figure 3. Incremental pore size distribution of carbon Norit initial and with adsorbed NaCl in amounts of 5, 10 and 20 wt.%.

The overall mass-transfer coefficient (β_e^*), and kinetic adsorption capacity (W_e) values, were calculated from the breakthrough curves of TBB and DMMP (Table 1) using a procedure described elsewhere.[4,5] The log-scaled TBB and DMMP breakthrough plots vs. time are presented in Figures 5 and 6. Addition of water vapour to the airstream with TBB leads to a decrease in the breakthrough time (Table 1), as the curves shift toward shorter time (Figure 5). Additionally, the shape and slope of the curves at a given NaCl concentration are similar at RH = 0% and 50% but significantly shift to shorter time at RH = 80%. This effect is observed at $C_{\mathrm{NaCl}} = 0$ and 5%, but at 15% the curve positions are proportional to the RH value. Results of the breakthrough dynamics are not shown at C_{NaCl} = 20 wt.% because of the strong negative effects of NaCl deposits.

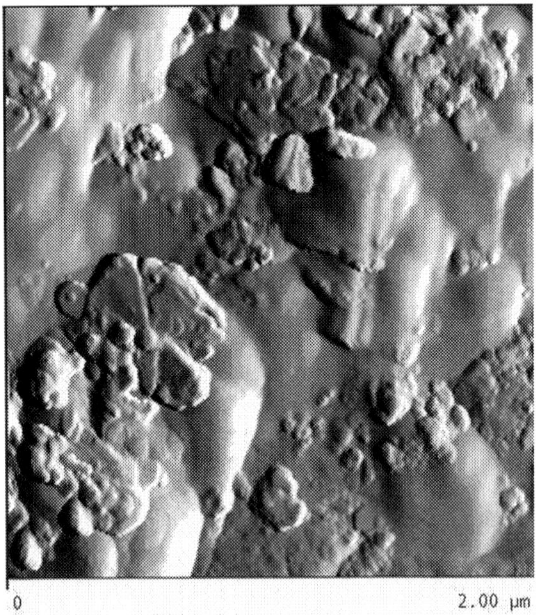

Figure 4. AFM image (2 × 2 × 1 µm, recorded using a NanoScope III (Digital Instruments, USA) apparatus with a tapping mode) of carbon Norit at C_{NaCl} = 20 wt.%.

Low inorganic salt amounts (5 wt.% or smaller) on the carbon surface lead to small changes in the TBB breakthrough plots at RH = 0%. However, for polar DMMP, addition of NaCl (even 5 wt.%) strongly lowers the breakthrough time (more than two times) and the effective overall adsorption rate coefficient at RH = 0 % (Table 1).

Comparison of the kinetic adsorption capacity for TBB (Table 1, W_e) (RH = 0%) shows little change at different amounts of NaCl. More significant changes are observed at higher relative humidity. Water and TBB molecules are not competitors for adsorption surface sites. These molecules should prefer different sites; polar for water, and nonpolar sites on the basal planes for TBB. However, the presence of NaCl changes the distribution of polar sites at the surface. This leads to formation of water droplets at the outer surface of carbon particles where large amounts of NaCl can locate (Figure 4). These droplets may plug pore entrances and inhibit penetration of TBB into these pores. Therefore, the breakthrough time becomes shorter. Clearly, these effects become stronger at high RH values.

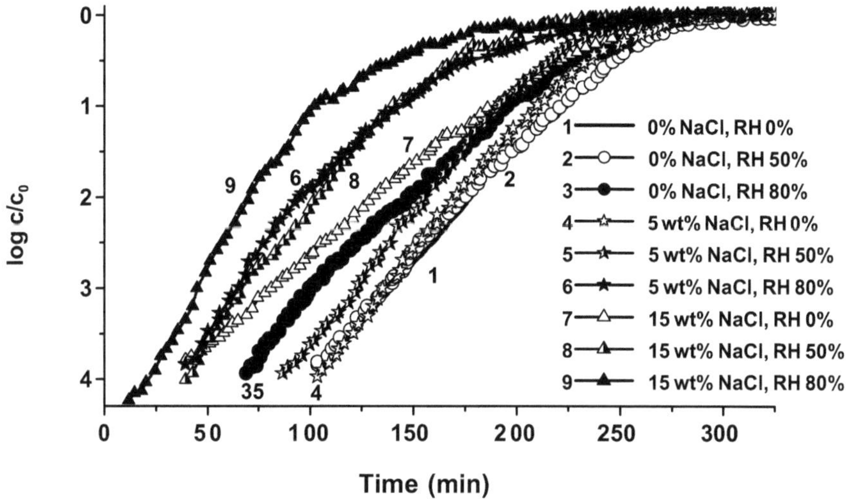

Figure 5. Log-scaled TBB breakthrough plots vs. time for beds of activated carbon with different amounts of pre-adsorbed NaCl (0, 5 and 15 wt.%) and different relative humidity of air RH = 0, 50, and 80%.

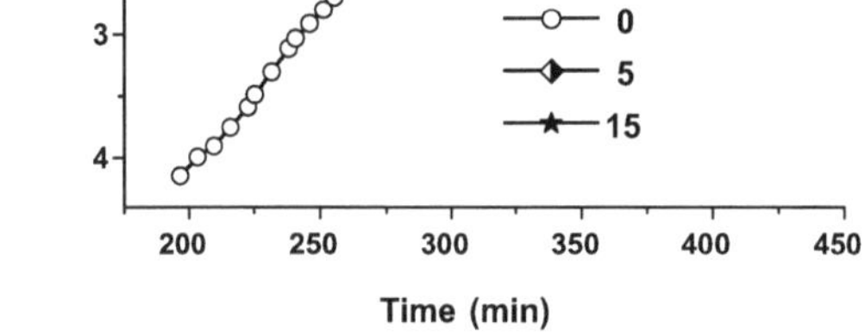

Figure 6. Log-scaled DMMP breakthrough plots vs. time for beds of activated carbon with different NaCl amount.

For DMMP, the effect of NaCl on the kinetic adsorption capacity is similar to TBB. However, the NaCl effect on the t_b for DMMP is stronger than that for TBB, especially at C_{NaCl} 5 wt.%. The DMMP breakthrough curves at 5 wt% and 15 wt% of NaCl are displaced toward shorter time (Table 1, Figure 5), in comparison with the curve for the sample without NaCl. It was not possible to measure the effect of RH because of experimental difficulties with the DMMP/water vapour mixtures.

4. Conclusion

Analysis of the breakthrough dynamics for TBB and DMMP on carbon beds with different amounts of pre-adsorbed NaCl and different air stream relative humidity, shows that NaCl exhibits deleterious effects at any humidity. However, at low amounts of NaCl these effects are insignificant. This can be caused by the distribution of NaCl at the entrances of pores, reducing penetration of organics into pores. At low NaCl contents, its negative effects are stronger for DMMP than for TBB.

References

1. V. B. Fenelonov, *Porous Carbon* (Nauka, Novosibirsk, 1995).
2. *Facilities of Individual and Collective Protection*, edited by K.M. Nikolaev (Military Academy of Chemical Protection, Moscow, 1976).
3. L. A. Jonas and J. A. Rehrmann, Predictive equations in gas adsorption kinetics, *Carbon* 11, 59-64 (1973).
4. D. Palijczuk, V. M. Gun'ko, R. Leboda, J. Skubiszewska-Zięba, S. Ziętek, Porous Structure of Activated Carbons and Tert-butylbenzene Breakthrough Dynamics, *J. Colloid Interface Sci.* 250, 5-17 (2002).
5. V. M. Gun'ko, D. Palijczuk, R. Leboda, J. Skubiszewska-Zięba, R. Szmigielski, S. Ziętek, Influence of pore structure and pretreatments of activated carbons and water effects on breakthrough dynamics of tert-butylbenzene, *J. Colloid Interface Sci.* 294(1), 53-68 (2006).
6. V. M. Gun'ko and S. V. Mikhalovsky, Evaluation of slitlike porosity of carbon adsorbents, *Carbon* 42, 843-849 (2004).
7. M. Melillo, V. M. Gun'ko, L. I. Mikhalovska, G. J. Phillips, J. G. Davies, A. W. Lloyd, S. R. Tennison, O. P. Kozynchenko, D. J. Malik, M. Streat, and S. V. Mikhalovsky, Ibuprofen Adsorption Affected by Bovine Serum Albumin, *Langmuir* 20, 2837-2851 (2004).
8. V. V. Turov, V. M. Gun'ko, R. Leboda, T. J. Bandosz, J. Skubiszewska-Zięba, D. Palijczuk, W. Tomaszewski, and S. Ziętek, Influence of Organics on Structure of Water Adsorbed on Activated Carbons, *J. Colloid Interface Sci.* 253, 23-34 (2002).
9. V. M. Gun'ko and T. J. Bandosz, Heterogeneity of Adsorption Energy of Water, Methanol and Diethyl Ether on Activated Carbons: Effect of Porosity and Surface Chemistry, *Phys. Chem. Chem. Phys.* 5, 2096-2103 (2003).
10. V. M. Gun'ko, V. V. Turov, J. Skubiszewska-Zięba, R. Leboda, M. D. Tsapko, and D. Palijczuk, Structural characteristics of a carbon adsorbent and influence of organic solvents on interfacial water, *Appl. Surf. Sci.* 214, 178-189 (2003).
11. S. V. Mikhalovsky, V. M. Gun'ko, V. V. Turov, R. Leboda, and W. R. Betz, Investigation of Structural and Adsorptive Characteristics of Various Carbons, *Adsorption* 11, 657-662 (2005).

INFLUENCE OF THE NATURE OF ACTIVE SURFACE SITES OF HIGHLY DISPERSE OXIDES ON ADSORPTION OF HEAVY METAL IONS

V. I. ZARKO, V. M. GUN'KO,[,*] L. S. ANDRIYKO, E. V. GONCHARUK,[1]
[1]*Institute of Surface Chemistry, 17 General Naumov Street, 03164 Kiev, Ukraine*
M. MATYSEK, E. SKWAREK, W. JANUSZ
[2]*Faculty of Chemistry, Maria Curie-Sklodowska University, 20031 Lublin, Poland*

Abstract. Different surface sites of silica, alumina, titania, and related mixed oxides, and pre-adsorbed poly(vinyl alcohol) (PVA) and poly(ethylene glycol) (PEG), affect the interaction with Ni(II), Cd(II), and Pb(II) studied using adsorption and potentiometric titration methods. Bridging hydroxyls (Brönsted acid sites) capable of taking part in ion-exchange reactions, and pre-adsorbed polymers, enhance the adsorption of metal ions at pH < 7. At pH > 7 PVA adsorbed on silica slightly enhances the adsorption of metal ions but pre-adsorbed PEG gives the opposite result. PEG has only electron-donor functionalities (C–O–C) which reduce the number of possible adsorption sites. PVA has COH groups with both electron-donor and proton-donor properties, which can effectively interact with metal ions and their hydroxy species.

Keywords: fumed silica; alumina/silica, titania/silica; alumina/silica/titania; Ni(II); Cd(II); Pb(II); poly(ethylene glycol); poly(vinyl alcohol); adsorption; potentiometric titration; surface charge density

1. Introduction

Nanooxides such as fumed silica, titania, alumina, etc. (with spherical primary nanoparticles) are used as fillers for complex drugs, enterosorbents, vaccine adjuvants, food additives, etc.[1,2] In general nonporous spherical nanoparticles can adsorb solutes in low amounts because the desolvation effect causes an

*To whom correspondence should be addressed. V. M. Gun'ko, Institute of Surface Chemistry, 17 General Naumov Street, 03164 Kiev, Ukraine; e-mail: gun@voliacable.com

J.P. Blitz and V.M. Gun'ko (eds.), Surface Chemistry in Biomedical and Environmental Science, 429–438.

increase in Gibbs free energy upon adsorption. However, nanoparticles form aggregates with relatively narrow gaps between particles where the adsorption potential for small molecules and ions is much higher than at the outer surface, adsorption isotherms exhibit hysteresis loops.[3,4] There are different active surface sites (e.g. bridging hydroxyls $M_1O(H)M_2$) at the surface of mixed oxides[5-7] which can take part in ion-exchange reactions with dissolved metal ions more effectively than silanols at the silica surface.[8] The adsorption of polymers or biomacromolecules on the surface of nanooxides can change the adsorption capacity with respect to small molecules and ions. Their adsorption onto nanooxides, unmodified and modified by adsorbed macromolecules, should be considered for a more detailed description of the behavior of drug composites with nanooxides in organisms. Therefore the aim of this work is to study the adsorption of several metal ions and related species onto unmodified and modified nanooxides (silica and titania) and mixed nanooxides (silica/alumina, silica/titania and alumina/silica/titania).

2. Experimental

2.1. MATERIALS

Fumed silica A-300 (S_{BET} = 232 m^2/g), titania (50 m^2/g), mixed oxides SiO_2/Al_2O_3 at alumina content C_{Al2O3} = 1, 3, 8, and 23 wt.% (labelled as SA1 (207 m^2/g), SA3 (188 m^2/g), SA8 (308 m^2/g), and SA23 (347 m^2/g) respectively), SiO_2/TiO_2 at C_{TiO2} = 2, 9, 14, 20, 63, 65, and 94 wt.% (labelled as ST2 (77 m^2/g), ST9 (188 m^2/g), ST14 (156 m^2/g), ST20 (86 m^2/g), ST63 (83 m^2/g), ST65 (34 m^2/g) and ST94 (30 m^2/g) respectively), and $SiO_2/Al_2O_3/TiO_2$ (AST82 (39 m^2/g) at C_{TiO2} = 82 wt.%); poly(vinyl alcohol) (PVA, at molecular weight $W_m \approx 43$ kDa) and poly(ethylene glycol) (PEG, $W_m \approx 35$ kDa) pre-adsorbed on oxides were used to study adsorption of metal ions. Aqueous suspensions with distilled water were sonicated by an ultrasonic disperser (44 kHz) for 5 min. The oxides were described in detail elsewhere.[3,9-15]

2.2. ADSORPTION OF METAL IONS

Cd(II), Pb(II) and Ni(II) adsorption on oxides and PVA (monolayer loading)/ oxide surfaces (oxide concentration C_{ox} = 0.2 wt%) was carried out from aqueous solution of CdX_2 PbX_2 or NiX_2 ($X = ClO_4$). The initial concentration was $10^{-3} - 10^{-6}$ M (concentration of radioactive species ^{115}Cd, $^{210}Pb(II)$ and ^{59}Ni was 10^{-6} M) with addition of a neutral electrolyte (10^{-3} M $NaClO_4$) using a

Teflon cell (50 cm^3) temperature-controlled at $T = 25 \pm 0.2$°C.[8] The pH value was varied by addition of 0.1 M HCl or NaOH solutions. The gamma radioactivity of the solution was determined using a Beckman Gamma 5500B counter.

2.3. SURFACE CHARGE DENSITY

To evaluate the surface charge density, potentiometric titrations were carried out using a Teflon thermostated vessel in a nitrogen atmosphere free from CO_2 at 25 ± 0.2°C. The solution pH was measured using a PHM240 Research pH-meter (G202C and K401 electrodes) coupled with an REC-61 recorder. The surface charge density was calculated using the potentiometric titration data for a blank electrolyte solution and oxide suspensions (C_{ox} = 0.2 wt.%), at a constant salinity of 10^{-3} M NaCl with and without addition of the Ni(II) solution. The difference of acid or base volume utilized to obtain the same pH value as that for the background electrolyte of the same ionic strength gives the surface charge density.[8]

3. Results and Discussion

Various surface sites such as strong Brönsted and Lewis acid sites are characteristic for mixed oxides with silica/alumina, silica/titania and alumina/silica/titania in contrast to silica having only weak amphoteric sites with surface silanols.[5-7] Changes in the surface content of titania in ST or alumina in SA affect the content of acidic bridges SiO(H)Ti[6] or SiO(H)Al[7], and these changes are not linear functions of the total content of these oxides. Features of ST are reflected in the amounts of Si atoms in the structures $Si(OM)_4$ (Q_4 in Table 1), $Si(OM)_3(OH)$ (Q_3) and $Si(OM)_2(OH)_2$ (Q_2), and their contributions change with increasing C_{TiO2} value.

In the case of a large titania content in ST94 or AST88 (C_{SiO2} = 6 and 8 wt.% respectively), silica does not represent a separate phase because it forms a solid solution in titania.[10] However, titania in ST9 or larger C_{TiO2} values forms its own phase, observed in XRD patterns. The distribution of alumina

Table 1. Position of peaks (ppm) and their area in the ^{29}Si NMR spectra of silica/titania.

Sample	Q_4			Q_3			Q_2		
	min –120	max –105	Peak area	min –104	max –95	Peak area	min –94	max –85	Peak area
ST9	–108.8		2015	–99.0		2148	–90.0		729
ST20	–109.7		2067	–100.4		2600	–91.7		525
ST65	–111.6		1405	–102.1		808	–92.1		33
	–106.5		1069	–99.8		1507	–90.2		246

in SA, and titania in ST, at the particle surface as a function of the total content of alumina or titania differs (smoother curve is for ST than for SA). The plateau adsorption of Pb(II) and Ni(II) has a shape corresponding to the surface concentrations of alumina or titania respectively.[11] These results are evidence of the direct influence of the surface composition of mixed oxides on their adsorption capacity with respect to metal ions.

Fumed silica possessing relatively large specific surface area can adsorb different metal ions up to 100% at pH > 7 (Figure 1), when the contribution of different hydroxy species of these ions increases and the surface charge density of silica becomes more negative. Therefore, a tenfold increase in the initial $C_{Pb(II)}$ value leads to a tenfold increase in the plateau adsorption (Figure 2).

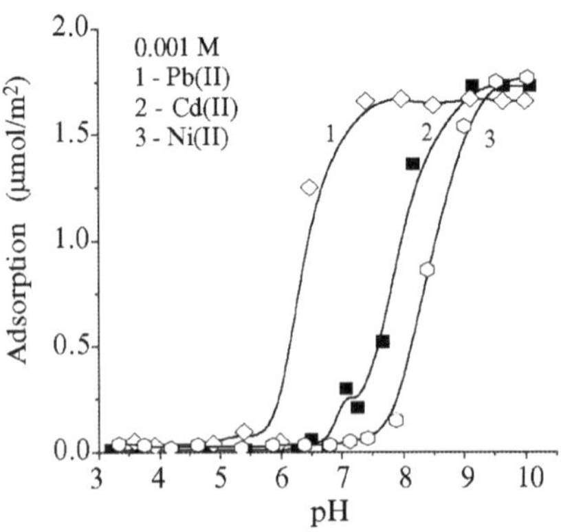

Figure 1. Adsorption of different ions on nanosilica A-300 as a function of pH at 0.001 M $NaClO_4$.

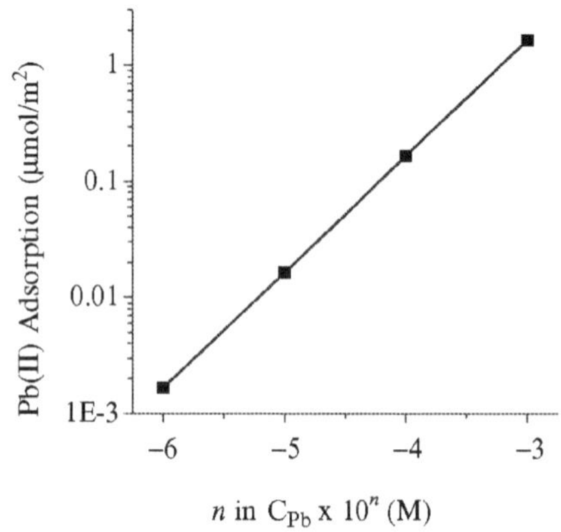

Figure 2. Plateau adsorption of Pb(II) on silica A-300 as a function of initial concentration at 0.001 M $NaClO_4$.

Pre-adsorption of polymers causes an increase in the adsorption of metal cations at low pH values (Figures 3 and 4). There is a difference in the effects of PVA (with COH functionalities) and PEG (C-O-C) because the adsorption of PEG diminishes the plateau adsorption of Cd(II) (Figure 4) in contrast to PVA (Figure 3). This can be caused by the absence of proton-donor groups in PEG reducing the probability of ion-exchange reactions, and diminishes the number of accessible surface silanols as proton-donors because of the strong interaction of silanols with PEG molecules. In effect PEG molecules shield silanols and do not create appropriate conditions for the adsorption of metal cations by ion-exchange reactions or complexation with C-O-C functionalities. This effect is seen in the difference between the surface charge density on silica with adsorbed PVA and PEG (Figure 5). The adsorption of PEG causes a noticeable reduction in the negative charging of particles, in contrast to adsorbed PVA which slightly enhance the negative charge with pH, in comparison with unmodified silica. Similar effects are observed for the adsorption of Cd(II) onto SA samples, both unmodified and modified by PVA or PEG (monolayer coverage) (Figure 6). The plateau adsorption of Cd(II) on A-300/PEG (Figure 4) is lower than on A-300/PVA (Figure 3).

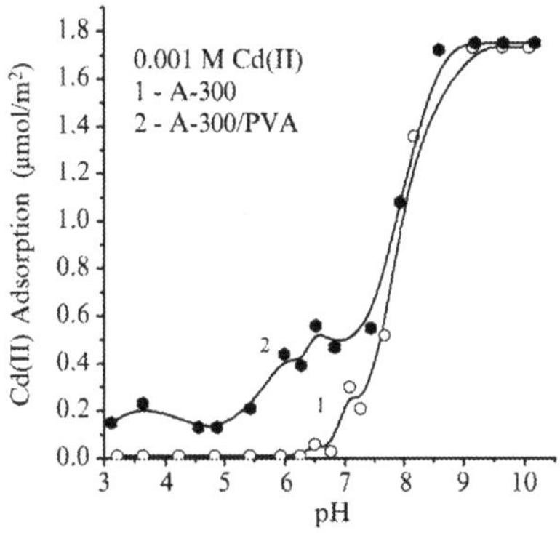

Figure 3. Adsorption of Cd(II) on silica A-300 (1) unmodified and (2) with pre-adsorbed PVA (monolayer coverage 0.15 g/g) as a function of pH at 0.001 M $NaClO_4$.

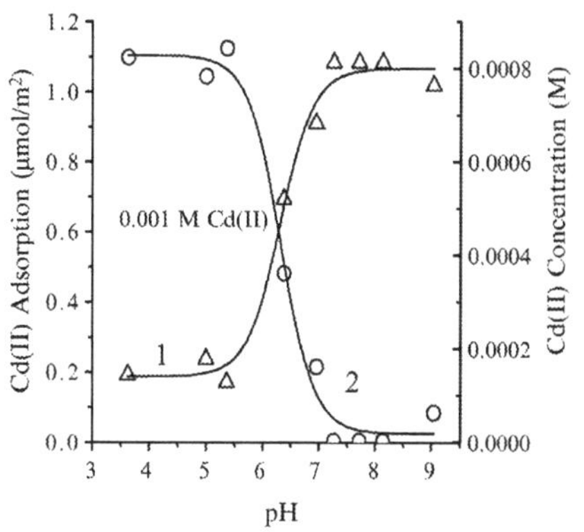

Figure 4. (1) Adsorption of Cd(II) on silica A-300 modified by PEG (monolayer coverage 0.125 g/g) and (2) equilibrium concentration of Cd(II) as a function of pH at 0.001 M $NaClO_4$.

At low pH values (pH < 6) the adsorption of all metal species is low not only for silica (Figure 1) but also for titania and ST interacting with Ni(II) (Figure 7). Adsorption on silica at 6 < pH < 8 is much lower than that on titania or ST94. At pH < 7 the surface charge density is practically independent of the adsorption of Ni(II) (Figure 8).

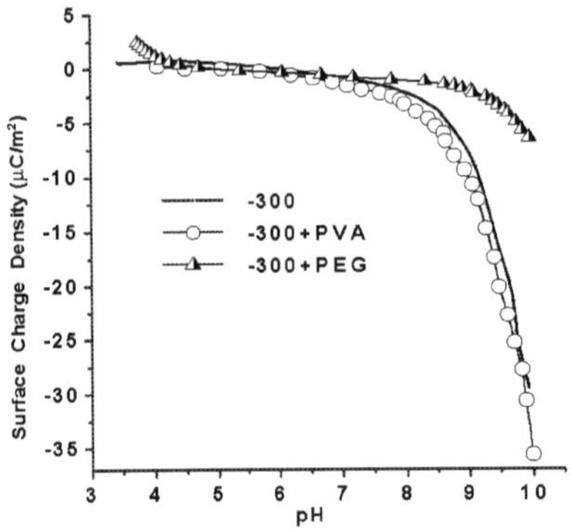

Figure 5. Surface charge density on silica unmodified and modified by adsorbed PVA (0.15 g/g) and PEG (0.125 g/g) as a function of pH at 0.001 M NaCl.

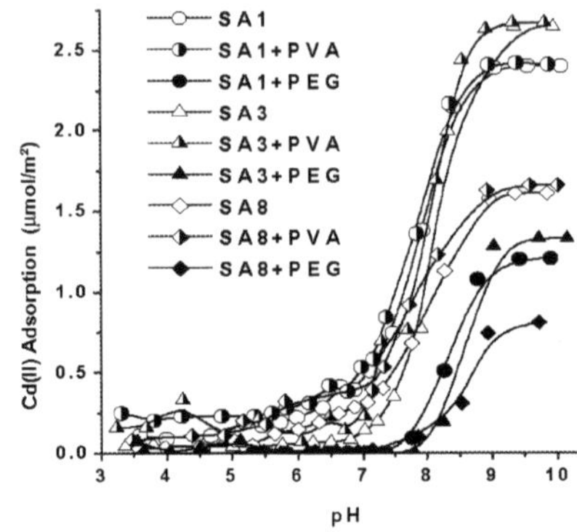

Figure 6. Adsorption of Cd(II) (initial 0.001 M) on SA unmodified and modified by adsorbed PVA (monolayer coverage 0.16, 0.14 and 0.1 g/g respectively) or PEG (0.125, 0.11 and 0.113 g/g) as a function of pH at 0.001 M $NaClO_4$.

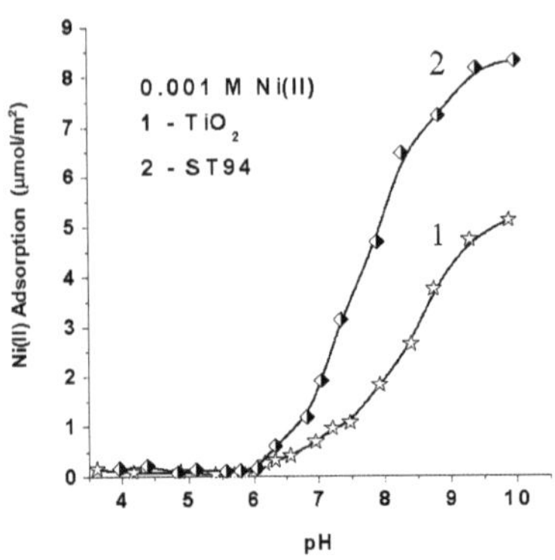

Figure 7. Adsorption of Ni(II) on (1) titania and (2) silica/titania ST94 as a function of pH at 0.001 M $NaClO_4$.

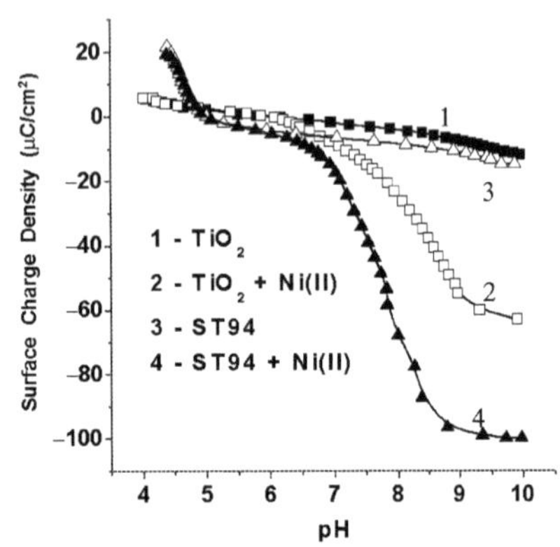

Figure 8. Surface charge density of (1) titania, (2) titania/Ni(II), (3) ST94, and ST94/Ni(II) at 0.001 M Ni(II) and 0.001 M NaCl.

However, at pH > 7 when the contribution of hydroxy species of Ni(II) (as well as other bivalent cations) increases, the negative charge of the surface grows.

This is accompanied by a noticeable increase in the adsorption of Ni(II) (Figure 7). A smaller effect with respect to the surface charge density is observed on the adsorption of Ni(II) on AST82 (Figure 9) because the surface charge density of the initial AST82 is much higher (Figure 10) than titania and ST94 (Figure 8). Additionally, the σ_0(pH) curves for AST82 and AST82/Ni(II) cross at a point close to the point of zero charge of alumina. However, the surface content of alumina ~14 wt.% is only slightly larger than its total content ~12 wt.% in AST82.

The effects of changes in the surface composition of mixed oxides SA and ST, and the initial concentration of dissolved metal ions, are revealed in changes of (i) the pH values corresponding to adsorption of 50% of the ions ($pH_{50\%}$) and (ii) the pH range of adsorption from 10 to 90% of the ions ($\Delta pH_{10-90\%}$) (Tables 2-4).

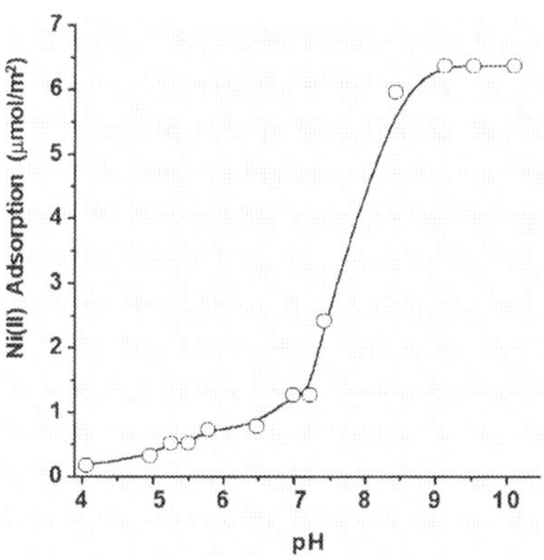

Figure 9. Adsorption of Ni(II) on AST82 as a function of pH at initial concentration 0.001 M Ni(II) and 0.001 M $NaClO_4$.

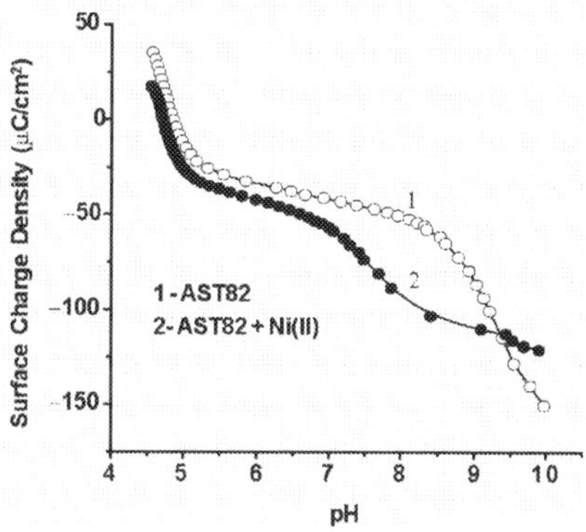

Figure 10. Surface charge density of (1) AST82 and (2) AST82/Ni(II) at 0.001 M Ni(II) and 0.001 M NaCl.

An increase in the C_M value from 10^{-6} to 10^{-3} leads to displacement of the point of adsorption of 50% of the ions toward higher pH values. However, the $\Delta pH_{10-90\%}$ range becomes narrower with exception of ST at great C_{TiO2} values. These effects can be explained by nonuniformity of the surface of mixed oxides characterized by a variety of active surface sites. At low C_M values, the number of maximum active surface sites is enough to adsorb (by the ion-exchange mechanism) the majority of ions at lower pH values.

Table 2. The $pH_{50\%}$ and $\Delta pH_{10\text{-}90\%}$ values on adsorption of Ni(II) on SA.

Oxide	Parameter	Initial concentration of Ni(II) (M)			
		10^{-3}	10^{-4}	10^{-5}	10^{-6}
SA1	$pH_{50\%}$	7.34	6.94	6.12	6.08
	$\Delta pH_{10\text{-}90\%}$	1.62	2.82	3.38	3.14
SA3	$pH_{50\%}$	7.99	8.20	7.60	7.51
	$\Delta pH_{10\text{-}90\%}$	1.30	1.51	2.51	2.47
SA8	$pH_{50\%}$	8.29	7.14	6.37	5.90
	$\Delta pH_{10\text{-}90\%}$	1.90	2.7	2.68	3.22
SA23	$pH_{50\%}$	8.37	7.23	6.95	5.93
	$\Delta pH_{10\text{-}90\%}$	1.07	2.53	3.08	4.67

Table 3. The $pH_{50\%}$ and $\Delta pH_{10\text{-}90\%}$ values on adsorption of Cd(II) on SA.

	Parameter	Initial concentration of Cd(II) (M)			
		10^{-3}	10^{-4}	10^{-5}	10^{-6}
SA1	$pH_{50\%}$	7.03	6.36	5.63	5.26
	$\Delta pH_{10\text{-}90\%}$	2.17	2.61	2.73	3.20
SA3	$pH_{50\%}$	8.04	7.41	7.27	6.73
	$\Delta pH_{10\text{-}90\%}$	1.44	1.84	2.48	2.10
SA8	$pH_{50\%}$	7.45	6.12	5.62	4.95
	$\Delta pH_{10\text{-}90\%}$	2.42	2.94	2.67	3.42

Table 4. The $pH_{50\%}$ and $\Delta pH_{10\text{-}90\%}$ values on adsorption of Ni(II) on ST.

Oxide	Parameter	Initial concentration of Ni(II) (M)			
		10^{-3}	10^{-4}	10^{-5}	10^{-6}
ST9	$pH_{50\%}$	7.54	7.33	6.66	6.34
	$\Delta pH_{10\text{-}90\%}$	1.78	2.58	3.19	3.29
ST14	$pH_{50\%}$	7.82	7.52	7.26	7.12
	$\Delta pH_{10\text{-}90\%}$	1.85	2.79	3.92	4.18
ST20	$pH_{50\%}$	8.23	7.47	6.52	6.27
	$\Delta pH_{10\text{-}90\%}$	1.12	2.74	2.57	2.59
ST63	$pH_{50\%}$	10.97	6.61	6.08	5.66
	$\Delta pH_{10\text{-}90\%}$	7.95	2.89	3.66	3.23
ST65	$pH_{50\%}$	8.72	6.87	6.25	
	$\Delta pH_{10\text{-}90\%}$	4.67	1.89	2.09	
ST94	$pH_{50\%}$	8.02	6.87	6.97	
	$\Delta pH_{10\text{-}90\%}$	3.53	2.68	1.90	

This corresponds to the displacement of the point of $pH_{50\%}$ and the increase in the $\Delta pH_{10\text{-}90\%}$ range. However, the latter is typically larger than the changes in the $pH_{50\%}$ value. In the case of ST63, ST65, and ST94, the opposite tendency is observed for changes in the $\Delta pH_{10\text{-}90\%}$ range which, increases with increasing C_M value. This result can be caused by an increase in uniformity of the ST samples at high titania content, which distributes nearly uniformly in the volume and on the surface of the ST samples.[11]

4. Conclusion

The adsorption of bivalent metal cations Ni(II), Cd(II), and Pb(II) and related hydroxy species onto nanosilica depends on surface modification by poly(vinyl alcohol) or poly(ethylene glycol). The surface composition of titania, silica/alumina, silica/titania, and alumina/silica/titania strongly affects the adsorption of metal ions mainly in basic solutions. In acidic solutions, the adsorption increases onto silica/polymer, e.g. at $pH < 6$ it reaches 25% for A-300/PVA and 18% for A-300/PEG. AST82 adsorbs approximately 15% Ni(II) at $pH \approx 6$. For other systems, the adsorption of cations is close to zero at $pH < 6$. Consequently, the use of nanooxides as components of drug composites including polymers or enterosorbents (interacting with macromolecules in native media) can lead to adsorption of metal ions from the liquid biomedia. This phenomenon can affect the interaction of oxide nanoparticles with biomacromolecules and other components of the organism, and, therefore, should be considered to understand mechanistic details of the action of nanooxides in biosystems.

References

1. *Silicas in Medicine and Biology*, edited by A. A. Chuiko (SMI, Stavropol, 1993).
2. *Medical Chemistry and Clinical Application of Silica*, edited by A. A. Chuiko (Naukova Dumka, Kiev, 2003).
3. V. M. Gun'ko, I. F. Mironyuk, V. I. Zarko, E. F. Voronin, V. V. Turov, E. M. Pakhlov, E. V. Goncharuk, Yu. M. Nychiporuk, T. V. Kulik, B. B. Palyanytsya, S. V. Pakhovchishin, N. N. Vlasova, P. P. Gorbik, O. A. Mishchuk, A. A. Chuiko, J. Skubiszewska-Zięba, W. Janusz, A. V. Turov, and R. Leboda, Morphology and surface properties of fumed silicas, *J. Colloid Interface Sci.* 289, 427-445 (2005).
4. V. M. Gun'ko, V. M. Bogatyrev, V. V. Turov, R. Leboda, J. Skubiszewska-Zięba, L. V. Petrus, G. R. Yurchenko, O. I. Oranska, and V. A. Pokrovsky, Composite powders with titania grafted onto modified fumed silica, *Powder Technology*, submitted for publication.
5. V. M. Gun'ko, V. I. Zarko, V. V. Turov, R. Leboda, E. Chibowski, E. M. Pakhlov, E. V. Goncharuk, M. Marciniak, E. F. Voronin, and A. A. Chuiko, Characterization of Fumed Alumina/Silica/Titania in the Gas Phase and Aqueous Suspension, *J. Colloid. Interface Sci.* 220, 302-323 (1999).
6. V. M. Gun'ko, V. I. Zarko, V. V. Turov, R. Leboda, E. Chibowski, L. Holysz, E. M. Pakhlov, E. F. Voronin, V. V. Dudnik, and Yu. I. Gornikov, CVD-Titania on Fumed Silica Substrate, *J. Colloid. Interface Sci.* 198, 141-156 (1998).

7. V. M. Gun'ko, V. V. Turov, V. I. Zarko, E. F. Voronin, V. A. Tischenko, V. V. Dudnik, E. M. Pakhlov, and A. A. Chuiko, Active Site Nature of Pyrogenic Alumina/Silica and Water Bound to Surfaces, *Langmuir* 13, 1529-1544 (1997).
8. V. M. Gun'ko, V. I. Zarko, I. F. Mironyuk, E. V. Goncharuk, N. V. Guzenko, M. V. Borysenko, P. P. Gorbik, O. A. Mishchuk, W. Janusz, R. Leboda, J. Skubiszewska-Zięba, W. Grzegorczyk, M. Matysek, and S. Chibowski, Surface Electric and Titration Behaviour of Fumed Oxides, *Colloids Surf. A* 240, 9-25 (2004).
9. V. M. Gun'ko, V. I. Zarko, R. Leboda, and E. Chibowski, Aqueous Suspensions of Fumed Oxides: Particle Size Distribution and Zeta Potential, *Adv. Colloid Interface Sci.* 91, 1-112 (2001).
10. V. M. Gun'ko, J. Skubiszewska-Zięba, R. Leboda, K. N. Khomenko, O. A. Kazakova, M. O. Povazhnyak, and I. F. Mironyuk, Influence of Morphology and Composition of Fumed Oxides on Changes in Their Structural and Adsorptive Characteristics on Hydrothermal Treatment at Different Temperatures, *J. Colloid Interface Sci.* 269, 403-424 (2004).
11. V. Gun'ko, R. Leboda, V. Turov, V. Zarko, A. Chuiko, Oxide, carbon, and carbon/mineral adsorbents for toxic, explosive, and narcotic compounds – synthesis, properties, and applications, in: Pure and Applied Surface Chemistry and Nanomaterials for Human Life and Environmental Protection, Proceedings of NATO ARW, (Kiev, Ukraine, September 14-17, 2005, NATO Science Series II: Mathematics, Physics and Chemistry, Springer, edited by J. P. Blitz and V. M. Gun'ko).
12. V. M. Gun'ko, V. I. Zarko, E. Chibowski, V. V. Dudnik, R. Leboda, and V. A. Zaets, Structure of Pyrogenic TiO_2 and TiO_2/SiO_2 and Influence of the Active Surface Site Nature on Interaction with Water, *J. Colloid. Interface Sci.* 188, 39-57 (1997).
13. V. M. Gun'ko, V. I. Zarko, V. V. Turov, R. Leboda, and E. Chibowski, The Effect of Second Phase Distribution in Disperse X/Silica (X = Al_2O_3, TiO_2, and GeO_2) on its Surface Properties, *Langmuir* 15(18), 5694-5702 (1999).
14. V. M. Gun'ko, V. I. Zarko, R. Leboda, M. Marciniak, W. Janusz, and S. Chibowski, Highly Dispersed X/SiO_2 and $C/X/SiO_2$ (X = Alumina, Titania, Alumina/Titania) in the Gas and Liquid Media, *J. Colloid Interface Sci.* 230, 396-409 (2000).
15. V. M. Gun'ko, I. F. Mironyuk, V. I. Zarko, V. V. Turov, E. F. Voronin, E. M. Pakhlov, E. V. Goncharuk, R. Leboda, J. Skubiszewska-Zięba, W. Janusz, S. Chibowski, Yu. N. Levchuk, and A. V. Klyueva, Fumed Silicas Possessing Different Morphology and Hydrophilicity, *J. Colloid Interface Sci.* 242, 90-103 (2001).

SUBJECT INDEX

^{1}H NMR spectroscopy 180, 188
4-Nitroanisole 19, 20
^{13}C NMR spectroscopy 109, 110

α-Chymotrypsin 171, 172

Acetonitrile 99
Activated carbons 94-97, 413-415, 419, 420, 421-427
Addition theorem 116
Adsorbed polymer 433, 434
Adsorbents 69-77, 91-100
Adsorption 307-313, 317-323, 429-437
Adsorptive characteristics 178-180
Aerogels 404
AFM 108, 139
Aggregation 246
Aggregation, degree 371
Ag-PEO 153, 155
Alkali metal promoted catalysts 359-367
Alkanes isomerization 350, 352, 354
Allydextran 150
Aluminium chloride, basic 371-373
Aluminium hydroxide 380
Aluminium salts 369, 370
Aluminium sulfate, basic 371
Amino acid 393
Amorphous silica 287-289, 293, 294
Amperometry 253
Antibacterial modification 152
Antifreeze protein, b-helix 5
Antioxidant power 307, 313
Aqueous suspension 181-189

Binding energy 362
Bioactive silica 191-204
Bio-application 265, 270, 271, 274
Biocompatibility 205-209, 265
Biofilm formation 156
Biofouling 164, 175
Biomaterials 145
Biosensors 146, 265, 271-274
Bovine serum albumin, BSA 154, 181-184, 277-285, 287, 290, 293-297
Breakthrough 94, 95
Breakthrough dynamics 414
Breakthrough dynamics, dimethylmethylphosphonate 421-427
Breakthrough plot 416, 417
Breakthrough time 415-420, 421, 423, 425, 426

Capillary flow porometry 215, 222-227
Carbon coating 408
Carbon monoxide 359-367
Carbon monoxide, adsorption 352, 353
Carbon-silica adsorbents 123-133, 135-143
Carbon-mineral adsorbents 98, 123-133, 135-143
Carboxylic acids 383
Catalyst activity 359-367
Catalysts 41
Cd(II), Cadmium ions 98-100, 143, 383-391, 431-437
Cell adhesion 160, 164, 175
Cell culture 205-213
Cell membrane 320-323
Cell surface 277, 278
Cells, E. coli 378
Cells, E. coli, removal 379
Charge neutralisation 369, 376
Chemically modified silicas 105-111
Cluster 395, 400, 401
Co-adsorption 383-391
Coagulant dose 370, 377
Coagulant dose, zones 377, 378
Coagulants 369
Collective diffusion 393, 398, 399, 401
Colloid 9
Colloidal particles 369
Colloidosomes 16
Condensation polymerization 13

Conformational changes 287, 288, 294
Contaminants, incorporation 370
Contrast matching 395, 400
Controlled uptake and release 11, 14, 19
Control release 25
Core-shell particles 15, 17, 18
Counter-ions, adsorption 376
Covalent functionalization 270
Cristobalite 288-296
Cross-flow membrane 17
Cross-linking 125, 133, 394-396
Cross-linking monomer 12, 19
Cyclohexane breakthrough 416, 419
Cs(I) 98-100
Cytostatic action 299, 300, 304

Debye-Huckel equation 114, 115
Desorption 308-313
Destructive sorbents 404
Detection of V-type nerve agent 253-262
Detoxication 197, 201, 202
Deuterated phenol 395, 400, 401
Diarrhea 194, 197-200
Differential charging 49-50
(Dimethylamino)azobenzene 180, 181
Direct electron transfer 265, 273, 274
Disinfection 378
Dispersion polymerization 12, 13
Divinylbenzene 104, 105
Donor plasma decontamination 301
DRIFT specroscopy 359, 362, 363
Drug immobilization 300
DSC 250
Dynamic adsorption 413, 418
Dynamic light scattering, DLS 182-187, 393, 395, 398, 400

Efferent therapy 191
Elastic modulus 398, 401
Electrical double layer, EDL 187
Electric sparks, discharge 380
Electrodeposition 253, 258-262
Electrokinetic potential 371, 373, 374
Electrolyte 114
Electron paramagnetic resonance 353, 354
Electrostatic interaction 114-116
Electrostatic stabilization 13
Emulsion polymerization 13
Ensemble of spherical particles 114
Enterosorption 191-204
Enzyme 15, 265, 272-274
ESR 248, 249
Explosives 96

Fibrinogen 154
First order transition 399, 401
Flocculation, sweep 377, 378
Folin-Ciocalteu method 312, 313
Friction coefficient 398, 399, 401
Fructose 277-285
FTIR spectroscopy, Infrared spectroscopy 107, 289, 290, 293, 294
Fullerene, hydration 3-6
Fumed alumina 93, 98, 99, 429-437
Fumed alumina/silica/titania 429-437
Fumed silica 93, 98, 99, 135-143, 177-189, 191-204, 277-285, 429-437
Fumed silica/alumina 93, 98, 99, 181, 429-437
Fumed silica/titania 93, 98, 99, 181, 429-437
Fumed titania 181, 429-437

Gibbs free energy 180, 188
Glass transition temperature 233
Glucose 277-285
Glutaraldehyde 151
Goethite 383-391

Halocarbons 406
Hamaker constant 13
Heavy/deuterated water 400
Heavy metal ions 325-328, 333, 343-347
Henry model 396
Homogenizer 17
Hydration, degree 371, 373
Hydrogels 230

Hydrogen bond 318, 319
Hydrolysis products 369, 370, 374
Hydrophilic surfaces 3
Hydrophilicity 243, 287, 288, 291, 294, 295, 296
Hydrophobic surfaces 2-3
Hydrothermal modification 125, 129-131
Highly disperse magnetite 299, 300, 305

Ice, crystal growth 4-5
Ice inhibitor 5
Image analysis 215, 221, 222, 224, 227
Immobilization 25, 265, 307, 313
Immunoglobulin G, IgG 173-175
Immunoassays 150
Impurities 369, 370
Impurities, inorganic, removal 370
Infectious diseases 196, 198, 199
Infinite systems 117, 118
Interfacial polymerization 16
Interfacial tension 17
Internal phase separation 17
Inverse emulsion polymerization 12
Iron oxide 247-249
Iron - promoter 352-355
Isoelectric point, IEP 181, 182

Janus particles 11

Layer-by-layer deposition 16
Layer, Stern 374
Liposomes 250
Low temperature oxidation 359-367
Lundelius rule 397

Macro-monomer 13
MAD-galactose 148
MAD maleimido-butyramide 146, 148
Magnetically sensitive carrier 299, 303-305
MALDI 277, 278, 283-285
Mass spectrometry 277, 278, 283
MCM-41 325-330
MCM-48 325-330
Medicinal chemistry 191
Medicinal preparation 192, 193, 195, 197, 199, 201
Mercury adsorption 325, 329, 331, 332
Mercury porosimetry 215, 222-227
Mesoporous organosilicas 325, 327
Metallic nanotube 269
Methacrylic monomers 104, 105
MgO 406
Microcalorimetry 288, 295
Microencapsulation 23-34
Microgel particles 12, 21
Migration 246
Model protein 394
Modified spherical Bessel functions 113-121
Moessbauer spectroscopy 353, 354
Molecular layering method, adsorbents 35-41
Molecular mobility 236
Molecular sieve 413-420, 423
Molybdate, hydration 8-9

N2 plasma 151
NaCl 421-427
Nanoceramics 43
Nanocomposites 104-112, 231, 307-313
Nanocrystalline oxides 403
Nanodrop 5-6
Nanomaterials 35-40
Nanoparticles 241-247, 250, 277, 278, 300, 302
Nanosilica 177-189
Nanotechnology 35-37, 299, 306
Nanotextured surfaces 211
Nanotoxicology 241, 242
Nanotubes 205, 206, 211, 246, 249, 265-273
Nanowheel 5-9
Neutravidin 150
Ni(II) 435, 436
Nitrogen adsorption 125, 126, 128
Non-covalent functionalization 270
n-Pentane isomerization 351, 354, 355

Organosilane 337-341
Ordered mesoporous silicas 325, 326
Oscillatory reactions 365
Osmotic modulus 398, 399, 401
Osmotic pressure 21
Ovalbumin 181, 182
Oxidation 359-367

Palladium oxide 359-367
Palladium species 359-367
Palygorskite 125, 132, 133
Particle aggregation 179, 182, 184, 187-189
Particle mobility 185, 186
Particle size distribution 183-185
Particles, hydrolysis product, surface, aggregation 371
Particles, primary, aluminium sulfate, hydrolysis product 375
Particles, primary, hydrolysis product, radius 371
Patchy particles 11
Pb(II) 98-100, 432
Permeability 19
Phenol 393-401
Phloroglucinol 393, 396-399
Phospholipids 320-322
Phosphoric acid 142
Photo-immobilization 146
Plasma 151
Platidiam 299-305
Platinum - promoter 350-355
p-Nitrophenol adsorption 126, 131, 137, 141
Pollutant 394, 395
Polyacrylamide 299-305
Poly(dimethylsiloxane) 12
Poly(ethylene glicol), PEG 183-185, 433, 434
Poly(ethylene oxide), PEO 14, 153-155, 165-170, 174, 175
Poly(hydroxyethyl acrylate) 230
Polyimides 61
Polymer 42
Polymer brush 164-166
Poly(methacrylic acid) 17
Poly(methylmethacrylate) 17
Poly(N-isopropylacrylamide), PNIPA 12, 393-401
Polyoxomolybdate, hydration 5-9
Polyoxotungstate clusters 351
Polypyrrole 253, 257, 260
Polysyrene, PS 149, 212
Poly(tetrahydrofuran) 17
Poly(vinyl alcohol), PVA 183-185, 433, 434
Polyvinylchloride, PVC 152-155
Poly(vinylpyridine) 16, 20
Poly(vinyl pyrrolidone), PVP 183-185
Potentiometric titration 431-437
Pore size distribution 93, 140, 342, 343
Pore volume 123-133, 135-143
Porosity 19
Porous copolymers 103-111
Porous materials 69-77, 123-133, 135-143
Porous structure 123-133, 135-143
Precipitate, aluminium hydroxide, solubility 370
Protein adsorption 151, 154, 160, 161, 163, 164, 166, 175, 181-184, 290, 296
Proteins 287, 288, 290, 293-297
Proteus mirabilis 186
Pseudomonas aeruginosa 152, 157
Purulent wound 197, 202, 203
Pyoinflammatory diseases 196, 197, 202
Pyrolysis 123-133, 135-143
Pyrroloquinoline quinone 253, 256

Quartz 242, 247, 287-297
Quantum chemical calculations 309, 311, 312, 315-323
Quantum dot 165, 269, 270, 272, 274

Raman spectroscopy 351
Rayleigh ratio 393, 398, 399
Reactive oxygen species, ROS 243
Reactivity 244, 246, 247
Reactivity ratio 13
Recovery rate 96

Red blood cell, RBC 186, 277, 280, 281, 284
Relaxation 238
Release kinetics 19
Resorcinol 393, 396-399

SBA-15 325-330, 333
Scanning electron microscopy, SEM 218
Selective uptake 14
Selectivity - alkane isomerization 354, 355
Semiconducting nanotube 270
Shell thickness 19
Silanols 288-296
Silica 123-133, 243, 244, 249-251, 287-297, 307-313, 337-347, 383-391
Silica gel 135-143
Silica shell 18
Silica surface 316-323
Siloxane 288, 293, 294, 296
Silver nitrate AgNO3 155, 156
Silylation 308, 310, 313
Slip plane 373
Small angle neutron scattering, SANS 13, 393, 395, 400, 402
Sodium oxopalladate 366
Sodium promotion 361, 363, 364
Solid phase extraction 96
Solvation 309, 311, 312
Sorbent 191, 194-203
Spermatozoa 277-281
Spherical functions 113-121
Spin-spin relaxation time 374
Spin trap 248, 249
Sr(II) 98-100
Stannic oxide 361-367
Steric stabilization 13
Streptavidin 150
Structural characteristics 177-182
Styrene 104, 105
sulphur-containing amino acid 316-323
Surface 243-248, 251, 287-297
Surface area 123-133, 135-143, 178-182, 244
Surface charge density 181, 431-437
Surface chemistry 191-204
Surface coverage 360
Surface modification 159-165, 300, 337-346
Surface pressure isotherm 79-88
Surface properties 287-297
Surface silanols 180-182
Surface sites 180-182
Sustained and triggered release 12, 14, 20, 21
Swelling 393-398, 401

Telomerase 205,209-213
TBB breakthrough 415, 416, 419, 420, 421-427
tert-Butylbenzene 95
Tetraethoxysilane 18
Texturology 69-77
Thermal modification 129-131
Thermodynamic model 79-89
Thiols 253-260
Time-of-Flight Secondary Mass Spectrometry, Tof-SIMS 145
Time-resolved XPS 51
Tin dioxide 361-367
Tissue scaffolds 215-227
Titania 61-66, 242, 250, 383-391, 407
Titanium oxyfluoride 410
Toxicity 287, 297
Translocation 244
Transmission electron microscopy, TEM 126, 130, 131, 353
Tridymite 288-295
Trimethylchlorosilane 308, 310
Tungsten oxide 351

Ultrafine particles 241, 242
Uptake isotherm 393, 397
UV spectroscopy 307, 308, 312, 313

Virus 299-305
Vitamin C 307-313
Vitamin E 307-313
Volume phase transition, VPT 393, 394, 397

Water 1-10, 413-418, 421-427
Water, binding at surfaces 2
Water, explicit model 9
Water, icosahedral cluster model 3
Water, nanodrop 5-6
Water sorption isotherm 234
Water treatment 369

X-ray absorption spectroscopy 353
X-ray diffraction, XRD 126, 133, 137, 142
X-ray fluorescence, XRF 126, 138
X-ray photoelectron spectroscopy, XPS 45-55, 145, 149, 359, 362-366

Yield 19

Zirconia, tungstated - acidity 352
Zirconia, tungstated - redox properties 353

9 781402 047398

MIX
Papier aus verantwortungsvollen Quellen
Paper from responsible sources
FSC® C105338

If you have any concerns about our products,
you can contact us on
ProductSafety@springernature.com

In case Publisher is established outside the EU,
the EU authorized representative is:
Springer Nature Customer Service Center GmbH
Europaplatz 3, 69115 Heidelberg, Germany

Printed by Libri Plureos GmbH
in Hamburg, Germany